International Review of
Cytology
A Survey of
Cell Biology

MOLECULAR MECHANISMS OF WATER TRANSPORT ACROSS BIOLOGICAL MEMBRANES

VOLUME 215

International Review of Cytology
A Survey of Cell Biology

Guest Edited by

Thomas Zeuthen
Institute of Medical Physiology
The Panum Institute
University of Copenhagen
Copenhagen, Denmark

Wilfred D. Stein
Department of Biological Chemistry
Silberman Institute of Life Sciences
Hebrew University
Jerusalem, Israel

MOLECULAR MECHANISMS OF WATER TRANSPORT ACROSS BIOLOGICAL MEMBRANES

VOLUME 215

ACADEMIC PRESS

An Elsevier Science Imprint

San Diego San Francisco New York Boston London Sydney Tokyo

This book is printed on acid-free paper. ∞

Academic Press
An Elsevier Science Imprint.
525 B Street, Suite 1900, San Diego, California 92101-4495, USA
http://www.academicpress.com

Academic Press
32 Jamestown Road, London NW1 7BY, UK
http://www.academicpress.com

International Standard Book Number: 0-12-364619-7

PRINTED IN THE UNITED STATES OF AMERICA
02 03 04 05 06 07 MM 9 8 7 6 5 4 3 2 1

CONTENTS

Water as It Applies to the Function of Enzymes

J. A. Kornblatt and M. J. Kornblatt

Aquaglyceroporins: Channel Proteins with a Conserved Core, Multiple Functions, and Variable Surfaces

Andreas Engel and Henning Stahlberg

Molecular Physiology of Aquaporins in Plants

C. Maurel, H. Javot, V. Lauvergeat, P. Gerbeau, C. Tournaire, V. Santoni, and J. Heyes

Osmotic Adaptation in Yeast—Control of the Yeast Osmolyte System

Stefan Hohmann

Adaptation of Kidney Medulla to Hypertonicity: Role of the Transcription Factor TonEBP

Seung Kyoon Woo and H. Moo Kwon

Passive Water Transport in Biological Pores

Thomas Zeuthen and Nanna MacAulay

Cell Volume Homeostasis: Ionic and Nonionic Mechanisms
The Sodium Pump in the Emergence of Animal Cells

Wilfred D. Stein

Cotransporters as Molecular Water Pumps

Thomas Zeuthen and Nanna MacAulay

General Models for Water Transport across Leaky Epithelia

Thomas Zeuthen

Paracellular Fluid Transport by Epithelia

Bruria Shachar-Hill and Adrian E. Hill

Osmoregulation and Contractile Vacuoles of Protozoa

Richard D. Allen and Yutaka Naitoh

Molecular Mechanisms of Water Transport in the Eye

Steffen Hamann

CONTRIBUTORS

Numbers in parentheses indicate the pages on which the authors' contributions begin.

Richard D. Allen (351), *Department of Microbiology, Pacific Biomedical Research Center, University of Hawaii, Honolulu, Hawaii 96822*

Andreas Engel (75), *M. E. Müller-Institute for Microscopic Structural Biology, Biozentrum, University of Basel, CH-4056 Basel, Switzerland*

P. Gerbeau (105), *Biochimie et Physiologie Moléculaire des Plantes, Agro-M/CNRS/INRA/UM2, F-34 060 Montpellier Cedex 1, France*

Steffen Hamann (395), *Department of Medical Physiology, The Panum Institute, University of Copenhagen, DK-2200 Copenhagen N, Denmark*

J. Heyes (105), *Biochimie et Physiologie Moléculaire des Plantes, Agro-M/CNRS/INRA/UM2, F-34 060 Montpellier Cedex 1, France*

Adrian E. Hill (319), *Physiological Laboratory, Cambridge University, Cambridge CB2 3EG, United Kingdom*

Stefan Hohmann (149), *Department of Cell and Molecular Biology, Microbiology, Göteborg University, S-405 30 Göteborg, Sweden*

H. Javot (105), *Biochimie et Physiologie Moléculaire des Plantes, Agro-M/CNRS/INRA/UM2, F-34 060 Montpellier Cedex 1, France*

J. A. Kornblatt (49), *Enzyme Research Group, Departments of Biology, Biochemistry and Chemistry, Concordia University, Montreal, Quebec, Canada H3G 1M8*

M. J. Kornblatt (49), *Enzyme Research Group, Departments of Biology, Biochemistry and Chemistry, Concordia University, Montreal, Quebec, Canada H3G 1M8*

H. Moo Kwon (189), *Division of Nephrology, Johns Hopkins University, Baltimore, Maryland 21205*

V. Lauvergeat (105), *Biochimie et Physiologie Moléculaire des Plantes, Agro-M/CNRS/INRA/UM2, F-34 060 Montpellier Cedex 1, France*

Nanna MacAulay (203, 259), *Institute of Medical Physiology, The Panum Institute, University of Copenhagen, DK 2200 Copenhagen N, Denmark*

C. Maurel (105), *Biochimie et Physiologie Moléculaire des Plantes, Agro-M/CNRS/INRA/UM2, F-34 060 Montpellier Cedex 1, France*

Yutaka Naitoh (351), *Department of Microbiology, Pacific Biomedical Research Center, University of Hawaii, Honolulu, Hawaii 96822*

V. Adrian Parsegian (1), *Laboratory of Physical and Structural Biology, National Institute of Child Health and Human Development, NIH, Bethesda, Maryland 20892*

R. P. Rand (33), *Department of Biological Sciences, Brock University, St. Catharines, Ontario, Canada L25 3A1*

V. Santoni (105), *Biochimie et Physiologie Moléculaire des Plantes, Agro-M/CNRS/INRA/UM2, F-34 060 Montpellier Cedex 1, France*

Bruria Shachar-Hill (319), *Physiological Laboratory, Cambridge University, Cambridge CB2 3EG, United Kingdom*

Henning Stahlberg (75), *M. E. Müller-Institute for Microscopic Structural Biology, Biozentrum, University of Basel, CH-4056 Basel, Switzerland*

Wilfred D. Stein (231), *Department of Biological Chemistry, Silberman Institute of Life Sciences, Hebrew University, Jerusalem 91904, Israel*

C. Tournaire (105), *Biochimie et Physiologie Moléculaire des Plantes, Agro-M/CNRS/INRA/UM2, F-34 060 Montpellier Cedex 1, France*

Seung Kyoon Woo (189), *Division of Nephrology, Johns Hopkins University, Baltimore, Maryland 21205*

Thomas Zeuthen (203, 259, 285), *Institute of Medical Physiology, The Panum Institute, University of Copenhagen, DK 2200 Copenhagen N, Denmark*

FOREWORD

Water is the most abundant substance in the body, yet, compared to ions and to many organic substrates, the least understood. The physical chemistry, the molecular biology, and the transport physiology of water are not well developed or integrated. The aim of this volume, therefore, is to present the state of the art in these disciplines and to bridge the gaps between them. It is our hope that this will help us understand the behavior of water to the same degree as we understand that of ions.

The first three chapters of this volume deal with the basic questions concerning water itself and its interactions with the components of the membrane. A central problem is the biophysics of protein–water interactions. Two chapters discuss how water interacts with the lipid components of the cell and how water plays a pivotal role in the function of enzymes. This first group of chapters forms a general framework for the later chapters. The concepts developed provide the basis for a clearer understanding of water movement through membrane-spanning channels and of the interaction of water and ions—in and around the membrane proteins—that provides the driving forces for water. Such understanding of water transport will give a clearer general picture of how the individual cell, be it a protozoa, a yeast, or an animal or plant cell, maintains its osmotic equilibrium. Despite a century of investigations, there are, as yet, no satisfactory models to explain how water crosses multicellular systems such as epithelia. Our authors suggest some new models, based on mechanisms of water–protein interactions. Finally, two specialized systems are considered: the osmotic regulation of those kidney cells that are exposed to great changes in their bathing osmoticum and the case of the water transport of the eye, in relation to its function.

We see these systems, which interact with and move water, as a hierarchy. Understanding each level of complexity—water with protein, water with enzyme, water with membrane protein, membrane transport with cell, cells within epithelia, external environment with expression level—will enable us to tackle the problems that arise at the next highest complexity level. This volume is meant to provide an aid to such studies.

Thomas Zeuthen, Copenhagen
Wilfred D. Stein, Jerusalem

Protein–Water Interactions

V. Adrian Parsegian
Laboratory of Physical and Structural Biology, National Institute of Child Health and Human Development, NIH, Bethesda, Maryland 20892

By recognizing the forces that drive water transport across cell membranes or across tissues, we can see how water is driven to and from proteins. We learn from examples. When a dissolved protein (bovine serum albumin) accumulates water relative to small solutes, it effectively withdraws a number of water molecules from the bath; the number of water molecules changes with the identity but not with the concentration of small solutes. When a large ionic channel (VDAC or alamethicin) opens, it withdraws water from its bathing solution; excluded solute stabilizes the closed state in proportion to activity of water, the osmotic stress created by the solute, rather than in proportion to the activity of the solute itself. Hemoglobin too acts like an osmometer whose loading of oxygen shifts with the chemical potential of water. Assemblies of many macromolecules (proteins, nucleic acids, polysaccharides, lipids), subjected to the osmotic stress of completely excluded solutes, fight dehydration with powerful, exponentially varying intermolecular forces. Should we speak of these sensitivities and responses as solute effects or water effects? Intuitive but rigorous thermodynamics, developed in a set of appendixes, provides a surprisingly practical guide to alternatives in language.

KEY WORDS: Gibbs adsorption isotherm, Hydration forces, Intermolecular interactions, Ionic channels, Molecular assembly, Osmotic stress, Surface excess/deficit, Water effect.

I. Introduction

A channel on the outer membrane of a mitochondrion opens to pass ATP, while opening it pulls in ~1000 waters. Hemoglobin relaxes upon the binding of oxygen; it draws ~65 waters away from the small solutes in its bathing solution. An enzyme skitters along a DNA double helix, pushes down onto its specific binding site, and squeezes out ~100 molecules of water.

We measure these numbers by watching the consequences of making it more or less difficult for the proteins to draw water from the bathing solution. Just as in bulk water transport, where water moves under the osmosis created by small thirsty solutes, so over the dimensions of macromolecules, water moves under stress. This can be on the millimeter scale of tissues, the micrometer scale of cells, or the nanometer scale of proteins (Fig. 1). Change stress; change flow. Proteins are osmometers. The same thermodynamic potentials that drive water transport also determine a single protein's swelling or a pair of proteins' expulsion of water in mating.

So linked are "water" and "protein" that a mere list with sketches of the techniques and ideologies connecting them would exceed reasonable limits. Surprisingly forgotten in such lists, at least until recent years, is the realization that proteins and other large molecules react directly to the activity of water much as they do to pH and to the activities of ligands. If there are two observable forms of a protein with different amounts of associated water, we can know the volume or number of waters transferred between the solution and the protein in its transit between states. When the protein has a more difficult time taking up water from the bath, the equilibrium ratio between states will shift to favor the drier state.

Why address protein–water interactions from the perspective of water effects? Water is a far more abundant ligand than any other species bathing a protein. Hydration forces have emerged as a major feature of the interactions between macromolecules nearing contact. Within the past few years, many systems have revealed responses that vary directly with osmotic stress.

Superficial interest in hydration deals with the differences in the amounts of water touched in various conformations and associations. More profound inquiry demands answers to questions such as:

- How much energy does it take for the protein to wrest water from the many other solutes in its milieu?

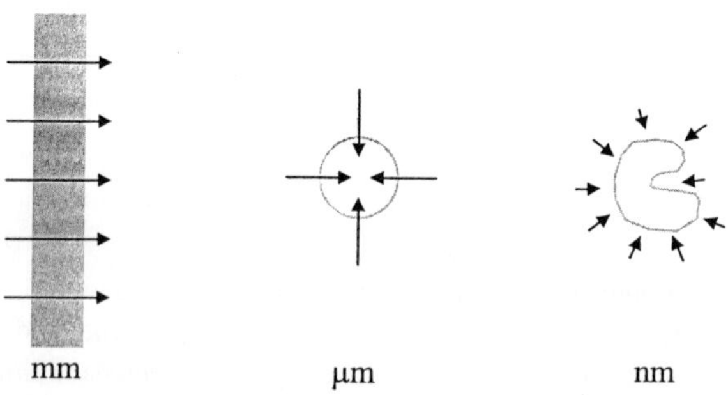

FIG. 1 Flow under osmotic stress, on the millimeter scale of tissues, the micrometer scale of cells, and the nanometer scale of proteins.

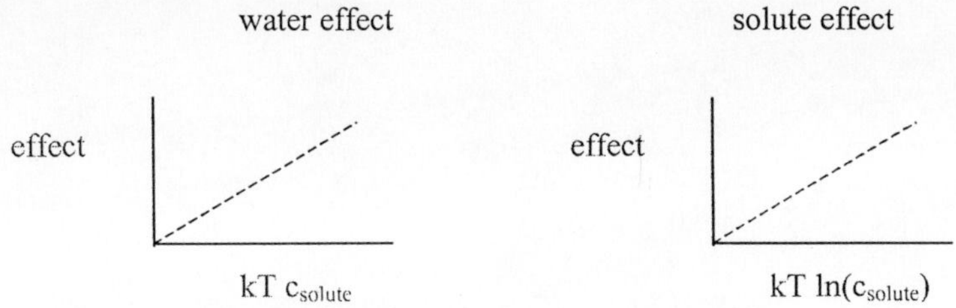

FIG. 2 The quantitative measure of hydration is the proportionality of effect to c_{solute}. The quantitative measure of solute effect is proportionality of effect to $kT\ \ln(c_{\text{solute}})$. The same effect cannot be linear in both c_s and $\ln(c_s)$. (It need not be linear in either.)

- How does one distinguish between the protein's work of wresting water and the protein's direct interactions with the solutes themselves?
- If we ignore the stress of water, do we miss essential features of the conditions under which proteins work?

We decide to speak explicitly of a "water effect" on a protein when the effect is proportional to the chemical potential (or osmotic pressure or stress)[1] of water rather than proportional to the chemical potential of small solutes (Fig. 2). Think of a solute s added to the solution bathing a protein. At low concentration c_s, its chemical potential μ_s varies as $kT\ln(c_s)$. At the same time, the chemical potential of water μ_w varies as $kT\ c_s$, the van't Hoff law for dilute solutions.

II. Baptism by Total Protein Immersion

By measuring solution vapor pressure vs added protein, Courtney *et al.* (2000) show how a protein takes water from small solutes that share its solution. Addition of each protein molecule increases solution osmolality by an increment that is virtually constant over a wide range of protein concentrations. Specifically, in a regime where the number density of proteins is negligible compared with the number density of the small solutes, each added protein surrounds itself

[1]Chemical potential denotes the work needed to transfer a number of molecules. In the particular case of water, equivalent to the chemical potential of water, osmotic stress (or pressure) refers to the energy available to move or to remove a volume of water (as opposed to a number of water molecules). The distinction is operational as well as linguistic. To know the work needed to transfer a volume V_w of water, multiply it by a quantity with units of pressure (Force/Area = Energy/Volume). If that pressure Π_{osmotic} is changed by a small amount, $d\Pi_{\text{osmotic}}$, there is a change $V_w d\Pi_{\text{osmotic}}$ in the amount of work needed to transfer the volume V_w. If that volume V_w corresponds to N_w molecules of water, multiply by chemical potential μ_{water} for the work of transfer and think of the change in chemical potential $d\mu_{\text{water}}$ for the change $N_w\ d\mu_{\text{water}}$ in work. In either case, the operational measure of an osmotic effect is change in proportion to the chemical potential μ_{water}.

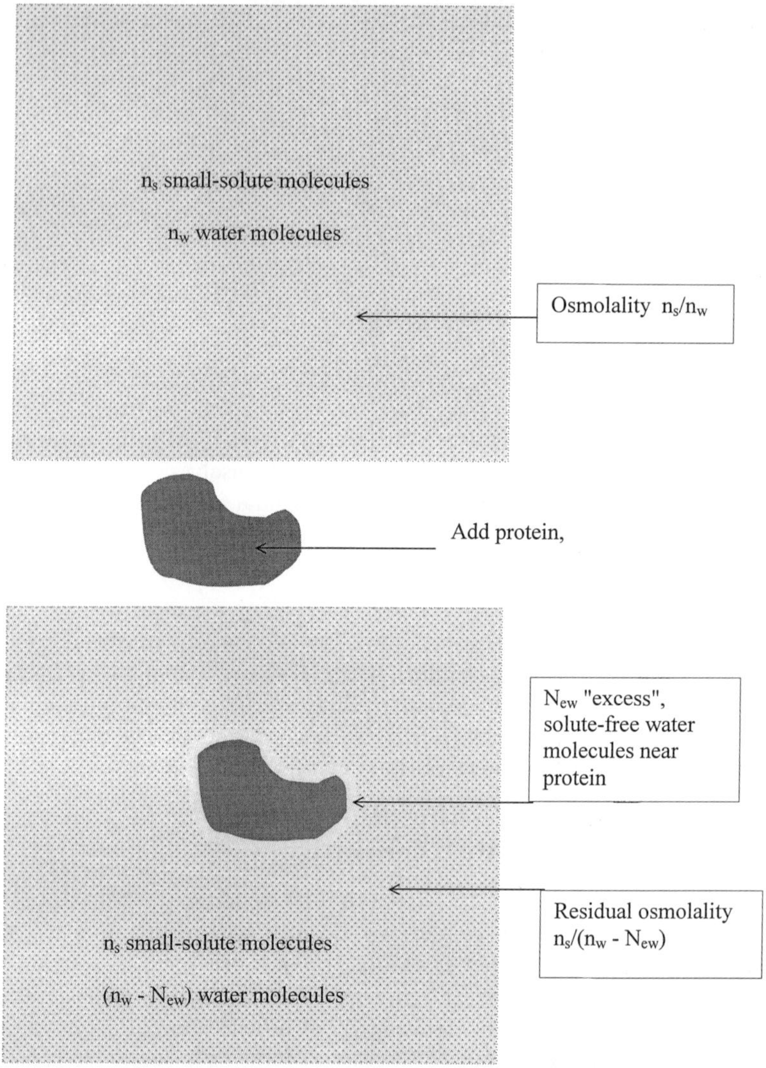

FIG. 3 Exclusion by the protein creates a thermodynamically defined excess number N_{ew} of solute-free waters near the protein (Appendix 1). It is possible to speak of the positive, excess N_{ew} in water molecules as a deficit (or negative excess) of solute, $N_{es} = -N_{ew}\,(n_s/n_w)$ [Appendix 2, Eq. (A2.12)].

with a region of solution depleted of solute. The consequence of the preferential attraction for water is to leave a bathing solution of higher solute concentration (Fig. 3).

More intriguing, the data (Fig. 4) show a factor of 7 difference in this preference for water, from glycerol—the least excluded from near the protein—to

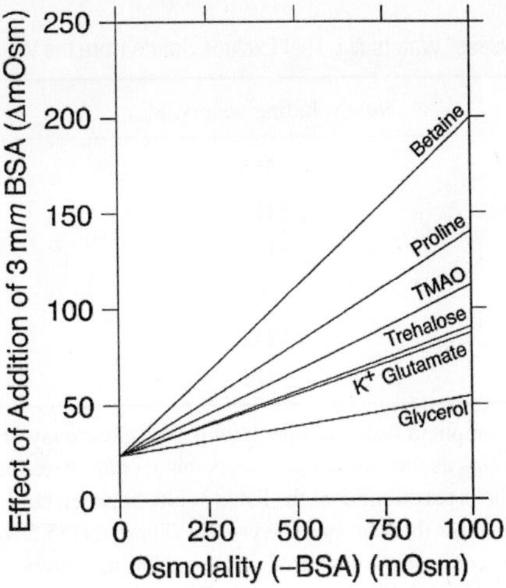

FIG. 4 Change in solution osmolality upon addition of 3 millimolal BSA vs osmolality of the two-component solution without BSA. The measured solution osmolality reflects the interplay of three species. (From Fig. 2 in Courtenay *et al.*, 2000.)

most-excluded betaine. That is, among these solutes, the amounts of water taken by the protein from the small-solute solutions differ by a factor of almost 7 (Table I).

There are many equivalent ways to express the three-way protein–solute–water competition. In the regime where the volume fraction of proteins is tiny, each protein molecule interacts independently with the surrounding solution.[2] In its fight for water (or solutes), the protein disturbs the small-molecule solution in its vicinity. This push–pull, excess–deficit ambiguity touches the limits on how we perceive. We can say "The protein repels solute more strongly than it repels water," or "The protein attracts water more strongly than it attracts solutes." Needless effort is spent dancing on such mental pinheads.

In the present case, there is less solute in this water around the protein than in a similar amount of water in the solution. Think "solute depletion" or "water excess"; it amounts to the same thing. The linearity between solution osmolality and protein molality shows that each added protein competes as effectively as its predecessors in fighting for water against the small solutes in the bath.

[2]The molecular mass of BSA is 66 kg/mol; protein density is ~4/3g/ml; 3 mmol BSA has a volume of 3×10^{-3}mol $\times \frac{66\text{kg}}{\text{mol}} \times \frac{3\text{L}}{4\text{kg}} = 0.15L$, compared to 1 L for 1 kg of water. There are 3×10^{-3} mol $\times$ 0.602×10^{24} molecules/mol $= 1.8 \times 10^{21}$ molecules per 1.15 L $= 1.15 \times 10^3\,\text{cm}^3 = 1.15 \times 10^{27}$ Å^3 of solution. Each protein is allowed an average volume ~ 0.64×10^6 Å^3 or ~ (86 Å)^3, a volume larger than the ~(44 Å)^3 volume of the protein. In fact (E. Courtenay, personal communication), in the actual experiments, there is some interaction between proteins; the relation between total osmolality is not strictly linear. "However, the preferential interaction coefficient at constant water chemical potential is not a function of protein concentration" (E.C.).

TABLE I

The Number of "Excess" Waters N_{ew} That Exclude Solute from the Vicinity of BSA Protein

Solute	No. excluding waters, N_{ew}	$\Gamma_{\mu 1,\mu 2}/m_3'$
Glycerol	555	−10
K^+ Glu^-	1110	−20
Trehalose	1170	−21
TMAO	1499	−27
Proline	1943	−35
Betaine	2775	−50

These N_{ew} are computed from column 2 Table 5 of Courtenay *et al.* (2000). Those authors list $\Gamma_{\mu 1,\mu 2}/m_3'$ as the number of solutes that *would be* excluded by the protein in a 1-molal solution, a partitioning of the listed solute between bulk water and a "local domain" one-water-layer thick around the protein. There are 55.5 waters per solute in a 1-*m* solution. The equivalent number of solute-excluding waters is therefore $55.5 \times \Gamma_{\mu 1,\mu 2}/m_3'$. Note that for those data, because $\Gamma_{\mu 1,\mu 2}/m_3'$ is said to be constant, N_{ew} is constant at all solute concentrations. Solute deficit $N_{es} = \Gamma_{\mu 1,\mu 2}$ is linear in solute concentration. See Appendix 3 for discussion of the different languages and Appendix 2 for the conversion between N_{ew} and N_{es}. The accessible surface area of BSA is given as 29,000 Å^2 (Courtenay *et al.*, 2000). Were the occupation area of a water molecule 9 Å^2, a monolayer of solute-excluding waters would contain $29{,}000/9 \approx 3225$ molecules.

The immersion of the protein raises the chemical potential of the water in the solution (Fig. 4). The magnitude of this consequence is linear in the chemical potential of the water in the pre-protein solution. This change with c_{solute} tells us that there is a change in the amount of work needed to insert the protein. Just as the Gibbs adsorption isotherm tells us that the energy to create an interface changes in proportion to the relative number of each component in its vicinity, so we can say that the change in proportion of solute to water near the protein is the key to connecting the work of immersion to the change in the ratio of small solute s and water w (Appendix 2, Eqs. (A2.7) and (A2.8), and Parsegian *et al.*, 2000):

$$d\mu_{\text{protein immersion}} = -N_{ew}\,d\mu_w = -N_{es}d\mu_s. \quad (1)$$

This formalism already asserts that if the numbers of s molecules and w molecules near the protein are in the same ratio as they are in the original bath, then there is no change on the immersion energy of the protein from adding s or w to the bath. We can ask, What makes this part of the solution around the protein different from the remainder of the solution? It is ratio of the total number of waters to the total number of solutes compared to the ratio (n_s/n_w) that matters.

N_{ew} and N_{es}, schematically portrayed in Fig. 3 and formally elaborated in the appendixes, have physical meaning. In the Gibbs spirit, N_{ew} is the number of waters around the protein that are in excess of the number of waters there would be if the protein did not redistribute w's and s's. We can say that the protein creates an excess number N_{ew} of waters free of solute. The protein might attract solutes relative to waters (or repel water relative to solutes), in which case N_{ew} is negative and N_{es} positive. The sensitivity of protein immersion to solute and solvent chemical potentials is a general idea, as general as the Gibbs isotherm for surface energy. It is the sensitivity of protein free energy to the activities of water and solutes that lets us measure these excesses and, sometimes, to gauge the shape of proteins.

If an experiment shows a linear response to μ_w, then we know that N_{ew} is a constant, and we can speak of a "water effect." In such a case, N_{es} cannot be a constant (Appendix 2). In their linear responses to solution molality, the Courtenay *et al.* (2000) data show precisely the number of water molecules that the bovine serum albumin (BSA) draws to itself from the various small-solute solutions.

The linearity of the data in Fig. 4 tells us that we have a water effect, a change in free energy proportional to changes in water activity. We could think of a solvent-accessible layer clear of solute, but it would be a layer of different thickness for different solutes. Its value depends on all the factors that drive solutes from or toward a surface. But there are many features of proteins that are not "surfaces." Rather than speak of data in terms of particular models, it is better to keep to the numbers N_{ew} that fall out neatly from basic definitions.

III. Transitions between States: Tipping Balances

Rather than look at proteins in only one state, it is more instructive to look at changes in protein–water interactions through transitions between functional states. In the simplest reaction scheme

$$\mathrm{a} \leftrightarrow \mathrm{b} \tag{2}$$

between two states a and b, the probabilities of their occurrence follow the same kind of dependence on free energies that determine the extent of a chemical reaction (Parsegian *et al.*, 1995). These probabilities depend on the difference $(G_b - G_a)$ in the free energy needed to go from a to b. Think of this difference as the work it would take if we had the hands to bring the molecule from one form to the other without relying on the kick of thermal energy to effect this transition. To connect this energy difference with Boltzmann thermal distribution for the probability of being in one state vs. the other, measure $(G_b - G_a)$ in units of the thermal energy kT per molecule (or RT per mole). In reaction language, the ratio of concentrations or probabilities goes as

$$K_{\mathrm{eq}} \equiv \frac{[\mathrm{b}]}{[\mathrm{a}]} = \frac{\Pr(\mathrm{b})}{\Pr(\mathrm{a})} = e^{-(G_b - G_a)/kT} \tag{3a}$$

or

$$\ln(K_{\text{eq}}) = -(G_b - G_a)/kT. \tag{3b}$$

If these two states have different volumes V_a and V_b of solute-excluding water, an added osmotic pressure Π_{osmotic} of the excluded nonbinding solute creates an additional work $\Pi_{\text{osmotic}}(V_b - V_a)$ to go between the two forms. Then K_{eq} changes with Π_{osmotic} as

$$\ln(K_{\text{eq}}) = -(G_b - G_a)/kT - \Pi_{\text{osmotic}}(V_b - V_a)/kT. \tag{4}$$

If measured $kT\ln(K_{\text{eq}})$ is linear in Π_{osmotic}, the slope is the difference $(V_b - V_a)$ in aqueous volumes inaccessible to added solute.

A. Inversion, the Outside-in Surfaces of Ionic Channels

As an immediate corrective to the purely surface thinking of immersion, we turn now to cavities. The hydration event is the opening of an ionic channel bathed by excluded solutes. Then the act of opening requires extra work to separate water from excluded solute, work that necessarily varies with excluded-solute concentration.

Specifically, imagine a cavity too small for entry of a large solute (Fig. 5). It is as though a semipermeable membrane bounded the small space. Creating such a space requires work to purify the water that must be drawn away from the bathing solution to fill that space. Increasing concentrations of excluded solute mean increasing works of purification to draw a volume of water $V_{\text{ew}}^{\text{open}} - V_{\text{ew}}^{\text{closed}}$. With addition of osmotic stress, the free energy of the open form will increase by $V_{\text{ew}}^{\text{open}} \times \Pi_{\text{osmotic}}$ while that of the closed form will go up as $V_{\text{ew}}^{\text{closed}} \times \Pi_{\text{osmotic}}$. The ratio of open vs. closed states will change with Π_{osmotic}:

$$\left.\frac{[\text{Open}]}{[\text{Closed}]}\right|_{\Pi_{\text{osmotic}}} = \left.\frac{[\text{Open}]}{[\text{Closed}]}\right|_{\Pi_{\text{osmotic}}=0} e^{-\Pi(v_{\text{ew}}^{\text{open}} - v_{\text{ew}}^{\text{closed}})/kT}. \tag{5}$$

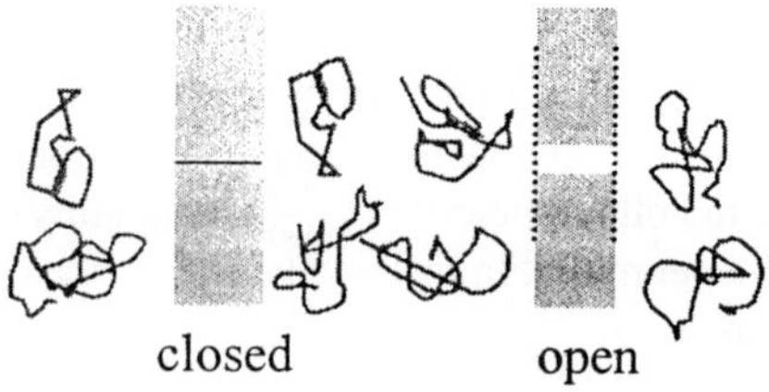

FIG. 5 Open vs closed channel under the osmotic stress of excluded solute.

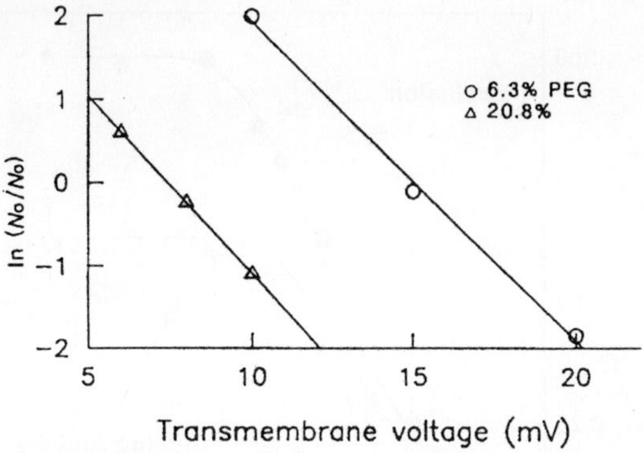

FIG. 6 Shift in [open]/[closed] ratio of the VDAC channel with addition of polyethylene glycol (PEG) polymer osmotic stress. Added PEG decreases the ratio of N_{open}/N_{closed} for the number of times the channel is periodically observed open vs closed. (Data from Vodyanoy *et al.,* 1993; Bezrukov and Vodyanoy, 1993.)

Plotting ln([Open]/[Closed]) vs $\Pi_{osmotic}$ will give a slope $-\left(V_{ew}^{open} - V_{ew}^{closed}\right)/KT$. That is, the open/closed ratio vs osmotic strength gives the difference in solute-inaccessible volumes of water associated with the two forms of the channel. Several mesoscopic channels, particularly VDAC (Zimmerberg and Parsegian, 1986) (Fig. 6) and alamethicin (Vodyanoy *et al.,* 1993; Bezrukov and Vodyanoy, 1993) (Figs. 7, 8), have been investigated.

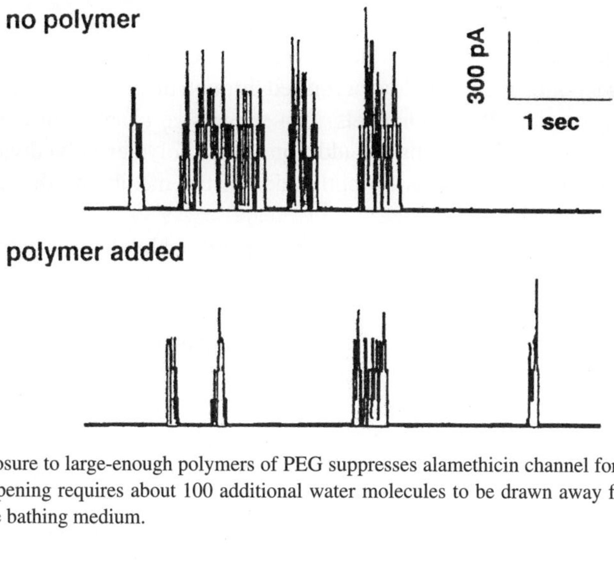

FIG. 7 Exposure to large-enough polymers of PEG suppresses alamethicin channel formation. Each successive opening requires about 100 additional water molecules to be drawn away from the large solutes in the bathing medium.

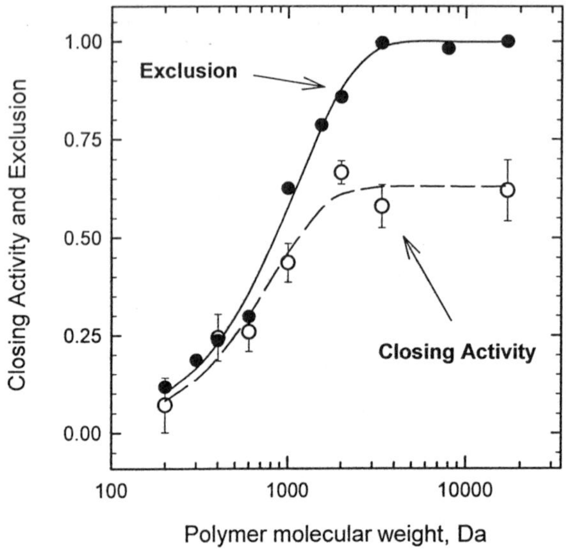

FIG. 8 Different-size PEGs reveal different volumes of solute-inaccessible water. Large PEGs are completely excluded from the open channel and exert maximum effect on closure. Smaller PEGs, able to enter the open channel, exert little power to close the channel. (Figure courtesy of S. Bezrukov, drawing on data from Bezrukov and Vodyanoy, 1993, and Vodyanoy *et al.,* 1993.) Closing activity (dashed line, open circles) is measured as $-\ln[([\text{open}]/[\text{closed}])_{\text{with PEG}}/([\text{open}]/[\text{closed}])_{\text{no PEG}}]$. Exclusion (solid circles, solid line) is derived from relative conductance of channels, those bathed in PEG solution vs channels in PEG-free solution.

With ionic channels, it is possible to watch the occurrence of open vs closed states by the amount of current flow permitted through them. Consider an electrical current burst of alamethicin channels with successive changes in conductance. Suppression of channel opening by addition of the polymer polyethylene glycol (PEG) to the bathing solutions on both sides of the membrane that houses the channel is seen in Fig. 7.

It is easy to visualize solute exclusion that operationally defines "large" solutes (Fig. 5). It is almost as easy to see that the degree of exclusion can vary with solute size. Large, completely excluded solutes are not equivalent to small solutes that enter the cavity. There will usually be a smaller apparent change in the difference $(V_{\text{open}} - V_{\text{closed}})$ in volume that is accessible to the smaller solutes. Figure 8 shows the variable power of PEGs to close alamethicin channels osmotically à la Eq. (5). Large PEGs exert a maximum osmotic stress and consequent closing activity (open circles, dashed lines). The power to close is almost lost with smaller PEGs. Measured open-channel conductance reveals that large polymers stay out of the channel, but smaller PEGs enter and obstruct ionic current (Vodyanoy *et al.,* 1993;

Bezrukov and Vodyanoy, 1993). Figure 8 shows that the power to exert osmotic stress and the degree of polymer exclusion are strongly correlated.

B. Protein Surfaces and Cavities at the Same Time: Hemoglobin

Crudely but instructively, we can distinguish two kinds of protein association with water (cf. Peter Rand's chapter, "The Lipid–Water Interface: Revelations by Osmotic Stress.") First the protein "surface" perturbs water in its vicinity; this perturbation can exclude solutes, as indicated in the BSA-immersion measurements. Second, aqueous cavities such as some ionic channels will exclude large solutes though the cavity is large enough that water properties are not strongly perturbed. Hemoglobin shows both kinds of association.

Water is an effector of hemoglobin (Colombo *et al.,* 1992). Much as with the change in open/closed ratios of channels with the addition of excluded thirsty solutes, loading of oxygen is shifted with the addition of small solutes as various as NaCl, glucose, and PEG.

This classic allosteric protein was subjected to different pressures, pO_2, of oxygen at constant temperature, (atmospheric) pressure, pH, ionic conditions, but the oxygen titration was conducted in solutions with different concentrations of a neutral solute "s." The idea was to see how the presence of these solutes changes the oxygen loading curve and, consequently, to learn about the competition between these neutral solutes and water interacting with the hemoglobin molecule.

The chemical potential of oxygen will go as $d\mu_{O2} = kT\, d\ln(pO_2)$ because the pressure of oxygen is low enough for it to be treated as an ideal gas. The work done on a hemoglobin molecule by oxygen pressure and water/solute activity (at constant temperature and pressure) can be written in a general but impractical form as

$$d\mu_{\rm Hb} = -N_{O_2}\,{\rm d}\mu_{O_2} - N_w\,d\mu_w - N_s\,d\mu_s. \tag{6}$$

Because the chemical potential of added solute is connected with the chemical potential of water by the Gibbs–Duhem constraint, $n_s\,d\mu_s + n_w\,d\mu_w = 0$ (Eq. (A2.3)),

$$-d\mu_{\rm Hb} = N_{O_2}\,d\mu_{O_2} + N_{\rm ew}\,d\mu_w = N_{O_2}\,d\mu_{O_2} + N_{\rm es}{\rm d}\mu_s. \tag{7}$$

Consider, for example, the observation that hemoglobin takes up less oxygen at a given pO_2 when neutral solutes are added to the solution (Fig. 9). What can we learn from this observed decrease in bound oxygen N_{O_2} with added solute? Using the cross derivative, $\partial N_{O_2}/\,\partial\mu_s|_{\mu_{O_2}}$ can be interpreted as

(*a*) The decrease in the relative amount $N_{\rm es}$ of solute as oxygen is loaded:

$$\partial N_{O_2}/\partial\mu_s\,|_{\mu_{O_2}} = \partial N_{\rm es}/\partial\mu_{O_2}\big|_{\mu_s}. \tag{8}$$

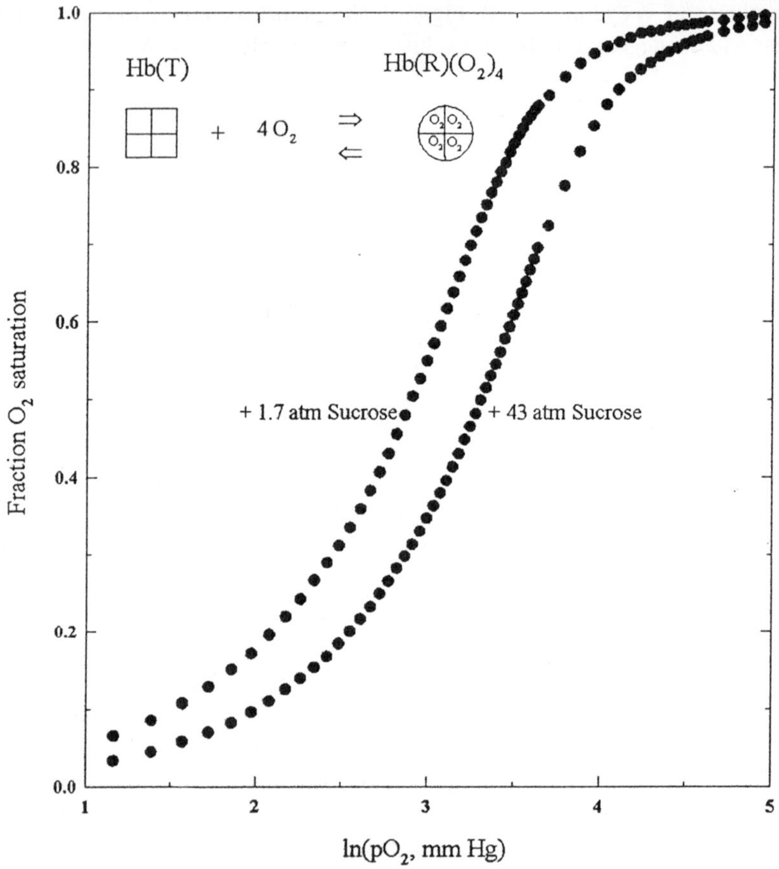

FIG. 9 Osmotic suppression of oxygen loading by tetrameric hemoglobin A.

(*b*) The increase in the relative amount of water $N_w\,[1-(N_s/N_w)(n_w/n_s)]$ associated with hemoglobin as it loads oxygen:

$$\partial N_{\mathrm{O}_2}/\partial \mu_w\,\big|_{\mu_{\mathrm{O}_2}} = \partial N_{\mathrm{ew}}/\partial \mu_{\mathrm{O}_2}\big|_{\mu_w}. \tag{9}$$

We can assign these changes to the population of two forms of hemoglobin, or we can treat them as averages over all states, as written here. That is all that pure thermodynamics tells us (but we are allowed to use our brains to decide which kind of picture makes more sense). In either case, we see the combined acts of adding oxygen and adding solute as ways to work on hemoglobin. From the comparative action of each, it is possible to infer an excess or deficit of water or solute “bound” to the hemoglobin with an energetic implication every bit as meaningful as the binding of oxygen itself. The shift in loading curves, $\partial N_{\mathrm{O}_2}/\partial \mu_w\,\big|_{\mu_{\mathrm{O}_2}}$, integrated

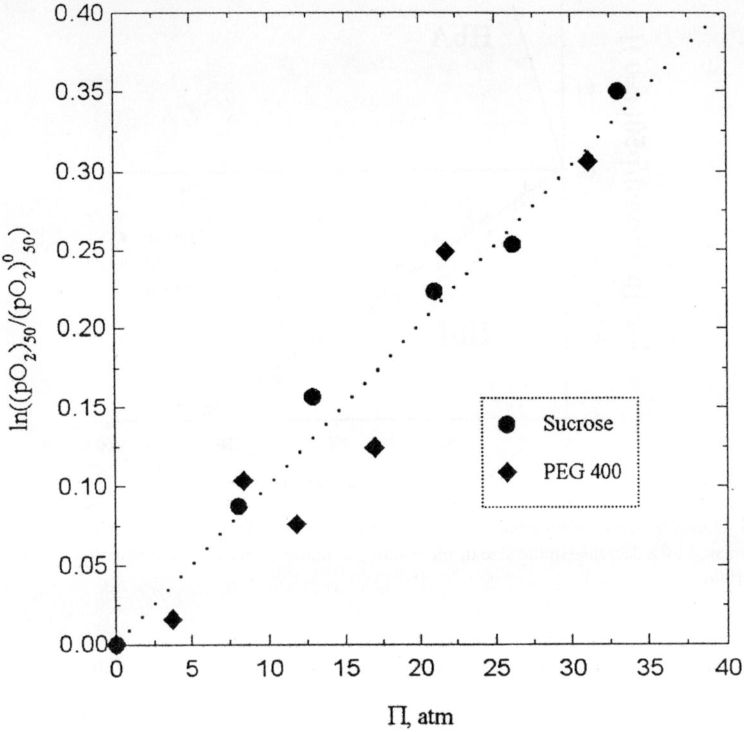

FIG. 10 Osmotic stress-induced shift in $p_{50}(O_2)$, the oxygen pressure required to half-load hemoglobin. Equivalent effects of two different osmotic agents.

over the full range of oxygen pressures, gives

$$N_{\text{ew}}^{\text{oxy}} - N_{\text{ew}}^{\text{deoxy}} = \int_{\mu_{O_2}=0}^{\mu_{O_2}=\text{“}\infty\text{”}} \left.\frac{\partial N_{\text{ew}}}{\partial \mu_{O_2}}\right|_{\mu_w} d\mu_{O_2}$$

$$= \int_{\mu_{O_2}=0}^{\mu_{O_2}=\text{“}\infty\text{”}} \left.\frac{\partial N_{O_2}}{\partial \mu_w}\right|_{\mu_{O_2}} d\mu_{O_2} \sim 65 \text{ waters.} \tag{10}$$

The same number emerges from the shift in $(pO_2)_{50}$, the O_2 pressure at which hemoglobin is 50% loaded with oxygen. In the spirit of allostery (Wyman and Gill, 1990), we imagine a step-like addition Δn_{O_2} of four oxygen molecules at this halfway point $(pO_2)_{50}$ (Fig. 10). The change in $d\mu_{O_2}^{50} = d\left[kT \ln\left(p_{O_2}^{50}\right)\right]$ with change $d\mu_w = -\bar{v}_w\, d\Pi$ in water activity looks like a Clausius–Clapeyron equation,

$$\frac{d\mu_{O_2}^{50}}{d\mu_w} = \frac{\Delta n_w}{\Delta n_{O_2}}. \tag{11}$$

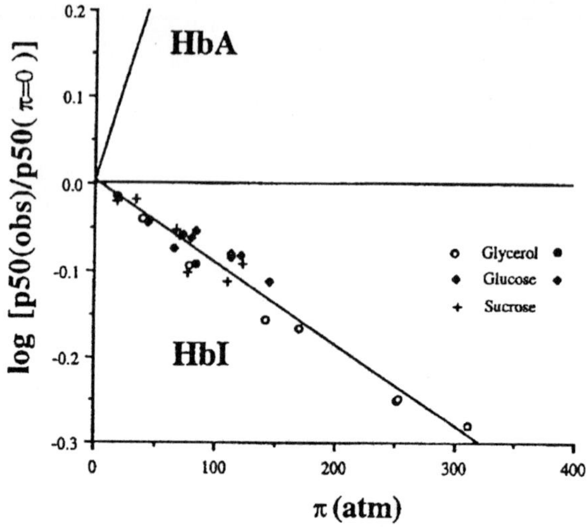

FIG. 11 Osmotic stress aids loading of oxygen by HbI, a dimeric form of hemoglobin whose response is dominated by a decrease in the size of the solute-excluding cavity upon oxygenation. (From Royer *et al.,* 1996.)

The measured slope $d\mu_{O_2}^{50}/d\mu_w$ and the known Δn_{O_2} give the change in excess waters,

$$\Delta n_w = \frac{d\mu_{O_2}^{50}}{d\mu_w}\Delta n_{O_2}. \tag{12}$$

Further analysis suggests that this gain in waters with oxygenation is a combination of two kinds of events. Hemoglobin has a cavity that is *smaller* in the oxy form; if osmotic sensitivity were due to this cavity, the oxy form would have fewer rather than more waters. Royer *et al.* (1996) have examined the osmotic sensitivity of a dimeric HbI form of hemoglobin that preserves this cavity but does not show any of the hydration changes that go with changes in tetrameric packing (Fig. 11). They nicely confirm that HbI loads oxygen more easily when it is subjected to osmotic stress. There is a decrease of ~6.2 waters per dimer upon oxygenation.

The net change of ~65 waters for the tetramer is likely due to a positive term from increased surfaces, as in the BSA immersion measurements, offset by a cavity term, similar to the solute exclusion seen in channels.

IV. Intermolecular Forces: Forces of Solvation

From the perspective of work on single molecules, it is almost too easy to proceed to the idea of doing work on configurations of molecules from which one

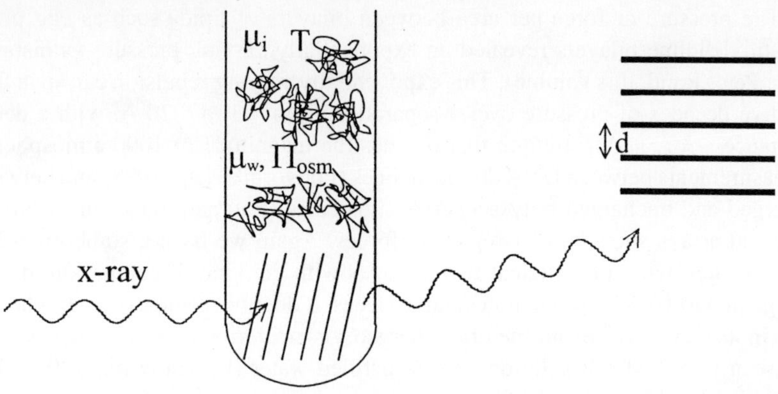

FIG. 12 Scheme of an array of macromolecules under stress of excluded polymer. The aggregate acts as a single unit in a reservoir of solutes i and water w of varied chemical potential μ_i and μ_w or Π_{osmotic}.

particular solute is completely excluded. Large polymers cannot enter the small spaces between molecules in the lamellar arrays of membranes, the hexagonal arrays of DNA, or any of a great number of macromolecular phases.

Here, the system observed is the array as a whole rather than a single molecule (Fig. 12). It responds to the varied stress of excluded polymers (as opposed to small solutes), to the osmotic stress of competition for water, as well as to all conditions in the bathing reservoir. The change in chemical potential of water is singled out for special consideration as the operative parameter for force measurement. A change $d\mu_w$ equals a change $d\Pi_{\text{osmotic}}$ except for a difference in sign convention and for the factor of molecular (or molar) volume of water $\bar{v}_w$,

$$d\mu_w = -\bar{v}_w \, d\Pi_{\text{osmotic}}, \; \bar{v}_w \, N_w = V_w, \tag{13}$$

where V_w is the total volume of water in the array, and N_w is the number of molecules or moles of water:

$$dG(T, p, \mu_w, \{\mu_i\}, ..) = -SdT + Vdp + V_w \, d\Pi_{\text{osmotic}} - \sum_i N_i \, d\mu_i. \tag{14}$$

X-ray scattering gives the spacing and the arrangement in the array as a function of the forces exerted by external polymer as well as of the chemical potentials of all species that can exchange between the array and the sample. Usually the measured spacings d can be immediately connected with the volume of aqueous solution, which is mostly the volume of water. The applied osmotic stress Π_{osmotic} coupled with d can often be interpreted in terms of the intermolecular force with which the array fights back against the compression of stress (Parsegian *et al.,* 1986). The forces that have been measured have been summarized in several places (e.g., Leikin *et al.,* 1993; Podgornik *et al.,* 1998).

The pressure or force per area between bilayers of lipids such as egg phosphatidylcholine bilayers revealed an exponentially varying pressure vs distance (see Peter Rand, this volume). This exponentially varying repulsion can span four or five decades of pressure over a separation from ~3 to ~20 Å, with a decay distance ~3 Å and a strength that reaches the magnitude of 1000 atmospheres. Measurements between DNA double helices (Podgornik *et al.,* 1998) and between charged and uncharged polysaccharide molecules (Rau and Parsegian, 1990) in ordered arrays show similar repulsive forces. Again we have a stubborn ~3-Å exponential, with a coefficient that changes with the kind of counterion, over a range of 4 to 10 Å between molecular surfaces. It does not seem to matter whether the molecular surfaces are neutral, charged, or zwitterionic. Similar decays can be seen in ~2 M salt solution and in distilled water (Leikin *et al.,* 1993). The commonality of exponential forces in so many different systems leads one to think in terms of H-bonding and water structuring. It appears as though a law of nature has been given us: *exponentially varying forces dominate over the important last 10 Å where molecules come into contact* (Fig. 13).

Proteins are more interesting. Structured molecular surfaces such as those of collagen also interact exponentially but with a decay constant that combines the ~3-Å λ_{hyd} of water with the natural lengths of the structured surface (Leikin *et al.,* 1995). For example, because of structural contributions, collagen triple helices show an exponential repulsion with a decay length of $\lambda_{\mathrm{exptl}} = 0.65$ Å (Fig. 14), a distance that can be rationalized as a coupling between natural lengths of water and of collagen (Leikin *et al.,* 1995). The disturbance of water near a protein has now been seen in X-ray diffraction by insulin (Badger and Caspar, 1991; Yu and Caspar, 1998).

And a new idea has emerged. Variation in the ordering of water near large molecules goes with spatial variation in the ability of that water to dissolve salts and neutral solutes. Recent work on polysaccharides and on DNA suggests that solutes often distribute near macromolecules with an exponential variation in concentration $\sim e^{-r/\lambda}$ whose λ is similar to that seen with hydration forces. The salt concentration gradient generated by the hydroxypropylcellulose (HPC) surface itself appears to be an exponential function with the same 3- to 4-Å decay length as the hydration force (Bonnet-Gonnet *et al.,* 2000). The implication is that salt interacts with HPC through hydration or water structuring forces. Having seen this partitioning of NaCl in progressively ordered water, we are now in a position to ask how solutes redistribute between all kinds of molecular surfaces including those of proteins. Such redistributions themselves create forces between molecules (Appendix 5).

Again, we are working with thermodynamic forces, spatial derivatives of free energy. The temperature dependence of forces automatically measures entropy vs separation. When such a measurement has been possible on cleanly ordered systems, the entropy that emerges is itself exponential with an ~3-Å exponential

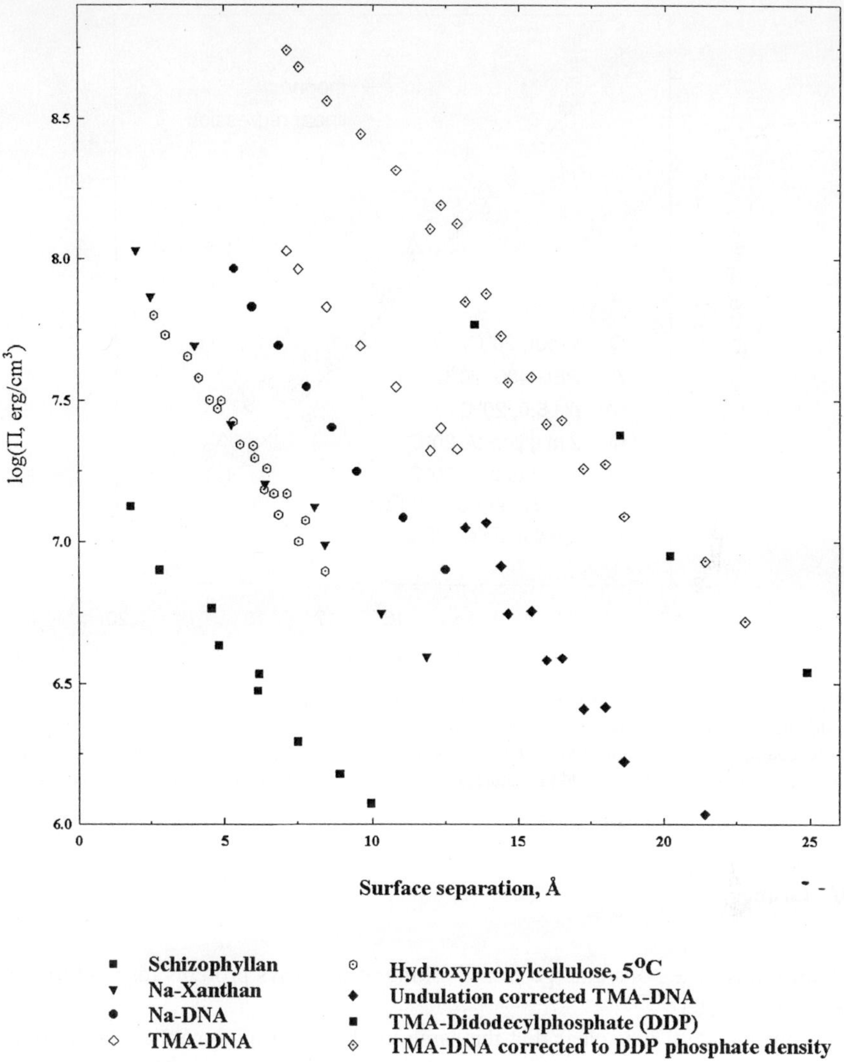

FIG. 13 Stress vs separation between macromolecules or membranes at different distances between surfaces. (From Leikin *et al.,* 1993.)

decay (Leikin *et al.,* 1991). For example, it is easier to push together HPC at higher temperatures; the entropy of the total system—HPC + bathing solution—increases as the molecules are brought together (Bonnet-Gonnet *et al.,* 2000). One possibility for the source of this entropy is the progressive release of increasingly ordered water as macromolecules are pushed together.

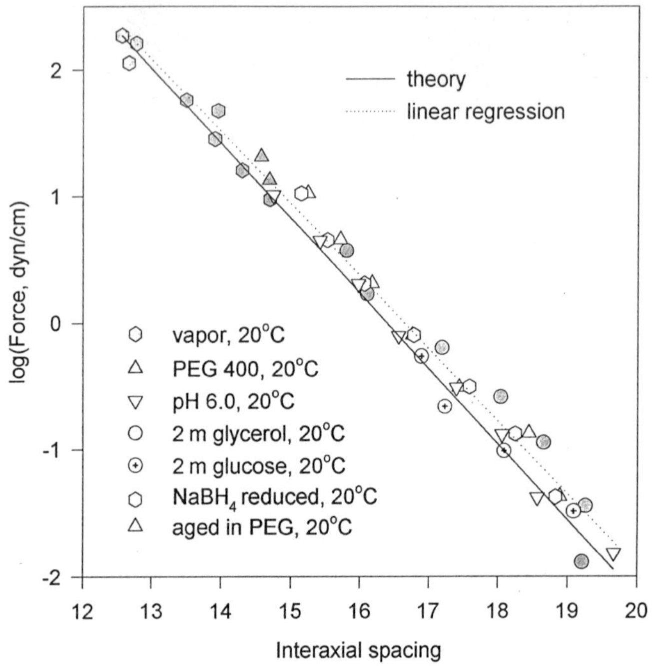

FIG. 14 Exponential force between collagen triple helices. In this case, the force extends over 5 decades. It is robust in the presence of a wide variety of solutes and conditions. The theory used to rationalize this exponential force combines the natural lengths of the water hydration force seen between unstructured surfaces and the natural length in the repeat period of the collagen triple helix.

V. Unity

What kinds of lessons emerge from looking at the ways proteins respond to the activity of water?

From the start, it is clear that the same kinds of thermodynamic forces drive water transport over all length scales. More difficult to see is that a protein is large enough that its attraction or repulsion of solutes can be spoken of in the same terms as attraction and repulsion of solutes by a macroscopic surface. This macroscopic-equivalence feature immediately liberates us to use classical concepts from the thermodynamics of surfaces. It also disciplines us to learn how to make the conceptual transition between smooth surfaces and molecular bodies.

The mind craves the certainty of structure, eschews disorder. Yet disorderly, continuously changing, diffusive molecules surround proteins and surfaces. The protein "surface" itself is rough and disorderly on the scale of molecular structure and atomic motion. We often speak of protein solvation in the geometric sense, for example, the numbers of water molecules that could coat the solvent-accessible area of a crystalline protein structure. In contrast, thermodynamics begs most structural questions, but it allows us to think about the cost of creation. It gives us procedures to ask, What composes this collection of waters and solutes that dwell in the protein's space? How do solutes and solvent drive proteins between different functionally distinct states?

In principle, a protein is in touch with everything around it. In practice, the protein's sensitivity to the price, or chemical potential, of all species depends only on the number of each component compared to the number that would be in that space were the protein not there. The cost of putting a protein into the solution or the cost of moving the protein between different forms depends on time-average, relative numbers of solutes, and solvent. Singular focus on structure misses essential mixing and averaging.

When it comes to numbers, at 55 M, water is the acknowledged winner. By any measure, there are more water molecules around a protein than any of the other small molecules in the solution. Numerical supremacy matters. The energetic consequence of a bathing species is proportional to its numerical excess or deficit (again compared with what would be there if there were no protein). That excess or deficit, multiplied by a change in chemical potential, gives the change in protein free energy. From this numbers perspective, too many labs still neglect the motive powers of water. With careful attention rightfully paid to the chemical potential of protons, ligands, small solutes, salts, etc., there is still relatively little attention given to the chemical potential of the most numerous "ligand," water. By intellectualizing first rather than looking at water effects in the same spirit as responses to pH or temperature or ligands, people have created unnecessary biases and headaches. Computation and discussion based on rigid structures miss the averaging that a living protein uses to respond to water and solute activity.

Several examples in this chapter and elsewhere in this book show that we can think profitably of water effects as well as of the effects of other solutes. Given that we vary the chemical potential of water by adding solutes, are we playing language games? Certainly there is some freedom in choice of language. We can say, "There are 55-glycerol-excluding waters" and speak of the effect of these waters in the context of free energy of solution. We can also say "There are 10 glycerols fewer than the number of glycerols that would be around the protein if it were in a freely mixing 1 molal glycerol solution." The 55 waters remains a pleasingly constant number for proteins in 0 to 1 molal solution. That number is the constant coefficient for sensitivity of protein immersion to water activity over that same range of added solute.

In any case, an ability to focus alternately on solute or solvent as the active agent presents an opportunity. Though it requires linguistic agility to speak alternatively in terms of solute or solvent action, we can escape the mental confines of either language. With this freedom to speak in the most general terms about protein–water interactions, we can go beyond thinking in terms of rigid models. We can see the dynamics of flexible proteins in solution not only in terms of energetics of contact with water but also in terms of a protein's ongoing competition for the water in the course its activities. We might see, too, how the energies that have been revealed in hydration forces can drive protein function.

Glossary

μ_i chemical potential of species i

n_i total number of molecules of species i

w water, species #1 in 1–2–3 notation

p protein, species #2

s small solute, species #3

$\bar{v}_w$ molecular (or molar) volume of water

Π, osmotic pressure or stress, sometimes written Π_{osmotic} for emphasis; $\bar{v}_w\, d\Pi = -n_w\, d\mu_w$.

$m_3 = m_s$, molality of small solutes, moles per kilogram of water, proportional to $n_3/n_1 = n_s/n_w$.

N_{ew}, number of waters in excess of what there would be were there no preferential interaction of solute and water with protein. $N_{\text{ew}} \equiv 0$ for no net redistribution of solute and water. ΔN_{ew} or $\Delta N_{\text{ew}}^{\text{ab}} = N_{\text{ew}}^{\text{b}} - N_{\text{ew}}^{\text{a}}$, difference in the number of excess waters for two different states a, b of a protein.

N_{es}, ΔN_{es} or $\Delta N_{\text{es}}^{\text{ab}} = N_{\text{es}}^{\text{b}} - N_{\text{es}}^{\text{a}}$, the same excesses and differences for a small solute s.

$V_{\text{ew}} = N_{\text{ew}}\, \bar{v}_w$, volume corresponding to excess number of water molecules; $\Delta V_{\text{ew}} = (V_{\text{ew}}^{\text{b}} - V_{\text{ew}}^{\text{a}}) = N_{\text{ew}}^{\text{ab}} \bar{v}_w$, difference in volume of excess or small-solute-inaccessible water between forms a and b.

$\Delta G_{\text{ab}} = G_b - G_a$, free energy difference or work necessary for an individual macromolecule to go between the two forms or states, a, b.

$K_{\text{eq}} = e^{-\Delta G_{\text{ab}}/kT} = \frac{[\text{b}]}{[\text{a}]} = \frac{\text{Pr(b)}}{\text{Pr(a)}}$, concentration ratio or probability ratio of b and a.

Also see Appendix 3.

Appendix 1

How to Connect the Change in Solution Osmolality by Immersion of a Protein with the Change in the Work of Protein Immersion as a Function of Solution Osmolality

Think numbers and number densities. Think of a solution consisting of n_w molecules of water and n_s molecules of a small solute s (glycerol, trehalose, proline, betaine, tetramethylamine N-oxide (TMAO), K^+ glutamate). Its osmolality is proportional to the ratio n_s/n_w. (Though osmolality is measured as moles solute per kilogram of mass solvent, it is essentially a ratio of numbers.) Next imagine that immersion of one protein molecule, the first protein added, withdraws N_{ew} water molecules from the solution. The operational reality of N_{ew} will be spelled out below. For now, imagine s being repelled from the protein relative to water or, the competitive equivalent, water attracted to protein relative to solute s.

After the protein removes N_{ew} waters from the original solution, there are $(n_w - N_{\text{ew}})$ waters available to mix with the n_s solutes. The new osmolality is proportional $n_s/(n_w - N_{\text{ew}})$; the change in osmolality is proportional to $\frac{n_s}{n_w - N_{\text{ew}}} - \frac{n_s}{n_w}$. In an effectively infinite bath, $n_w >>> N_{\text{ew}}$, the change in osmolality of the remaining solution, after the first protein's immersion, is $\frac{n_s}{n_w - N_{\text{ew}}} - \frac{n_s}{n_w} = \frac{n_s}{n_w}\frac{N_{\text{ew}}}{n_w}$. Two protein molecules will withdraw $2N_{\text{ew}}$ waters and change osmolality by $\frac{n_s}{n_w}\frac{2N_{\text{ew}}}{n_w}$. As long as the solution is dilute in the number of proteins, n_p ("dilute" defined as $n_p N_{\text{ew}} <<< n_w$ and n_s), added proteins will withdraw $n_p N_{\text{ew}}$ waters and the solution osmolality will go as $\frac{n_s}{n_w}\frac{n_w}{n_w} n_p$, i.e., proportional to the number density of proteins.

More strictly speaking, we should say that the result of adding proteins is to create solutions whose vapor pressure, or chemical potential of water, is that which is measured for a solution of s and w consisting of n_s small-solute molecules and $n_w - n_p N_{\text{ew}}$ water molecules. It is this vapor pressure that is measured in the exacting Courtenay *et al.* (2000) measurements.

The osmolar consequence of added protein is linear in the original osmolality n_s/n_w. It is also reportedly, almost surprisingly, effectively linear in the number of added proteins n_p, or equivalently, linear in n_p/n_w. In other words, these two proportionalities are what is seen in $\frac{n_s}{n_w}\frac{N_{\text{ew}}}{n_w} n_p = \frac{n_s}{n_w}\frac{n_p}{n_w} N_{\text{ew}}$. The linear master plot, the effect on osmolality of adding a given amount of protein, vs original osmolality n_s/n_w immediately gives the numbers of excluding waters N_{ew}. Its slope is $\frac{n_p}{n_w} N_{\text{ew}}$; the data are for a given, known n_p/n_w. Voila N_{ew}!

Appendix 2

What Does It Mean to a Protein That It Sequesters N_{ew} Excess Waters When Immersed into the n_s, n_w Solution?

For learning purposes only, just for the next few paragraphs, imagine a spherical protein of radius a plunged into an effectively infinite bath. Its push/pull on waters and solutes creates local number densities $\nu_s(r)$ and $\nu_w(r)$. At large enough radial position r, far from the protein, these number densities go to those of the bathing solution. But let us not restrict ourselves *a priori* to saying how far is "far." In principle, the domain of perturbation can extend indefinitely, and we can think of the protein and an absurdly infinite number of neighbors, $\mathbf{N}_w$ and $\mathbf{N}_s$ (for which I have chosen an absurdly shaped $\mathbf{N}$),

$$\mathbf{N}_w = \int_a^\infty \nu_w(r)4\pi r^2 dr,\ \mathbf{N}_s = \int_a^\infty \nu_s(r)4\pi r^2 dr. \qquad \text{(A2.1)}$$

From the protein's point of view, not only is it situated among an infinite number of s's and w's, but also these s's and w's come at a price. Recall that the chemical potential μ_s or μ_w is the energy it takes to add a molecule of s or a molecule of w to the original bathing solution. An increase in μ_s means it takes more work to add a molecule of s *to* the bathing solution; an increase in μ_w, more work to add a molecule of water to the bathing solution.

If we think of the protein and its surrounding molecules as an entity, if we think of these neighbors as belonging to the dissolved protein, then raising the chemical potential μ_s or μ_w in which the protein is bathed makes it easier for the protein to acquire s or w *from* the bath.

It is as though the protein "owns" $\mathbf{N}_w$ waters and $\mathbf{N}_s$ solutes. If μ_w goes up, it is harder to transfer w to the bathing solution, easier to move w to the protein. It is as though the protein has paid less for its collection of magnitude $\mathbf{N}_w$. Ditto with an increase of μ_s and the protein's possession of $\mathbf{N}_s$ solutes. The act of protein immersion would change by an amount $-\mathbf{N}_w\, d\mu_w$ and $-\mathbf{N}_s\, d\mu_s$:

$$dG_{\text{protein}}^{\text{immersion}} = -\mathbf{N}_w \mathrm{d}\mu_w - \mathbf{N}_s d\mu_s. \qquad \text{(A2.2)}$$

The reason for the minus sign should be obvious. The greater the number $\mathbf{N}_s$ or $\mathbf{N}_w$ in the domain of the protein, the less the free energy of collecting those w's and s's with an increase in μ_s or μ_w in the surrounding solution.

Now we can eliminate the absurdity of infinitely many $\mathbf{N}_w$ waters and $\mathbf{N}_s$ solutes. Think of two different immersions into two different solutions whose chemical potentials differ slightly, by $d\mu_w$ and $d\mu_s$. From the Gibbs–Duhem relation, any change in chemical potentials suffers the constraint

$$n_w d\mu_w + n_s d\mu_s = 0. \qquad \text{(A2.3)}$$

Chemical potentials cannot change independently.[3] We can say

$$d\mu_w = -(n_s/n_w)d\mu_s \text{ or } d\mu_s = -(n_w/n_s)d\mu_w \qquad \text{(A2.4)}$$

to allow us to say

$$dG_{\text{protein}}^{\text{immersion}} = -[\mathbf{N}_w - (n_w/n_s)\mathbf{N}_s]d\mu_w$$

or

$$dG_{\text{protein}}^{\text{immersion}} = -[\mathbf{N}_s - (n_s/n_w)\mathbf{N}_w]d\mu_s. \qquad \text{(A2.5)}$$

The energy of protein immersion changes with chemical potential of the bath only to the extent that the ratio of s's to w's, $\mathbf{N}_s/\mathbf{N}_w$, differs from their ratio (n_s/n_w)

[3]Recall that, for $n_s << n_w$, this $n_w d\mu_w + n_s d\mu_s = 0$ condition gives the van't Hoff law, osmotic pressure due to s added to w: $d\mu_s = d(kT \ln(n_s/n_w)$, $d\mu_w = -\, v_w\, d\, \Pi_{\text{osmotic}}$, where v_w is the volume of a water molecule.

in the original solution. If the protein did nothing to the distribution of s and w, then there would be no consequence of changing the amount of s compared to w in the bathing solution.

- If the protein accumulates s's relative to w's, if $\mathbf{N}_s/\mathbf{N}_w > n_s/n_w$, then increased μ_s (or decreased μ_w) lowers the work to put the protein into the solution.
- If the protein accumulates w's relative to s's, if $\mathbf{N}_s/\mathbf{N}_w < n_s/n_w$, then increased μ_s (decreased μ_w) increases the work of putting the protein into the solution. With more s in the bath relative to w, it is harder to get the desired w's out of the n_s, n_w bath.

Return now to the distributions $\nu_s(r)$ and $\nu_w(r)$ where we had

$$dG_{\text{protein}}^{\text{immersion}} = -\mathbf{N}_w d\mu_w - \mathbf{N}_s d\mu_s = \int_a^\infty [-\nu_w(r)d\mu_w - \nu_s(r)d\mu_s]\, 4\pi r^2 dr, \tag{A2.6}$$

but add the necessary constraint between $d\mu_w$ and $d\mu_s$,

$$dG_{\text{protein}}^{\text{immersion}} = -\left\{\int_a^\infty [\nu_w(r) - \nu_s(r)\,(n_w/n_s)]\, 4\pi r^2 dr\right\} d\mu_w \equiv -N_{\text{ew}} d\mu_w \tag{A2.7}$$

or, equivalently,

$$dG_{\text{protein}}^{\text{immersion}} = -\left\{\int_a^\infty [\nu_s(r) - \nu_w(r)\,(n_s/n_w)]\, 4\pi r^2 dr\right\} d\mu_s \equiv -N_{\text{es}} d\mu_s. \tag{A2.8}$$

Now the numbers multiplying the $d\mu_w$ and $d\mu_s$ are reasonable quantities. Despite the integration to "infinity," N_{ew} and N_{es} are finite numbers. The local number density $\nu_w(r)$ is reduced by a factor $[1 - (\nu_s(r)/\nu_w(r))(n_w/n_s)]$ that goes to zero at large enough distances from the protein (where $\nu_w(r)/\nu_s(r) \to (n_w/n_s)$); similarly $\nu_s(r)$.

The excess numbers of waters and solutes are then

$$N_{\text{ew}} = \int_a^\infty [\nu_w(r) - \nu_s(r)\,(n_w/n_s)]\, 4\pi r^2 dr \tag{A2.9a}$$

and

$$N_{\text{es}} = \int_a^\infty [\nu_s(r) - \nu_w(r)\,(n_s/n_w)]\, 4\pi r^2 dr. \tag{A2.9b}$$

Departing from the fantasy of a spherically symmetric protein, we may write these excesses as general integrals over the entire volume. Speak of a vector position $\mathbf{r}$ and integrals over all locations:

$$N_{\text{ew}} = \int_{\text{Volume}} [\nu_w(\mathbf{r}) - \nu_s(\mathbf{r})\,(n_w/n_s)]\, d\mathbf{r},$$

$$N_{\text{es}} = \int_{\text{Volume}} [\nu_s(\mathbf{r}) - \nu_w(\mathbf{r})\,(n_s/n_w)]\, d\mathbf{r}. \tag{A2.10}$$

Because

$$dG_{\text{protein}}^{\text{immersion}} = -N_{\text{ew}}d\mu_w = -N_{\text{es}}d\mu_s \tag{A2.11}$$

and

$$n_w d\mu_w + n_s d\mu_s = 0, \tag{A2.3}$$

we can write

$$N_{\text{ew}} = -N_{\text{es}}(n_w/n_s) \text{ or } N_{\text{es}} = -N_{\text{ew}}(n_s/n_w). \tag{A2.12}$$

An excess in one species is equivalent to a deficit in the other, evaluated in exact proportion to the n_s/n_w ratio in the bathing solution.

Appendix 3

1–2–3 and w–p–s, Translations

Notation can be a problem. Solute–water–protein interactions are frequently described in a language of "preferential interaction coefficients." These coefficients are based on the same thermodynamics as those of osmotic stress analyses that directly look at the effects of water on proteins and other macromolecules. Because of occasional confusion in the literature (Timasheff, 1998), it is worth pointing out explicitly the connection between the two kinds of notation (Parsegian *et al.*, 2000). The single most difficult feature of preferential interaction coefficients is to remember that

1 = water = w
2 = protein = p
3 = small solute or "cosolvent" = s

For a three-component solution, the Gibbs–Duhem relation is

$$n_1 d\mu_1 + n_2 d\mu_2 + n_3 d\mu_3 = 0 \text{ or } n_w d\mu_w + n_p d\mu_p + n_s d\mu_s = 0, \tag{A3.1}$$

where, by definition, the n_i are the total numbers of molecules of each component. To create the language of osmolalities in the 1–2–3 notation, the numbers of each species are divided by an amount of substance 1 (water) to create the requisite "per 1 kg of water[4],"

$$m_1 d\mu_1 + m_2 d\mu_2 + m_3 d\mu_3 = 0. \tag{A3.2}$$

Preferential interaction coefficients are defined in terms of changes that take place under imposed constant-chemical-potential conditions. For a constant water

[4]But for the historical convention of units "per kg water" rather than "per water molecule", m_1 would be a number ratio, $n_1/n_1 = n_w/n_w = 1$; $m_2 = n_2/n_1 = n_p/n_w$; $m_3 = n_3/n_1 = n_s/n_w$.

chemical potential, define

$$\Gamma_{\mu_1} \equiv \left.\frac{\partial m_3}{\partial m_2}\right|_{T,p,\mu_1}. \tag{A3.3}$$

The idea is, "If I add protein '2' to a solution while holding constant the chemical potential of water '1,' how does the molality of '3' (small solute molecules) change?" Conceptually, the equation assumes that the change in the number of small solutes can be measured while the water chemical potential is held fixed. At least in principle, one can imagine equilibrium dialysis between a protein-containing solution and a small-solute solution. In that spirit, it is possible to speak of the small-solute molality $m_3{}^{\Delta}$ on the two-component side of the membrane under the constraint that the water chemical potential is the same on both sides of the membrane. The difference in small-solute population because of the presence of protein "2" is then a key to the "iso-osmolal" preferential interaction coefficient,

$$\Gamma_{\mu_1} \equiv \left.\frac{\partial m_3}{\partial m_2}\right|_{T,p,\mu_1} \cong \frac{m_3 - m_3^{\Delta}}{m_2}. \tag{A3.4}$$

A similar coefficient can be constructed using the small-solute molality m_3^* under the constraint of constant chemical potential of small solute "3",

$$\Gamma_{\mu_3} \equiv \left.\frac{\partial m_3}{\partial m_2}\right|_{T,p,\mu_3} \cong \frac{m_3 - m_3^*}{m_2}. \tag{A3.5}$$

And, of course, there must be a derivative for the practical case of immersion into a reservoir, or between a three-component 1–2–3 solution separated by a dialysis membrane from a 1–3 solution, wherein both μ_1 and μ_3 are held fixed by virtue of an effectively infinite bath:

$$\Gamma_{\mu_1\mu_3} \equiv \left.\frac{\partial m_3}{\partial m_2}\right|_{T,p,\mu_1,\mu_3} \cong \frac{m_3 - m_3'}{m_2}. \tag{A3.6}$$

Imagine a protein that repels solute 3 compared to its repulsion of water 1 (or attracts water more than it does small solute). Add a protein to one side of a dialysis membrane dividing two solutions: m_3 the amount of solute 3 per amount of water 1 on the protein side becomes less than m_3^* or $m_3{}^{\Delta}$ or m'_3 the amount of 3 per amount of 1 on the protein-free side. In fact, in a two-component solution, a constant chemical potential of one component automatically enforces a constant chemical potential of the other, $m_1 d\mu_1 + m_3 d\mu_3 = 0$!

The last preferential interaction coefficient expression pleasingly transforms to the N_{ew} expressions used in Appendix 1. The protein-free side of the dialysis bag is, in the actual experiments, the infinite bath whose reservoir-power keeps fixed the chemical potentials of solute and water despite the intrusion of the protein. Better, recognize that in well-controlled lab experiments, chemical potentials are enforced in the form of stipulated pH, salts, temperature, pressure, etc. Those constraints are the *de facto* reservoir.

In the language of Appendix 1, $m^*{}_3$ or $m_3{}^\Delta$ or m'_3 equals $\frac{n_s}{n_w}$ without the protein and $m_3 = \frac{n_s}{n_w - n_p N_{ew}}$ in the three-component solution.

The requisite derivative of immersion, tantamount to the preferential interaction coefficient in the dilute limit, is

$$\left.\frac{\partial m_3}{\partial m_2}\right|_{T,p,n_1,n_3} = \partial\left(\frac{n_s}{n_w - n_p N_{ew}}\right) \Big/ \partial\left(\frac{n_p}{n_w}\right)\Bigg|_{T,p,n_1,n_3}$$
$$= \frac{n_s}{\left(n_w - n_p N_{ew}\right)^2} n_w N_{ew}. \tag{A3.6a}$$

Experimentally, in the dilute-protein, linear limit of the Courtenay *et al.* (2000) experiments, where $n_w >>> n_p N_{ew}$, this preferential coefficient reduces to

$$\left.\frac{\partial m_3}{\partial m_2}\right|_{T,p,n_1,n_3} = \frac{n_s}{n_w} N_{ew}. \tag{A3.6b}$$

The change in $\left.\frac{\partial m_3}{\partial m_2}\right|_{T,p,n_1,n_3}$ with original osmolality $\frac{n_s}{n_w}$ immediately gives N_{ew}. There is thus happy equivalence in languages.

Appendix 4

An Example of Water- and Solute-Counting

To see how a spatially varying distribution of solute and solvent density converts into excess quantities, think in a language where there is an energy $E_s(\mathbf{r})$ to put a molecule s at a position $\mathbf{r}$ relative to an energy of the same kind of molecule s at an infinite distance from the protein. As a consequence of this interaction, there are spatially varying number densities $\nu_s(\mathbf{r})$ of solute and $\nu_w(\mathbf{r})$ of water. Specifically, there are $\nu_s(\mathbf{r})$ solutes per unit volume and $\nu_w(\mathbf{r})$ waters per unit volume at a position $\mathbf{r}$. If the volume of a solute is $\bar{\upsilon}_s$ and volume of a water is $\bar{\upsilon}_w$, then by definition of number densities (per unit volume) and molecular volumes,

$$\nu_w(x)\bar{\upsilon}_w + \nu_s(x)\bar{\upsilon}_s = 1. \tag{A4.1}$$

The Boltzmann density distribution in solute number density $\nu_s(\mathbf{r}) = \nu_s^\infty e^{-E_s(\mathbf{r})/kT}$, then automatically enforces a water density distribution

$$\nu_w(\mathbf{r}) = \left(1 - \nu_s(\mathbf{r})\bar{\upsilon}_s\right)/\bar{\upsilon}_w. \tag{A4.2}$$

Note the consistency with the total volume of a solution containing n_s solutes of volume $\bar{v}_s$ and n_w waters of volume $\bar{v}_w$ is $n_w\bar{v}_w + n_s\bar{v}_s$. The number density of solutes in the infinite bathing reservoir is $\nu_s^\infty = \frac{n_s}{n_w\bar{v}_w+n_s\bar{v}_s}$ and of waters $\nu_w^\infty = \frac{n_w}{n_w\bar{v}_w+n_s\bar{v}_s}$, so that $\nu_w^\infty\bar{v}_w + \nu_s^\infty\bar{v}_s = 1$.

For the excess number of solutes, recall (Appendix 2)

$$N_{\text{es}} = \int_{\text{Volume}} [\nu_s(\mathbf{r}) - \nu_w(\mathbf{r})(n_s/n_w)]\, d\mathbf{r}. \tag{A4.3}$$

Introducing $\nu_w(\mathbf{r}) = (1 - \nu_s(\mathbf{r})\bar{v}_s)/\bar{v}_w$ gives [5]

$$\nu_s(\mathbf{r}) - \nu_w(\mathbf{r})(n_s/n_w) = \left(e^{-E_s(\mathbf{r})/kT} - 1\right)(n_s/\bar{v}_w n_w). \tag{A4.4}$$

Then

$$N_s = (n_s/\bar{v}_w n_w)\int_{\text{Volume}} \left(e^{-E_s(\mathbf{r})/kT} - 1\right) d\mathbf{r}. \tag{A4.5}$$

For simplicity in illustration, consider a flat surface. This is a one-dimensional situation wherein the excess number is really an excess number per unit area. We will use the symbol Γ_s to preserve the essential connection with the Gibbs adsorption isotherm. The total number N_s will then be Area x $\Gamma_s = A\Gamma_s$, where

$$\begin{aligned} N_{\text{es}} &= \int_{x=0}^{\infty} [\nu_s(x) - \nu_w(x)(n_s/n_w)]dx = \int_{x=0}^{\infty} [\nu_s(x) - \nu_w(x)(\nu_s^\infty/\nu_w^\infty)]dx \\ &= (n_s/\bar{v}_w n_w)A\int_{x=0}^{\infty} \left(e^{-E_s(x)/kT} - 1\right) dx. \end{aligned} \tag{A4.6}$$

This is a flat surface that interacts with solute and water in such a way that there is an exponentially varying change in energy to place a solute molecule a distance x from the surface. $E_s(x) = E\,e^{-x/\lambda}$. For a weak interaction, $E \ll kT$, $e^{-E_s(x)/kT} \approx 1 - E_s(x)/kT$ so that

$$N_{\text{es}} = -(n_s/\bar{v}_w n_w)\,A(E/kT)\int_{x=0}^{\infty} e^{-x/\lambda}dx = -(n_s/\bar{v}_w n_w)\,A(E/kT)\lambda. \tag{A4.7}$$

For a dilute solution, $n_s \ll n_w$, $(n_s/\bar{v}_w n_w)$ is the number density of solute. If solutes are repelled relative to water, then $E > 0$. The solute deficit is the bulk solution concentration times a volume $A\lambda(E/kT)$. Note that this deficit is linear in solute concentration. At the same time, the excess N_{ew} of water is (Appendix 2)

$$\begin{aligned} N_{\text{ew}} &= -N_{\text{es}}(n_w/n_s) = (n_w/n_s)(n_s/\bar{v}_w n_w)\,A(E/kT)\lambda \\ &= (A\lambda/\bar{v}_w)(E/kT). \end{aligned} \tag{A4.8}$$

This N_{ew} is independent of solute concentration.

[5] $\nu_s(\mathbf{r}) - \nu_w(\mathbf{r})(n_s/n_w) = \nu_s(\mathbf{r}) - \nu_w(\mathbf{r})(\nu_s^\infty/\nu_w^\infty) = \nu_s(\mathbf{r}) - \bar{v}_w\nu_w(\mathbf{r})(\nu_s^\infty/\bar{v}_w\nu_w^\infty) = \nu_s(\mathbf{r}) - (1 - \bar{v}_s\nu_s(\mathbf{r}))(\nu_s^\infty/\bar{v}_w\nu_w^\infty) = \nu_s(\mathbf{r})(1 + (\bar{v}_s\nu_s^\infty/\bar{v}_w\nu_w^\infty)) - (\nu_s^\infty/\bar{v}_w\nu_w^\infty) = e^{-E(\mathbf{r})/kT}(\nu_s^\infty/\bar{v}_w\nu_w^\infty) - (\nu_s^\infty/\bar{v}_w\nu_w^\infty) = (e^{-E(\mathbf{r})/kT} - 1)(\nu_s^\infty/\bar{v}_w\nu_w^\infty) = (e^{-E(\mathbf{r})/kT} - 1)(n_s/\bar{v}_w n_w)$.

Appendix 5

Pushing Proteins

What if two solute-excluding surfaces come near? Even as the free energy of one such surface changes with the concentration of excluded solute, so will the strength of interaction between two surfaces. For simplicity, consider an energy of exclusion that decays exponentially with distance from both surfaces. $E_s(x) = E\,e^{-x/\lambda}$ for an isolated surface (Appendix 4) becomes

$$E_s(x) = E\left(e^{-x/\lambda} + e^{-(D-x)/\lambda}\right) \quad \text{(A5.1)}$$

between two flat parallel surfaces a distance D apart. Now the exclusion of s between two surfaces of area A is

$$\begin{aligned} N_{\text{es}} &= \int_{x=0}^{D} [\nu_s(x) - \nu_w(x)(n_s/n_w)]dx \\ &= (n_s/\bar{v}_w n_w)\,A \int_{x=0}^{D} \left(e^{-E_s(x)/kT} - 1\right) dx. \end{aligned} \quad \text{(A5.2)}$$

For a weak interaction, $E \ll kT$, $e^{-E_s(x)/kT} \approx 1 - E_s(x)/kT$ so that

$$\begin{aligned} N_{\text{es}} &= -(n_s/\bar{v}_w n_w)\,A(E/kT) \int_{x=0}^{D} [e^{-x/\lambda} + e^{-(D-x)/\lambda}]dx \\ &= -2(n_s/\bar{v}_w n_w)\,A(E/kT)\,\lambda\left[1 - e^{-D/\lambda}\right], \end{aligned} \quad \text{(A5.3)}$$

which varies with separation as

$$\frac{\partial N_{\text{es}}(D, \mu_s)}{\partial D}\Big|_{\mu_s} = -2(n_s/\bar{v}_w n_w)\,A(E/kT)e^{-D/\lambda}. \quad \text{(A5.4)}$$

Alternatively, recalling that $N_{\text{ew}} = -N_{\text{es}}\,(n_w/n_s)$ (Appendix 4),

$$\begin{aligned} N_{\text{ew}} &= (2/\bar{v}_w)\,A(E/kT)\,\lambda[1 - e^{-D/\lambda}] \text{ and } \frac{\partial N_{\text{ew}}(D, \mu_s)}{\partial D}\Big|_{\mu_s} \\ &= (2/\bar{v}_w)\,A(E/kT)e^{-D/\lambda}. \end{aligned} \quad \text{(A5.5)}$$

N_{ew} is independent of solute concentration, but N_{es} is not.

At infinite separation, N_{es} and N_{ew} are simply twice those of two isolated surfaces. At contact, all the w's and s's have been expelled so that there is no "excess" or deficit of one over the other. And in between? N_{es} and N_{ew} vary exponentially. The change in interaction energy with added solute has an exponentially varying coefficient. We can speak of energy of interaction that depends on changes in separation D and on changes in solute chemical potential $d\mu_s = kT\, d\ln(n_s)$,

$$dG(D,\mu_s) = \left.\frac{\partial G}{\partial \mu_s}\right|_D d\mu_s + \left.\frac{\partial G}{\partial D}\right|_{\mu_s} dD = -N_{\mathrm{es}}(D,\mu_s)\,d\mu_s - P(D,\mu_s)\,A\,dD. \tag{A5.6}$$

Because $\dfrac{\partial^2 G}{\partial \mu_s\,\partial D} = \dfrac{\partial^2 G}{\partial D\,\partial \mu_s}$,

$$\left.\frac{\partial N_{\mathrm{es}}(D,\mu_s)}{\partial D}\right|_{\mu_s} = A\left.\frac{\partial P(D,\mu_s)}{\partial \mu_s}\right|_D = -2(n_s/\bar{v}_w n_w)\,A(E/kT)e^{-D/\lambda}. \tag{A5.7}$$

The change in pressure $P(D,\mu_s)$ between two parallel flat surfaces, or the change in force $AP(D,\mu_s)$ between two flat parallel faces of area A, equals the change in solute excess vs separation. In this weak perturbation example, sensitivity to solute is linear in solute concentration and exponentially varying in separation.

We can also see how the interaction free energy changes with solute concentration,

$$G(D,\mu_s) - G(D=\infty,\mu_s) = -\int_\infty^D P(D,\mu_s)\,A\,dD, \tag{A5.8}$$

so that

$$\left.\frac{\partial(G(D,\mu_s) - G(D=\infty,\mu_s))}{\partial \mu_s}\right|_D = -\int_\infty^D A\left.\frac{\partial P(D,\mu_s)}{\partial \mu_s}\right|_D dD$$

$$= -2(n_s/\bar{v}_w n_w)\,A\lambda(E/kT)e^{-D/\lambda}. \tag{A5.9}$$

The change in the work to bring the two surfaces to contact is

$$\frac{\partial(G(D=0,\mu_s) - G(D=\infty,\mu_s))}{\partial \mu_s}\Big|_D = -2(n_s/\bar{v}_w n_w)\,A\lambda(E/kT), \tag{A5.10}$$

exactly twice the change in the energy of an isolated surface.

Expressed in terms of the change in water chemical potential,

$$\frac{\partial(G(D,\mu_w) - G(D=\infty,\mu_w))}{\partial \mu_w}\Big|_D = -(2/\bar{v}_w)\,A\lambda(E/kT)e^{-D/\lambda}, \tag{A5.11}$$

these results can be converted into the interaction between oppositely curved surfaces. According to the Derjaguin approximation (Verwey and Overbeek 1999), the interaction force F_{ss} between two spheres of radius R near contact connects with the free energy per area G_{pp} between parallel planes of the same material by

$$F_{\mathrm{ss}} = \pi R G_{\mathrm{pp}}, \tag{A5.12}$$

where G_{pp} is given the reference value $G_{pp}(D\rightarrow\infty) = 0$. The change in sphere–sphere force with change in solute- or water chemical potential is

$$\left.\frac{\partial F_{ss}(D, \mu_s)}{\partial \mu_s}\right|_D = -2\pi R(n_s/\bar{v}_w n_w)\lambda(E/kT)e^{-D/\lambda} \tag{A5.13}$$

and

$$\left.\frac{\partial F_{ss}(D, \mu_w)}{\partial \mu_w}\right|_D = -(2\pi R/\bar{v}_w)\lambda(E/kT)e^{-D/\lambda}. \tag{A5.14}$$

Excluded solute, $E > 0$, creates an added attraction between spheres.

Acknowledgments

Sergey Bezrukov, Per Lyngs Hansen, Jack Kornblatt, Rudi Podgornik, Peter Rand, and Don Rau have contributed their reliably perceptive thoughts. I also thank Elizabeth Courtenay and Thomas Record for generous access to their preprints, for instructive correspondence, and for providing me with a copy of a figure.

References

Badger, J., and Caspar, D. L. D. (1991). Water-structure in cubic insulin crystals. *Proc. Natl. Acad. Sci. USA* **88,** 622–626.

Bezrukov, S. M., and Vodyanoy, I. (1993). Probing alamethicin channels with water-soluble polymer: Effect on conductance of channel states. *Biophys. J.* **64,** 16–25.

Bonnet-Gonnet, C., Leikin, S. L., Chi, S., Rau, D. C., and Parsegian, V. A. (2000). Measurement of forces between hydroxypropylcellulose polymers: Temperature favored assembly and salt exclusion. *J. Phys. Chem. B* **105**(9), 1877–1886.

Colombo, M. F., Rau, D. C., and Parsegian, V. A. (1992). Protein solvation in allosteric regulation: A water effect in hemoglobin. *Science* **256,** 655–659.

Courtenay, E., Capp, M. W., Anderson, C. F., and Record, M. T. (2000). Vapor pressure osmometry studies of osmolytes–protein interactions: Implications for the action of osmoprotectants in vivo and for the interpretations of "osmotic stress" experiments in vitro. *Biochemistry* **39,** 4455–4471.

Leikin, S., Rau, D. C., and Parsegian, V. A. (1991). Measured entropy and enthalpy of hydration as a function of distance between DNA double helices. *Phys. Rev. A* **44,** 1156–1162.

Leikin, S., Parsegian, V. A., Rau, D. C., and Rand, R. P. (1993). Hydration forces. *Annu. Rev. Phys. Chem.* **44,** 369–395.

Leikin, S., Rau, D. C., and Parsegian, V. A. (1995). Temperature-favored assembly of collagen is driven by hydrophilic not hydrophobic interactions. *Nature Struct. Biol.* **2**(3), 205–210.

Parsegian, V. A., Rand, R. P., Fuller, N. L., and Rau, D. C. (1986). Osmotic stress for the direct measurement of intermolecular forces. *In* "Methods in Enzymology; Biomembranes: Protons and Water: Structure and Translocation" (L. Packer, Ed.), pp. 400–416. Academic Press, New York.

Parsegian, V. A., Rand, R. P., and Rau, D. C. (1995). *In* "Methods in Enzymology" (M. L. Johnson and G. K. Ackers, Eds.), Vol. **259,** pp. 43–94. Academic Press, New York.

Parsegian, V. A., Rand, R. P., and Rau, D. C. (2000). Osmotic stress, crowding, preferential hydration, and binding. A comparison of perspectives. *Proc. Natl. Acad. Sci. USA* **97**(8), 3987–3992.

Podgornik, R., Strey, H. H., and Parsegian, V. A. (1998). Colloidal DNA. *Curr. Opin. Colloid Interface Sci.* **3,** 534–539.

Rau, D. C., and Parsegian, V. A. (1990). Direct measurement of forces between linear polysaccharides xanthan and schizophyllan. *Science* **249,** 1278–1281.

Royer, W. E., Jr., Pardanani, A., Gibson, Q. H., Peterson, E. S., and Friedman, J. M. (1996). Ordered water molecules as key allosteric mediators in a cooperative dimeric hemoglobin. *Proc. Natl. Acad. Sci. USA* **93,** 14,526–14,531.

Selected osmotic stress data, http://aqueous.labs.brocku.ca/osfile.html and http://dir.nichd.nih.gov/Lpsb/docs/osmdata/osmdata.html.

Timasheff, S. N. (1998). In disperse solution, "osmotic stress" is a restricted case of preferential interactions. *Proc. Natl. Acad. Sci. USA* **95,** 7363–7367.

Verwey, E. J. W., and Overbeek, J. Th. G. (1999). "Theory of the Stability of Lyophobic Colloids." Dover Publications, Mineola, NY (originally published by Elsevier, 1948).

Vodyanoy, I., Bezrukov, S. M., and Parsegian, V. A. (1993). Probing alamethicin channels with water-soluble polymers. Size modulated osmotic action. Biophys. *J.* **65,** 2097–2105.

Wyman, J., and Gill, S. J. (1990). "Binding and Linkage, Functional Chemistry of Biological Macromolecules." University Science, Mill Valley, CA.

Yu, B., and Caspar, D. L. D. (1998). Structure of cubic insulin crystals in glucose solutions. *Biophys. J.* **74,** 616–622.

Zimmerberg, J. J., and Parsegian, V. A. (1986). Polymer inaccessible volume changes during opening and closing of a voltage-dependent ionic channel. *Nature* **323,** 36–39.

The Lipid–Water Interface: Revelations by Osmotic Stress

R. P. Rand
Department of Biological Sciences, Brock University, St. Catharines, Ontario, Canada L25 A31

Lipids at the bilayer–water interface are highly disordered and mobile, and large areas of the bilayer undergo thermal undulations. Water penetrates significantly down to the hydrocarbon chain level. This water, and water out to about 10 Å from the surface, is perturbed by the lipid surface in a way that produces a strong hydration repulsion and precludes molecular contact between bilayers. Its removal costs work, but most of this water is neither a permeable barrier nor unavailable to solvate other solutes. All hydrophilic surfaces show this "hydration force." Most lipids have an excess higher free energy when packed within a bilayer membrane since in isolation they pack into high curvature assemblies with polar groups on the concave side. Osmotic stress measurements of those curved assemblies yield a measure of monolayer elastic parameters and the excess higher free energy, which likely controls embedded proteins. Osmotic stress experiments can determine whether water is energetically significant, or not, in almost any system. The osmotic effect of solutes, independent of specific binding, is to compete with lipids and proteins for water. Solute affinity for water can modify lipid packing and protein conformation, coupling lipid and protein structure and function to osmolality at the molecular level.

KEY WORDS: Amphiphiles, Lipid assemblies, Hydration forces, Membrane curvature, Osmotic stress

I. Introduction

It is the love–hate relationship between water and hydrophobic components that drives much of the interesting energetics of membrane phenomena. Membrane assembly, membrane interactions, internal membrane activity, and transbilayer movements are all dictated by minimizing the hate, maximizing the love, seeking some compromise, relieving frustrations.

Lipid molecules do not come alone; their solubility in water is nearly immeasurable (Parsegian and Rand, 1992). And their ability even to protrude out of the membrane plane is poor (Parsegian and Rand, 1991). To that we can attribute the enabling stability of membranes in cells and the requirement for sophisticated molecular mechanisms to distribute lipids among their many compartments.

Global lipid insolubility drives lipid assembly in water. The assembly that forms is the energetic compromise of solvating the hydrophilic parts of the lipid yet protecting the hydrophobic parts from exposure to water. A common assembly that reaches that compromise is the lipid bilayer. But individual lipids in "flat" membrane assemblies are likely to be energetically frustrated since most of them prefer, in isolation, to assemble into other quite unrelated highly curved geometries. Those geometries better satisfy, energetically, the hydrophobic/hydrophilic dilemma (Luzzati and Husson, 1962; Seddon, 1990) and geometrically, steric packing stresses.

Constraint into planar membranes, required by the biological cell, becomes a configuration of frustration for many of its constitutive membrane lipids and proteins. The latter must interact with the warring factions of water and the conflicting parts of the lipid. One believes that such frustrations of the lipids and proteins are required and used energetically in many membrane functions (Gruner, 1985).

Once assembled, membranes are prevented from interacting nonspecifically with each other because the love of the hydrophilic coating of the assembly for water is so strong it is hard to break (LeNeveu *et al.,* 1976; Parsegian *et al.,* 1979; Rand and Parsegian, 1989). That "vicinal" water forms part of the membrane and dominates their interactions; it has to be breached for membranes to make functional contact, and especially for membranes to fuse. On the other hand that vicinal water is a trivial barrier to solutes passing through membranes.

Solutes, of any kind, whether or not they interact specifically with lipids or proteins, inevitably affect water activity. And water activity necessarily affects lipid–water and protein–water interactions. The circle is complete. "osmotic stress," the nonspecific effect of solutes on molecular systems operating in an aqueous environment, affects lipid and protein assembly and function in membranes.

Osmotic stress takes on new meaning when it is recognized that all solutes act osmotically on any compartment from which they are excluded. Importantly and nonintuitively, exclusion, with resultant osmotic stress, does not need a membrane! It may result from something as simple as the steric exclusion of a solute from a molecular water-filled cavity, such as the cleft in a protein, or a membrane channel (Parsegian *et al.,* 1995; Rand *et al.,* 2000; see Parsegian, this volume).

What follows are empirical examples illustrating the principles expressed above. The references and attributions are necessarily far from comprehensive but should act as a guide to the literature. The age-old and deceptively simple principles of amphiphilicity applied to an increasingly diverse array of participating molecular species, coupled to the osmotic effects of solutes, continue to be the origins of much rich experimental phenomena and understanding of molecular assembly and function.

II. "Boundary" Water

Although the lipid–water interface is disordered and mobile, much has been learned about the water at the bilayer interface through measurements of bilayer interactions. Bilayer membranes within multilayer assemblies can be forced together osmotically. The work against interbilayer forces that is required to do this can be measured directly using a technique developed three decades ago. That method, osmotic stress, has been extended to many other systems (Parsegian *et al.,* 1995; Rand *et al.,* 2000) and as a widely applicable experimental strategy is illustrated in Fig. 1.

Osmotic stress applied to multilamellar lipid systems shows that all lipid bilayers experience strong mutual repulsion when within 10 to 20 Å of each other. That repulsion grows exponentially with a decay distance of about 2 Å for all lipids, and it prohibits molecular contact between bilayer membranes. Typical force–distance plots are shown in Fig. 2.

Measurements have now been made on most lipid species, both by osmotic stress and by mechanical means (Israelachvili, 1985). They reveal the universality of the interbilayer hydration force at small separation. All the quantitative structural and energetic parameters that characterize these many lipid systems are provided on a website (http://aqueous.labs.brocku.ca/osfile.html). Species differences reside in the prefactor of the exponential force. For example, the differences between the most common membrane phospholipids, phosphatidylcholine (PC) and phosphatidylethanolamine (PE), and the effects of methylation have been explored in detail (Rand *et al.,* 1988; Rand and Parsegian, 1989; see Parsegian, this volume). Methylation of PE, even once, largely destroys the ability of the polar group to form intermolecular bonds and results in a higher level of hydration like the triply-methylated PC. The common charged membrane lipid phosphatidylserine, once divested of its effective charge by screening with monovalent ions or bound with divalent cations, shows a hydration repulsion more like that of PE than PC. Always the decay distance is small and the resultant force large, precluding molecular contact with other bilayers. Similar force measurements have been made between many very different kinds of hydrophilic surfaces—proteins, polysaccharides, coated mica, myelin membanes, DNA, etc. Universally all show a charge-independent, common, short-range exponential force, called a hydration force. Comprehensive reviews of measurement details have been published (Parsegian *et al.,* 1995; Rand *et al.,* 2000).

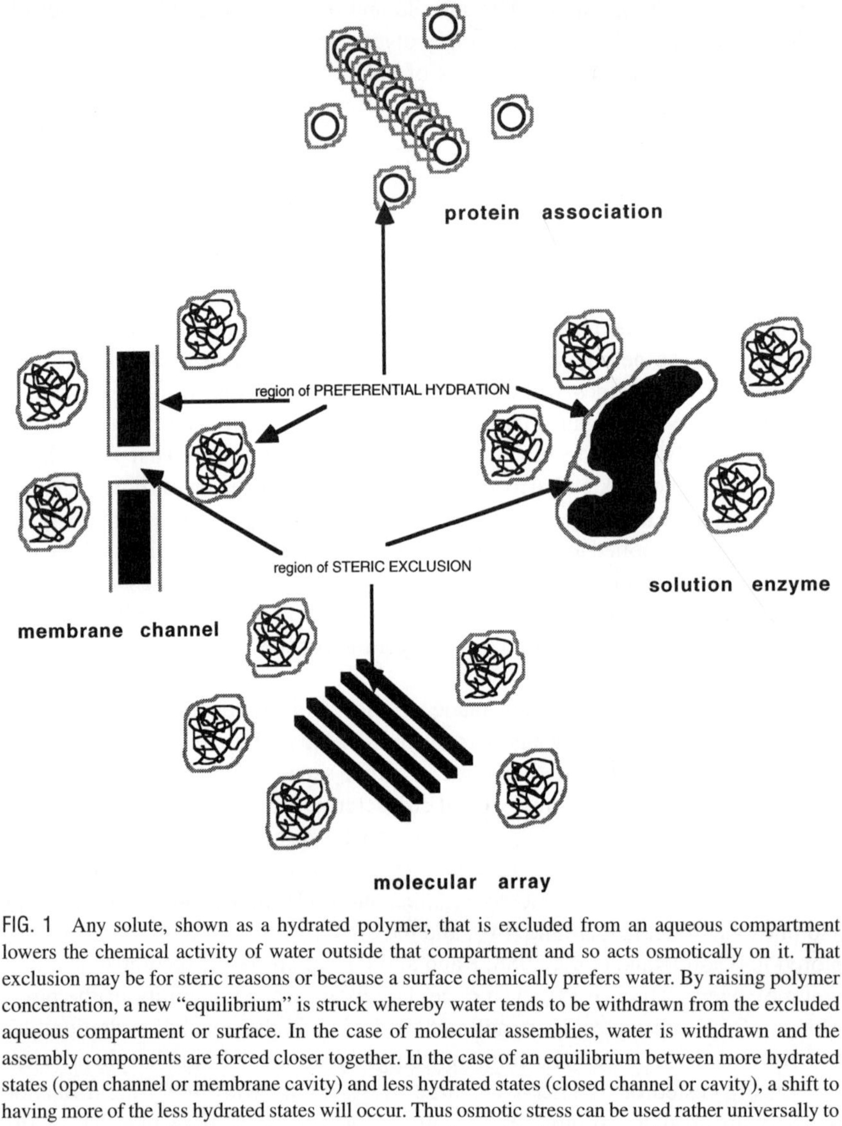

FIG. 1 Any solute, shown as a hydrated polymer, that is excluded from an aqueous compartment lowers the chemical activity of water outside that compartment and so acts osmotically on it. That exclusion may be for steric reasons or because a surface chemically prefers water. By raising polymer concentration, a new "equilibrium" is struck whereby water tends to be withdrawn from the excluded aqueous compartment or surface. In the case of molecular assemblies, water is withdrawn and the assembly components are forced closer together. In the case of an equilibrium between more hydrated states (open channel or membrane cavity) and less hydrated states (closed channel or cavity), a shift to having more of the less hydrated states will occur. Thus osmotic stress can be used rather universally to measure quantitatively the role of water both in assemblies and during conformational changes between states of different hydration.

The fundamental nature and physical chemistry of the hydration force is buried deep in the physics of water (Leikin *et al.*, 1993). But the empirical fact of hydration repulsion is now well established and can be viewed at a simple level by considering it as the surface perturbation of water dipoles (Fig. 3).

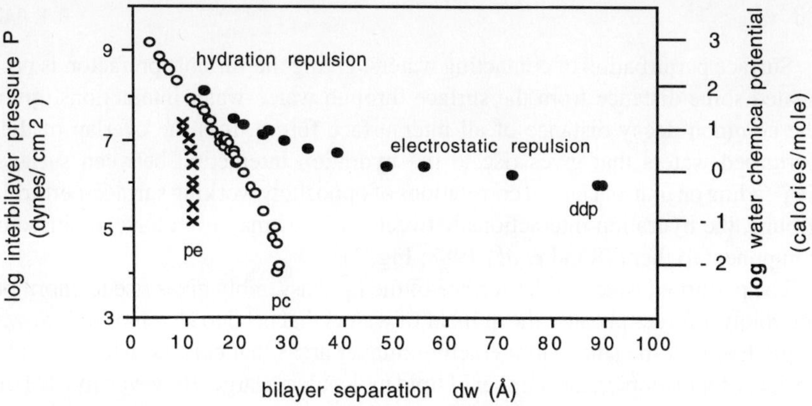

FIG. 2 As osmotic pressure P is increased outside a multilamellar phase of lipids, water is withdrawn from the array, and bilayer separation decreases. Neutral lipids start from finite separation in bulk water; phosphatidylethanolamine (pe) closer than phosphatidylcholine (pc). Charged lipids, didodecyl phosphate (ddp), start from indefinitely large separation. All lipids, and in fact all hydrophilic surfaces (Parsegian *et al.,* 1995), experience large repulsive forces within 10–20 Å separation that grow exponentially and rapidly. The universal, charge- and salt-independent nature of this force points to the work of dehydrating the surface groups as its origin. The chemical potentials of water near the lipid surface, measured as the difference from bulk water (right-hand axis), fall off continuously and as rapidly from the surface. Water at the surface has chemical potentials (0.1–1.0 kcal/mol) similar to the hydration energies of ions. Water at about 10 Å from the surface has chemical potentials of about 10^{-3}–10^{-4} kcal/mol that would behave as bulk water to solutes.

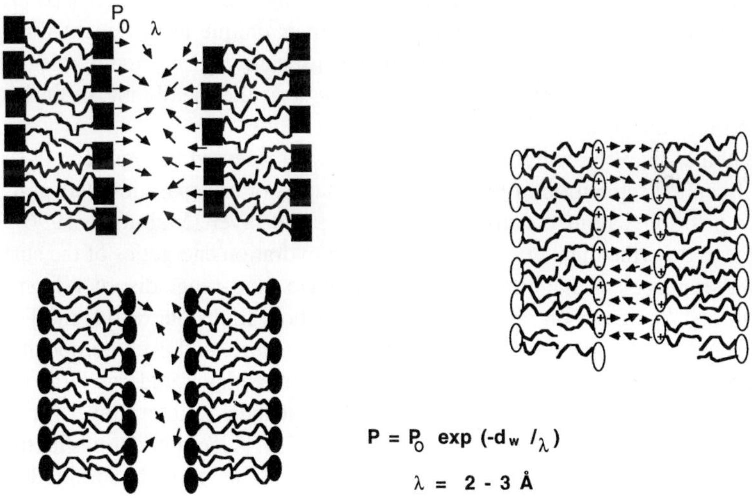

FIG. 3 The universal hydration force can be naively viewed as having (i) its origin in the surface perturbation of water dipoles, which sets the magnitude of the preexponential factor, and (ii) its decay depending on water–water interactions, accounting for the commonality of its decay length. Laterally correlated inverse perturbations, possible in the case of PE and not PC (methylated PE), for example, can give rise to an attractive component of the hydration force, also seen in other molecular interactions (Rand *et al.,* 1988).

Surface perturbation of contacting waters, giving the variable prefactor, is propagated some distance from the surface through water–water interactions, giving the common decay distance of all intersurface forces. It is the overlap of these perturbed waters that gives rise to the hydration interaction between surfaces. Depending on unusual lateral correlations of oppositely working surface-perturbing groups, the hydration interactions between surfaces may even have an attractive component to them (Rand *et al.,* 1988; Fig. 3).

The perturbed water on the surface of the lipid assembly gives rise to enormous interbilayer forces because the number of waters that need to be removed is always large. Even for the smallest interacting bilayer areas that exist (small unilamellar vesicles, for example), the number of lipid molecules is large. However, most of that perturbed water is very little different from bulk water and should not be considered "bound" and unavailable for other "solutes." The osmotic stress technique used to measure these forces also gives a direct measure of the chemical potential of these waters. Plotted in Fig. 2 is the difference in chemical potential between bulk water and the surface-perturbed waters as it varies from the surface. Those waters in direct contact with the surface polar groups have chemical potentials near those resembling hydration potentials for solutes (kcal/mol) and might be considered "unavailable" to solvate other solutes near the surface. In fact, the polar groups are held close packed in the bilayer assembly at a concentration far beyond what they would be if free in saturated solution. But the chemical potential of the water decays rapidly with distance from the first layer to that of bulk water, and most of it can solvate other species. In other words, the surface polar groups are like any solute in changing water activity; it is their unique arrangement that gives rise to the huge hydration force that can be so sensitively detected by the osmotic stress method. But the perturbed water layers near the membrane surface would not represent a permeability barrier to a small solute.

Whether bilayer separations are described as presented in Figs. 2 and 3, commonly called separation of the Luzzati interfaces, or as separation between some other position within the disordered polar group layer (McIntosh, 1996), each serves to describe the thermodynamics of the hydration energetics of the surfaces. However, it is clear that bilayer membranes have short-range disorder, from polar group and hydrocarbon chain mobility and whole molecule diffusion. This has prevented good understanding of the structural nature, and even composition, of the interface with water. In addition, large areas of a bilayer undergo large thermal undulations perpendicular to its surface that contribute to interbilayer forces and further hinder structural definition (Helfrich, 1973). Accommodating fluctuations into interbilayer forces (Evans and Parsegian, 1983) has brought a refined picture of bilayer interactions but has not changed the fundamental understanding or evaluation of the short-range hydration forces and polar group hydration (Tristram-Nagle *et al.,* 1998). Systematic X-ray studies of specifically labeled lipids and their mixtures have produced a better description of the distribution of molecular constituents, phosphate groups, glycerol backbone, and water, particularly within the polar group region (Wiener and White, 1992). Those studies combined with nuclear magnetic resonance and molecular simulation studies (Feller *et al.,* 1997) produce a more detailed description of the structural and dynamic disorder within the bilayer. Wide Gaussian distributions of every molecular constituent appear. Water is seen with high probability down to the hydrocarbon layer, and with low probability within it.

Special circumstances lead to dehydration of lipid surfaces, cutting through the work of taking off the water, at least in the case of interacting membranes. These include primarily ionic interactions of divalent cations or proteins with anionic lipids. Formation of salt bridges between phosphatidylserine molecules on neighboring bilayers and subsequent displacement of water is a powerful force driving membranes together. It can even lead to the dehydration of strongly hydrated neutral lipids in the neighborhood of the anionic lipids (Coorssen and Rand, 1995), as shown in Fig. 4. Whether similar effects occur on isolated bilayers is much less clear. Divalent cation affinity for the isolated anionic surface is orders of magnitude less than for interacting layers (McLaughlin *et al.,* 1981). Many studies attempting to understand the interaction of peptides with the lipid water surface are currently being undertaken (Jacobs and White 1989; http://blanco.biomol.uci.edu/peptide_bilayer_energetics.html#ref_1). How hydration of the isolated lipid surface is overcome awaits more detailed knowledge of its structure.

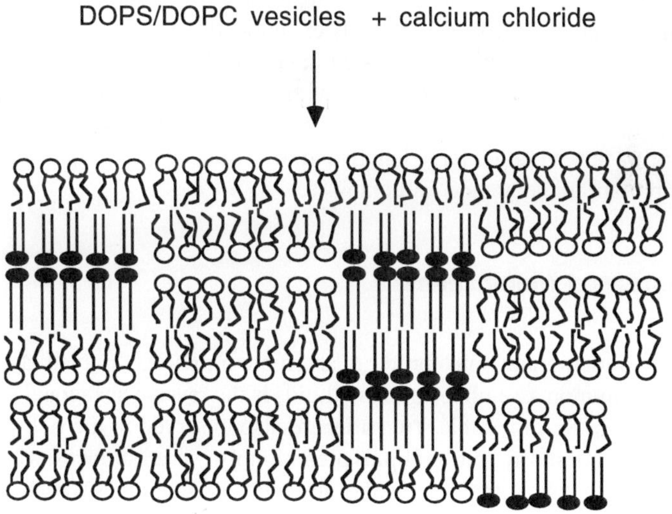

FIG. 4 Mixtures of the anionic lipid dioleoylphosphatidylserine (DOPS) and the neutral lipid dioleoylphosphatidylcholine (DOPC) form a single dehydrated multilamellar phase in the presence of divalent cations, with DOPS forming large mosaic aggregates in the gel state. The strong dehydrating effect of the divalent cation anionic lipid complex dehydrates the strongly hydrated neighboring DOPC molecules (Coorssen and Rand, 1995; Feigenson, 1989).

III. Lateral Segregation and Mosaic Structure

Membranes contain a large variety of lipids that differ both in the combination of hydrocarbon chains, which themselves vary in length and degree of unsaturation, and in the species of polar group. At some length and time scale, the surface of the lipid layer cannot be homogeneous. While some particular lipids act as specific activators of membrane enzymes, the specific roles of the majority are unclear. The question of lateral lipid segregation and the formation of a mosaic structure of microdomains within the cell membrane surface through lateral interactions of lipids remains an enigma. Earlier studies in model systems focused on, and demonstrated, the segregation of lipids according to the mobility state of their hydrocarbon chains (Silvius *et al.,* 1996). Lipids with different hydration interactions would be expected to segregate on contact with neighboring bilayers (Rand, 1981). Divalent cation interactions with acidic lipids can drive lateral interaction (Feigenson 1986, 1989; Coorssen and Rand, 1995). More recently, the segregation of lipids into "rafts" containing sphingolipids and cholesterol has been clearly demonstrated (Brown and London, 1998; Reitveld and Simons, 1998). The nature of the interactions that lead to lipid aggregation include primarily intermolecular hydrogen bonding between polar groups, especially possible with PE and sphingolipids. van der Waals interactions between hydrocarbon chains, particularly enhanced for longer and more saturated chains, also drives lipid segregation (Brown, 1998). Lateral segregation into regions enriched in particular lipids containing particular membrane proteins is invoked in models of intracellular trafficking. While lipid segregation is more clearly demonstrated in compositionally simpler model systems of isolated lipids, its demonstration within intact membranes is complicated by the complexity of lateral interactions (Tian-Yu and Silvius, 2000). Nevertheless, it is emerging that on rather large length and time scales there is lipid segregation that is important in membrane trafficking.

IV. Frustrated Membranes

Most isolated membrane lipids prefer to assemble into highly curved layers. One of the most common assemblies of diacyllipids is the reverse hexagonal structure, shown in Fig. 5. Several cubic structures are also observed in the rich polymorphism of phospholipid systems. If lipids prefer highly curved surfaces, what are the implications of constraining them within flat bilayer membranes (Gruner, 1985)?

A lipid's curvature preference, or lowest free energy assembly, can be determined by providing all the water needed to quench the thirst of the polar groups, and oil to fill spaces between the intrinsically curved monolayers where the lipid chains cannot reach (Kirk and Gruner, 1985). The curved monolayers are considered unstressed under these conditions, expressing the intrinsic curvature of the assembled lipids. The usefulness of examining assemblies such as the hexagonal

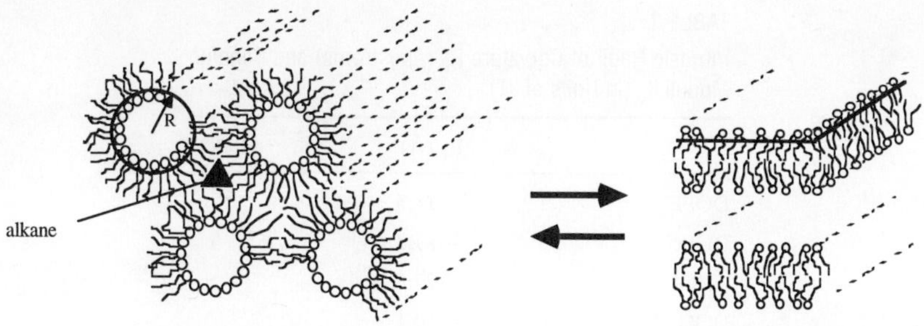

FIG. 5 One common nonlamellar assembly of diacylphospholipids is the reverse hexagonal phase. Most membrane lipids form it under conditions of full hydration and in the presence of alkane, which is required to fill the interstices where the lipid chains cannot reach. Under these circumstances, it is considered that the monolayers assemble into their stress-free geometry and express their intrinsic curvature. A measure of this curvature, combined with a measure of the modulus of bending this monolayer, give a quantitative estimate of the curvature free energy excess of maintaining such lipid in a flat bilayer. That free energy excess is thought to be important in modulating both membrane protein conformations (see Fig. 6) and membrane fusion activity.

structure is not that they exist in membranes, but that their intrinsic curvature, and the energy required to unbend their monolayers, give some measure of the free energy state of the lipid constrained in a flat membrane (Helfrich, 1973; Rand *et al.,* 1990). The farther the intrinsic curvature is from flat, and the harder it is to unbend the monolayer, the more frustrated and higher the free energy that lipid has in the bilayer membrane. That free energy can be used in membrane "reactions."

More than curvature energy contributes to the total free energies of lipid assemblies. Different lipids assemblies can be very delicately balanced energetically. For example, the coexistence of the multilamellar and hexagonal phases is common, and calorimetric measurements of their transition enthalpies indicate a very small difference in their free energies. In a comprehensive study of a PE system (Kozlov *et al.,* 1994), the complete complex of structural transitions required four contributions to the free energy of the system: (i) hydration, (ii) bending, (iii) van der Waals, and (iv) interstitial or chain stress energies. Very small differences between large values of the four component free energies determine the structure, accounting for the rich polymorphism. Polar group hydration and monolayer bending energies were particularly delicately balanced in the facile transition from flat to curved. That study highlighted the variety of contributions to the free energy of these systems, accounting for their structural lability.

Intrinsic curvatures and bending moduli have been measured for a number of membrane lipids (Rand and Parsegian, 1997). Again, osmotic stress is used to deform the hexagonal structure by removing water from the structure in an energetically controlled way, and the structural consequences are followed by, for example, X-ray diffraction. Table I gives examples of the intrinsic or spontaneous curvatures and elastic parameters measured for a number of different lipid species. Molecules like the second messengers diacyglycerol and lysolipids contribute

TABLE I
Intrinsic Radii of Curvature R_0 (Angstroms) and Bending Moduli K_c (in Units of kT)

	R_0 (Å)	K_c/kT
DOPE	− 28.5	12
DOPC	− 87.3	9
CHOL	− 22.8	
DOG	− 10.1	
O-LYSOPE	> 400	11
O-LYSOPC	+ 38	12

DOPE, dioleoylphosphatidylethanolamine; DOPC, dioleoylphosphatidylcholine; CHOL, cholesterol; DOG, dioleoylglycerol; *O*-LYSOPE/PC, oleoyllysophosphatidylethanolamine/choline.

particularly anomolous, and opposite, ratios of polar to hydrocarbon volumes, affecting monolayer curvature the most. It is these intrinsic curvatures and bending energies that can be used quantitatively in comprehensive models and mechanisms both of protein conformational change in single membranes (May and Ben-Shaul, 1999) and of membrane fusion (Siegel, 1993).

The excess free curvature energy of lipids in bilayers is considered fundamental to the energetics of membrane structure and function. The conceptual

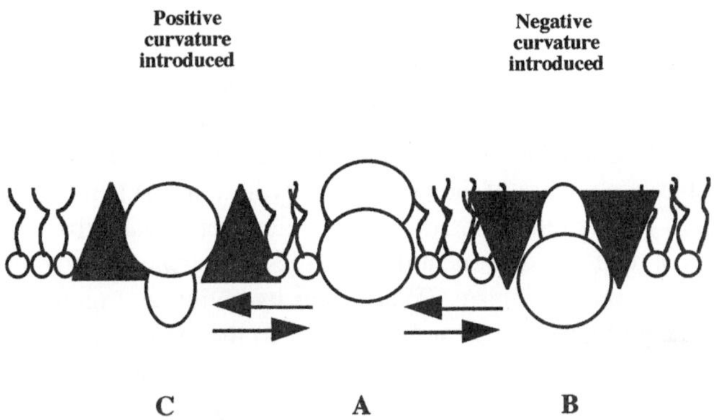

FIG. 6 Conceptual connection between excess curvature free energy of membrane lipids and membrane protein conformation. One lipid monolayer is shown with its complement of an integral membrane protein with three different conformations. (A) Protein conformation within an unstressed monolayer, or one whose intrinsic curvature is effectively zero. (B) Protein conformation within a monolayer with negative curvature stress, induced for example by the production of second messengers such as diacylglycerols or by decreased polar group area. (C) Protein conformation within a monolayer with positive curvature stress, induced for example by the production of second messengers such as lysolipids or by an increase in polar group area by electrostatic interactions.

connection between curvature energy and membrane protein conformation is illustrated in Fig. 6. A growing number of studies correlate membrane enzyme and channel activity with the species characteristics of their neighboring lipids. The most dramatic illustration of the strong dependence of membrane protein conformation on lipid composition is that of membrane channel gating kinetics. Both alamethecin (Bezrukov *et al.,* 1998, see Parsegian, this volume) and gramicidin (Lundbaek *et al.,* 1994) channel lifetimes depend dramatically on lipid composition, often correlated with intrinsic curvature and/or hydrophobic thickness. A parallel hypothesis, that of the hydrophobic mismatch between lipid and protein hydrophobic thicknesses (Mouritsen and Bloom, 1984; Lewis and Engelmann, 1983), involves concepts that likely contain stresses directly related to curvature and the other components of free energy mentioned above. Interactions at the polar group level are particularly enigmatic.

V. Water, Solutes, and Osmotic Stress at the Molecular Level

One widespread cellular function that centers directly on water activity is cellular osmoregulation. All cells, but microorganisms in particular, respond to changes in water activity, however it changes. For microorganisms, that change can go quickly from rain to desiccating sunlight. An osmosensory mechanism, which eventually regulates cellular osmolarity through solute transport mechanisms, must function not on specific solute interactions, but on any condition that changes water activity. Is the osmotic sensor found in the cellular membrane stressed by volume or pressure change? Does the osmotic change even need a membrane across which it is sensed? Or can molecules themselves detect osmotic changes? Some of these basic questions are beginning to be delineated and answered for microorganisms (Wood, 1999). One such study suggests clearly that the osmotic sensor resides within the molecule (MacMillan *et al.,* 1999).

The experimental strategy of osmotic stress, so instructive in revealing the role of water at molecular interfaces, has been extended to single molecules (Parsegian *et al.,* 1995; Rand *et al.,* 2000). This has brought new perspectives to the possible role of osmotic effects on molecules within cells. Given that any aqueous compartment that excludes a solute is necessarily osmotically stressed by that solute, nonspecific solute/compartment interactions become a potential source of conformational free energy change. Figure 7 illustrates the general case of the mutual interactions of solute, water, and macromolecule that can form the basis of a mechanism for osmotic control of protein function.

The interplay between solute, water activity, and membrane channel gating is dealt with in Parsegian's chapter, this volume. The role of water in protein structure is treated in the chapter by Kornblatt and Kornblatt, this volume. One specific example that illustrates the very general effects that excluded solutes and subsequent

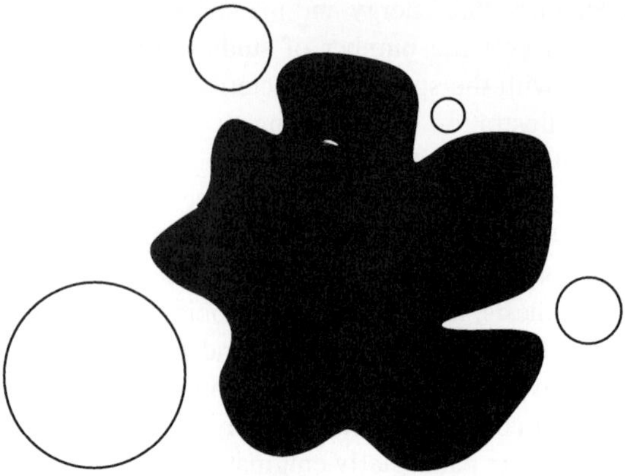

FIG. 7 The general case of the cross section of a protein with various size grooves and cavities and exposed to different size molecules, including water, is shown. These molecules are variously excluded from the grooves and cavities, either simply for steric reasons or because of differential affinity of the surface for water and solute (Parsegian *et al.,* 2000; Timasheff, 1998). Polyethylene glycols, for example, are preferentially excluded, giving the preferential hydration of such surfaces, measured directly by Timasheff and used widely in osmotic stress experiments. The possibility arises, then, that different intracellular solutes, regulated by the cell, stress different protein compartments osmotically, regulating the conformational states of the protein. Membrane channels would be one such compartment (see Parsegian, this volume).

osmotic stress can have on a functioning protein structure is summarized in Fig. 8. It is the strong osmotic dependence of glucose binding to the enzyme hexokinase.

1. Large solutes, above a certain minimum size, are excluded from all aqueous compartments around hexokinase. Their osmotic pressure, or equivalently the resultant reduction of the chemical potential of water "outside" the protein, increases the affinity of hexokinase for glucose: quantitatively, the effect shows that about 350 water molecules are lost when one glucose is bound.
2. Large osmolytes can apparently dehydrate hexokinase to the extent of about 300 water molecules even without substrate binding.
3. Solutes of decreasing size gain increasing access to, and act osmotically on, ever smaller compartments around the protein.

This one example shows how osmotically sensitive a protein can be. Increasing numbers of proteins show osmotic sensitivity, and this provides reason to think that it is the triumvirate of solute–macromolecule–water that functions in concert in controlling many macromolecular reactions. It is extremely unlikely that water activity itself varies significantly within the cell. However, the role of water is invoked by the species of solute and the aqueous compartments around the protein that exclude it (Fig. 7). Different osmotic stresses of a macromolecule to different solutes, which the cell can control, lend considerable versatility to the system. It

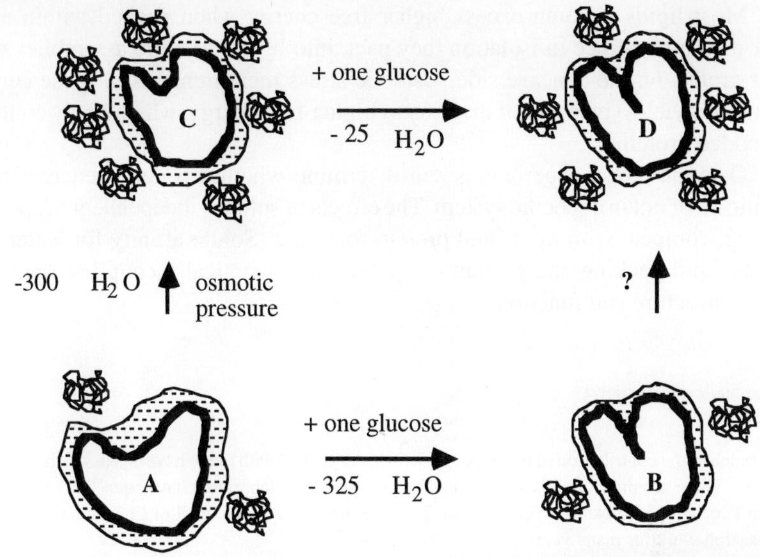

FIG. 8 The osmotic sensitivity of glucose binding to hexokinase shows that water inhibits binding and hexokinase is associated with ΔN_w fewer water molecules when glucose does bind (Reid and Rand, 1997). Glucose binding induces the detectable conformational change required for catalysis (A–B, C–D). Using polyethyleneglycol (PEG) molecules of MW 1000 to 10,000 at low osmotic pressures, ΔN_w for that conformational change is 325 $\pm$ 10% (A–B). However, ΔN_w decreases with increasing osmotic pressure to about 25 water molecules (C–D). These larger MW PEGs are apparently excluded from the same maximum size aqueous compartment around the protein, and act to osmotically shrink that compartment by 300 waters (A–C). Nothing is known about any osmotically driven ΔN_w of the bound state (B–D).

would be intriguing to investigate such osmotic effects on the very protein that regulates water permeability itself, aquaporin.

VI. In Sum

1. The lipids at the bilayer–water interface are highly disordered and mobile, and large areas of the bilayer undergo thermal undulations. Water likely penetrates significantly down to the hydrocarbon chain level.

2. This water, and water out to about 10 Å from the surface, is perturbed by the lipid surface in a way that produces a strong hydration repulsion between bilayers. Its removal costs work, but most of this water is neither a permeable barrier nor unavailable to solvate other solutes. All hydrophilic surfaces show this hydration force.

3. Most lipids have an excess higher free energy when packed within a bilayer membrane since in isolation they pack into high curvature assemblies with polar groups on the concave side. Osmotic stress measurements of those curved assemblies yield a measure of that excess higher free energy, which likely controls embedded proteins.

4. Osmotic stress experiments can determine whether water is energetically significant or not in a specific system. The effects of solutes, independent of specific binding, compete with lipids and protein for water. Solute affinity for water can modify lipid packing and protein conformation, osmotically coupling lipid and protein structure and function

Acknowledgments

I owe much to the contribution of many people who have worked with me or have discussed these issues with me. There are many, but two valuable close long-time collaborators have been Nola Fuller and Adrian Parsegian. The Natural Sciences and Engineering Research Council of Canada has supported my research work for many years.

References

Bezrukov, S. M., Rand, R. P., Vodyanoy, I., and Parsegian, V. A. (1998). Lipid packing stress and polypeptide aggregation: Alamethicin channel probed by proton titration of lipid charge. *Faraday Discuss.* **111,** 173–183.

Brown, D. A., and London, E. (1998). Structure and origin of lipid domains in biological membranes. *J. Membr. Biol.* **164,** 103–114.

Brown, R. E. (1998). Structure and origin of ordered lipid domains: What physical studies of model membranes reveal *J. Cell Sci.* **111,** 1–9.

Coorssen, J. R., and Rand, R. P. (1995). Structural effects of neutral lipids on divalent cation-Induced interactions of phosphatidylserine-containing bilayers. *Biophys. J.* **68,** 1009–1018.

Evans, E., and Parsegian, V. A. (1983). Energetics of membrane deformation and adhesion in cell and vesicle aggregation. *Ann. NY Acad. Sci.* **416,** 13–33.

Feigenson, G. W. (1986). On the nature of calcium ion binding between phosphatidylserine lamellae. *Biochemistry* **25,** 5819–5825.

Feigenson, G. W. (1989). Calcium ion binding between lipid bilayers: The 4-component system of phosphatidylserine, phosphatidylcholine, calcium chloride, and water. *Biochemistry* **28,** 1270–1278.

Feller, S. E., Yin, D., Pastor, R. W., and MacKerrell, A. J. (1997). Molecular dynamics simulation of unsaturated lipids at low hydration: Paramaterization and comparison with diffraction studies. *Biophys. J.* **73,** 2269–2279.

Gruner, S. M. (1985). Intrinsic curvature hypothesis for biomembrane lipid composition: A role for nonbilayer lipids. *Proc. Natl. Acad. Sci. USA* **82,** 3665–3669.

Helfrich, W. (1973). Elastic properties of lipid bilayers: Theory and possible experiments. *Z. Naturforsch.* **28C,** 693–703.

Israelachvili, J. N. (1985). Measurements of hydration forces between macroscopic surfaces. *Chem. Scripta* **25,** 7–14.

Jacobs, R. E., and White, S. H. (1989). The nature of the hydrophobic binding of small peptides at the bilayer interface: Implications for the insertion of transbilayer helices. *Biochemistry* **28,** 3421–3427.

Kirk, G. L., and Gruner, S. M. (1985). Lyotropic effects of alkanes and headgroup composition on the L_{α}-H_{II} lipid liquid crystal phase transition: Hydrocarbon packing versus intrinsic curvature. *J. Phys.* **46,** 761–769.

Kozlov, M. M., Leikin, S., and Rand, R. P. (1994). Bending, hydration and void energies quantitatively account for the hexagonal-lamellar-hexagonal reentrant phase transition in dioleoylphosphatidylethanolamine. *Biophys. J.* **67,** 1603–1611.

Leikin, S., Parsegian, V. A., Rau, D. C., and Rand, R. P. (1993). Hydration forces. *Annu. Rev. Phys. Chem.* **44,** 369–395.

LeNeveu, D. M., Rand, R. P., and Parsegian, V. A. (1976). Measurement of forces between lecithin bilayers. *Science* **259**(19), 601–603.

Lewis, B. A., and Engelmann, D. M. (1983). Bacteriorhodopsin remains dispersed in fluid phospholipid bilayers over a wide range of bilayer thicknesses. *J. Mol. Biol.* **166,** 203–210.

Lundbaek, J. A., and Anderson, O. S. (1994). Lysophospholipids modulate channel function by altering the mechanical properties of lipid bilayers. *J. Gen. Physiol.* **104,** 645–673.

Luzzati, V., and Husson, F. (1962). The structure of the liquid-crystalline phases of lipid–water systems. *J. Cell. Biol.* **12,** 207–219.

MacMillan, S. V., Alexander, D. A., Culham, D. E., Kunte, H. J., Marshall, E. V., Rochon, D., and Wood, J. M. (1999). The ion coupling and organic substrate specificities of osmoregulatory transporter ProP in *Escherichia coli. Biochim. Biophys. Acta* **1420,** 30–44.

May, S., and Ben-Shaul, A. (1999). Molecular theory of lipid–protein interaction and the L-HII transition. *Biophys. J.* **76,** 751–767.

McIntosh, T. J. (1996). Hydration properties of lamellar and non-lamellar phases of phosphatidylcholine and phosphatidylethanolamine. *Chem. Phys. Lipids* **81**(2), 117–131.

McLaughlin, S., Mulrine, N., Gresalfi, T., Vaio, G., and McLaughlin, A. (1981). Adsorption of divalent cations to bilayer membranes containing phosphatidylserine *J. Gen. Physiol.* **77,** 445–473.

Mouritsen, O. G., and Bloom, M. (1984). Mattress model of lipid protein interactions in membranes. *Biophys. J.* **46,** 141–153.

Parsegian, V. A., Fuller, N. L., and Rand, R. P. (1979). Measured work of deformation and repulsion of lecithin bilayers. *Proc. Natl. Acad. Sci. USA* **76**(6), 2750–2754.

Parsegian, V. A., and Rand, R. P. (1991). On molecular protrusion as the source of hydration forces. *Langmiur* **7,** 1299–1301.

Parsegian, V. A., and Rand, R. P. (1992). Comparisons of solubilities predicted by the protrusion model with experimental critical micelle concentration estimates from equations of Cevc and Marsh. *Langmuir* **8,** 1502.

Parsegian, V. A., Rand, R. P., and Rau, D. C. (1995). Macromolecules and water: Probing with osmotic stress. *Methods Enzymol.* **259,** 43–94.

Parsegian, V. A., Rand, R. P., and Rau, D. C. (2000). Osmotic stress, crowding, preferential hydration, and binding: A comparison of perspectives. *Proc. Natl. Acad. Sci. USA* **97,** 3987–3992.

Rand, R. P. (1981). Interacting phospholipid bilayers: Measured forces and induced structural changes. *Annu. Rev. Biophys. Bioeng.* **10,** 277–314.

Rand, R. P., Fuller, N., Parsegian, V. A., and Rau, D. C. (1988). Variation in hydration forces between neutral phospholipid bilayers: Evidence for hydration attraction. *Biochemistry* **27,** 7711–7722.

Rand, R. P., Fuller, N. L., Gruner, S. M., and Parsegian, V. A. (1990). Membrane curvature, lipid segregation, and structural transitions for phospholipids under dual-solvent stress. *Biochemistry* **29,** 76–87.

Rand, R. P., and Parsegian, V. A. (1989). Hydration forces between phospholipid bilayers. *Biochim. Biophys. Acta* **988,** 351–376.

Rand, R. P., and Parsegian, V. A. (1997). Hydration, curvature, and bending elasticity of phospholipid bilayers. *In* "Lipid Polymorphism and Membrane Properties," (R. Epand, Ed.). Academic Press, New York.

Rand, R. P., Parsegian, V. A., and Rau, D. C. (2000). Intracellular osmotic action. *Cell. Mol. Life Sci.* **57,** 1018–1032.

Reid, C., and Rand, R. P. (1997). Probing protein hydration and conformational states in solution. *Biophys. J.* **72,** 1022–1030.

Reitveld, A., and Simons, K. (1998). The differential miscibility of lipids as the basis for the formation of functional membrane rafts. *Biochim. Biophys. Acta* **1396,** 467–479.

Seddon, J. M. (1990). Structure of the inverted hexagonal (HII) phase and non-lamellar phase transitions of lipids. *Biochim. Biophys. Acta* **1031,** 1–69.

Siegel, D. P. (1993). Energetics of intermediates in membrane fusion: Comparison of stalk and inverted micellar intermediate mechanisms. *Biophys. J.* **65,** 2124–2140.

Silvius, J. R., del Giudice, D., and Laleur, M. (1996). Cholesterol at different bilayer concentrations can promote or antagonize segregation of phospholipids of differing chain length. *Biochemistry* **35,** 569–572.

Tian-Yu, W., and Silvius, J. R. (2000). Different sphingolipids show differential partitioning into sphingolipid/cholesterol-rich domains in lipid bilayers. *Biophys. J.* **79,** 1478–1489.

Timasheff, S. N. (1998). In disperse solution, "osmotic stress" is a restricted case of preferential interactions. *Proc. Natl. Acad. Sci. USA* **95,** 7363–7367.

Tristram-Nagle, S., Petrache, H. I., and Nagle, J. F. (1998). Structure and interactions of fully hydrated dioleoylphosphatidylcholine bilayers. *Biophys. J.* **75,** 917–925.

Wiener, M. C., and White, S. H. (1992). Structure of a fluid dioleoylphosphatidylcholine bilayer determined by joint refinement of X-ray and neutron diffraction data. III. Complete structure. *Biophys. J.* **61,** 437–447.

Wood, J. M. (1999). Osmosensing by bacteria: Signals and membrane-based sensors. *Microbiol. Mol. Biol. Rev.* **63,** 230–262.

Water as It Applies to the Function of Enzymes

J. A. Kornblatt and M. J. Kornblatt
Enzyme Research Group, Departments of Biology, Biochemistry and Chemistry, Concordia University, Montreal, Quebec, Canada H3G 1M8

Escherichia coli and *Saccharomyces cerevisiae* can metabolize, grow, and divide over osmotic pressures ranging from 0.24 atm to about 100 atm [Record, T. M. *et al.* (1999). *Trends Biochem. Sci.* **23,** 143–148, 190–194; Wood, J. M. (1999). *Microbiol. Mol. Biol. Rev.* **63,** 230–262; Marachal, P. A., and Gervais, P. (1994). *Appl. Microbiol. Biotechnol.* **42,** 617–622]. At the higher end of the range, they perform their functions with difficulty, but they can survive. Over the full span of pressures, the activity of water goes from 0.9998 to 0.93. Neither of the authors can survive at anything like these extremes; some of their enzymes and enzymatic complexes would "fall apart," would either cease to function or would denature. We would very much like to know just how the two microbes manage.

KEY WORDS: Water, Enzyme function, Enzyme structure.

I. Introduction

If one surveys the total number of enzymes known, there are probably only a few, if any, that do not involve water in one way or another. How important is the water? At one extreme, one posits that water is relatively unimportant in maintaining enzyme structure and function; at the other extreme, one would argue that it is water that does most of the work and that a protein serves one purpose only: it provides a scaffold on which water molecules can be ordered to carry out biochemical reactions. Does the truth lie somewhere between the extremes? The problem before us is to limit our discussion so as not to lose sight of what is essential for understanding.

Where does an enzymologist find water? The question does not have a unique answer since it depends on the methodology used. It is for this reason that we dedicate a substantial portion of the review to the tools that can and have been used to define the location and function of water in and around proteins.

The protein cytochrome P450cam, a heme enzyme, catalyzes the reaction

$$\text{camphor} + O_2 + 2\,\text{putidaredoxin (reduced)} \rightarrow \\ \text{5-exohydroxycamphor} + H_2O + 2\,\text{putidaredoxin (oxidized)} \qquad (1)$$

Its structure is known to high resolution. We use it here to attempt an illustration of the goal of the review: "Water as it applies to the function of enzymes." Figure 1 shows two structures. Figure 1A shows the x-ray crystallographic coordinates of the protein and the heme when camphor is not present (Poulos *et al.*, 1986); note the six waters (greenish) that sit as a cluster just in front of the heme. Figure 1B shows the cavity after camphor has gained access (Schlichting *et al.*, 2000). In the presence of camphor, there is no water in the pocket. In the absence of camphor, the cavity is filled with water, the polarity of the cavity is high, and the dielectric coefficient of the region is relatively high. These conditions favor the Fe(III) form of the heme over Fe(II). The iron cannot easily be reduced by its reducing substrate, putidaredoxin. Binding of substrate, camphor, drives the water out of the pocket (Fig. 1B) with the result that the polarity of the cavity is lower and the dielectric coefficient in the vicinity of the heme is lowered. This favors the Fe(II) form of the heme over Fe(III); the heme is now capable of being reduced by putidaredoxin, and the catalytic cycle can continue.

What is the role of water in the function of this enzyme?

1. It takes up space thereby facilitating good packing and reducing the flexibility of the protein in the vicinity of the heme.
2. It keeps electrons from being transferred to the cytochrome P450 from the reduced putidaredoxin. The reduced protein is capable of reacting with

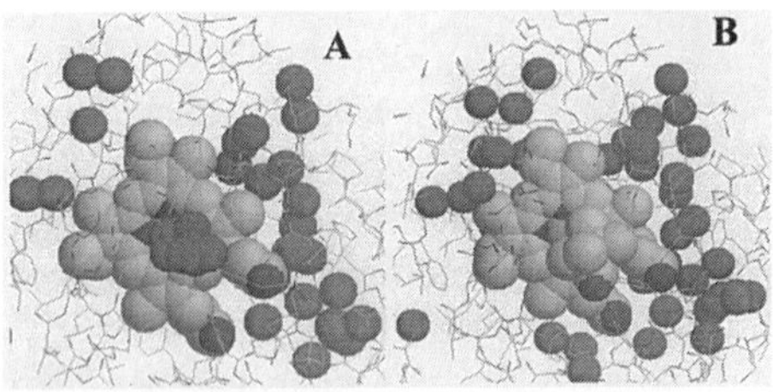

FIG. 1 (A) (**1PHC**). [PDB files are shown in (**Bold**).] Cytochrome P450CAM showing the group of six waters in the heme pocket when camphor is absent. The waters are shown in greenish blue. The carbons of the heme are shown as gray spacefilling structures. The iron of the heme is pink and is almost totally obscured by the cluster of waters. (B) (**1DZ6**). In this depiction of cytochrome P450CAM, camphor has displaced the waters shown in (A). The carbons of the camphor are shown as spacefilling gray. The oxygen of the camphor is red. (See also color insert.)

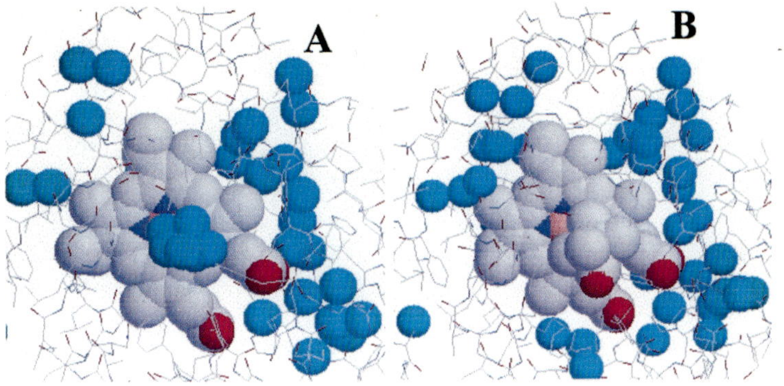

FIG 3.1. (A) **(1PHC).** [PDB files are shown in bold.] Cytochrome P450CAM showing the group of six waters in the heme pocket when camphor is absent. The waters are shown in greenish blue. The carbons of the heme are shown as gray spacefilling structures. The iron of the heme is pink and is almost totally obscured by the cluster of waters. (B) **(1DZ6).** In this depiction of cytochrome P450CAM, camphor has displaced the waters shown in (A). The carbons of the camphor are shown as spacefilling gray. The oxygen of the camphor is red.

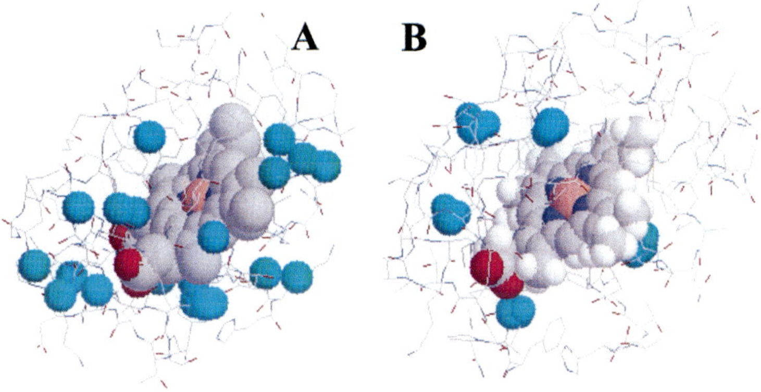

FIG 3.2. (A) **(1HRC).** [PDB files are in bold.] The structure of horse cytochrome *c* as determined by x-ray crystallography. Only those portions of the molecules less than 10 Å from the heme are shown. This results in hiding most of the waters shown in the complete structure. Water number 166 is rigorously conserved in all naturally occurring cytochromes c. It is part of three waters at 9 o'clock. The color scheme is the same as in Fig. 1. (B) **(1OCD).** The structure of horse heart cytochrome c as determined by NMR. The color scheme is the same as in Fig. 1. Only those portions of the molecules within 10 Å of the heme are shown. NB: One of the waters at 9 o'clock is water 166 and is rigorously conserved in all naturally occurring cytochromes c. (A) and (B) indicate that some waters can be seen by both NMR and x-ray crystallography.

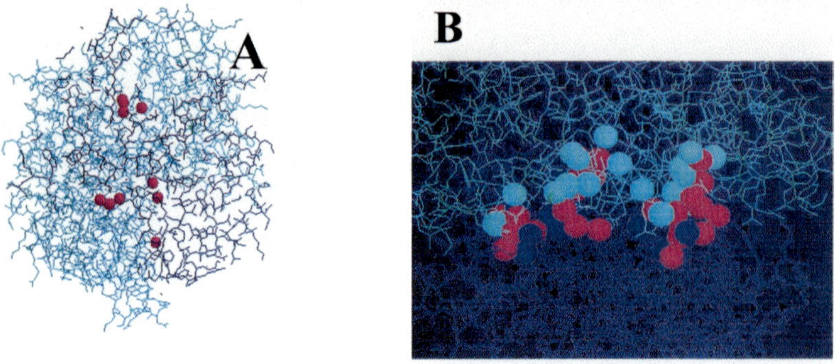

FIG 3.4. (A) **(1ONE).** [PDB files are in bold.] The X-ray structure of yeast enolase looking down on the dimer interface is shown. The enzyme is composed of two identical subunits. About 60% of the top subunit (dark blue) has been removed so that the rough surface of the interface can be envisioned. The 10 reddish balls are water that are at the interface. (B) **(1ONE).** The molecule has been rotated by 90° so that one is looking through the interface. The dark blue lines and balls constitute atoms of one subunit, while the light blue balls and lines correspond to the other. The reddish balls are water molecules that are situated within 3 Å of the dark blue and light blue balls; they are within hydrogen bonding distance.

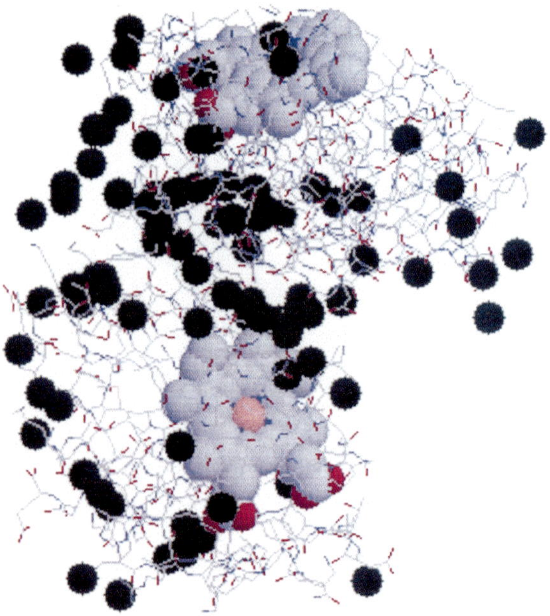

FIG 3.5. **(2PCC).** [PDB files are in bold.] The complex of horse cytochrome *c* and yeast cytochrome *c* peroxidase as determined by X-ray crystallography. The peroxidase heme is shown at the top and the cytochrome *c* heme at the bottom. Only those portions of the complex that are within 29 Å of the heme of cytochrome *c* are shown. The black balls represent water molecules. Pay particular attention to the leaf-like cluster situated between the two hemes; this is part of the interface between the peroxidase and the cytochrome *c*.

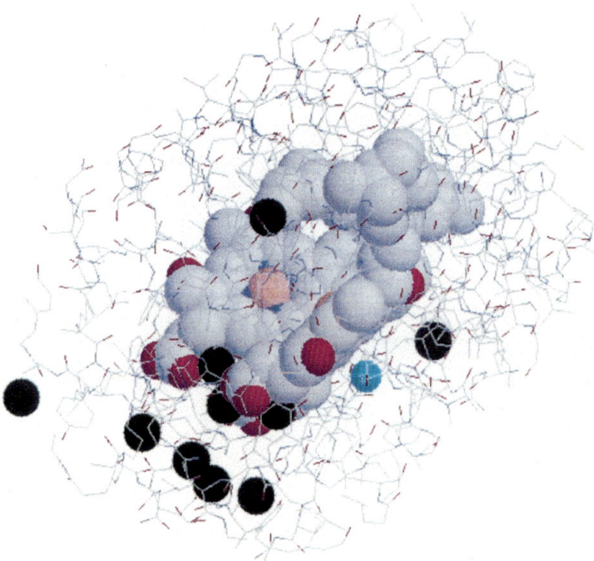

FIG 3.7. **(1AR1).** [PDB files are in bold.] The cytochrome *c* oxidase showing the two cytochromes: the *a* cytochrome is almost a full face view, while the a_3 cytochrome is shown end on. The bluish ball to the right of the a_3 cytochrome is CuB, part of the binuclear center. The black balls represent water molecules that are close to the proposed proton pumping channel.

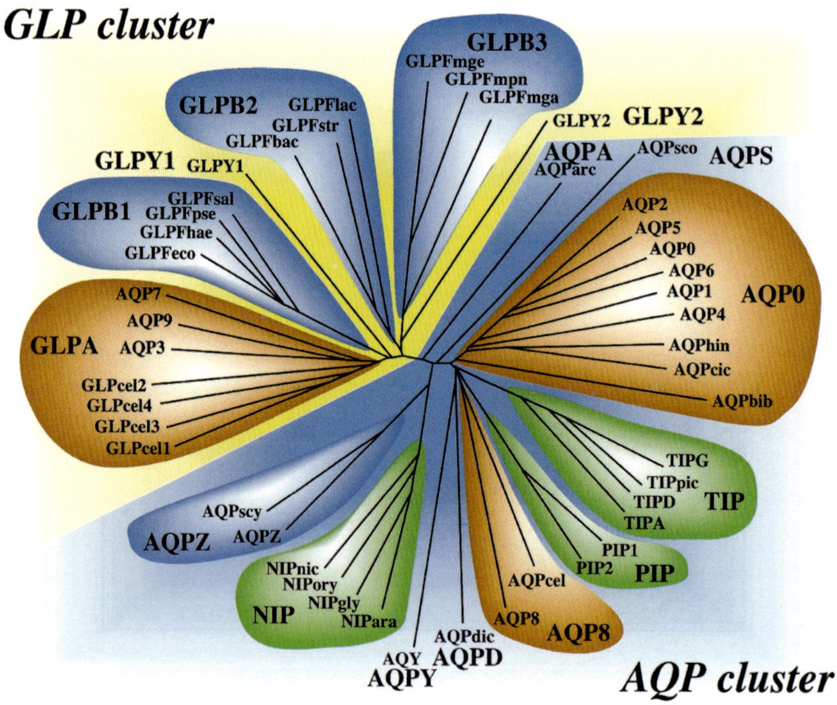

FIG 4.1. Phylogenetic analysis of the aquaglyceroporin (MIP) family suggests a classification into two clusters, an AQP0-like cluster (blue area) and a GlpF-like cluster (yellow area). Further distinction is made between subfamilies, which include sequence types. The types are a 46-sequence subset considered to be representative of the whole family of ~160 sequences.

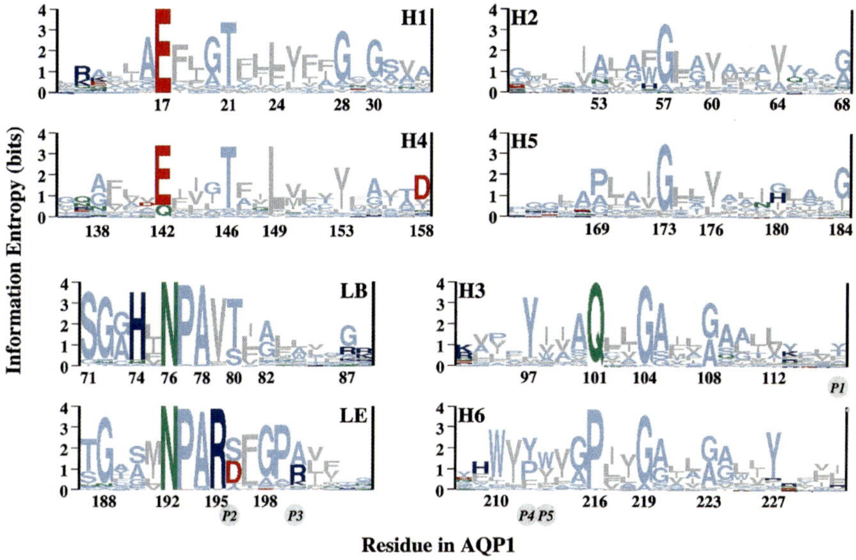

FIG 4.2. Multiple alignment of 160 AQP/GLP sequences and subsequent phylogenetic analyses yielded 46 subtypes (Heymann and Engel, 2000). To identify critical residues, the 46 characteristic sequences were aligned and the conservation of each residue calculated (Heymann and Engel, 2000). Sequence logos, whose heights are a measure of conservation, are drawn with the residue numbers for AQP1. They reveal the conservation patterns of helices H1–H6 and the highly conserved loops, LB and LE. Helices are grouped in pairs according to the sequence similarity between the first and second half of the protein. The five positions (P1–P5) that were found by Froger *et al.* (1998) to be different between the AQP and GLP clusters are given in italics in circles. Colors: gray, hydrophobic; light blue, polar; green, amide; red, acidic; dark blue, basic.

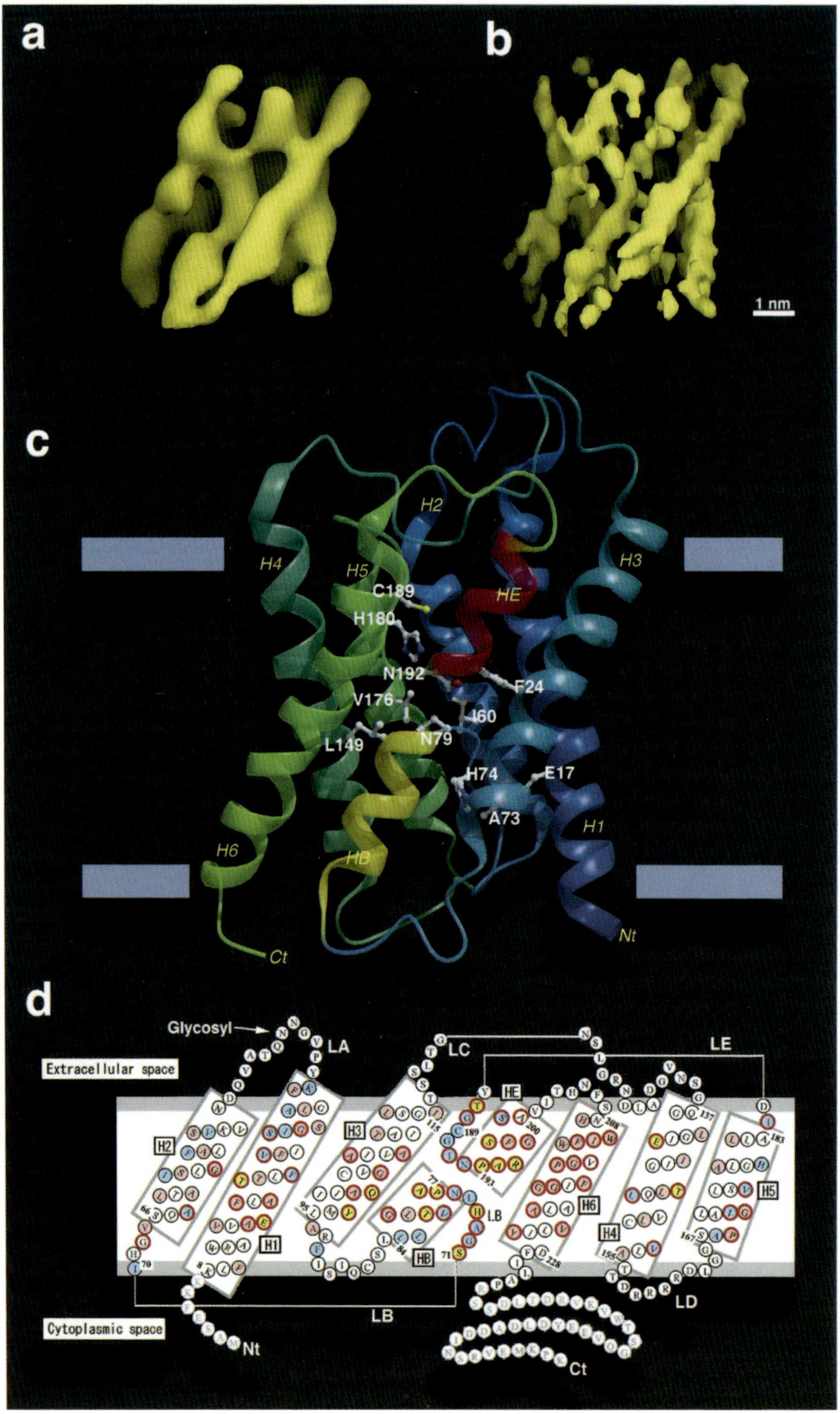

FIG 4.7. The structure and sequence of AQP1. (a) The structure from cryoelectron microscopy, rendered at 6.0 Å resolution, reveals rod-like densities, which correspond to highly tilted α-helices in a right-handed helix arrangement. (b) At 3.8 Å resolution, the α-helices reveal a right-handed twist. These data were the structural basis for the model-building (Murata *et al.,* 2000). (c) Ribbon representation of the atomic model of AQP1, showing the extracellular side up. Helices are labeled in yellow italics, channel-lining residues are labeled in white. (d) The amino acid sequence of AQP1.

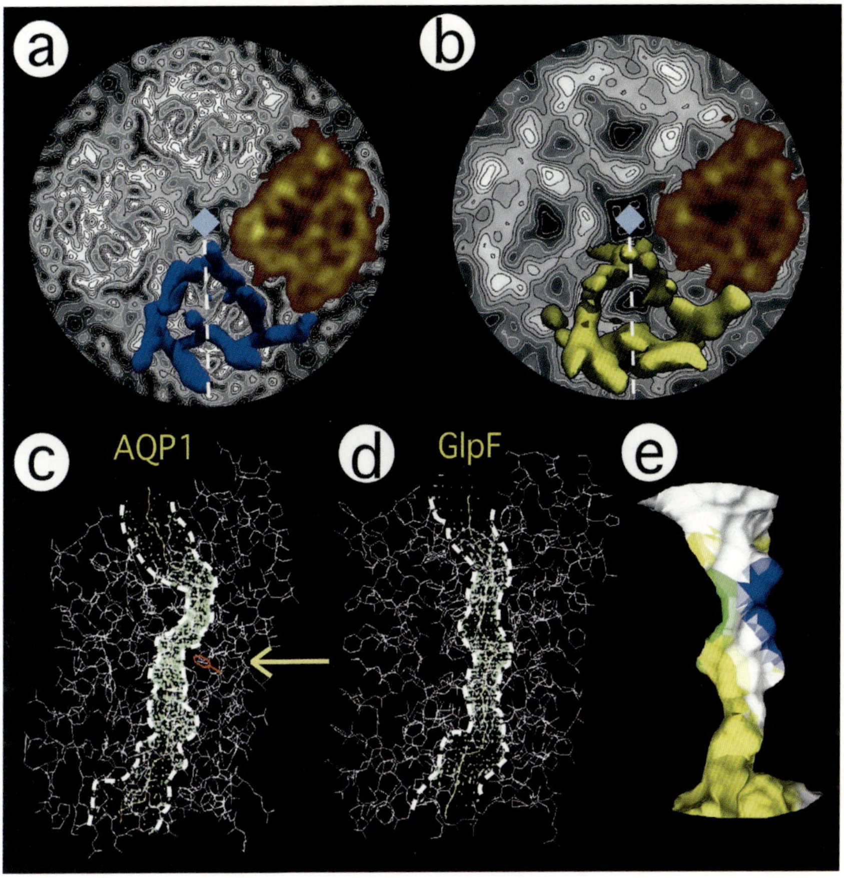

FIG 4.8. Atomic modeling of the GlpF channel based on the structure and sequence homology of GlpF and AQP1 and the atomic structure of AQP1 (Murata *et al.,* 2000). (a) The 3.5 Å AQP1 projection structure with the 3D map of the monomer at 6.9 Å resolution (blue) and the projection of the atomic model (brown). (b) The 3.7 Å GlpF projection structure with the 3D map of the monomer at 6.9 Å resolution (yellow) and the projection of the constructed model of GlpF. The dotted line indicates the pseudo-two-fold symmetry axis of the monomer. (c, d) Channels of AQP1 (c) and GlpF (d), calculated by HOLE (Vriend, 1990). (e) The amphiphilic nature of the GlpF channel is revealed by color-coding the accessible surfaces: green, aromatic residues; yellow, hydrophobic residues; white, polar residues; blue, positively charged residues.

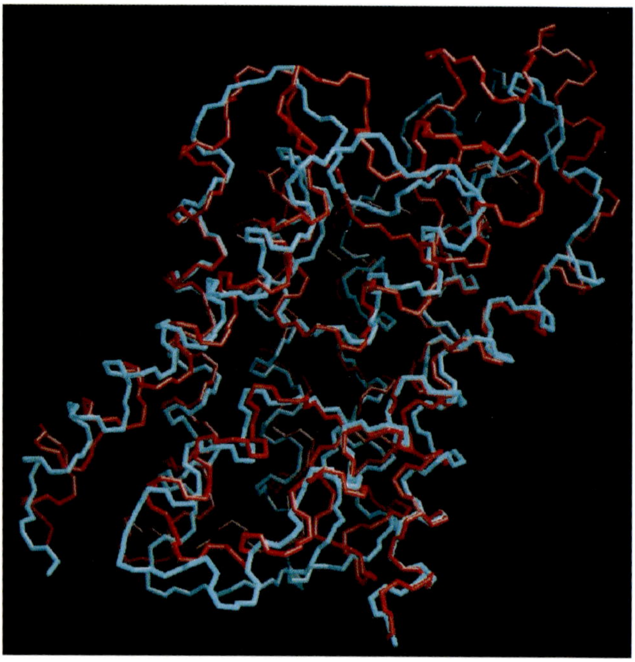

FIG 4.9. Superposition of the backbone traces of the X-ray structure of GlpF (red) and a previous EM structure of AQP1 (cyan, PDB entry 1FQY) (Murata *et al.,* 2000).

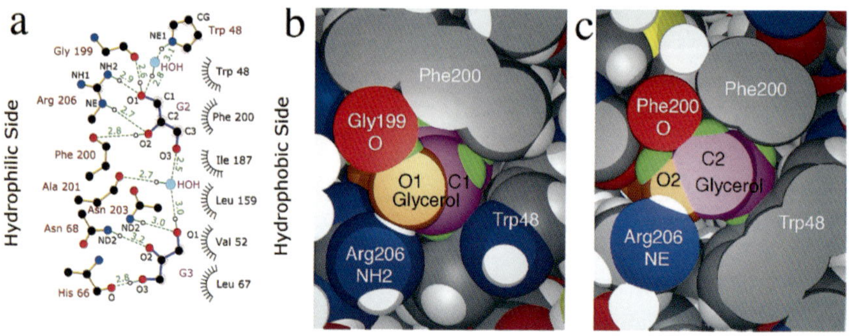

FIG 4.10. The structure of the GlpF reveals three glycerol molecules at specific sites in the channel. (a) The hydrogen bonding network (dotted lines) showing interactions of G2, G3, and associated water molecules with the selectivity filter. Hydrophobic contacts are indicated as radial lines around atoms or residues. Distances (in Å) between heavy atoms are indicated. (b, c) Cross-section through the channel (space-filling rendering) showing the interactions with G2OH2, O2 (orange) and its associated CH group (C2; purple), and hydrogen (green). The cross-section shows the donor NEH of Arg 206, acceptor carbonyl oxygen of Phe 200, and hydrophobic corner formed by Trp 48 and Phe 200 viewed edge-on to the aromatic rings.

oxygen, in the absence of camphor, to produce activated oxygen species that are released in solution. The presence of water in the pocket reduces this activity, thus helping to avoid damage to the bacterium's DNA and membranes.

3. When camphor finally does gain access to the binding cavity, the midpoint potential of the protein is altered so that electron transfer from the putidaredoxin is facilitated.
4. Water, by its presence or absence, determines whether the enzyme is a good catalyst with high specificity or a poor catalyst that leaks noxious chemicals into the cytosol.

The above is obviously not true for other enzymes, but each enzyme will follow a similar pattern; water will be present at various sites, and this water will modify the properties of the protein and/or participate in the reactions being catalyzed. The decision as to how water and enzyme will interact will be made on the basis of energetics. Where the return outweighs the energetic cost, water will fill spaces that might otherwise form large voids; water here will fill a structural role. In many cases, the active site will be filled with water that must be displaced by substrate or product; water here will contribute to the overall dynamics of substrate binding and product release. In many cases, water will contribute to catalysis by modulating the characteristics of the groups at the active site. In cytochrome P450, it modulates the midpoint potential of the iron. In enzymes that rely on the amino acids aspartate, glutamate, histidine, tyrosine, arginine, and lysine to initiate chemistry or transport, water modulates the pKs of those amino acids.

II. Tools for Studying Water

Figure 2A shows the structure of horse cytochrome *c* as determined by x-ray crystallography (Bushnell *et al.,* 1990). Figure 2B shows the same protein determined by nuclear magnetic resonance (NMR; Qi *et al.,* 1996). Waters are shown in green. The NMR and x-ray structures of the protein portion are much the same. Close inspection of A and B would show that the waters detected by NMR are also detected by x-ray crystallography, but that most of the waters detected in the x-ray structure are not detected by NMR. Recent results from molecular dynamics calculations indicate that it is possible, based on the x-ray structure of the protein portion of the structure, to predict the presence of many of the waters detected by x-ray crystallography (Garcia and Hummer, 2000). These differences between the techniques do not indicate that one is right while others are not; the differences do indicate the ephemeral nature of the subject (and the subject of this article). Water is there; what do we have to do to see it? What does it do? What do we have to do to see it in action?

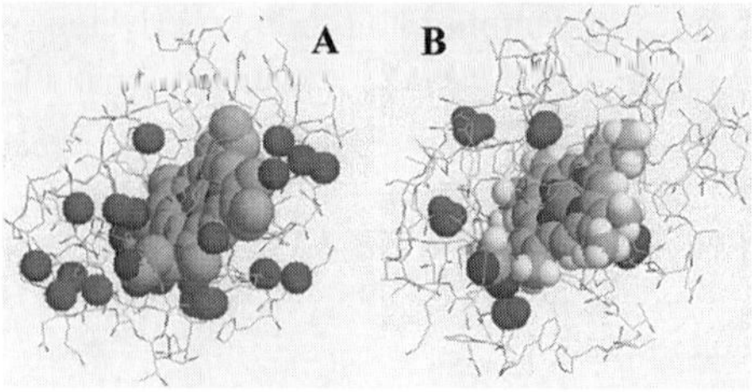

FIG. 2 (A) **(1HRC)**. [PDB files are in (**Bold**).] The structure of horse cytochrome *c* as determined by x-ray crystallography. Only those portions of the molecule less than 10 Å from the heme are shown. This results in hiding most of the waters shown in the complete structure. Water number 166 is rigorously conserved in all naturally occurring cytochromes *c*. It is part of the cluster of three waters at 9 o'clock. The color scheme is the same as in Fig. 1 (B) **(1OCD)**. The structure of horse heart cytochrome *c* as determined by NMR. The color scheme is the same as in Fig. 1. Only those portions of the molecules within 10 Å of the heme are shown. NB: One of the waters at 9 o'clock is water 166 and is rigorously conserved in all naturally occurring cytochromes *c*. (A) and (B) indicate that some waters can be seen by both NMR and x-ray crystallography. (See also color insert.)

The main methods can be divided into three types—those that permit the direct observation of water molecules, those that permit one to infer the involvement of water (thermodynamics), and those that permit the researcher to perturb water and observe the consequences.

A. Direct Observation

High resolution structure determination, by x-ray diffraction or by NMR, reveals some of the water molecules that are bound to proteins. For a water molecule to be "seen" by x-rays, the site must have a reasonably high occupancy, but it need not be occupied all the time. The rate at which the water is exchanging with bulk solvent is not important as long as there is often a water at that site and the water molecule is ordered, that is, it always has the same orientation. The result of the latter requirement is that the water molecules seen by x-rays are those that are hydrogen bonded to the protein or are part of a group of water molecules that are hydrogen bonded to each other and, ultimately, to the protein. Another limitation of the x-ray method is that the conditions used to crystallize the protein—high concentrations of salts or polyethylene glycol, for example—are conditions of osmotic stress (see below) and thus may have perturbed (removed) some of the bound water molecules. Waters that are "seen" by NMR do not have to be ordered; they can be rotating, changing position slightly, etc. However, they must remain in the general neighborhood long enough for transfer of magnetization to occur. Estimates of the lifetimes of waters that can be seen range from 1–2 ns to 100–200 μs (Ernst *et al.,* 1995). The result of these differences in what is seen by the two methods is that one does not see all of the same molecules by the two methods (Billeter, 1995; Ernst *et al.,* 1995). It is important to remember that the failure to "see" water at a given site does not mean that it is not present. These two techniques show us some, but not necessarily all, of the water associated with the protein. Another technique with the potential for direct observation of water molecules is neutron diffraction; however, to date, it has not been extensively used.

B. Thermodynamics

When two solvated molecules come together and bind, there is a decrease in entropy due to the loss of degrees of freedom of both partners. To understand how the entropy of the complete system changes, we must ask what happens to the water molecules bound to the surfaces that interact. Are the surfaces desolvated (Fig. 3, path a)? In this case, solvent entropy will increase due to the release of bound water. Is water trapped at the interface (Fig. 3, path b)? This water is more ordered and immobilized, hence solvent entropy decreases. The magnitude of the $T\Delta S$ term can give us some indication as to whether or not binding involves a net release of water molecules.

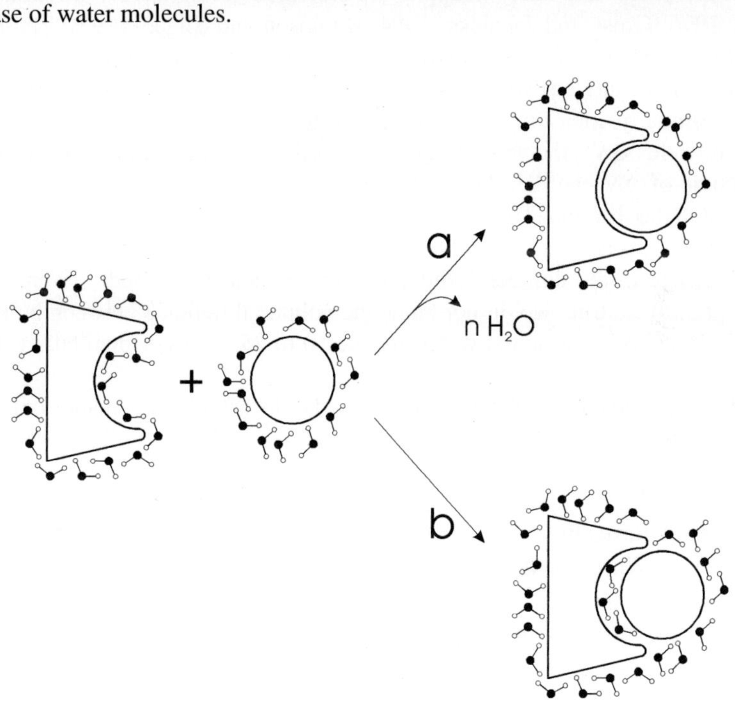

FIG. 3 Two possible modes of binding a ligand to a macromolecule. In path a, desolvation of the two surfaces occurs; binding is accompanied by a net release of water. In path b, some water molecules are trapped at the interface; these waters may mediate some of the interactions between the two molecules.

C. Perturbation Methods

Proteins and their ligands function in aqueous solution; almost anything that a protein does—binding a ligand, catalyzing a reaction, transmitting a signal—will involve changes in solvation of the protein and/or ligand. The converse is also true—anything that perturbs solvation will probably perturb the normal functioning of the protein. Perturbation of solvation can be achieved by the application of hydrostatic or osmotic pressure. Consider a simple reaction, such as A $\leftrightarrows$ B. If A and B differ in their volumes, then the application of hydrostatic pressure will drive the reaction (displace the equilibrium) in the direction of the species which occupies the smaller volume. Bound water occupies a smaller volume than does bulk water. Thus, exposing buried surface to solvent, as in the dissociation of an oligomeric protein, results in that surface being hydrated. Bulk water becomes bound water, and the system volume decreases. Similarly, the disruption of a salt bridge results in both partners being hydrated, with a corresponding decrease in volume; this phenomenon is called electrostriction. Elimination of voids or packing defects could also decrease the volume. The specialized equipment required for these experiments and many examples are described in various reviews (Moshaev *et al.,* 1996; Gross and Jaenicke, 1994; Robinson and Sligar, 1995). The basic experiment consists of subjecting the reaction of interest to hydrostatic pressure; the variation of ln(Keq) with pressure yields ΔV, the volume change for the reaction. Similarly, measuring the variation of the rate of the reaction (ln(k)) with pressure yields $\Delta V^{\ddagger}$, the volume change for formation of the transition state. The magnitude of ΔV can then be used to evaluate the number of water molecules involved in the reaction. There are several serious limitations to this analysis. The first is that one is assuming that the only factor contributing to the volume changes is the change in the number of bound water molecules. Secondly, even if that assumption were true, we do not know the volume difference between bulk and bound. The density of bound water may be 10 to 25% greater than bulk (Zhang *et al.,*1996).

Let us return to our simple reaction. If A and B do differ in their degree of hydration, then decreasing the water activity, by adding cosolvents, will shift the equilibrium toward the less hydrated species. This technique has also been extensively reviewed (Parsegian *et al.,* 1995, 2000; Robinson and Sligar, 1995; see Parsegian, this volume). Basically, the reaction of interest is performed in solutions of varying water activity; ln(K) or ln(k) is plotted as a function of osmotic pressure, yielding the *number* of water molecules involved. Unfortunately, the cosolvents that are used—glycerol, sucrose, polyethylene glycol, etc.—can have multiple effects on reactions; these have been extensively commented on (Parsegian *et al.,* 1995, 2000; Timasheff, 1998; Robinson and Sligar, 1995). In addition to lowering water activity, they also change the dielectric constant and increase the viscosity, among other things. Sucrose, glycerol, and ficoll all decrease the rate of carbonic anhydrase activity. The data can be satisfactorily explained by an effect of solvent viscosity on the rate of the reaction or by an effect of water activity, with two water molecules participating in the reaction (Pocker and Janjic, 1989). Much of the effort in experimental design and data analysis lies in trying to determine which change in solvent properties is responsible for the observed effects on the reaction being studied.

A related technique is that of dilatometry, in which the actual volume change of a reaction is directly and physically measured (reviewed in Kornblatt *et al.*, 1998). The volume changes are usually small; this means that highly sensitive and precise equipment must be used, as well as large amounts of protein. For example, the ΔV for dissociation of the dimeric enzyme, enolase, measured by dilatometry, is 110 mL mol^{-1} (unpublished data). This measurement required 0.7 μmol (66 mg) of the protein; the actual volume change that was measured was only 0.03 μL.

III. Structural Water

Where is the water that is associated with enzymes and what is it doing there? Before 1960, we thought of it as nothing but a convenient solvent. We then realized just how little water there was inside the "typical" cell, and water was elevated to a privileged position with properties unique to the proteins, membranes, nucleic acids, and carbohydrates with which it was associated. Our thinking has since moderated but is certain to change again. Current thinking can be easily summarized:

A typical water molecule spends a certain percentage of its time being just water (the second-to-hour time scale), a certain percentage of its time very tightly associated with a given protein (the second-to-microsecond time scale), another portion tightly associated (the microsecond-to-picosecond time scale), and another portion weakly associated with proteins (Garcia and Hummer, 2000; Otting *et al.*, 1991).

In what follows, we focus on several examples which illustrate the principles of where we find water in small, monomeric enzymes, multimeric enzymes and, finally, enzyme–substrate complexes. We will look at surface water, internal water, and water at interfaces.

A. Surface Water

Proteins in solution are surrounded (hydrated) by one or more layers of water or other solvent. The organization of water around polar groups may not be the same as that around nonpolar groups. The first layer of water has a higher density than that of bulk water. Evidence for this layer comes from analysis of x-ray results and from direct observation by neutron scattering (Svergun *et al.*, 1998). Although one is tempted to view this layer as purely passive, changes in hydration accompany changes in quaternary structure or conformation of a protein. As discussed in Section IV.B, perturbation of this layer can modify the behavior of the protein.

B. Internal Water

1. Cytochrome *c*

Let us look again at the waters of cytochrome *c*, as shown in Fig. 2. Of these, 120 have been detected in the horse protein using x-ray diffraction. By molecular dynamics, one sees much the same picture as one does using x-ray diffraction, but here one can distinguish between those waters that spend a great deal of time inside the protein and those that are ephemeral. Based on the two structures of Fig. 2, it is clear that water 166 (yeast numbering; found at about 9 o'clock in both panels of Fig. 2) is tightly associated with the protein. This internal water molecule is near the heme group and is part of a hydrogen bond network that includes three residues—asn52, tyr67, and thr68—which are directly hydrogen bonded to this water. Does it fulfill a known function? Conventional wisdom would say that it should since wat166 is highly conserved throughout evolutionary time. Wat166 may move slightly as the protein goes through an oxidation–reduction cycle; x-ray and NMR studies come to different conclusions on this point. If, using site-directed mutagenesis, one mutates the residues in the vicinity of wat166, one of two things happens; in the mutants N52I and Y67F/N52I, wat166 is removed. In the mutants N52A and Y67F, a new water molecule, wat300, is introduced alongside wat166. The midpoint potentials of these mutant forms of cytochrome *c* change, and their stability to denaturation changes. *The altered cytochromes c continue to be totally functional as far as one can tell.* Cells containing cytochrome *c* missing wat166 grow normally. What role can we then assign to wat166 so that we can explain the evolutionary conservation of this water? So far, an explanation eludes us. One can say that when mutations are carried out that result in the water being removed, the space usually gets occupied by other groups. In the case of cytochrome *c*, mutation does not result in the introduction of empty space. Accordingly, wat166 appears to function in much the same way as an amino acid not necessary for function. It takes up space; it does not introduce any large energetic instabilities; it is sufficiently mobile that the protein can undergo any necessary conformational changes; in short, it allows the remainder of the protein to "easily" maintain its folding pattern. (The relevant data have been reviewed in Lett *et al.,* 1996).

2. Bovine Trypsin Inhibitor (BPTI)

BPTI is a low molecular weight protein that inhibits trypsin and trypsin-like proteins. It does so by acting as a very poor substrate that, once cleaved, does not release from the trypsin. Both x-ray (Huber *et al.,* 1974) and NMR (Otting and Wuthrich, 1989) structures are known. Both techniques agree on number and position of the four tightly bound waters. The exchange rates for these buried waters have been estimated by NMR and by molecular dynamics (Otting and Wuthrich, 1989; Otting *et al.,* 1991; Denisov *et al.,* 1995). Three of the waters are in a pore-like cavity and are hydrogen bonded to each other and to the protein; they exchange rapidly, with residence times of 15 ns to 1 μs. The fourth water, wat122, is in a small cavity between two loop segments; it is hydrogen bonded to four groups, two in each loop segment, and is near the disulfide bond that connects the two loops. It has a slower rate of exchange, with a residence time of 10–200 μs. The differences in exchange rates probably reflect differences in the flexibility of the protein. Exchange of a buried water with bulk water requires a channel in the protein that opens transiently. This occurs as a result of random conformational fluctuations in the protein. The more rigid environment of wat122 limits its rate of exchange. Wat122 is adjacent to gly36. Mutation of this residue to serine expels the water; as expected, the hydroxyl of the serine can make two of the four hydrogen bonds made by wat122. Does the mutation change anything vis-à-vis the function of the molecule? The stability of the protein is reduced by less than 1 kcal mol^{-1}. More significantly, the glycine to serine mutant is a poorer inhibitor of the proteases (Berndt *et al.,* 1993); this probably results from wat122 being in the region of the protease binding site. Is the water necessary for good inhibition or does the water contribute to the rigidity of this region, which in turn is necessary for inhibition? Or have minor structural rearrangements occurred? We do not yet have answers to these questions.

Both of these examples illustrate the ability of water to fill spaces. If one makes a mutation that creates a space for water, by replacing a tyrosine with a phenylalanine for example, one often finds, in the x-ray structure, that the space is occupied by water. Similarly, eliminate a space by substituting a large residue for a small and the water is lost. Does this rule hold all the time? Are all internal cavities occupied by water?

3. T4 Lysozyme

This is an interesting small protein; its function is to help the bacteriophage T4 get out of the bacterium. The structure of the protein has been known for many years. As part of a study on the role of hydrophobic interactions in protein stability, the Matthews group has produced and done structural analysis on many mutants of T4 lysozyme (Xu *et al.,* 1998). The relevant mutations are located in the hydrophobic core of the protein and change a large residue—leucine, isoleucine, valine, methionine, and phenylalanine—to a small one (alanine). In most cases, this creates a new cavity inside the protein. Fifteen cavities were observed; 13 of them, ranging in size from 11 to 123 Å^3, were empty. Since the average size of a water molecule in bulk water is about 30 Å^3, at least some of these cavities were big enough to accommodate water. Of the mutations that produced cavities, all were destabilizing (they made the protein more susceptible to denaturation), and only two incorporated waters into the newly formed cavities. Why haven't more of the mutant proteins incorporated waters? The answer has to come from energetics: The substitution of a small side chain for a large one is destabilizing. This results from two factors that include transfer energies and loss of van der Waals contacts. To insert a water molecule into the hole would alleviate some of the loss of van der Waals energy but would increase other energetic costs by having a polar residue in a hydrophobic milieu. To the extent that the water might/can hydrogen bond to other residues in the vicinity, the energetic cost can be reduced and even reversed (Dunitz, 1994; Ernst *et al.,* 1995). In the two cavities where waters were observed, the waters were hydrogen bonded to polar groups on the protein.

There is a caveat: It has been pointed out that positionally disoriented water will not be seen by x-ray analysis. Interleukin-1β contains a large hydrophobic cavity, large enough to accommodate two to four water molecules. This cavity is empty in the crystal structures. However, NMR data (Ernst *et al.,* 1995) indicate that the hole is filled with water; this water is disordered and not hydrogen bonded to the protein.

In some enzymes, internal water molecules are present at the active site. Their possible contributions to catalysis will be considered in Section IV.

C. Water at Interfaces

1. Enolase

Enolase, a dimeric enzyme, catalyzes the interconversion of 2-phosphoglycerate and phosphoenolpyruvate. In yeast, the enzyme is composed of two identical subunits. From the point of view of one surface recognizing another, of one surface making good contacts with another, this presents a special problem. It means that there are constraints operating on the structure of homodimeric enzymes that are not operating on other proteins.

The x-ray crystallographic structure of yeast enolase is shown in Fig. 4A (Larsen *et al.,* 1996). We show the rugged surface of the protein that has to make contact with an identical surface on the other monomer. The problem, needless to say, is how

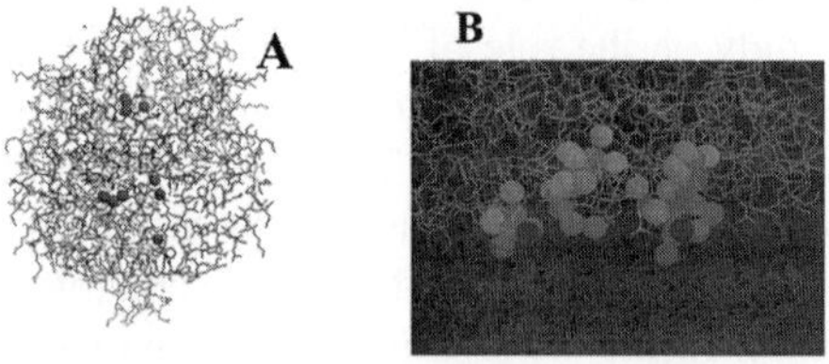

FIG. 4 (A) (**1ONE**). [PDB files are in (**Bold**).] The x-ray structure of yeast enolase looking down on the dimer interface is shown. The enzyme is composed of two identical subunits. About 60% of the top subunit (dark blue) has been removed so that the rough surface of the interface can be envisioned. The 10 reddish balls are waters that are at the interface. (B) (**1ONE**). The molecule has been rotated by 90° so that one is looking through the interface. The dark blue lines and balls constitute atoms of one subunit, while the light blue balls and lines correspond to the other. The reddish balls are water molecules that are situated within 3 Å of the dark blue and light blue balls; they are within hydrogen bonding distance. (See also color insert.)

to maximize the good contacts and avoid creating large, potentially destabilizing voids. Yeast enolases solve the problem by having a relatively small surface that makes good contacts on the two monomers and having an auxiliary surface that makes poorer contacts that are linked by water molecules. This is shown in Fig. 4B, where we emphasize the region where the two monomers come together. In the figure, one monomer is green; one monomer is blue; the water molecules at the interface are red spheres. The solid blue and green spheres are atoms of those residues of the protein that are within 3.0 Å of one of the interfacial waters, that is, the blue and green spheres represent protein atoms that can hydrogen bond to the waters.

What is significant about these waters is that in some instances they make good contacts with monomer A, in other instances with monomer B, and in some instances with other water molecules. They are far from the active site. These interfacial waters form a cushion—they take up space and improve van der Waals contacts and hydrogen bonding—that improves otherwise poor contacts.

The dissociation of enolase into subunits is a simple one-step, reversible reaction. Osmotic and hydrostatic pressure have opposite effects on the dissociation (Paladini and Weber, 1981; Kornblatt *et al.,* 1982, 1993a). Hydrostatic pressure favors dissociation, as would be predicted if the subunit surfaces which are buried in the dimer become hydrated upon dissociation. Osmotic pressure favors association. Volume changes measured by osmotic pressure, hydrostatic pressure, and dilatometry are consistent (unpublished data). The measured volume changes and the effects of both hydrostatic and osmotic pressure reflect the net balance, upon dissociation, between buried surface that becomes hydrated and interfacial waters that are liberated. Mammals have three enolase genes and both the homo- and heterodimers form. One would predict that the regions of the subunits that make close contact are highly conserved between the isozymes, while some variation would be tolerated on the surfaces that are bridged by water. Structures are not yet available for any mammalian enolase.

2. Antigen–Antibody Complexes

The binding of lysozyme to two different monoclonal antibodies was studied as a function of osmotic pressure and temperature (Goldbaum *et al.,* 1996). The binding of antibody D1.3 was weakened by increasing concentrations of glycerol; this suggests that binding requires the presence of water—i.e., water is present in the interface between the two macromolecules. The binding of antibody D44.1, on the other hand, was strengthened by glycerol, consistent with dehydration of the two surfaces occurring upon complex formation. For D1.3, $T\Delta S$ and ΔH both became smaller (less negative) as glycerol concentration increased, while with D44.1 they became more negative. Thus, although the effects of glycerol could be due to other factors, the thermodynamic results are consistent with water being trapped at the lysozyme/D1.3 interface but excluded from the lysozyme/D44.1 interface. The x-ray structures of these complexes suggest that more water molecules are in the interface with D1.3 than with D44.1 and that these water molecules form hydrogen bonds with the antibody and the antigen, thus contributing to the energetics of the interaction (Bhat *et al.,* 1994).

3. Cytochrome *c*/Protein Complexes

Cytochrome *c* serves as an electron donor or acceptor with a number of proteins. How similar are the interactions with the various partners? Three complexes have been studied by a combination of hydrostatic and osmotic pressure.

a. Cytochrome c/Cytochrome c Peroxidase Complex Hydrostatic pressure studies showed that this complex (yeast peroxidase/horse cytochrome *c*) was not very sensitive to pressure (Kornblatt *et al.,* 1986). Cytochrome *c* contains many positive charges necessary for binding to the oppositely charged peroxidase. As the complex forms, many of these charges must be dehydrated. If that were the whole story, hydrostatic pressure should promote the dissociation by promoting the electrostriction of these charges. It does not. The binding of substrate to enzyme must therefore be more complex. The easiest explanation is that water gets trapped in the interface. The crystal structure (Pelletier and Kraut, 1992) showed that the interface between the yeast peroxidase and cytochrome *c* is filled with water, as is shown in Fig. 5. Horse cytochrome *c* is not the normal partner for yeast peroxidase.

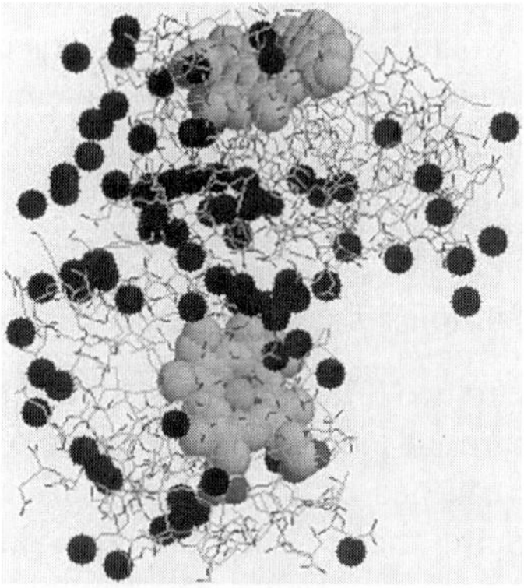

FIG. 5 (**2PCC**). [PDB files are in (**Bold**).] The complex of horse cytochrome *c* and yeast cytochrome *c* peroxidase as determined by x-ray crystallography. The peroxidase heme is shown at the top and the cytochrome *c* heme at the bottom. Only those portions of the complex that are within 29 Å of the heme of cytochrome *c* are shown. The black balls represent water molecules. Pay particular attention to the leaf-like cluster situated between the two hemes; this is part of the interface between the peroxidase and the cytochrome *c*. (See also color insert.)

However, the crystal structure of the yeast cytochrome *c*/yeast peroxidase complex was also determined (see PDB file 2pcc); this interface is also filled with water.

b. Cytochrome c/Cytochrome b5 Complex The complex has been subjected to hydrostatic and osmotic pressures. The two give complementary information and are in agreement. Hydrostatic pressure causes the complex to dissociate. Since both members are highly charged, this would indicate that the charges become hydrated on dissociation and that they are naked in the complex. The amount of water trapped in the interstices of the complex is not enough to offset electrostriction. Osmotic pressure experiments also indicate that water is excluded in the main. The excess of water in the free components vs the bound is 3 ± 1 (Kornblatt *et al.,* 1993b). There is no crystal structure for this complex.

c. Cytochrome c/Cytochrome c Oxidase The cytochrome *c*/cytochrome *c* oxidase complex has not been crystallized. There is, nonetheless, a wealth of information about the interactions between the two proteins. The significant aspects of the binding of the two proteins are:

1. There is a ring of positive charge on cytochrome *c* that makes contact with a negatively charged region on subunit II of the oxidase. The dissociation coefficient is extremely sensitive to ionic strength as expected. It indicates that the product of the charges on the two proteins is about −25 (Kornblatt *et al.,* 1992).
2. The positive surface charges of cytochrome *c* have been chemically modified. The modified protein will still bind to the oxidase but its affinity is much reduced.
3. The interaction is not sensitive to hydrostatic pressure (Kornblatt *et al.,* 1984, 1992).
4. It is sensitive to osmotic pressure (Kornblatt *et al.,* 1993b).

When these four aspects of the binding phenomenon are put together, they indicate that the interfacial region is highly charged and that the charges have, in the main, been stripped of water. In the regions between the charges, there is hydrophobic surface that makes poor contact between the two proteins. This surface sequesters water, perhaps a total of 12 ± 1 (Kornblatt *et al.,* 1993b).

Two points emerge from studying the above complexes. (1) Water has a role in specificity; cytochrome *c* can recognize and bind to three different partners by varying the number and position of water molecules at the interface. (2) Since osmotic pressure measures the sensitivity of a system to changes in water activity, and hydrostatic pressure measures the sensitivity to volume changes, the results of the two techniques are not always comparable.

IV. Catalysis and Conformation

In this unit, we will try to show that the static picture discussed in Section III is insufficient if we are to understand how enzymes, transporters, and channels work. A more dynamic approach is necessary. Changing water activity provides the enzymologist with a powerful and gentle tool for perturbing a protein and studying the consequences.

A. Water Involved in Specificity and Binding

Protein–ligand interactions do not always involve a perfect fit. In addition, specificity is not absolute; proteins may accept multiple ligands, they may accommodate mutations. Water molecules are the universal substitute—they are "both the cement that fills crevices . . . and the lubricant that allows motion" (Levitt and Park, 1993). How does a major histocompatibility (MHC) molecule bind any of a thousand different peptides? Each MHC molecule has specificity for one or two positions of the nine-residue peptides that are bound; this specificity is due to one or two pockets with the required size, shape, and charge. The rest of the binding site is lined with water molecules. Any of a number of polar side chains can be accommodated. A large side chain may interact directly with side chains of the MHC molecule, while interactions of smaller side chains will be mediated by one or more water molecules (Smith *et al.,* 1996). The number and positions of the water molecules will vary from one peptide ligand to another.

The bacterial L-arabinose binding protein binds L-arabinose (a five-carbon sugar) and D-galactose (a six-carbon sugar) with almost equal affinities. Both sugars are bound in the 6-member pyranose form (Fig. 6). In this form, arabinose has two hydrogens as substituents on carbon-5, while galactose has a -CH_2OH group. The reader will have surmised that changes in number or position of water molecules will be observed in the structures of the protein with the two different sugars (Quiocho *et al.,* 1989). There are two water molecules—310 and 311—near C-5 of arabinose. Wat310 forms four hydrogen bonds—to the sugar, to the protein, and to wat311, which is in turn hydrogen-bonded to the protein. The -CH_2OH on C-5 of galactose overlaps the space that was occupied by wat311. This water is displaced, resulting in the loss of three hydrogen bonds. Wat310 shifts position slightly. Some hydrogen bonds are maintained; some are lost, and both wat310 and the C-6 OH of galactose form new bonds. The net result is that only one hydrogen bond is lost, and all of the hydrogen-bonding groups in this region of the protein remain paired.

The immunochemistry literature contains many other examples of binding reactions where a change of a single amino acid in either antigen or antibody is accommodated by a gain or loss of water molecules.

Restriction enzymes cleave DNA at specific base sequences; that is what makes them such powerful tools for the molecular biologists. Many of them have what

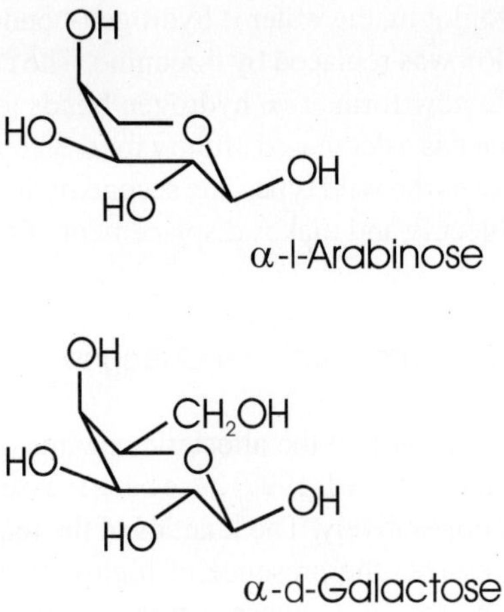

FIG. 6 α-L-Arabinose and α-D-galactose. When bound to the arabinose binding protein, both sugars are in the 6-membered, pyranose ring form shown here. In the complex with the protein, there are two water molecules near C-5 of arabinose. When galactose is bound, the -CH_2OH group on C-5 displaces one of the water molecules and forces the other to shift slightly.

is called "star" activity: cleavage activity at alternative sequences. This activity usually shows up when solution conditions are altered—changes in pH, ionic strength, glycerol concentration, etc. Since at least some of these conditions would decrease water activity, Robinson and Sligar (1994, 1998) studied the specificity of several restriction enzymes as a function of hydrostatic and osmotic pressure. Their results showed that the binding of the restriction enzyme EcoRI to DNA involves the release of large amounts of water. Some (about 75) are released upon binding to nonspecific sites. An additional 70 molecules are released upon binding at the canonical (specific) site and still more upon binding to alternative (star activity) sites. The differences in the numbers of waters are probably too large to be accounted for by simple desolvation of the DNA and protein. Spolar and Record (1994), on the basis of thermodynamic and structural studies, have suggested that site-specific binding of proteins to DNA involves local folding, which may be extensive. Folding, dimerization, conformational changes, and changes in the curvature of the DNA—all could involve major changes in hydration. A further complication is that changes in hydration affect the specificity of cleavage of DNA, as well as specificity of binding.

Some active sites contain water that is displaced by the substrate. The substrate binding pocket of P450cam contains six water molecules; none of these are present when the substrate is bound. The heme pocket of myoglobin contains a water molecule which is displaced when oxygen binds (Cameron *et al.*, 1993). How do these waters get out of the active site as the substrate is entering? In neither example is there an obvious escape path for the water. In the case of myoglobin, it has been suggested that displacement of the water molecule could be a slow step in the overall reaction. In myoglobin, the water is hydrogen-bonded to his64. A mutation was made in which val68 was replaced by threonine. The crystal structure showed that the water molecule now forms two hydrogen bonds to the protein—to his64 and to thr68. The mutant has a decreased affinity for oxygen and a decreased rate of oxygen binding, relative to the wild type. The second hydrogen bond in the mutant stabilizes the water molecule and makes displacement of the water more difficult.

B. Water Involved in Conformational Changes

The effects of osmotic pressure on the allosteric enzyme aspartate transcarbamylase (ATCase) (LiCata and Allewell, 1997) have been measured. This enzyme binds its substrate aspartate cooperatively. The kinetics of the reaction were studied as a function of osmotic pressure; the presence of high molecular weight osmolytes increased the K_m for aspartate. Somewhere between 180 and 250 water molecules are taken up by the protein as it binds substrate. Ever since the classic ultracentrifuge experiments by Schachman's group (Gerhart and Schachman, 1968), it has been known that the substrate free (T) and fully saturated (R) states of this protein differ in shape. There is a good correlation between the increase in bound water and the increase in surface area for the T to R transition.

The binding of glucose to hexokinase results in a conformational change that closes a cleft between two domains of the protein. Oxygen binding to hemoglobin produces a conformational change in that protein . The binding of ligands (glucose for hexokinase and oxygen for hemoglobin) to both proteins have been studied as a function of osmotic pressure. In both cases, the observed effects were consistent with those predicted from the crystal structures. The cleft closing in hexokinase is associated with a release of water from the cleft; osmotic pressure increases the affinity of the enzyme for its substrate (Rand *et al.,* 1993). The T to R transition in hemoglobin results in an increase in exposed surface area. Accordingly, the oxygen affinity of hemoglobin is decreased by osmotic pressure (Colombo *et al.,* 1992). A striking feature of all three of these studies is that the effects of osmotic pressure are observed in the range of osmotic pressures that bracket physiological conditions. Allosteric enzymes are finely poised to respond to changes in substrate concentrations and therefore may be quite sensitive to other changes.

C. Water and Catalysis

Water may play a specific role in an enzymatic mechanism. The most obvious case is that of a hydrolytic enzyme. The water molecule that is the substrate, although it ultimately comes from the bulk solvent, may be bound at the active site. β-Lactamase (Zawadzke *et al.,* 1996) cleaves its substrates by a two-step, acyl enzyme mechanism. Based on the x-ray structure, a water molecule that is interacting with asn170 is proposed to be the water that hydrolyzes the acyl enzyme. Asn170 was changed to gln170; the authors reasoned that the larger side chain of glutamine would displace the water molecule. The crystal structure confirmed this prediction. What were the kinetic consequences? Depending on the substrate, k_{cat} was decreased by a factor of 10^3 to 10^4. In the wild-type enzyme, formation of the acyl enzyme is slow and is followed by a rapid deacylation; this is the step that requires a water molecule. In the mutant, deacylation is slow, consistent with the loss of the catalytic water molecule. Why then does any cleavage of substrate occur? It is proposed that there is some flexibility of the protein (alternate conformations for gln170 or of the acyl group of the substrate) that gives water access to the active site.

Many reactions involve the net release or uptake of protons. The protons are proposed to travel from bulk solvent to the active site via a network of proton acceptors/donors. The network may include one or more water molecules. For several enzymes of this type—triose phosphate isomerase (Zhang *et al.,* 1999), carbonic anhydrase (Tu *et al.,* 1989), Mn superoxide dismutase (Ramilo *et al.,* 1999)—efforts have been made to interrupt this hydrogen-bonded network via site-directed mutagenesis. The kinetic properties and crystal structures of the mutants have been determined; in all cases, the hydrogen-bonded network was disrupted by the mutation. The mutations decreased the rates of the enzymatic reactions, but only by about an order of magnitude (10- to 50-fold); this may be because proton transfer is not a slow step in these reactions for the wild-type enzyme and/or because alternative pathways for proton transfer exist or are readily established when needed. A further complication in trying to interpret experiments such as these is that water at an active site may modulate activity in a number of ways. This has already been discussed in the Introduction, in reference to P450cam, but it is worth repeating. Water may assist in holding residues in the correct position or orientation. Water may modify the hydrophobicity or polarity or dielectric constant of the active site, thus "titrating" the pKs of residues or substrates, or modifying the tightness of binding via charge–charge interactions. Water at any given active site is probably assisting catalysis in multiple ways.

D. Water and Pumps

1. Photoreaction Centers: The Cytochrome b_6f of Higher Plants and the *Rhodobacter sphaeroides* Reaction Centers

Structural and functional studies on these two enzymes (Martinez *et al.,* 1996; Ponamarev and Cramer, 1998; Ermler *et al.,* 1994; Baciou and Michel, 1995) have, in many ways, formed the basis for our thinking on how water is involved in energy transduction by proton pumping enzymes. The structural work on both systems has shown that there is a chain of water molecules that either traverses the entire membrane spanning region of the protein or comes close to spanning it. In the case of cytochrome b_6f, this chain is five molecules long, whereas it contains about 12 waters in the *Rhodobacter* enzyme. The chains contain waters arranged in single file. It has been proposed that this arrangement would permit two very important things to occur: (1) the very rapid transfer of a proton from one space to another via the so-called Grotthuss mechanism (Pomès and Roux, 1998) and (2) the control of the proton translocation and electron transfer, since there would be an activation energy for the process of about 8 kcal/mol. Mutagenesis studies on the *Rhodobacter* enzyme have clearly shown that the integrity of the water chain is essential for activity of the enzymes. For these two proteins as well as several others, the Grotthuss mechanism is clearly attractive. The waters are aligned with their dipoles sharing the same orientation. When the energy barrier is overcome, the dipoles reorient so that a proton that was exposed on an internal face is now exposed on the external. This requires overcoming the high activation energy and finally some mechanism for getting the waters back to their original orientation. It is essential to note that the waters involved in the Grotthuss mechanism correspond more—in our thinking—to essential amino acid residues than they do to waters in the bulk. The waters in the chain move on the short time scale, but the motions are very restricted and correspond to rotations rather than transpositions.

2. Bacteriorhodopsin

Like the photoreaction centers above, this protein has helped to crystallize our thinking about how proton pumps work. The protein sits in the (purple) membrane of *Halobacterium salinarium* (formerly *halobium*) where it absorbs visible light via its all-trans-retinal chromophore. If the chromophore absorbs a photon, it is converted to 13-cis retinal; the protein then undergoes many small but regulated conformational changes, which result in the reorientation of retinal, amino acid residues, and water. The x-ray structure of the unexcited molecule is known to about 1.5 Å (Luecke *et al.,* 1999a) and the structures of several intermediates are also known to high resolution (Luecke *et al.,* 1999b; Edman *et al.,* 1999).

The ground state is known as bR570. It contains a chain consisting of about seven waters and five key amino acids. Between the waters and the five residues, a complete jagged chain is formed. The intermediate states are J, K, L, M, N, and O. The K state is formed within about 4 ps of excitation, and represents the conversion of the retinal to 13-cis. During this 4 ps, wat402, the top water, is moved out of its ground state position; this allows residue asp85 to move into position so that it can accept the proton from the destabilized Schiff base. This is the start of proton transfer toward the exterior. In the ground state, wat402 acts as a blocker so that there is no transfer to asp85. The K to L conversion requires about 1 μs. There is proton transfer along the amino acid–water "wire"; the conversion of L to M is accompanied by the delivery of the proton to the surface. This results from the reorientation of several amino acids and waters, the pKs of which change substantially with the altered orientation. It is probable but not certain that it is the waters, stabilized by glu204 and arg82, that act as the transfer agents by a Grotthuss mechanism (Rammelsberg *et al.*, 1998).

3. Cytochrome *c* Oxidase

This protein is central to the bioenergetics of nonphotosynthetic eukaryotes. The mitochondrial oxidase accepts electrons one at a time from cytochrome *c* and stocks them in four metal centers. It then transfers four electrons to molecular oxygen and, during the same cycle, transfers four protons across the mitochondrial membrane. The oxidase consumes about 95% of the oxygen that we use and supplies us with about 30% of our total useful energy. The enzyme both *in vivo* and *in vitro* exhibits what is known as respiratory control. It turns over more rapidly when the need for energy is great and more slowly when there is little need. For the oxidase, "need" is sensed through the proton gradient across the membrane through which it is inserted. In the absence of a membrane, the oxidase senses pH but not a gradient. The current feeling is that the oxidase has an obligate requirement to transfer protons through the protein during the process of electron transfer. How is water involved in this process?

Early studies showed that the fully oxidized protein was insensitive to hydrostatic pressure, as was the fully reduced protein. During electron transfer, however, the protein became exquisitely sensitive to pressure, so that by about 200 MPa (2000 atm), the protein would no longer transfer electrons from cytochrome *c* to oxygen (Kornblatt *et al.*, 1988). It was blocked at an internal electron transfer step. We reasoned that the volume change was probably far too large to have resulted from a conformational change in the protein; that it must instead be due to solvent fluxes. The application of osmotic stress showed that enzymatic activity was also inhibited by high concentrations of osmolytes (Kornblatt and Hui Bon Hoa, 1990). The conclusion that followed was relatively straightforward. Hydrostatic pressure exerts its action by, among other things, promoting hydration of ionic charges, polar surfaces, and hydrophobic surfaces. To promote hydration of a surface inhibits the dehydration step. If water movement is involved in the catalytic cycle of the protein, hydrostatic pressure must be promoting the hydration of the oxidase and blocking the dehydration step. In just the opposite fashion, the application of osmotic pressure should be promoting the dehydration of the protein and inhibiting its hydration. We proposed that water was acting as a carrier for protons and that it and the protons had to gain access to a privileged compartment. This compartment was gated on both sides of the protein, thereby providing a mechanism by which the proton gradient across the mitochondrial membrane could be maintained. In the intervening time, several crystal structures of the oxidase have been published (Tsukihara *et al.,* 1996; Iwata *et al.,* 1995). Of the several present in the Protein Data Bank only one (1ar1) shows the waters that were present in the original publications. The other publications have, for one reason or another, left them out of the deposited structures.

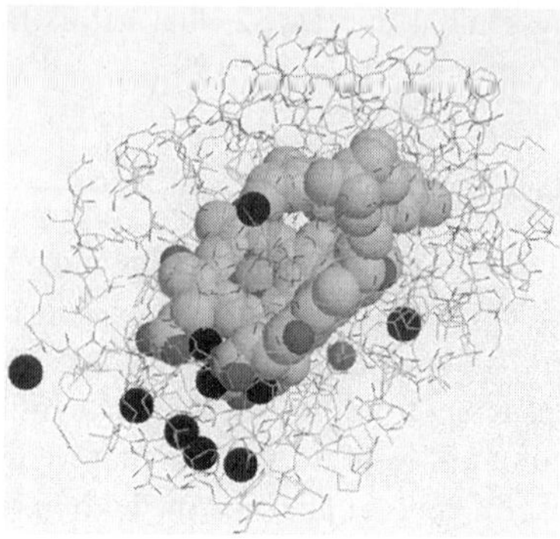

FIG. 7 **(1AR1)**. [PDB files are in (**Bold**).] The cytochrome *c* oxidase showing the two cytochromes: the *a* cytochrome is almost a full face view, while the a_3 cytochrome is shown end on. The bluish ball to the right of the a_3 cytochrome is CuB, part of the binuclear center. The black balls represent water molecules that are close to the proposed proton pumping channel. (See also color insert.)

Figure 7 shows the crystal structure of the oxidase in the region of the two hemes of the protein: the black solids are internal waters; the blue ball is copper. What is clear is that the interior of the protein contains several water molecules and that some of these are close to the presumed proton pump path. Our data, based on the dynamics of electron transfer, indicate that between 4 and 10 waters move across the membrane during a catalytic cycle (Kornblatt and Hui Bon Hoa, 1990; Kornblatt and Kornblatt, 1992; Kornblatt, 1998) . The interior of the protein, near the pumping path, shows 5 or 6 (Iwata *et al.,* 1995; Ostermeir *et al.,* 1997).

V. Conclusions

One thing that becomes clear from these studies is that almost anything a protein does involves bound water. The very important corollary is that anything we do that changes water activity will probably affect the process we are studying. It has been estimated that the normal osmotic pressure for *E. coli* is 7.6 atm (LiCata and Allewell, 1997); 8% sucrose has an osmotic pressure of 6 atm, while that of 12% glycerol is 36 atm. Thus, the conditions under which we study enzymes—dilute solutions of protein in buffers or concentrated solutions in the high salts often used for crystallization—are far from physiological and may produce significant alterations in the behavior of the protein. Many systems are probably poised so that slight changes in conditions will perturb the equilibrium significantly. Kd for the dissociation of yeast enolase is about 10^{-9} M. It will always be dimeric unless we subject it to extreme conditions. However, if the Kd were 10^{-7} M, the enzyme might be monomeric under the conditions of an enzyme assay. Changing assay conditions might change water activity (as with the restriction enzymes) and shift the equilibrium. There are a number of systems—DNA binding proteins, receptors, allosteric enzymes—in which the response of the protein involves dimerization or a conformational change, both of which are sensitive to water activity. Probably the most important question is whether or not changes in water activity occur within a cell. If so, do they alter the behavior of the proteins within that cell and thereby alter the behavior of the cell? That question falls in the domain of the rest of this volume.

Appendix

Abbreviations, PDB (Protein Data Bank) Files, and Conventions
10 atmospheres = 1 MPa
ATCase, aspartate transcarbamylase
BPTI, bovine trypsin inhibitor
1PHC = cytochrome P450CAM without camphor
1DZ6 = cytochrome P450CAM with camphor
1HRC = horse heart cytochrome *c*, the x-ray structure
1OCD = cytochrome *c*, the NMR structure
1ONE = yeast enolase;
2PCB = yeast cytochrome *c* peroxidase complexed with horse cytochrome *c*
2PCC = yeast cytochrome *c* peroxidase complexed with yeast iso-1-cytochrome *c*
1AR1 = *Paracoccus denitrificans* cytochrome *c* oxidase.
Mutant forms of proteins: The general form used is as follows. An alanine at position 10 that is changed to a glycine is designated as A10G. A phenylalanine at position 20 that is changed to a histidine is F20H.

Acknowledgments

We thank the Natural Science and Engineering Research Council (Canada) for financial support. Drs. Claude Balny and Franck Travers generously provided us with an agreeable space during the writing of this review. We dedicate this work to the memory of Dr. Pierre Douzou, a friend who knowingly or unknowingly contributed to some of the ideas that we have expressed.

References

Baciou, L., and Michel, H. (1995). Interruption of the water chain in the reaction center from *Rhodobacter sphaeroides* reduces the rates of the proton uptake and of the second electron transfer to Q_B. *Biochemistry* **34,** 7967–7972.

Berndt, K. D., Beunink, J., Schroder, W., and Wuthrich, K. (1993). Designed replacement of an internal water molecule in BPTI: Structural and functional implications of a glycine to serine mutation. *Biochemistry* **32,** 4564–4570.

Bhat, T. N., Bentley, G. A., Boulot, G., Greene, M. I., Tello, D., Dall'Acqua, W., Souchon, H., Schwartz, F. P., Mariuzza, R. A., and Poljak, R. J. (1994). Bound water molecules and conformational stabilization help mediate an antigen–antibody association. *Proc. Natl. Acad. Sci. USA* **91,** 1089–1093.

Billeter, M. (1995). Hydration water molecules seen by NMR and by x-ray crystallography. *Prog. Nucl. Magn. Reson. Spectrosc.* **27,** 635–645.

Bushnell, G. W., Louie, G. V., and Brayer, G. D. (1990). High-resolution three dimensional structure of horse heart cytochrome *c. J. Mol. Biol.* **214,** 585.

Cameron, A. D., Smerdon, S. J., Wilkinson, A. J., Habash, J., Helliwell, J. R., Li, T., and Olson, J. S. (1993). Distal pocket polarity in ligand binding to myoglobin: deoxy and carbonmonoxy forms of a threonine ^{68}(E11) mutant investigated by x-ray crystallography and infrared spectroscopy. *Biochemistry* **32,** 13,061–13,070.

Colombo, M. F., Rau, D. C., and Parsegian, V. A. (1992). Protein solvation in allosteric regulation: A water effect on hemoglobin. *Science* **256,** 655–659.

Denisov, V. P., Halle, B., Peters, J., and Horlein, H. D. (1995). Residence times of the buried water molecules of BPTI and its G36S mutant. *Biochemistry* **34,** 9046–9051.

Dunitz, J. D. (1994). The entropic cost of bound water in crystals and biomolecules. *Science* **264,** 670.

Edman, K., Nollert, P., Royant, A., Beihali, H., Pebay-Peyroula, E., Hadju, J., Neutz, R., and Landau, E. M. (1999). High resolution structure of an early intermediate in the bacteriorhodopsin photocycle. *Nature* **401,** 822–825.

Ermler, U., Fritzsch, G., Buchanan, S. K., and Michel, H. (1994). Structure of the photosynthetic reaction centre from *Rhodobacter sphaeroides* at 2.65 Å resolution: Cofactors and protein–cofactor interactions. *Structure* **2,** 925–936.

Ernst, J. A., Clubb, R. T., Zhou, H.-X., Gronenborn, A. M., and Clore, G. M. (1995). Demonstration of a positionally disordered water within a protein hydrophobic cavity by NMR. *Science* **267,** 1813–1817.

Garcia, A. E., and Hummer, G. (2000). Water penetration and escape in proteins. *Proteins Struct. Funct. Genet.* **38,** 261–272.

Gerhart, J. C., and Schachman, H. K. (1968). Allosteric interactions in aspartate transcarbamylase. II. Evidence for different conformational states of the protein in the presence and absence of specific ligands. *Biochemistry* **7,** 538–552.

Goldbaum, F., Schwarz, F. P., Eisenstein, E., Cauerhff, A., Marriuzza, R. A., and Poljak, R. J. (1996). The effect of water activity on the association constant and the enthalpy of reaction between lysozyme and the specific antibodies D1.3 and D44.1. *J. Mol. Recognition* **9,** 6–12.

Gross, M., and Jaenicke, R. (1994). Proteins under pressure. The influence of high hydrostatic pressure on structure, function and assembly of proteins and protein complexes. *Eur. J. Biochem.* **221,** 617–630.

Huber, R., Kukla, D., Bode, W., Schwager, P., Bartels, K., Deisenhofer, J., and Steigemann, W. (1974). Structure of the complex formed by bovine trypsin and bovine pancreatic trypsin inhibitor. II. Crystallographic refinement at 1.9 Å resolution. *J. Mol. Biol.* **89,** 73–101.

Iwata, S., Ostermeier, C., Ludwig, B., and Michel, H. (1995). Structure at 2.8 Å resolution of cytochrome *c* oxidase from *Paracoccus denitrificans*. *Nature* **376,** 660–669.

Kornblatt, J. A. (1998). The water channel of cytochrome *c* oxidase: Inferences from inhibitor studies. *Biophys. J.* **75,** 3127–3134.

Kornblatt, J. A., and Hui Bon Hoa, G. (1990). A non-traditional role for water in the cytochrome *c* oxidase reaction. *Biochemistry* **29,** 9370–9376.

Kornblatt, J. A., and Kornblatt, M. J. (1992). Cytochrome *c* oxidase: The presumptive channel holds at least four water molecules. *Biochim. Biophys. Acta* **1099,** 182–184.

Kornblatt, J., Kornblatt, J., and Hui Bon Hoa, G. (1982). The pressure-induced, reversible inactivation of mouse brain enolases. *Eur. J. Biochem.* **128,** 577–581.

Kornblatt, J. A., Hui Bon Hoa, G., and English, A. M. (1984). Volume changes associated with the cytochrome *c* oxidase/porphyrin *c* equilibrium. *Biochemistry* **23,** 5906–5911.

Kornblatt, J. A., English, A. M., and Hui Bon Hoa, G. (1986). The effects of pressure on yeast cytochrome *c* peroxidase. *Eur. J. Biochem.* **159,** 39–43.

Kornblatt, J. A., Hui Bon Hoa, G., and Heremans, K. (1988). Effects of pressure on cytochrome oxidase: The aerobic steady state. *Biochemistry* **27,** 5122–5128.

Kornblatt, J. A., Theordorkis, J., Hui Bon Hoa, G., and Margoliash, E. (1992). Cytochrome *c* and cytochrome *c* oxidase interactions: The effects of ionic strength and hydrostatic pressure studied with site-specific modifications of cytochrome *c*. *Biochem. Cell. Biol.* **70,** 539–547.

Kornblatt, M. J., Kornblatt, J. A., and Hui Bon Hoa, G. (1993a). The role of water in the dissociation of enolase, a dimeric enzyme. *Arch. Biochem. Biophys.* **306,** 495–500.

Kornblatt, J. A., Kornblatt, M. J., Hui Bon Hoa, G., and Mauk, A. G. (1993b). Responses of two protein–protein complexes to solvent stress: Does water play a role at the interface?. *Biophys. J.* **65,** 1059–1065.

Kornblatt, J. A., Kornblatt, M. J., Rajotte, I., Hui Bon Hoa, G., and Kahn, P. C. (1998). Thermodynamic volume cycles for electron transfer in the cytochrome *c* oxidase and for the binding of cytochrome *c* to cytochrome *c* oxidase. *Biophys. J.* **75,** 435–444.

Larsen, T. M., Wedekind, J. E., Rayment, I., and Reed, G. H. (1996). A carboxylate oxygen of the substrate bridges the magnesium ions at the active site of enolase: Structure of the yeast enzyme complexed with the equilibrium mixture of 2–phosphoglycerate and phosphoenolpyruvate at 1.8 Å resolution. *Biochemistry* **35,** 4349.

Lett, C. M., Berghuis, A. M., Frey, H. E., Lepock, J. R., and Guillemette, J. G. (1996). The role of a conserved water molecule in the redox-dependent thermal stability of iso-1-cytochrome *c*. *J. Biol. Chem.* **271,** 29,088–29,093.

Levitt, M., and Park, B. H. (1993). Water: Now you see it, now you don't. *Structure* **1,** 223–226.

LiCata, V. J., and Allewell, N. M. (1997). Functionally linked hydration changes in *Escherichia coli* aspartate transcarbamylase and its catalytic subunit. *Biochemistry* **36,** 10,161–10,167.

Luecke, H., Schobert, B., Richter, H.-T., Cartailler, J.-P., and Lanyi, J. K. (1999a). Structure of bacteriorhodopsin at 1.55 Å resolution. *J. Mol. Biol.* **291,** 899–911.

Luecke, H., Schobert, B., Richter, H.-T., Cartailler, J.-P., and Lanyi, J. K. (1999b). Structural changes in bacteriorhodopsin during ion transport at 2 Å resolution. *Science* **286,** 255–260.

Marachal, P. A., and Gervais, P. (1994). Yeast viability related to water potential variation. *Appl. Microbiol. Biotechnol.* **42,** 617–622.

Martinez, S. E., Huang, D., Ponomarev, M., Cramer, W. A., and Smith, J. L. (1996). The heme linked redox center of chloroplast cytochrome *f* is linked to a buried five water chain. *Protein Sci.* **5,** 1081–1092.

Moshaev, V. V., Heremans, K., Frank, J., Masson, P., and Balny, C. (1996). High pressure effects on protein structure and function. *Proteins Struct. Funct. Genet.* **24,** 81–91.

Ostermeir, C., Harrenga, A., Ermler, U., and Michel, H. (1997). Structure at 2.7 Å resolution of the *Paracoccus denitrificans* two subunit cytochrome *c* oxidase complexed with an antibody Fv fragment. *Proc. Natl. Acad. Sci. USA* **94,** 10,547–10,553.

Otting, G., and Wuthrich, K. (1989). Studies of protein hydration in aqueous solution by direct NMR observation of individual protein-bound water molecules. *J. Am. Chem. Soc.* **111,** 1871–1875.

Otting, G., Liepinsh, E., and Wuthrich, K. (1991). Protein hydration in aqueous solution. *Science* **254,** 974–980.

Paladini, A. A., and Weber, G. (1981). Pressure-induced reversible dissociation of enolase. *Biochemistry* **20,** 2587–2593.

Parsegian, V. A., Rand, R. P., and Rau, D. C. (1995). Macromolecules and water: Probing with osmotic stress. *Methods Enzymol.* **259,** 43–94.

Parsegian, V. A., Rand, R. P., and Rau, D. C. (2000). Osmotic stress, crowding, preferential hydration and binding: A comparison of perspectives. *Proc. Natl. Acad. Sci. USA* **97,** 3987–3992.

Pelletier, H., and Kraut, J. (1992). Crystal structure of a complex between electron transfer partners cytochrome *c* peroxidase and cytochrome *c*. *Science* **258,** 1748–1755.

Pocker, Y., and Janjic, N. (1989). Molecularity of water in enzymic catalysis. Application to carbonic anhydrase II. *J. Am. Chem. Soc.* **111,** 731–733.

Pomès, R., and Roux, B. (1998). Free energy profiles for H^+ conduction along hydrogen-bonded chains of water molecules. *Biophys. J.* **75,** 33–40.

Ponamarev, M. V., and Cramer, W. A. (1998). Perturbation of the internal water chain of cytochrome *f* in oxygenic photosynthesis: Loss of concerted reduction of cytochromes *f* and *b6*. *Biochemistry* **37,** 17,199–17,208.

Poulos, T. L., Finzel, B. C., and Howard, A. J. (1986). Crystal structure of substrate free *Pseudomonas putida* cytochrome P-450. *Biochemistry* **25,** 5314–5322.

Qi, P. X., Beckman, R. A., and Wand, A. J. (1996). Solution structure of horse heart ferricytochrome *c* and detection of redox-related structural changes by high resolution ^{1H}NMR. *Biochemistry* **35,** 12,275–12,286.

Quiocho, F. A., Wilson, D. K., and Yvas, N. K. (1989). Substrate specificity and affinity of a protein modulated by bound water molecules. *Nature* **340,** 404–407.

Ramilo, C. A., Leveque, V., Guan, Y., Lepock, J. R., Tainer, J. A., Nick, H. S., and Silverman, D. N. (1999). Interrupting the hydrogen bond network at the active site of human manganese superoxide dismutase. *J. Biol. Chem.* **274,** 27,711–27,716.

Rammelsberg, R., Huhn, G., Lubben, M., and Gerwert, K. (1998). Bacteriorhodopsin's intramolecular proton-release pathway consists of a hydrogen bonded network. *Biochemistry* **37,** 5001–5009.

Rand, R. P., Fuller, N. L., Butko, P., Francis, G., and Nicholls, P. (1993). Measured change in protein solvation with substrate binding and turnover. *Biochemistry* **32,** 5925–5929.

Record, T. M., Courtenay, E. S., Cayley, D. S., and Guttman, H. J. (1999a). Response of *E. coli* to osmotic stress. *Trends Biochem. Sci.* **23,** 143–148.

Record, T. M., Courtenay, E. S., Cayley, D. S., and Guttman, H. J. (1999b). Biophysical compensation mechanisms buffering *E. coli* protein–nucleic acid interactions against changing environments. *Trends Biochem. Sci.* **23,** 190–194.

Robinson, C. R., and Sligar, S. G. (1994). Hydrostatic pressure reverses osmotic pressure effects on the specificity of *Eco*RI-DNA interactions. *Biochemistry* **33,** 3787–3793.

Robinson, C. R., and Sligar, S. G. (1995). Hydrostatic and osmotic pressure as tools to study macromolecular recognition. *Methods Enzymol.* **259,** 395–427.

Robinson, C. R., and Sligar, S. G. (1998). Changes in solvation during DNA binding and cleavage are critical to altered specificity of the *Eco*RI endonuclease. *Proc. Natl. Acad. Sci. USA* **95,** 2186–2191.

Schlichting, I., Berendzen, J., Chu, K., Stock, A. M., Maves, S. A., Benson, D. E., Sweet, R. M., Ringe, D., Petsko, G. A., and Sligar, S. G. (2000). Catalytic pathway of cytochrome P450cam at atomic resolution. *Science* **287,** 1615–1622.

Smith, K. J., Reid, S. W., Harlos, K., McMichael, A. J., Stuart, D. I., Bell, J. I., and Jones, E. Y. (1996). Bound water structure and polymorphic amino acids act together to allow the binding of different peptides to MHC class I HLA-B53. *Immunity* **4,** 215–228.

Spolar, R. S., and Record, M. T. (1994). Coupling of local folding to site-specific binding of proteins to DNA. *Science* **263,** 777–784.

Svergun, D. I., Richard, S., Koch, M. H. J., Sayers, Z., Kuprin, S., and Zaccai, G. (1998). Protein hydration in solution: Experimental observation by x-ray and neutron scattering. *Proc. Natl. Acad. Sci. USA* **95,** 2267–2272.

Timasheff, S. N. (1998). In disperse solution, "osmotic stress" is a restricted case of preferential interactions. *Proc. Natl. Acad. Sci. USA* **95,** 7363–7367.

Tsukihara, T., Aoyama, H., Yamashita, E., Tomazaki, T., Yamaguchi, H., Shinzawa-Itoh, K., Nakashima, R., Yano, R., and Yoshikawa, S. (1996). The whole structure of the 13-subunit oxidized cytochrome *c* oxidase at 2.8 Å. *Science* **272,** 1136–1144.

Tu, C., Silverman, D. N., Forsman, C., Jonnson, B.-H., and Lindskog, S. (1989). Role of histidine 64 in the catalytic mechanism of human carbonic anhydrase II studied with a site-specific mutant. *Biochemistry* **28,** 7913–7918.

Xu, J., Baase, W. A., Baldwin, E., and Matthews, B. W. (1998). The response of T4 lysozyme to large-to-small substitutions within the core and its relation to the hydrophobic effect. *Protein Sci.* **7,** 158–177.

Wood, J. M. (1999). Osmosensing by bacteria. *Microbiol. Mol. Biol. Rev.* **63,** 230–262.

Zawadzke, L. E., Chen, C. C. H., Banerjee, S., Li, Z., Wäsch, S., Kapadia, G., Moult, J., and Herzberg, O. (1996). Elimination of the hydrolytic water molecule in a class A β-lactamase mutant: Crystal structure and kinetics. *Biochemistry* **35,** 16,475–16,482.

Zhang, W., Capp, M. W., Bond, J. P., Anderson, C. F., and Record, M. T. (1996). Thermodynamic characterization of native BSA with highly excluded (glycine betaine) and moderately accumulated (urea) solutes by a novel application of vapor pressure osmometry. *Biochemistry* **35,** 10,506–10,516.

Zhang, Z., Komives, E. A., Sugio, S., Blacklow, S. C., Narayana, N., Xuong, N. H., Stock, A. M., Petsko, G. A., and Ringe, D. (1999). The role of water in the catalytic efficiency of triosephosphate isomerase. *Biochemistry* **38,** 4389–4397.

Aquaglyceroporins: Channel Proteins with a Conserved Core, Multiple Functions, and Variable Surfaces

Andreas Engel and Henning Stahlberg
M. E. Müller-Institute for Microscopic Structural Biology, Biozentrum, University of Basel, CH-4056 Basel, Switzerland

Membrane channels for water and small nonionic solutes are required for osmoregulation in bacteria, plants, and animals. Aquaporin-1, the water channel of human erythrocytes, is the first channel demonstrated to conduct water, by expression in *Xenopus* oocytes. Phylogenetic analyses reveal the existence of two clusters of subfamilies, the aquaporins (AQPs) and glycerol facilitators (GLPs). Sequence-based structure prediction provided a model comprising six membrane-spanning helices, while sequence analyses suggested strategic residues that are important for structure and function. The surface topography of several AQPs has been mapped by atomic force microscopy, revealing different features that correlate with differences in the loops connecting transmembrane helices. The 3D structures of AQP1 and GlpF have been determined by electron cryomicroscopy. The 3.8-Å density map allowed the first atomic model of AQP1 to be built, taking into account data from sequence analyses. This model provides some insight into the permeation of water through a channel that blocks the passage of protons. GlpF has been resolved to 6.9 Å, revealing helices that are similar to those of AQP1. Homology modeling shows the channel region of these distant aquaglyceroporins to be similar, as confirmed by the 2.2-Å structure of GlpF from X-ray crystallography.

KEY WORDS: AQP0, AQP1, AqpZ, GlpF, 2D-crystallization, Atomic force microscopy, Electron crystallography, Sequence homology of aquaglyceroporins

I. Introduction

Osmoregulation in bacteria, plants, and animals requires the presence of highly specific membrane channels that allow the passage of water and small nonionic solutes. The existence of such channels was shown by the pioneering work of Peter Agre and collaborators (Agre, 2000). Aquaporin-1 (AQP1), a 28-kDa protein discovered in human erythrocytes, was shown to function as a water-selective membrane pore by expression in *Xenopus* oocytes (Preston *et al.,* 1992). These experiments initiated many studies of the rapidly growing aquaglyceroporin superfamily. By sequence-based structure prediction, a model of the channel forming proteins was proposed that comprises six membrane-spanning helices and two long loops (Gorin *et al.,* 1984; Preston and Agre, 1991). Later, multiple alignments of a large number of aquaglyceroporin sequences provided phylogenetic trees that suggested the existence of two large subfamilies (Park and Saier, 1996; Heymann and Engel, 1999), the aquaporins (AQPs) and glycerol facilitators (GLPs), and allowed strategic residues to be identified which are likely to be important for structure and function (Heymann and Engel, 2000).

Biological membranes must control the entry and exit of ions from cells, because ion concentrations determine the electrochemical potential that energizes many vital processes. Therefore, aquaglyceroporins must be highly specific for water or small solutes to prevent ions from crossing the membrane. However, it is unclear how the passage of ions is prevented, especially protons which are able to "tunnel" through the matrix of hydrogen bonds in water (Pomès and Roux, 1996). An exception to this has been discovered: AQP6, the only water channel so far identified exclusively in internal membranes, was found to be a pH regulated chloride ion channel (Yasui *et al.,* 1999b).

The structurally best characterized water channel is AQP1 (Murata *et al.,* 2000), and the best characterized glycerol channel is GlpF of *E. coli* (Fu *et al.,* 2000). Three dimensional (3D) density maps of AQP1 have first been established to a resolution of 6 Å by three groups (Cheng *et al.,* 1997; Li *et al.,* 1997; Walz *et al.,* 1997), and subsequently to atomic resolution (Murata *et al.,* 2000; Ren *et al.,* 2000). These maps show a right-handed bundle of six highly tilted transmembrane helices that surround a central density formed by long loops. The atomic structures confirm the hourglass model of Jung *et al.* (1994), who proposed these two long loops to fold back into a frame of six α-helices. More recently, X-ray analyses provided a 2.2-Å model of GlpF, showing the channel with three glycerol molecules held in place by an extended network of hydrogen bonds and precise hydrophobic interactions (Fu *et al.,* 2000). The helix arrangement of GlpF is similar to that of AQP1, and homology modeling demonstrated the channel region of these distant aquaglyceroporins to be amazingly similar.

The surface topography of several AQPs has been mapped by atomic force microscopy (Scheuring *et al.,* 1999; Fotiadis *et al.,* 2000) and by metal-shadowing of freeze-dried 2D crystals (Walz *et al.,* 1996), revealing different features that could be correlated with differences in the loops connecting the transmembrane helices. These experiments provided insight into the cell-to-cell adhesion function provided by the major intrinsic protein of lens fiber cells, MIP, now also termed AQP0 (Fotiadis *et al.,* 2000).

In this chapter, the structural findings concerning aquaglyceroporins are reviewed and discussed in view of current sequence data and permeability measurements.

II. Sequence Homology

Aquaglyceroporins are ubiquitous proteins, whose sequences are found in most organisms. Therefore, the number of proteins belonging to this family increases rapidly as the sequences of new genomes become available. For instance, ProDom (http://protein.toulouse.inra.fr/prodom) delivers close to 300 sequences of aquaglyceroporins. At the close of the 1990s, 160 of them were retrieved from Genbank, SWISS_PROT, EMBL and the genome database, and submitted to multiple alignment and phylogenetic analysis (Heymann and Engel, 1999). The quality of alignment and phylogeny has been assessed in terms of consistency and interpretability by comparing the phylogenetic tree with all functional and biochemical data available. This tree is similar to those obtained by previous analyses, and it shows two major clusters of aquaglyceroporins, the AQPs (water channels) and GLPs (nonionic solute channels). The two major families emerged billions of years ago from an ancient bacterial channel protein (Fig. 1). In addition, classification of the two clusters revealed 9 AQP subfamilies comprising 27 AQP subtypes, and 6 GLP subfamilies comprising 19 GLP subtypes.

This analysis strongly suggests a classification scheme along taxonomic lines. It also showed that there is large diversity in aquaporin sequences not only between organisms but also within the same species. The occurrence of the number and types of aquaporin isoforms within the same organism is therefore instructive as to what we can expect to emerge in the near future, especially for completely sequenced genomes. *Haemophilus influenzae* has only one member of the AQP family, a GlpF type from the GLPB1 subfamily (Fig. 1), compared to the two sequences for *E. coli* (AqpZ and GlpF). Yeast has four sequences in three types, with one in the AQP cluster and two in the GLP cluster. *Arabidopsis* appears to have about 10 types, unexpectedly all found in the AQP cluster, accounting for 37 genes, reflecting the importance of water channels in plants. Four GLP sequences and two AQP sequences representing five types have been found in *Caenorhabditis elegans*. Mammals are represented mainly by human and rat sequences, with seven in the AQP cluster and three in the GLP cluster. The trend of increasing number of sequences with organism complexity is evident, with the distribution between

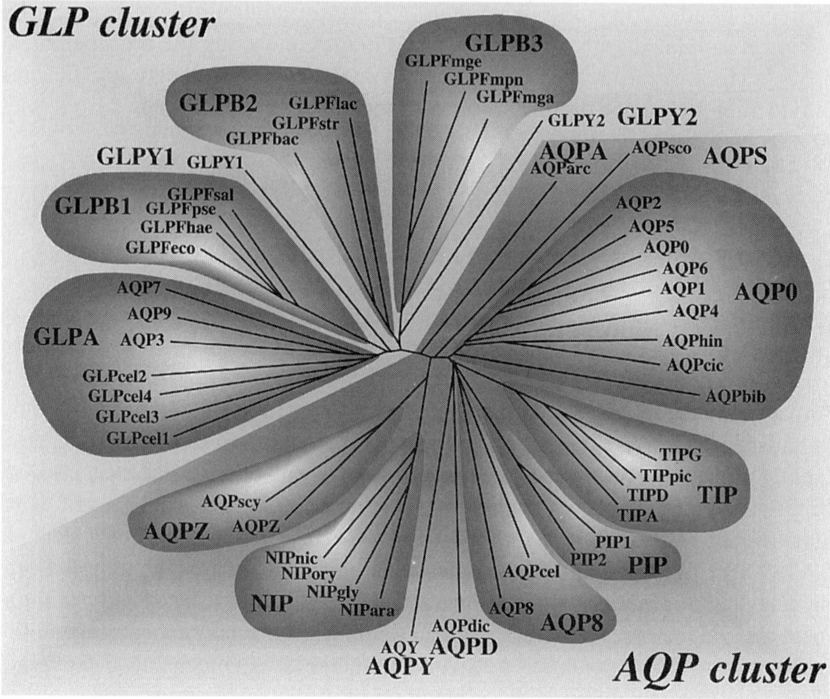

FIG. 1 Phylogenetic analysis of the aquaglyceroporin (MIP) family suggests a classification into two clusters, an AQP0-like cluster (blue area) and a GlpF-like cluster (yellow area). Further distinction is made between subfamilies, which include sequence types. The types are a 46-sequence subset considered to be representative of the whole family of ~160 sequences. (See also color insert.)

the AQP and GLP clusters still obscure. Keeping in mind the vast number of organisms for which no sequence information is available, the aquaglyceroporin family appears to consist of tens of subfamilies and over a hundred sequence types.

Multiple alignment of 46 sequences, selected to represent each subtype, revealed well-defined helical regions, allowing the ranking of the six membrane-spanning helices according to their length. The longest are helices 3 and 6, while helices 2 and 5 are the shortest. Both conservation and hydrophobicity of residues showed a distinct periodicity of approximately 4 in certain helical regions. A high periodic signal of conserved residues that is out of phase by 180° with respect to the periodic signal of hydrophobic residues suggests such helical regions to be in contact with the bilayer, with conserved residues buried. Length ranking and periodicity assessment allowed the fold of aquaglyceroporins to be predicted strictly from sequence analysis (Heymann and Engel, 2000). Such a prediction was attempted based on the 6-Å 3D map alone (Walz *et al.*, 1997). However, it turned out to be wrong. The analysis of periodic features in the aligned sequences also confirmed the unexpected half-helices formed by loop B and loop E (Mitsuoka *et al.,* 1999). A most convenient way to display the results of such sequence alignments is Tom Schneider's sequence logos (Schneider and Stephens, 1990), whose heights are proportional to the "certainty" ($-\log p_i$) of finding a residue at a particular position. As shown in Fig. 2, many subtle features can now be immediately extracted from such a representation. While the NPA motifs of loops B and E have been identified early as critical residues and hence conserved (>98%), Fig. 2 shows that all the residues between the N-termini of helices 2 and 5 and the NPA motifs in loops B and E, respectively, are almost as highly conserved, indicating their important structural or functional role. Two pairs of buried charged amino acids are also conserved (>90%), suggesting the possibility of ion-pair formation. Finally, the motif G(A)xxxG(A)xxG(A) is distinct in helices 3 and 6. This motif is characteristic of crossings in right-handed helical bundles (Russ and Engelman, 2000).

Periodicity analysis defines the buried faces of helical regions (Heymann and Engel, 2000). Compatible with this are the charged or polar residues on helices 1 and 4 that point toward the center of the molecule, possibly providing the hydrophilic surface expected for a water or a nonionic solute channel. However, two residues in the middle of helices 1 and 4 contradict this simple picture: Phe24 in helix 1 and Leu149 in helix 4. In GlpF, the corresponding sites are occupied with Leu. In fact, in 17 out of 21 GLP subfamilies, these positions harbor the L-L pair; in two subfamilies, it is an M-L pair. In contrast, 16 out of 28 AQP subfamilies exhibit an F-L, F-M, or L-F pair, while AQP0, AQP6, and BIB_DROME have a Tyr in position 24. However, 7 AQPs possess the L-L pair characteristic for GLPs. Among them are 3 NIPs, 2 archaeal AQPs, and 2 fly AQPs, the latter including AQPcic. AQPcic is noteworthy, as this is a water channel protein that has apparently been converted to a glycerol channel by a substitution of a tyrosine and a tryptophan by a proline and a leucine, in the sixth transmembrane helix (Froger *et al.,* 1998; Lagrée *et al.,* 1999). Because positions 24 and 149 are in the middle of these helices, this analysis suggests that the Leu and Phe may line the pore. In this case, these residues could be involved in determining the size and specificity of the pore, in turn indicating that several of the channels classified as AQPs may well be permeable to small nonionic molecules.

III. Physiology

Because the aquaporin family has only recently been discovered, physiological and functional analyses are quite limited. Of particular interest is the high flux combined with high selectivity for water found for the best-studied member, AQP1. The function of aquaglyceroporins is assessed by expressing proteins in oocytes (Preston *et al.,* 1992; Maurel *et al.,* 1993, 1994; Jung *et al.,* 1994; Kushmerick

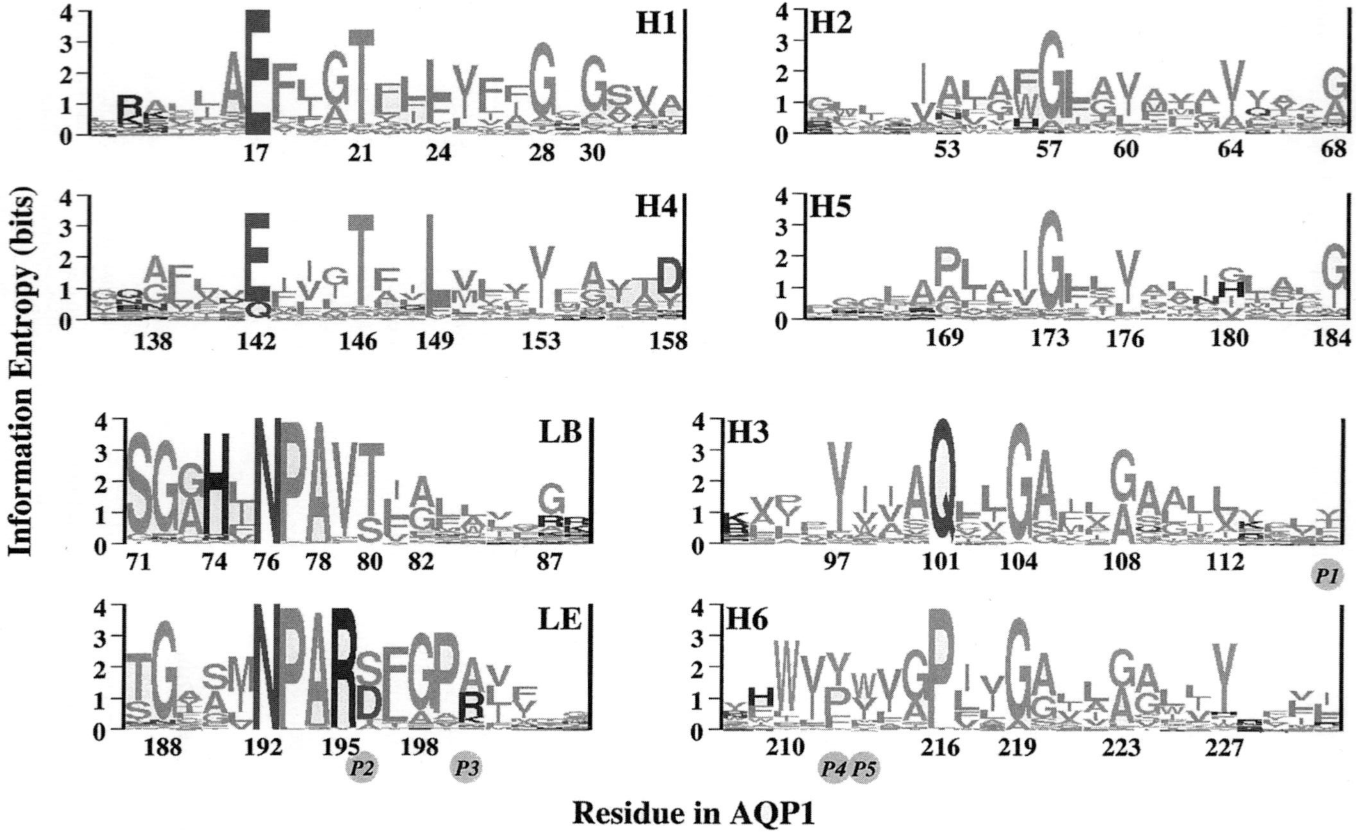

H1
H2
H4
H5
LB
H3
LE
H6
P1
P2
P3
P4 P5
Information Entropy (bits)
Residue in AQP1

et al., 1995; Le Cahérec *et al.*, 1996a; Yang and Verkman, 1997); by reconstituting proteoliposomes from purified proteins (Zeidel *et al.*, 1992, 1994), or by analyzing purified native membrane fractions (Gerbeau *et al.*, 1999). Water permeation induced by application of osmotic gradients can conveniently be observed by video microscopy (Preston *et al.*, 1992), stopped-flow fluorescence quenching (Zeidel *et al.*, 1992; Gerbeau *et al.*, 1999), or by light scattering measurements (Werten *et al.*, 2001). Permeation of small anionic solutes has been monitored using radioactive solutes (Maurel *et al.*, 1994) or by the kinetics of proteoliposome shrinkage and reswelling of proteoliposomes upon their exposure to concentration steps of the respective solutes (Fu *et al.*, 2000). Such experiments are carried out at different temperatures, thus allowing determination of the activation energy (Zeidel *et al.*, 1992). Ion conductance experiments demonstrating the lack of ion transport for most aquaglyceroporins have been achieved by reconstitution of the channel proteins in black lipid membranes (Yasui *et al.*, 1999a; Saparov *et al.*, 2001) or by assessing pH changes within proteoliposomes (Zeidel *et al.*, 1994).

Water channel proteins are defined as lowering the activation energy of a membrane for water permeation from 10–20 kcal/mol to below 5 kcal/mol (Chrispeels and Agre, 1994). The molar flux of water driven by osmotic pressure is given by

$$\Phi_w = P_f A \Delta c_s$$

where A is the area per channel and Δc_s is the difference in osmolarity across the membrane (Finkelstein, 1987). For water diffusing through a pure lipid membrane, the permeability coefficient, P_f, ranges from 10^{-5} to 0.01 cm/s, depending on the composition of the membrane (Finkelstein, 1987, p. 104). Normal red blood cells with AQP1 exhibit a P_f ~0.02 cm/s, while for AQP1-deficient cells, P_f ~0.003 cm/s [50]. Aquaporins expressed in *Xenopus* oocytes give a P_f ~0.02 cm/s for AQP1 compared to P_f ~0.001 cm/s for controls (Jung *et al.*, 1994; Mulders *et al.*, 1995), similar to that for red blood cells. This allows semi-quantitative measurements, but the uncertainty in the level of expression of particular proteins makes interpretation difficult.

Two-dimensional crystallization of AQP1 in the presence of lipids into closed vesicles offered a system where the protein density in the membrane is known to high accuracy (Walz *et al.*, 1994b). In these crystalline vesicles of about 3 μm

---

FIG. 2 Multiple alignment of 160 AQP/GLP sequences and subsequent phylogenetic analyses yielded 46 subtypes (Heymann and Engel, 2000). To identify critical residues, the 46 characteristic sequences were aligned and the conservation of each residue calculated (Heymann and Engel, 2000). Sequence logos, whose heights are a measure of conservation, are drawn with the residue numbers for AQP1. They reveal the conservation patterns of helices H1–H6 and the highly conserved loops, LB and LE. Helices are grouped in pairs according to the sequence similarity between the first and second half of the protein. The five positions (P1–P5) that were found by Froger *et al.* (1998) to be different between the AQP and GLP clusters are given in italics in circles. Colors: gray, hydrophobic; light blue, polar; green, amide; red, acidic; dark blue, basic. (See also color insert.)

TABLE I

Unit Permeability Coefficients for Water Permeation through Aquaporins

Aquaporin type	P_f ($\times 10^{-14}$cm^3/s $\cdot$ *channel*)	Reference[a]
AQP1	1–16	(1)
	11.7	(2)
	3.6	(3)
	5.43[b]	(4)
	4.6	(5)
	3.2	(6)
	4.9	(7)
	1.2	(8)
AQP0	0.028	(8)
	0.25	(7)
AQP2	3.3	(7)
AQP3	2.1	(7)
AQP4	24	(7)
	15	(9)
AQP5	5.0	(7)
AQY1-1p	4.32	(10)
NIPgly	0.32	(11)

[a]References: (1) Preston *et al.* (1992); (2) Zeidel *et al.* (1992); (3) Verbavatz *et al.* (1993); (4) Walz *et al.* (1994b); (5) Zeidel *et al.* (1994); (6) van Hoek *et al.* (1995); (7) Yang and Verkman (1997); (8) Chandy *et al.* (1997); (9) Yang *et al.* (1997); (10) Laize *et al.* (1999); (11) Rivers *et al.* (1997).

[b]From crystalline vesicles.

in diameter, the osmotic permeation coefficient for water flowing through the AQP1 channel was found to be $P_f = 0.472$ cm/s (Walz *et al.,* 1994b). For a crystalline unit cell of 9.6×9.6 nm containing eight AQP1 monomers (11.5 nm^2/channel), this translates into a unit osmotic permeability coefficient of $P_f \approx 5.43 \cdot 10^{-14} \frac{\text{cm}^3}{\text{s} \cdot channel} = 1.8 \cdot 10^9 \frac{water\ molecules}{\text{s} \cdot channel}$ (Table I). Figure 3 compares stopped-flow fluorescence decay measurements accomplished under identical conditions with liposomes, proteoliposomes, and 2D crystals of AQP1.

An attractive and simple model for the selectivity of the aquaporins is a size-exclusion mechanism, originally proposed by Heller *et al.* (1980) for GlpF from *E. coli*. It was concluded that the AQP1 channel specificity is also based on size-exclusion (Gutiérrez *et al.,* 1995), with the channel showing decreasing permeability for water, formamide, and ethylene glycol, and no permeability for urea or glycerol (Le Cahérec *et al.,* 1996b) (Table II). This indicated that the channel has a narrow part with a diameter about the size of a water molecule, ~3 Å. Reports on

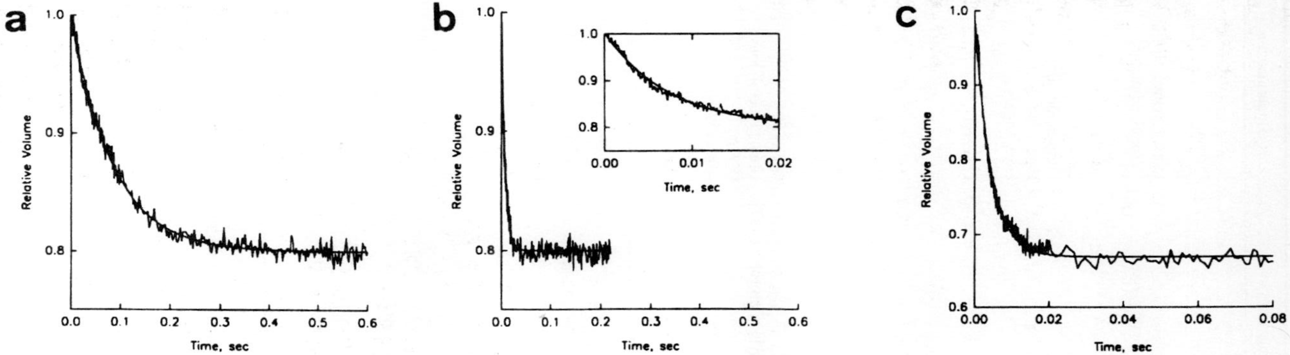

FIG. 3 Water permeation measured by stopped-flow fluorescence decay. (a) Liposomes, (b) proteoliposomes carrying a few AQP1 tetramers, (c) vesicle-shaped 2D AQP1-crystals. These were incubated overnight with fluorescein, washed briefly, and submitted to a sucrose gradient of 0.5 osmol by rapid mixing in a stopped-flow apparatus. The osmotic pressure-induced water flux induces a volume reduction yielding fluorescence quenching.

TABLE II

Selectivity of the Aquaporins to Compounds Other Than Water[a]

Type	Permeant to	Impermeant to	Reference[b]
AQP	**Cluster**		
AQP1		Protons	(1)
		Ions	(2)
		Urea, glycerol	(3)
	Formamide, ethylene glycol	Urea, glycerol, ions	(4)
	Ions[c]		(5)
		Glycerol	(6)
		Ions	(7)
	Glycerol	Formamide, urea	(8)
AQP0		Ions	(2)
	Glycerol	Urea, sugars	(9)
	Ions[c]		(10)
		Glycerol	(6)
	Glycerol		(11)
AQP2		Glycerol	(6)
	Glycerol		(8)
AQP4		Glycerol	(6)
	Glycerol		(8)
AQP5		Glycerol	(6)
		Glycerol	(8)
AQP6	Ions (at pH <5.5)		(12)
AQP8		Glycerol	(13)
AQPcic	Formamide, ethylene glycol	Urea, glycerol, ions	(4)
AQPZ		Glycerol	(14)
AQY1-1p		Urea, glycerol	(15)
γTIP		Glycerol, ions	(16)
NIPgly	Formamide, glycerol	Urea, acetamide	(17)
Nt-TIPa	Urea, glycerol		(18)
GLP	**Cluster**		
	Urea, glycerol		(8)
AQP3	Urea, glycerol		(19)
	Glycerol	Urea, sugars, lactate, ions	(20)
	Urea, glycerol		(3)
	Glycerol		(6)
	Urea, glycerol		(15)
AQP7	Urea, glycerol		(21)
	Glycerol		(22)
AQP9	Urea	Glycerol	(23)
	Urea, polyols, glycerol		(24)
GLPY2	Glycerol		(25)
GLPFeco	Urea, polyols	Sugars	(26)
	Glycerol	Ions	(27)
	Glycerol		(14)
	Glycerol	Urea	(18)

[a]It is taken for granted that a channel that allows passage of a molecule such as glycerol will allow at least the same rate of permeation of water, which may not be much higher than the water permeability of the membrane in general.

[b]References: (1) Zeidel *et al.* (1994); (2) Mulders *et al.* (1995); (3) Echevarria *et al.* (1996); (4) Le Cahérec *et al.* (1996b); (5) Yool *et al.* (1996); (6) Yang and Verkman (1997); (7) Agre *et al.* (1997); (8) Meinild *et al.* (1998); (9) Kushmerick *et al.* (1995); (10) Modesto *et al.* (1996); (11) Kushmerick *et al.* (1998); (12) Yasui *et al.* (1999a); (13) Ishibashi *et al.* (1997b); (14) Calamita *et al.* (1995); (15) Laize *et al.* (1999); (16) Maurel *et al.* (1993); (17) Rivers *et al.* (1997); (18) Gerbeau *et al.* (1999); (19) Ishibashi *et al.* (1994); (20) Ma *et al.* (1994); (21) Ishibashi *et al.* (1997a); (22) Kuriyama *et al.* (1997); (23) Ishibashi *et al.* (1998); (24) Tsukaguchi *et al.* (1999); (25) Luyten *et al.* (1995); (26) Heller *et al.* (1980); (27) Maurel *et al.* (1994).

[c]Controversial; see Agre *et al.* (1997).

the low water permeability of glycerol-specific channels (Echevarria *et al.,* 1996) challenge the simple size-exclusion model. However, more recent studies revealed GlpF to possess significantly higher water permeability than had been previously determined (Borgnia and Agre, 2001).

The high selectivity of the AQP1 channel suggests that it functions according to the single-file model, where a single chain of water molecules has to pass through the narrowest part of the pore (Finkelstein, 1987). This allows the determination of the length of this narrow part of the channel from the ratio of osmotic to diffusive flow of water (the file number). Studies on red blood cells from Colton-null patients (who lack aquaporin) compared to normal red blood cells yielded a length shorter than the membrane thickness, ~36 Å (Mathai *et al.,* 1996). Analysis of the low permeability of formamide indicated that this constriction might even be shorter, $L \sim 9$–13 Å (Whittembury *et al.,* 1997; Meinild *et al.,* 1998). Assuming the narrow part of the channel to have a length L of 10–20 Å, and considering $D_w = P_f \cdot L/A$ (Finkelstein, 1987), the apparent diffusion coefficient of water in the channel is $D_w \approx 0.4 \cdot 10^{-5}$ cm^2/s to $0.8 \cdot 10^{-5}$ cm^2/s. Compared to bulk water diffusion, $D_w = 2.1 \cdot 10^{-5}$ cm^2/s (Hille, 1992), diffusion in the channel is therefore about 3–5 times lower. Other measurements for AQP1 gave unit permeability coefficients in the range $P_f = 1–16 \cdot 10^{-14}$ cm^3/s $\cdot$ *channel* (Table I). The high variability in the determined permeabilities is likely due to uncertainty about the number of channels per unit membrane area. The determination for the crystalline vesicles is therefore taken as definitive, with several other measurements yielding comparable permeabilities.

Some aquaporins have low water permeability, $P_f = 0.03-0.3 \cdot 10^{-14}$ cm^3/s $\cdot$ *channel*, such as that for AQP0 (Chandy *et al.,* 1997; Yang and Verkman, 1997) and for NIPgly (Rivers *et al.,* 1997) (Table III). On the other end of the scale, AQP4 has been reported to have a $P_f = 15-24 \cdot 10^{-14}$ cm^3/s $\cdot$ *channel* (Yang

TABLE III
Similarity of Aquaglyceroporin Structures

Structure pair	Helical backbone RMSD (Å)
1FX8-1FQY	1.75
1FX8-1IH5	2.67
1FX8-*re*	1.36
1FQY-1IH5	2.45
1FQY-*re*	1.41
1IH5-*re*	2.24

Root mean square deviations (RMSD) between pairs of aquaglyceroporin structures, calculated over the heavy backbone atoms of all residues in helical conformation. 1FX8 denotes the PDB code of the x-ray structure of GlpF (Fu *et al.,* 2000); 1FQY and 1IH5, the two AQP1 models (Murata *et al.,* 2000; Ren *et al.,* 2000); and *re* denotes our refined AQP1 structure (de Groot *et al.,* 2001).

et al., 1997). While AQP4 remains selective for water (Yang and Verkman, 1997), the higher flux through AQP4 may be associated with multiple paths for water permeation to yield a diffusion coefficient significantly lower than in bulk water. Most aquaporins, however, have permeabilities comparable to that of AQP1 ($2-5 \cdot 10^{-14}$ cm^3/s · *channel*).

Aquaglyceroporins are channels permeated by small nonionic molecules in the size range of water to glycerol (Table II). The largest compound reported to pass through the *E. coli* GlpF is xylitol, while GlpF was impermeant to pentoses (Heller *et al.,* 1980). The linear xylitol molecule has a width similar to that of glycerol, while the sugars are ring structures with a larger cross section. More recent studies revealed a stereo- and enantioselectivity of GlpF for the homologous series of linear polyalcohols called alditols (Fu *et al.,* 2000).

The functional difference between the two clusters of the family is apparent in Table II, although the distinction is not clear cut. For instance, the AQP cluster proteins AQP0, NIPgly, and Nt-TIPa transport glycerol, while AQP9 of the GLP cluster appeared impermeant to glycerol in one study (Ishibashi *et al.,* 1998), but permeant in another (Tsukaguchi *et al.,* 1999). It is interesting to relate this to the critical sites Phe24 and Leu149 in AQP1. AQP0 is indeed special, as it reveals two bulky residues, Tyr and Phe, at these sites, explaining at least the low water permeability. NIPgly and related NIPs would fall into the category of glyceroporins, as they carry a Leu at both positions. Németh-Cahalan and Hall (2000) reported a 3-fold increase in water permeability of AQPO at pH 6.5 compared to pH 7.5. A similar increase was measured in absence of Ca^{2+} ions. Reports of AQP0 being permeant (Kushmerick *et al.,* 1998) or impermeant (Yang and Verkman, 1997) to glycerol are based on how experiments were done. In the latter case, the 10-min incubation of the oocytes with labeled glycerol proved to be too short to reveal the relatively slow uptake. The permeability of glycerol and urea in AQP3, AQP7, and AQP9 is 3 to 4 orders of magnitude lower than the water permeability (Echevarria *et al.,* 1996; Ishibashi *et al.,* 1994, 1997a,b, 1998). This is consistent with the glycerol permeability coefficient for AQP0 of $3.1 \cdot 10^{-19}$ cm^3/s · *channel,* 3–4 orders lower than the water permeability coefficient (Table I) (Kushmerick *et al.,* 1998). This suggests that in these channels a large flux of water accompanies solute permeation, in stark contrast to GlpF.

As demonstrated in several laboratories (Agre *et al.,* 1997; Zeidel *et al.,* 1994), aquaglyceroporins have no intrinsic ion conductance ability, with the exception of AQP6 (Yasui *et al.,* 1999b) (Table II). If the flow of water through the channel is accompanied by ions without selection, the conductivity in a 100 mM salt would be $\frac{0.1\ M}{55\ M} \cdot 2 \cdot 10^9 \frac{waters}{\mathrm{s} \cdot channel} = 4 \cdot 10^6 \frac{ions}{\mathrm{s} \cdot channel}$, which is comparable to the single channel current observed for ion channels (Hille, 1992, p. 330). Water permeation through other channels has been demonstrated, such as that for the Na^+/glucose cotransporter of the small intestine brush border membranes (Loo *et al.,* 1996; see Zeuthen and MacAuley, Cotransporters as Molecular Water Pumps, this volume). In this case, the water passage is only 6000 molecules/s · channel, with 260 water molecules accompanying each Na_2-sugar cotransport cycle.

Many aquaglyceroporin family members show a strong concentration of positive charge on the cytoplasmic side (especially loop D) and negative charge on the extracellular side. Perhaps these strongly charged entrances to the channel form efficient electrostatic filters preventing the passage of ions. Nevertheless, the inability of the AQP1 channel to conduct ions (Agre *et al.,* 1997; Zeidel *et al.,* 1994) is likely to be also related to the size of hydrated ions. To eliminate the hydration shell, the potassium channel KscA provides a network of hydrogen bond acceptors and donors to allow for energy-neutral stripping of water molecules (Doyle *et al.,* 1998). In the absence of such a network, ions could not permeate, whereas protons should be conducted along a water chain (Pomès and Roux, 1996). However, a pH shift experiment did not show any proton conductance through AQP1, while the control, gramicidin A, dissipated the proton gradient efficiently (Zeidel *et al.,* 1994).

An examination of the mechanisms of proton conductance through the gramicidin A channel suggested that clusters of waters bound to the protein slow down proton transfer, while a single-file hydrogen-bonded chain of water molecules conducts a proton more efficiently (especially when electrostatic interactions with the protein wall of the channel are switched off; Pomès and Roux, 1996). Hence, the structure of the water at the charged channel entrances may prevent proton flow. Furthermore, the orientations of the water molecules required for the transmission of the defect along the water chain may be restricted or disallowed within the channel. Thus, the specific interactions and arrangement of the water in the pore may provide a barrier to proton conduction.

Another mechanism might rely on the requirement of the transmission of a defect along the water chain to reorient the water molecules correctly for proton translocation. A gap too large to bridge by significant tunnelling of a proton might provide a barrier to proton conductance. In this respect, it is interesting to note that a gap in the ladder of consecutive main-chain carbonyls that form the polar line in the channel interior has been identified in the 2.2-Å GlpF structure. This feature is expected to be common to all AQPs, and determines distances that closely correspond to the distances between hydrogen bonded water molecules in the water channels (Nollert *et al.,* 2001).

IV. Progress in Structure Determination

Native and recombinant aquaglyceroporins appear to exist as tetramers in the native membranes, as initially shown by freeze-fracture electron microscopy (Verbavatz *et al.,* 1993). In most cases, solubilized and purified oligomers exhibit a square-shaped appearance and dimensions compatible with a tetrameric form (Fig. 4). A hypothesis that glyceroporins may exist as native monomers (Lagrée *et al.,* 1998) has been supported by the finding that the affinity of the GlpF tetramer associations were much lower than for AqpZ, the water channel of *E. coli* (Borgnia and Agre,

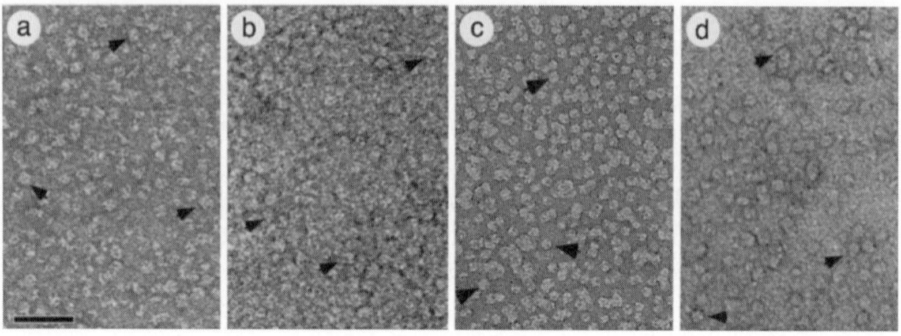

FIG. 4 All water channel proteins shown here are isolated as tetramers (arrows) after solubilization of the respective membranes with octyl glucoside. (a) AQP1 from erythrocyte ghosts (Walz *et al.*, 1994a). (b) AQP0 from ovine lens fiber cells (Hasler *et al.*, 1998). (c) AqpZ from *E. coli* cells over-expressing this water channel (Ringler *et al.*, 1999). (d) GlpF from *E. coli* cells over-expressing the glycerol facilitator (Braun *et al.*, 2000). Scale bar = 50 nm.

2001). However, this was not confirmed by the 2.2-Å GlpF structure determined by X-ray crystallography (Fu *et al.*, 2000) and the results from electron microscopy studies (Braun *et al.*, 2000). It was probably the oligomeric nature of GlpF and the stability of aquaglyceroporins that fostered their crystallization mainly into 2D but more recently also into highly ordered 3D crystals.

Examples of 2D crystals document differences between aquaglyceroporin subtypes and dependence on reconstitution conditions. In the presence of $MgCl_2$ at a concentration of a few mM, AQP1 forms vesicle-type crystals that are amenable to functional assays (Walz *et al.*, 1994b). At higher $MgCl_2$ concentrations, a significant fraction of sheet-like 2D crystals emerges (Walz *et al.*, 1994a) that shows single-layered electron diffraction patterns that extend to a resolution of 3 Å (Fig. 5a). GlpF assembles into extended crystalline layers that are not highly ordered. However, large tubular crystals that flatten upon adsorption to the carbon film are often found in the same preparations (Braun *et al.*, 2000). These crystals are highly ordered, as documented by electron diffraction pattern extending to 3.5 Å resolution, and exhibit the typical twinning of double-layered crystals (Fig. 5b). AQP0 also tends to form double-layered crystals, but in contrast to GlpF, the layers are stacked in register (Fotiadis *et al.*, 2000), yielding untwined diffraction patterns as illustrated by the IQ plot shown in Fig. 5c. Projection maps of bacterial, plant, and animal aquaglyceroporins have been published (Jap and Li, 1995; Mitra *et al.*, 1995; Walz *et al.*, 1995; Daniels *et al.*, 1999; Ringler *et al.*, 1999). All exhibit a pronounced similarity, as also documented by the examples shown in Fig. 5. These maps cannot show the surface variability of the aquaglyceroporins because the contribution of loops to the contrast of the projection maps is small compared to that of the helical protein core. Because the latter is highly conserved (see Fig. 2), the similarity of projection maps is not unexpected. However, even in projection, distinct differences emerge, as illustrated in Fig. 5. The 5.5-Å AQP0 projection

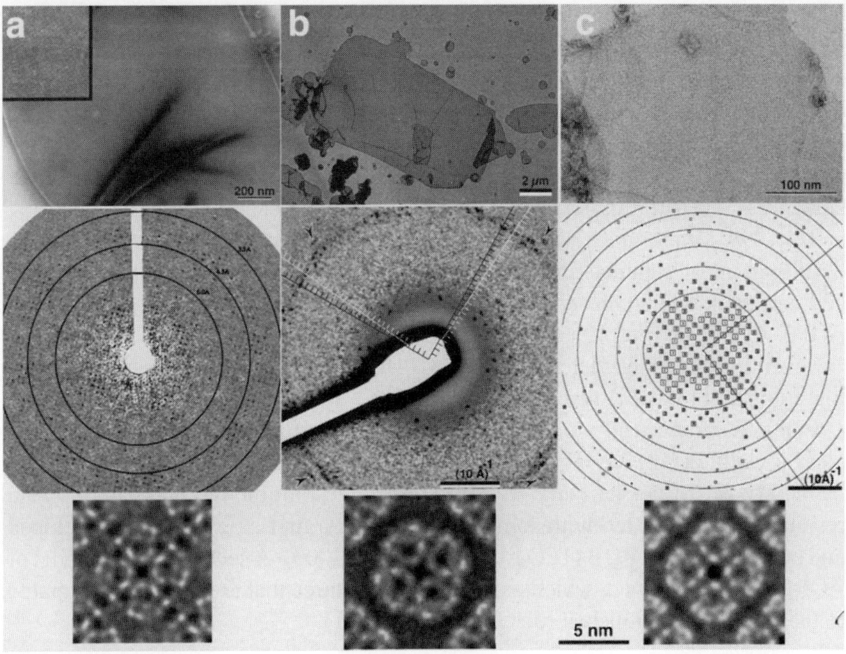

FIG. 5 Highly ordered 2D crystals of aquaglyceroporins allow structural information to be extracted to atomic resolution by electron crystallography. Reproduced are overview images of 2D crystals (top), diffraction pattern (center), and reconstructed unit-cells (bottom). (a) Single layered AQP1 crystals diffract to 3 Å resolution. The projection map calculated from electron images and diffraction patterns exhibits a resolution of 3.5 Å (Walz *et al.*, 1995). (b) Large tubular GlpF crystals collapse by adsorption to the carbon film. They were usually monocrystalline and well ordered and had a size of several micrometers in diameter. Scale bar = 2 μm. The electron diffraction pattern of a frozen hydrated 2D crystal demonstrates the excellent crystallinity of such rectangular crystals. Diffraction orders marked by arrows correspond to a resolution of 3.3 Å (Braun *et al.*, 2000). (c) AQP0 mostly forms double-layered crystals, but these crystals frequently reveal monolayered 2D crystals at their edges. The calculated power spectrum (IQ-plot (Henderson *et al.*, 1990)) demonstrates the order of these crystals (Hasler *et al.*, 1998). The data from AQP1 (a) and GlpF (b) were recorded with a He-cooled field-emission-gun microscope, while the data from AQP0 and the diffraction pattern of GlpF were recorded with a conventional microscope.

map from single-layered crystal edges is highly similar to the 3.5-Å map of AQP1, both revealing a handedness and a trapezoid monomer with an almost identical density distribution. In contrast, the 3.7-Å map of GlpF exhibits a significant density minimum within the monomer. We have speculated that this minimum is the channel, and have related the larger diameter of the density minimum to those critical amino acids that determine the pore size (Stahlberg *et al.*, 2000). The 2.2-Å GlpF structure now proves this correct. Interestingly, the projection maps all show a prominent low-density region about the four-fold axis. In AQP0 and AQP1, this minimum is more pronounced than that of the glycerol channel.

Most sensitive in detecting surface variations resulting from loop variability is the atomic force microscope (AFM). Not only does this instrument allow membrane proteins to be observed in buffer solutions, but it also allows conformational changes of single loops connecting α-helices or antiparallel β-strands to be followed (Engel and Müller, 2000). Using the AFM, it has been possible to visualize the large variation of aquaglyceroporin surfaces, and to identify the sidedness of bovine AQPO (Fotiadis *et al.,* 2000) and bacterial AqpZ (Scheuring *et al.,* 1999). Importantly, such studies have provided clues to understanding the development of cataracts in mice bearing mutations in AQPO in their lens fiber cell membranes (Fotiadis *et al.,* 2000). Figure 6 summarizes the surface reliefs of AQP0, AQP1, and AqpZ as measured with the AFM.

Several groups have determined the 3D structure of AQP1 by electron crystallography, first to 6–7 Å resolution (Cheng *et al.,* 1997; Li *et al.,* 1997; Walz *et al.,* 1997), and subsequently to atomic resolution. At 6.0 Å, the helices of AQP1 are clearly visible (Fig. 7a) (Walz *et al.,* 1997). At 4.5 Å, the helices reveal a right-handed twist (Mitsuoka *et al.,* 1999). At 3.8 Å resolution, the bulky amino acid residues could be fitted with some certainty, and a first atomic model developed and deposited in the PDB (1FQY) (Murata *et al.,* 2000). A second atomic model of AQP1 (1IH5) followed, which exhibits some features that are different compared to the first model published (de Groot *et al.,* 2001).

V. Structural Clues for Pore Specificity

The fold of AQP1, as determined by sequence analyses (Heymann and Engel, 2000) and by fitting of short helices to the 4.5 Å 3D density map (de Groot *et al.,* 2000), was confirmed. Beginning at the four-fold axis of the tetramer, the helices are arranged: 2-1-3 (1st repeat), 5-4-6 (2nd repeat). In the first repeat, helices 1 and 2 run almost parallel to each other, and are tilted to approximately the same angle (Fig. 7c). Helix 3 is oriented almost perpendicular to the axis roughly defined by the top-on view of the first two helices. The arrangement of first and second repeats exhibits the pseudo-two-fold axis running parallel to the membrane plane at the center of the molecule (Fig. 7d).

The short N-terminus of AQP1 is located on the cytoplasmic side of the membrane (Fig. 7d). Helix 1 then crosses the membrane and is connected to helix 2 by the extended extracellular loop A with the glycosylation site Asn41 forming a weak density in our map. Helix 2 returns back through the membrane adjacent to helix 1 near the four-fold axis of the AQP1 tetramer. At the cytoplasmic end of helix 2, loop B folds back into the membrane, positioning the NPA motif midway through the membrane. Following the NPA motif, loop B forms HB, the short pore helix located adjacent to helix 6. At the cytoplasmic surface farthest from the four-fold axis, HB emerges from the membrane and connects to helix 3, which crosses the

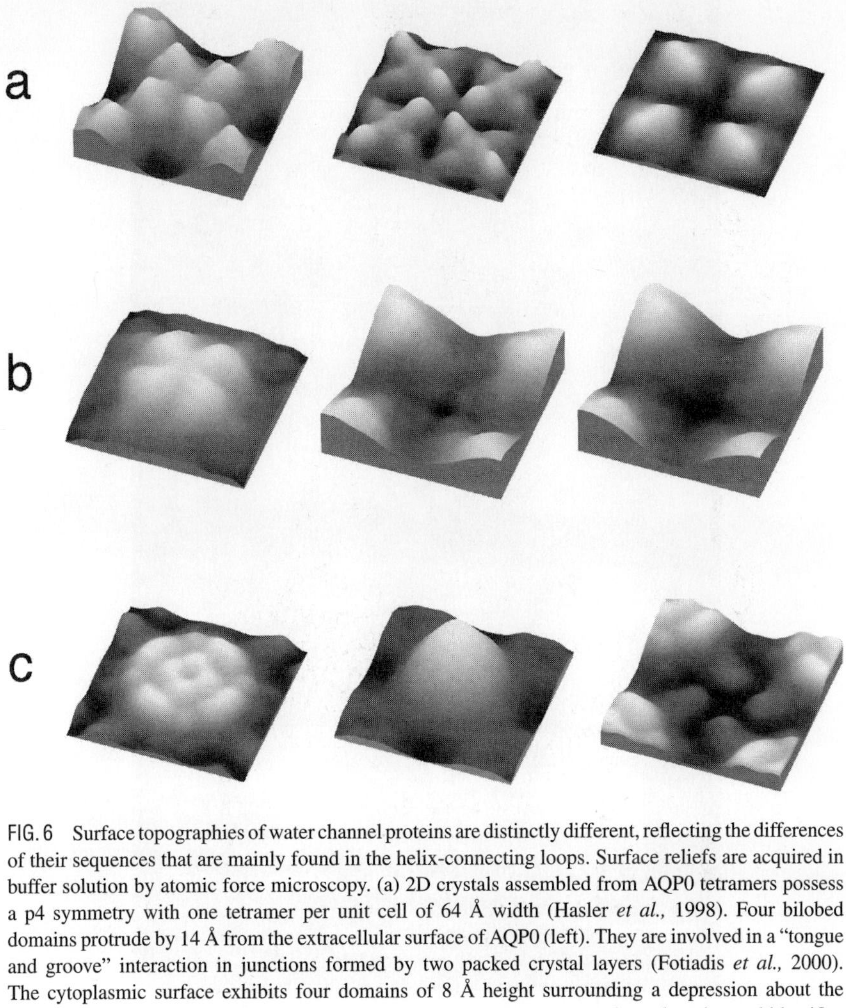

FIG. 6 Surface topographies of water channel proteins are distinctly different, reflecting the differences of their sequences that are mainly found in the helix-connecting loops. Surface reliefs are acquired in buffer solution by atomic force microscopy. (a) 2D crystals assembled from AQP0 tetramers possess a p4 symmetry with one tetramer per unit cell of 64 Å width (Hasler *et al.,* 1998). Four bilobed domains protrude by 14 Å from the extracellular surface of AQP0 (left). They are involved in a "tongue and groove" interaction in junctions formed by two packed crystal layers (Fotiadis *et al.,* 2000). The cytoplasmic surface exhibits four domains of 8 Å height surrounding a depression about the four-fold axis (center). Carboxypeptidase Y treatment *on mica* removes these domains and identifies them as carboxylic (Fotiadis *et al.,* 2000) (right). (b) AQP1 crystals have a $p42_12$ symmetry and a corrugated extracellular surface with four protrusion of 12 Å height (left). The cytosolic surface with four elongated peripheral and four small central domains produces a windmill-shaped structure with a height of 6 Å (center). Preliminary data from carboxypeptidase Y treatment *on mica* indicate that the carboxyl terminus is located at a position similar to that of AQP0 (Müller and Engel, unpublished data; right). (c) AqpZ has an extracellular surface with four elongated peripheral protrusions and four central protrusions of 7 Å height (left). The poly-His tag used for purification produces a massive protrusion on the cytosolic surface (center). After trypsin cleavage, the cytosolic surface exhibits four small protrusions (right). The elongated extracellular protrusion in AqpZ has been identified as loop C based on its volume and flexibility (Scheuring *et al.,* 1999). Correlation averages displayed comprise two tetramers and have a side length of approximately 95 Å. Horizontal and vertical scalings are identical, but the grey level range is adapted to go from black to full white within each topograph.

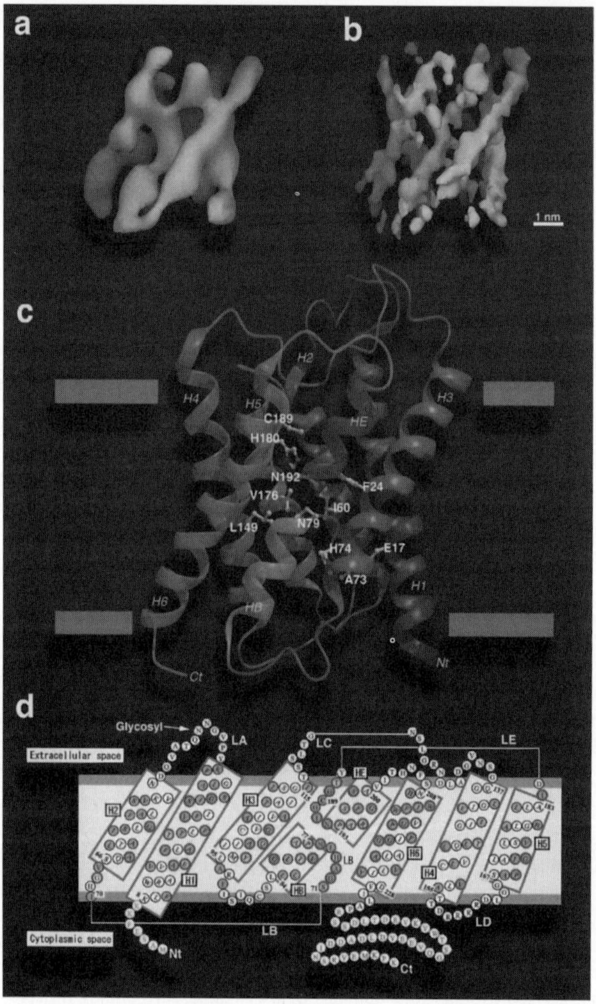

FIG. 7 The structure and sequence of AQP1. (a) The structure from cryoelectron microscopy, rendered at 6.0 Å resolution, reveals rod-like densities, which correspond to highly tilted α-helices in a right-handed helix arrangement. (b) At 3.8 Å resolution, the α-helices reveal a right-handed twist. These data were the structural basis for the model-building (Murata *et al.,* 2000). (c) Ribbon representation of the atomic model of AQP1, showing the extracellular side up. Helices are labeled in yellow italics, channel-lining residues are labeled in white. (d) The amino acid sequence of AQP1. (See also color insert.)

membrane adjacent to helix 1 and terminates on the extracellular surface. The 1st and 2nd repeats of each monomer are connected by the 18-residue loop C which spans the entire extracellular surface of the monomer joining helix 3 to helix 4. Oriented opposite to the first repeat, the second repeat begins at the extracellular surface as helix 4, which crosses the membrane and is connected to helix 5 by the short loop D near the four-fold axis. Helix 5 returns back through the membrane between helix 4 and helix 2 to the extracellular surface. Loop E folds back into the membrane from the extracellular side, and the NPA motif is positioned near the center of the molecule, where it crosses the NPA motif in loop B at a right angle and forms the short pore helix (HE). Helix 6 crosses the membrane adjacent to helix 4, with the C-terminus emerging on the cytoplasmic side of the membrane (Fig. 7c,d).

Three large helical crossing angles stabilize the right-handed coiled-coil interactions. Helices 1 and 3 (36.7°), helices 4 and 6 (40.9°), and helices 2 and 5 (28.5°) all have highly conserved glycine residues at the contact sites. The glycine residues at the contact regions are Gly 104 for helices 1 and 3, and Gly 219 for helices 4 and 6, and the crossing of helix 2 (1st repeat) and helix 5 (2nd repeat) is maintained by interactions involving highly conserved residues Gly 57 and Gly 173 (Heymann and Engel, 2000). The fitting of ridges into grooves in helices 2 and 5 locks the two AQP1 helical bundles together closely at the four-fold axis of the tetramer, opposite to the site where HB and HE interact with helices 6 and 3 (Fig. 7c). The conserved glycines in helices 3 and 6 reveal the GxxxG motif that is a framework for transmembrane helix–helix association (Russ and Engelman, 2000) (A100, G104, A108 in helix 3; G215, G219, A223 in helix 6; glycine can sometimes be replaced by alanine in the motif). Thus, the α-helices in both repeats are held together near the interior and exterior surfaces of the monomer. Gly 82 is the contact in pore helix HB with helix 6, and Gly 198 in pore helix HE with helix 3.

Additional highly conserved residues are concentrated in and around the two functional loops B and E that are closely positioned in the interior of the monomer (Heymann and Engel, 2000). Loops B and E are held together by van der Waals interactions between the prolines in the two NPA motifs (Pro 77 and Pro 193) that cross each other at an angle of approximately 90° (Fig. 7c,d). The asparagines in the NPA motifs cap the N-terminal end of the pore helices by hydrogen bonding with the main-chain NH groups at Val 79 and Arg 195. The positions of functional loops B and E are stabilized through ion pairs and hydrogen bonds. His 74 (loop B) forms an ion pair with the conserved Glu 17 (helix 1) anchoring the position of pore loop B, while Arg 195 in pore helix HE is connected by a salt bridge to the conserved Glu 142 (helix 4). These ion pairs are stabilized by conserved polar residues that surround them within the membrane; Thr 80, Gln 101, and Ser 196. Secondly, the positions of the NH groups in the main chain of the pore loops are stabilized by hydrogen bonds to highly conserved threonine residues, Thr 21 (helix 1) and Thr 146 (helix 4). In addition, the side chain of Ser 71 (loop B) forms a hydrogen bond with Tyr 97 (helix 3), further stabilizing the position of this loop.

Parallel to the progress in determining the structure of AQP1, the highly ordered GlpF 2D crystals allowed projections of tilted lattices to be recorded and a 3D map to 6.9 Å to be established in short time (Stahlberg *et al.,* 2000). Compared to the AQP1 structure, additional domains were found in this map that

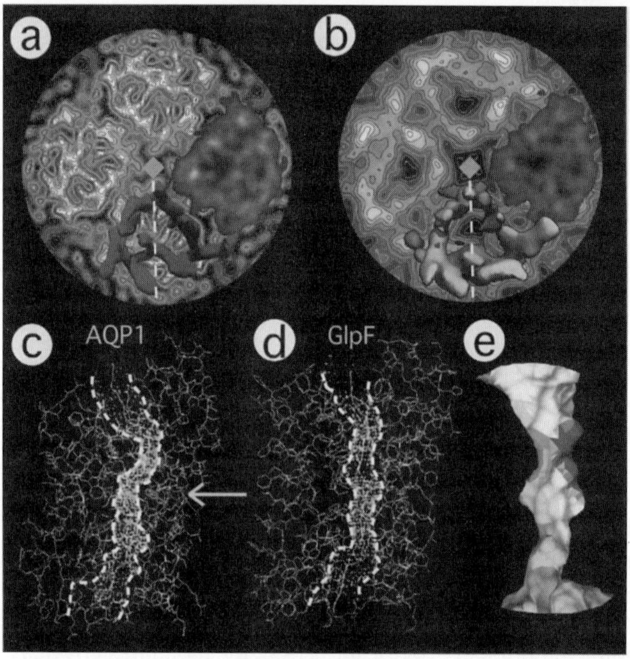

FIG. 8 Atomic modeling of the GlpF channel based on the structure and sequence homology of GlpF and AQP1 and the atomic structure of AQP1 (Murata *et al.,* 2000). (a) The 3.5 Å AQP1 projection structure with the 3D map of the monomer at 6.9 Å resolution (blue) and the projection of the atomic model (brown). (b) The 3.7 Å GlpF projection structure with the 3D map of the monomer at 6.9 Å resolution (yellow) and the projection of the constructed model of GlpF. The dotted line indicates the pseudo-two-fold symmetry axis of the monomer. (c,d) Channels of AQP1 (**c**) and GlpF (d), calculated by HOLE (Vriend, 1990). (e) The amphiphilic nature of the GlpF channel is revealed by color-coding the accessible surfaces: green: aromatic residues; yellow: hydrophobic residues; white: polar residues; blue: positively charged residues. (See also color insert.)

could be fully interpreted by alignment of the AQP1 and the GlpF sequences. The similarity of the density maps fostered modeling of the GlpF structure starting from the atomic model of AQP1, revealing the critical importance of the bulky residue, Phe24, that determines the constriction in AQP1 but is absent in GlpF (Fig. 8) (Stahlberg *et al.,* 2000). The 2.2-Å structure of GlpF (PDB code 1FX8) now available allows the value of lower resolution structures to be scrutinized. In particular, the GlpF model produced is in excellent agreement with the 2.2-Å atomic model. Although separated in early stages of evolution, AQP1 and GlpF reveal a striking structural homology, which parallels the high sequence homology (Fig. 9). Their structures are highly similar, exhibiting a root mean square deviation between the peptide backbones within the α-helical core of about 1.5 Å (Table III). This similarity and the success in modeling the GlpF channel, starting from the AQP1 structure, now fostered a refinement of the AQP1

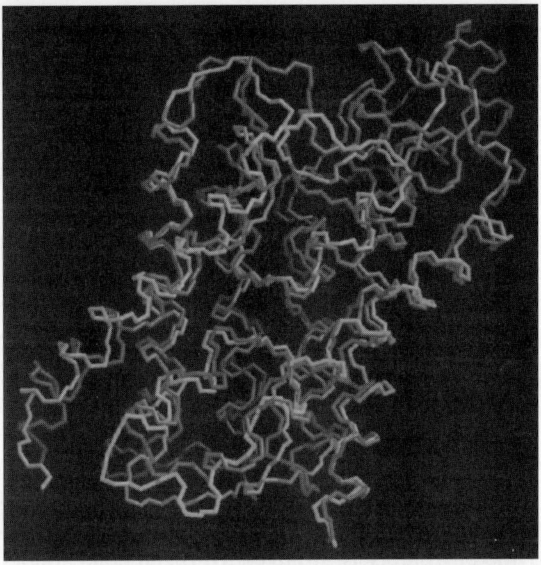

FIG. 9 Superposition of the backbone traces of the X-ray structure of GlpF (red) and a previous EM structure of AQP1 (cyan, PDB entry 1FQY) (Murata *et al.*, 2000). (See also color insert.)

structure, starting from that of GlpF. This refinement improved the overall quality of the model, and it increased its stability when tested by molecular dynamics calculations.

The specificity of the respective pores for water or nonionic solutes is due to the different side chains facing the channel. The critical function of AQP1 is its exceptional water permeability, 3×10^9 molecules per monomer per second (Zeidel *et al.,* 1992, 1994; Walz *et al.,* 1994b). The 3-Å diameter of the narrowest part of the AQP1 pore measured by van der Waals distance (the 6.5 Å of carbon–carbon distance was measured as the narrowest part of the pore) is only slightly larger than the 2.8-Å diameter of a water molecule. Although the span of this constriction is only one amino acid residue, this would block passage of larger solutes and ions. Unlike the potassium channel (Doyle *et al.,* 1998), AQP1 does not contain a structure to liberate ions from their hydration shell. Thus, the atomic structural model of AQP1 agrees with biophysical measurements, which showed no solute, ion, or proton permeability (Zeidel *et al.,* 1994; Gutiérrez *et al.,* 1995; Yang and Verkman, 1997).

The 3-Å constriction in the pore is located in the center of the membrane just beside the region where loops B and E interact with each other. Ile 60 (helix 2), Phe 24 (helix 1), Leu 149 (helix 4), and Val 176 (helix 5) form a hydrophobic surface lining the inside of the pore adjacent to Asn 76 and Asn 192 of the NPA motifs (Fig. 7c). Two histidines are located within 5 Å of the pore constriction.

His 74 is highly conserved and forms an ion pair with Glu 17. His 180 may play a specific role in water conductance, since this residue is conserved in water-selective members of the aquaporin family, but the residue is predominantly Gly in homologs permeated by glycerol (Heymann and Engel, 2000). Hydrophobicity of the inner pore surface and small axial dimension (5 Å) of the constriction are probably of crucial importance for facilitating the exceptionally high water permeability of AQP1.

The free sulfhydryl of Cys 189 in loop E projects into the pore from the extracellular side of the narrowing (Fig. 7c), where mercurial inhibition of the AQP1 water permeation (Preston *et al.,* 1993; Zhang *et al.,* 1993) must result from occlusion of the aqueous pathway. As expected for a symmetrical structure, Ala 73, the corresponding residue in loop B, resides at a comparable location on the cytoplasmic side of the narrowing, explaining the mercurial inhibition of the double mutant C189S, A73C (Jung *et al.,* 1994) (Fig. 7c).

X-ray crystallography provided the best-defined structure of an aquaglyceroporin (Fu *et al.,* 2000). At 2.2 Å resolution, three glycerol molecules are found to be positioned within the GlpF channel by an extraordinary hydrogen bonding network that is provided by precisely arranged hydrogen bond donors and acceptors lining the channel walls (Fig. 10). This network explains the stereospecificity of the GlpF pore, as well as the much lower permeation rate of glycerol in GlpF compared to water in AQP1. In addition, the GlpF pore exhibits a striking amphipathic nature with polar side chains or backbone carbonyls on one side and hydrophobic residues on the other side of the channel. Glycerol molecules are oriented therefore not only by hydrogen bonds, but also by a snug fit of their hydrophobic backs to the hydrophobic slide provided by the channel wall. It is tempting to compare GlpF with maltoporin, a bacterial outer membrane pore that allows passage of large sugars such as maltodextrins. These have been shown to permeate the outer membrane efficiently along a "greasy slide" formed by a set of aromatic residues within the maltoporin pore, itself revealing a pronounced amphipathicity (Schirmer *et al.,* 1995).

Water molecules traverse the AQP1 channel at a rate of $2 \cdot 10^9$/s, when driven by a sucrose gradient of 0.5 osmol. While the channel of AQP1 is also amphipathic, the hydrophilic surface area is smaller than in GlpF, providing fewer opportunities for hydrogen bonding. Compared to glycerol, water molecules move therefore less hindered, and in a single file, forming a hydrogen bond network mainly perpendicular to the membrane plane. The temperature dependence of the water permeability yields an activation energy of 3–5 kcal/mol, which is close to the energy of a hydrogen bond, suggesting that a water molecule moves through the channel as in bulk water, where hydrogen bonds must be broken and reformed as well. Quantum mechanical modeling of a single water molecule file in gramicidin has given proton transfer rates of 10^9/s, when hydrogen bonding to the channel walls was switched off (Pomès and Roux, 1996). Therefore, specific hydrogen bonds with the pore must be considered in order to understand the effective inhibition

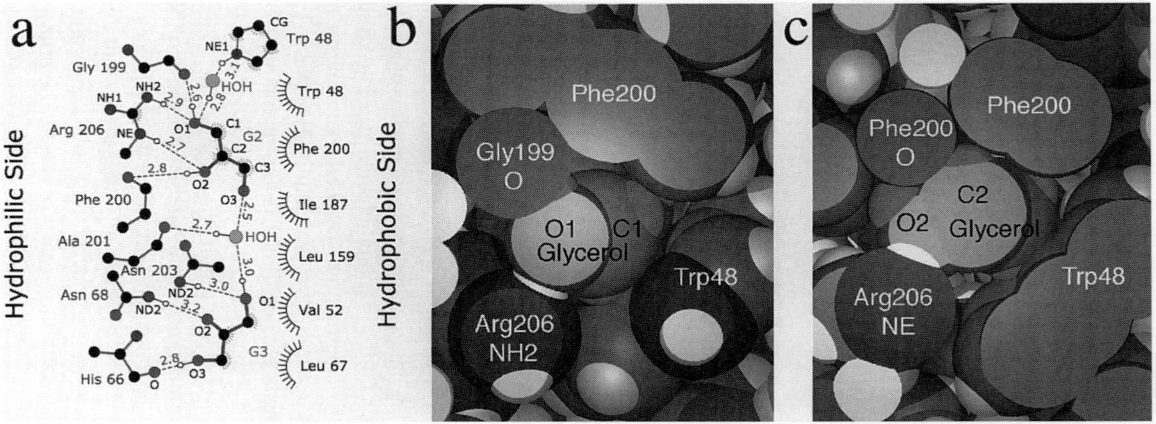

FIG. 10 The structure of GlpF reveals three glycerol molecules at specific sites in the channel. (a) The hydrogen bonding network (dotted lines) showing interactions of G2, G3, and associated water molecules with the selectivity filter. Hydrophobic contacts are indicated as radial lines around atoms or residues. Distances (in Å) between heavy atoms are indicated. (b,c) Cross section through the channel (space-filling rendering) showing the interactions with G2OH2, O2 (orange) and its associated CH group (C2; purple), and hydrogen (green). The cross section shows the donor NEH of Arg 206, acceptor carbonyl oxygen of Phe 200, and hydrophobic corner formed by Trp 48 and Phe 200 viewed edge-on to the aromatic rings. (See also color insert.)

of proton currents along single files of water molecules. We have speculated that the conserved asparagines of the NPA motifs, held in place in the center of the channel by the stacking interaction of the prolines, are the most likely hydrogen bonding sites (Murata *et al.,* 2000). Stabilized by hydrogen bonds of the carbonyls with backbone NH groups, these asparagines extend their amido groups into the channel, similar to the corresponding arrangement of the asparagines in the 2.2-Å GlpF structure (Fu *et al.,* 2000). However, at the resolution currently available, no water can be resolved in the AQP1 channel, and the confirmation of this hypothesis must await higher resolution structural data.

The role of the few charged residues in the pore as electrostatic filters that form a barrier for ions is not established yet. However, the available structures of AQP1 and GlpF do not show residues that would allow removing the hydration shell at low energy cost. Both mechanisms are hypothesized to be important as selectivity filters excluding the passage of ions (Doyle *et al.,* 1998).

VI. Regulation

At least three mammalian aquaporins appear to be directly regulated by pH. Thus, structural pH sensors must reside in these proteins. AQPO is expressed in the lens fiber cells of mammalian eye lenses. Its water permeability is reported to increase 3-fold at a change in pH from 7.5 to 6.5, which is the normal pH range found in the lens (7.2–6.5) (Németh-Cahalan and Hall, 2000). AQP3 is expressed in basolateral membranes of water-reabsorbing principal cells in renal collecting duct, airway epithelia, and secretory glands. AQP3 is permeated by water and glycerol at neutral pH, but the channel appears gated shut at pH <6 (Zeuthen and Klaerke, 1999). AQP6 resides in acid-secreting intercalated cells of renal collecting duct (Yasui *et al.,* 1999b) and undergoes a conformational change at pH <5.5, during which the channel opens for selective permeation by water and chloride ions (Yasui *et al.,* 1999a). AQP6 belongs to the AQP cluster, but it possesses a Tyr at position 24 instead of a Phe, implying a role of Tyr24 in the regulation of the channel. However, AQP3 is a typical GLP exhibiting the characteristic L–L pair, suggesting that different mechanisms of pH driven regulation might exist.

However, from the two atomic structures available, it seems unlikely that a pH change could force the highly specific pore formed by the six helices and the loops with the NPA motifs to conduct ions. As recently suggested from the structure of GlpF (Fu *et al.,* 2000), the large pore about the four-fold axis of the tetramer has dimensions and electrostatics resembling those of KcsA, the potassium channel (Doyle *et al.,* 1998). Indeed, all aquaglyceroporins exhibit a puzzling low density about the four-fold axis (see Fig. 5) that is related to a central cavity lined mostly with hydrophobic residues in AQP1 and GlpF. It is tempting to speculate that the pH induced chloride channel in AQP6 is related to this central pore of the tetramer (Nollert *et al.,* 2001).

Phosphorylation of specific residues in certain aquaporins may induce protein trafficking or channel gating. Several plant aquaporins exhibit a water permeability that depends on the level of phosphorylation. The water activity of the vacuolar membrane aquaporin α-TIP (Maurel *et al.,* 1995) as well as that of the plant plasma membrane aquaporin PM28A is regulated by phosphorylation (Johansson *et al.,* 1998). Putative phosphorylation sites of PM28A have been identified in the C-terminus of PM28 and in the first cytoplasmic loop (loop B) (Johansson *et al.,* 1998). AQP2 is expressed in collecting duct principal cells and contains a protein kinase A phosphorylation consensus at the C-terminus (Fushimi *et al.,* 1993). In response to vasopressin, AQP2 migrates from intracellular membrane vesicles to the apical membrane, thereby raising the surface water permeability (Nielsen *et al.,* 1995). The gene encoding AQP2 contains mutations in some forms of nephrogenic diabetes insipidus (Deen *et al.,* 1994), and AQP2 expression is perturbed in several other disorders of either water loss or water retention (reviewed in Marples *et al.,* 1999). To understand the transport mechanism responsible for fusion of the internal AQP2, carrying vesicles with the apical membrane, the structural difference of phosphorylated and dephosphorylated AQP2 tetramers must be established.

VII. Conclusions and Perspectives

The structures of AQP1 and GlpF represent a significant step toward a full understanding of aquaglyceroporin channels, and a higher resolution structure of AQP1 is expected soon. Nevertheless, the structures of further members of this protein family are required. Since crystallographic methods provide information about static structures, the flow of solutes within these channels cannot easily be measured. Molecular dynamics simulations are now in progress to obtain further clues about the dynamics of permeation.

Acknowledgments

Most of the research presented here was supported by grants from the Japan Society for the Promotion of Science (JSPS-RFTF96L00502), from the U.S. National Institutes of Health (HL33991, HL48268, EY11239), the Maurice E. Müller Foundation, the Swiss National Foundation for Scientific Research (4036–44062), and the European Union (BIO4-CT98-0024). The authors are indebted to Dr. R.M. Stroud for permission to reproduce the contents of Fig. 10, and to Dr. H. Grubmüller for Fig. 9.

References

Agre, P., Lee, M. D., Devidas, S., and Guggino, W. B. (1997). Aquaporins and ion conductance. *Science* **275,** 1490.

Agre, P. (2000). Homer W. Smith award lecture. Aquaporin water channels in kidney. *J. Am. Soc. Nephrol.* **11**(4), 764–777.

Borgnia, M. J., and Agre, P. (2001). Reconstitution and functional comparison of purified GlpF and AqpZ, the glycerol and water channels from *Escherichia coli. Proc. Nat. Acad. Sci. USA* **98,** 2888–2893.

Braun, T., Philippsen, A., Wirtz, S., Borgnia, M. J., Agre, P., Kühlbrandt, W., Engel, A., and Stahlberg, H. (2000). The 3.7 Å projection map of the glycerol facilitator GlpF: a variant of the aquaporin tetramer. *EMBO Rep.* **2,** 183–189.

Calamita, G., Bishai, W., Preston, G., Guggino, W., and Agre, P. (1995). Molecular cloning and characterization of AqpZ, a water channel from *Escherichia coli. J. Biol. Chem.* **270,** 29,063–29,066.

Chandy, G., Zampighi, G. A., Kreman, M., and Hall, J. E. (1997). Comparison of the water transporting properties of MIP and AQP1. *J. Membr. Biol.* **159,** 29–39.

Cheng, A., van Hoek, A. N., Yeager, M., Verkman, A. S., and Mitra, A. K. (1997). Three-dimensional organization of a human water channel. *Nature* **387,** 627–630.

Chrispeels, M., and Agre, P. (1994). Aquaporins: water channel proteins of plant and animal cells. *Trends Biochem. Sci.* **19,** 421–425.

Daniels, M. J., Chrispeels, M. J., and Yeager, M. (1999). Projection structure of a plant vacuole membrane aquaporin by electron cryo-crystallography. *J. Mol. Biol.* **294,** 1337–1349.

de Groot, B. L., Heymann, J. B., Engel, A., Mitsuoka, K., Fujiyoshi, Y., and Grubmüller, H. (2000). The fold of human aquaporin 1. *J. Mol. Biol.* **300,** 987–994.

de Groot, B. L., Engel, A., and Grubmüller, H. (2001). A refined structure of Aquaporin 1. *FEBS Lett.* **504,** 206–211.

Deen, P., Verdijk, M., Knoers, N., Wieringa, B., Monnens, L., van Os, C., and van Oost, B. (1994). Requirement of human renal water channel aquaporin-2 for vasopressin-dependent concentration of urine. *Science* **264,** 92–95.

Doyle, D. A., Cabral, J. M., Pfuetzner, R. A., Kuo, A., Gulbis, J. M., Cohen, S. L., Chait, B. T., and MacKinnon, R. (1998). The structure of the potassium channel: Molecular basis of K^+ conduction and selectivity. *Science* **280,** 69–77.

Echevarria, M., Windhager, E. E., and Frindt, G. (1996). Selectivity of the renal collecting duct water channel aquaporin-3. *J. Biol. Chem.* **271,** 25,079–25,082.

Engel, A., and Müller, D. J. (2000). Observing single biomolecules at work with the atomic force microscope. *Nat. Struct. Biol.* **7,** 715–718.

Finkelstein, A. (1987). "Water Movement through Lipid Bilayers, Pores, and Plasma Membranes, Theory and Reality." John Wiley and Sons, New York.

Fotiadis, D., Hasler, L., Müller, D. J., Stahlberg, H., Kistler, J., and Engel, A. (2000). Surface tongue-and-groove contours on lens MIP facilitate cell-to-cell adherence. *J. Mol. Biol.* **300,** 779–789.

Froger, A., Tallur, B., Thomas, D., and Delamarche, C. (1998). Prediction of functional residues in water channels and related proteins. *Prot. Sci.* **7,** 1458–1468.

Fu, D., Libson, A., Miercke, L. J., Weitzman, C., Nollert, P., Krucinski, J., and Stroud, R. M. (2000). Structure of a glycerol-conducting channel and the basis for its selectivity. *Science* **290,** 481–486.

Fushimi, K., Uchida, S., Hara, Y., Hirata, Y., Marumo, F., and Sasaki, S. (1993). Cloning and expression of apical membrane water channel of rat kidney collecting tubule. *Nature* **361,** 549–552.

Gerbeau, P., Guclu, J., Ripoche, P., and Maurel, C. (1999). Aquaporin Nt-TIPa can account for the high permeability of tobacco cell vacuolar membrane to small neutral solutes. *Plant J.* **18,** 577–587.

Gorin, M. B., Yancey, S. B., Cline, J., Revel, J.-P., and Horwitz, J. (1984). The major intrinsic protein (MIP) of the bovine lens fiber membrane: Characterization and structure based on cDNA cloning. *Cell* **39,** 49–59.

Gutiérrez, A. M., Gonzáles, E., Echevarría, M., Hernández, C. S., and Whittembury, G. (1995). The proximal straight tubule (PST) basolateral cell membrane water channel: Selectivity characteristics. *J. Membr. Biol.* **143,** 189–197.

Hasler, L., Walz, T., Tittmann, P., Gross, H., Kistler, J., and Engel, A. (1998). Purified lens major intrinsic protein (MIP) forms highly ordered tetragonal two-dimensional arrays by reconstitution. *J. Mol. Biol.* **279,** 855–864.

Heller, K. B., Lin, E. C., and Wilson, T. H. (1980). Substrate specificity and transport properties of the glycerol facilitator of *Escherichia coli*. *J. Bacteriol.* **144,** 274–278.

Henderson, R., Baldwin, J. M., Ceska, T. A., Zemlin, F., Beckmann, E., and Downing, K. H. (1990). Model for the structure of *Bacteriorhodopsin* based on high-resolution electron cryo-microscopy. *J. Mol. Biol.* **213,** 899– 929.

Heymann, B., and Engel, A. (2000). Structural clues in the sequences of the aquaporins. *J. Mol. Biol.* **295,** 1039–1053.

Heymann, J. B., and Engel, A. (1999). Aquaporins: Phylogeny, structure and physiology of water channels. *News Physiol. Sci.* **14,** 187–194.

Hille, B. (1992). "Ionic Channels of Excitable Membranes." Sinauer Associates Inc., Sunderland, Massachusets.

Ishibashi, K., Sasaki, S., Fushimi, K., Uchida, S., Kuwahara, M., Saito, H., Furukawa, T., Nakajima, K., Yamaguchi, Y., Gojobori, T., and Marumo, F. (1994). Molecular cloning and expression of a member of the aquaporin family with permeability to glycerol and urea in addition to water expressed at the basolateral membrane of kidney collecting duct cells. *Proc. Natl. Acad. Sci. USA* **91,** 6269–6273.

Ishibashi, K., Kuwahara, M., Gu, Y., Kageyama, Y., Tohsaka, A., Suzuki, F., Marumo, F., and Sasaki, S. (1997). Cloning and functional expression of a new water channel abundantly expressed in the testis permeable to water, glycerol, and urea. *J. Biol. Chem.* **272,** 20,782–20,786.

Ishibashi, K., Kuwahara, M., Kageyama, Y., Tohsaka, A., Marumo, F., and Sasaki, S. (1997b). Cloning and functional expression of a second new aquaporin abundantly expressed in testis. *Biochem. Biophys. Res. Commun.* **237,** 714–718.

Ishibashi, K., Kuwahara, M., Gu, Y., Tanaka, Y., Marumo, F., and Sasaki, S. (1998). Cloning and functional expression of a new aquaporin (AQP9) abundantly expressed in the peripheral leukocytes permeable to water and urea, but not to glycerol. *Biochem. Biophys. Res. Commun.* **244,** 268–274.

Jap, B. K., and Li, H. (1995). Structure of the osmo-regulated H_2O-channel, AQP-CHIP, in projection at 3.5 Å resolution. *J. Mol. Biol.* **251,** 413–420.

Johansson, I., Karlsson, M., Shukla, V., Chrispeels, M., Larsson, C., and Kjellbom, P. (1998). Water transport activity of the plasma membrane aquaporin PM28A is regulated by phosphorylation. *Plant Cell* **10,** 451–459.

Jung, J., Preston, G., Smith, B., Guggino, W., and Agre, P. (1994). Molecular structure of the water channel through aquaporin CHIP. The hourglass model. *J. Biol. Chem.* **269,** 14,648–14,654.

Kuriyama, H., Kawamoto, S., Ishida, N., Ohno, I., Mita, S., Matsuzawa, Y., Matsubara, K., and Okubo, K. (1997). Molecular cloning and expression of a novel human aquaporin from adipose tissue with glycerol permeability. *Biochem. Biophys. Res. Commun.* **241,** 53–58.

Kushmerick, C., Rice, S., Baldo, G., Haspel, H., and Mathias, R. (1995). Ion, water and neutral solute transport in *Xenopus* oocytes expressing frog lens MIP. *Exp. Eye Res.* **61,** 351–362.

Kushmerick, C., Varadaraj, K., and Mathias, R. (1998). Effects of lens major intrinsic protein on glycerol permeability and metabolism. *J. Membr. Biol.* **161,** 9–19.

Lagrée, V., Froger, A., Deschamps, S., Pellerin, I., Delamarche, C., Bonnec, G., Gouranton, J., Thomas, D., and Hubert, J. F. (1998). Oligomerization state of water channels and glycerol facilitators. Involvement of loop E. *J. Biol. Chem.* **273,** 33,949–33,953.

Lagrée, V., Froger, A., Deschamps, S., Hubert, J. F., Delamarche, C., Bonnec, G., Thomas, D., Gouranton, J., and Pellerin, I. (1999). Switch from an aquaporin to a glycerol channel by two amino acids substitution. *J. Biol. Chem.* **274,** 6817–6819.

Laize, V., Gobin, R., Rousselet, G., Badier, C., Hohmann, S., Ripoche, P., and Tacnet, F. (1999). Molecular and functional study of AQY1 from *Saccharomyces cerevisiae:* Role of the C-terminal domain. *Biochem. Biophys. Res. Commun.* **257,** 139–144.

Le Cahérec, F., Bron, P., Verbavatz, J., Garret, A., Morel, G., Cavalier, A., Bonnec, G., Thomas, D., Gouranton, J., and Hubert, J. (1996a). Incorporation of proteins into (*Xenopus*) oocytes by proteoliposome microinjection: Functional characterization of a novel aquaporin. *J. Cell Sci.* **109,** 1285–1295.

Le Cahérec, F., Deschamps, S., Delamarche, C., Pellerin, I., Bonnec, G., Guillam, M.-T., Thomas, D., Gouranton, J., and Hubert, J.-F. (1996b). Molecular cloning and characterization of an insect aquaporin. Functional comparison with aquaporin 1. *Eur. J. Biochem.* **241,** 707–715.

Li, H., Lee, S., and Jap, B. K. (1997). Molecular design of aquaporin-1 water channel as revealed by electron crystallography. *Nat. Struct. Biol.* **4,** 263–265.

Loo, D. D. F., Zeuthen, T., Chandy, G., and Wright, E. M. (1996). Cotransport of water by the Na^+/glucose cotransporter. *Proc. Natl. Acad. Sci. USA* **93,** 13,367–13,370.

Luyten, K., Albertyn, J., Skibbe, W. F., Prior, B. A., Ramos, J., Thevelein, J. M., and Hohmann, S. (1995). Fps1, a yeast member of the MIP family of channel proteins, is a facilitator for glycerol uptake and efflux and is inactive under osmotic stress. *EMBO J.* **14,** 1360–1371.

Ma, T., Frigeri, A., Hasegawa, H., and Verkman, A. S. (1994). Cloning of a water channel homolog expressed in brain meningeal cells and kidney collecting duct that functions as a stilbene-sensitive glycerol transporter. *J. Biol. Chem.* **269,** 21,845–21,849.

Marples, D., Frokiaer, J., and Nielsen, S. (1999). Long-term regulation of aquaporins in the kidney. *Am. J. Physiol.* **276,** F331–F339.

Mathai, J. C., Mori, S., Smith, B. L., Preston, G. M., Mohandas, N., Collins, M., van Zijl, P. C. M., Zeidel, M. L., and Agre, P. (1996). Functional analysis of aquaporin-1 deficient red cells. *J. Biol. Chem.* **271,** 1309–1313.

Maurel, C., Reizer, J., Schroeder, J., and Chrispeels, M. (1993). The vacuolar membrane protein γ-TIP creates water specific channels in *Xenopus* oocytes. *EMBO J.* **12,** 2241–2247.

Maurel, C., Reizer, J., Schroeder, J. I., Chrispeels, M. J., and Saier, M. H. Jr. (1994). Functional characterization of the *Escherichia coli* glycerol facilitator, GlpF, in *Xenopus* oocytes. *J. Biol. Chem.* **269,** 11,869–11,872.

Maurel, C., Kado, R. T., Guern, J., and Chrispeels, M. J. (1995). Phosphorylation regulates the water channel activity of the seed-specific aquaporin α-TIP. *EMBO J.* **14,** 3028–3035.

Meinild, A.-K., Klaerlee, D. A., and Zeuthen, T. (1998). Bidirectional water fluxes and specificity for small hydrophilic molecules in aquaporins 0-5. *J. Biol. Chem.* **273,** 32446–32451.

Mitra, A. K., van Hoek, A. N., Wiener, M. C., Verkman, A. S., and Yeager, M. (1995). The CHIP28 water channel visualized in ice by electron crystallography. *Nat. Struct. Biol.* **2,** 726–729.

Mitsuoka, K., Murata, K., Walz, T., Hirai, T., Agre, P., Heymann, J. B., Engel, A., and Fujiyoshi, Y. (1999). The structure of aquaporin-1 at 4.5-A resolution reveals short α-helices in the center of the monomer. *J. Struct. Biol.* **128,** 34–43.

Modesto, E., Lampe, P. D., Ribeiro, M. C., Spray, D. C., and Campos de Carvalho, A. C. (1996). Properties of chicken lens MIP channels reconstituted into planar lipid bilayers. *J. Membr. Biol.* **154,** 239–249.

Mulders, S. M., Preston, G. M., Deen, P. M. T., Guggino, W. B., van Os, C. H., and Agre, P. (1995). Water channel properties of major intrinsic protein of lens. *J. Biol. Chem.* **270,** 9010–9016.

Murata, K., Mitsuoka, K., Hirai, T., Walz, T., Agre, P., Heymann, J. B., Engel, A., and Fujiyoshi, Y. (2000). Structural determinants of water permeation through aquaporin-1. *Nature* **407,** 599–605.

Németh-Cahalan, K. L., and Hall, J. E. (2000). pH and calcium regulate the water permeability of aquaporin O. *J. Biol. Chem.* **275,** 6777–6782.

Nielsen, S., Chou, C.-L., Marples, D., Christensen, E. I., Kishore, B. K., and Knepper, M. A. (1995). Vasopressin increases water permeability of kidney collecting duct by inducing translocation of aquaporin-CD water channels to plasma membrane. *Proc. Natl. Acad. Sci. USA* **92,** 1013–1017.

Nollert, P., Harries, W. E. C., Fu, D., Miercke, L. J. W., and Stroud, R. M. (2001). Atomic structure of a glycerol channel and implications for substrate permeation in aqua(glycero)porins. *FEBS Lett.* **504,** 112–117.

Park, J. H., and Saier, M. H. (1996). Phylogenetic characterization of the MIP family of transmembrane channel proteins. *J. Membr. Biol.* **153,** 171–180.

Pomés, R., and Roux, B. (1996). Structure and dynamics of a proton wire: A theoretical study of H^+ translocation along the single-file water chain in the gramicidin A channel. *Biophys. J.* **71,** 19–39.

Preston, G., Jung, J., Guggino, W., and Agre, P. (1993). The mercury-sensitive residue at cysteine 189 in the CHIP28 water channel. *J. Biol. Chem.* **268,** 17–20.

Preston, G. M., and Agre, P. (1991). Isolation of the cDNA for erythrocyte integral membrane protein of 28 kilodaltons: member of an ancient channel family. *Proc. Natl. Acad. Sci. USA* **88,** 11,110–11,114.

Preston, G. M., Carroll, T. P., Guggino, W. B., and Agre, P. (1992). Appearance of water channels in *Xenopus* oocytes expressing red cell CHIP28 protein. *Science* **256,** 385–387.

Ren, G., Cheng, A., Reddy, V., Melnyk, P., and Mitra, A. K. (2000). Three-dimensional fold of the human AQP1 water channel determined at 4 Å resolution by electron crystallography of two-dimensional crystals embedded in ice. *J. Mol. Biol.* **301,** 369–687.

Ringler, P., Borgnia, M. J., Stahlberg, H., Maloney, P. C., Agre, P., and Engel, A. (1999). Structure of the water channel AqpZ from *Escherichia coli* revealed by electron crystallography. *J. Mol. Biol.* **291,** 1181–1190.

Rivers, R. L., Dean, R. M., Chandy, G., Hall, J. E., Roberts, D. M., and Zeidel, M. L. (1997). Functional analysis of nodulin 26, an aquaporin in soybean root nodule symbiosomes. *J. Biol. Chem.* **272,** 16,256–16,261.

Russ, W. P., and Engelman, D. M. (2000). The GxxxG motif: A framework for transmembrane helix–helix association. *J. Mol. Biol.* **296,** 911–919.

Saparov, S. M., Kozono, D., Rothe, U., Agre, P., and Pohl, P. (2001). Water and ion permeation of aquaporin-1 in planar lipid bilayers: Major differences in structural determinants and stoichiometry. *J. Biol. Chem.* **15,** 15.

Scheuring, S., Ringler, P., Borgnia, M., Stahlberg, H., Müller, D. J., Agre, P., and Engel, A. (1999). High resolution AFM topographs of the *Escherichia coli* water channel aquaporin Z. *EMBO J.* **18,** 4981–4987.

Schirmer, T., Keller, T. A., Wang, Y. F., and Rosenbusch, J. P. (1995). Structural basis for sugar translocation through maltoporin channels at 3.1 Å resolution. *Science* **267,** 512–514.

Schneider, T. D., and Stephens, R. M. (1990). Sequence logos: A new way to display consensus sequences. *Nucleic Acids Res.* **18,** 6097–6100.

Stahlberg, H., Braun, T., de Groot, B., Philippsen, A., Borgnia, M. J., Agre, P., Kühlbrandt, W., and Engel, A. (2000). The 6.9 Å structure of GlpF: A basis for homology modeling of the glycerol channel from *Escherichia coli. J. Struct. Biol.* **132,** 133–141.

Tsukaguchi, H., Weremowicz, S., Morton, C. C., and Hediger, M. A. (1999). Functional and molecular characterization of the human neutral solute channel aquaporin-9. *Am. J. Physiol.* **277,** F685–F696.

van Hoek, A. N., Wiener, M., Verbavatz, J.-M., Brown, D., Lipniunas, P. H., Townsend, R. R., and Verkman, A. S. (1995). Purification and structure-funtion analysis of native, PNGase F-treated, and endo-β-galactosidase-treated CHIP28 water channels. *Biochemistry* **34,** 2212–2219.

Verbavatz, J.-M., Brown, D., Sabolic, I., Valenti, G., Ausiello, D. A., Van Hoek, A. N., Ma, T., and Verkman, A. S. (1993). Tetrameric assembly of CHIP28 water channels in liposomes and cell membranes: A freeze-fracture study. *J. Cell Biol.* **123,** 605–618.

Vriend, G. (1990). WHAT IF: A molecular modeling and drug design program. *J. Mol. Graph.* **8,** 52–56.

Walz, T., Smith, B. L., Agre, P., and Engel, A. (1994a). The three-dimensional structure of human erythrocyte aquaporin CHIP. *EMBO J.* **13,** 2985–2993.

Walz, T., Smith, B. L., Zeidel, M. L., Engel, A., and Agre, P. (1994b). Biologically active two-dimensional crystals of aquaporin CHIP. *J. Biol. Chem.* **269,** 1583–1586.

Walz, T., Typke, D., Smith, B. L., Agre, P., and Engel, A. (1995). Projection map of aquaporin-1 determined by electron crystallography. *Nat. Struct. Biol.* **2,** 730–732.

Walz, T., Tittmann, P., Fuchs, K. H., Müller, D. J., Smith, B. L., Agre, P., Gross, H., and Engel, A. (1996). Surface topographies at subnanometer resolution reveal asymmetry and sidedness of aquaporin-1. *J. Mol. Biol.* **264,** 907–918.

Walz, T., Hirai, T., Murata, K., Heymann, J. B., Mitsuoka, K., Fujiyoshi, Y., Smith, B. L., Agre, P., and Engel, A. (1997). The 6Å three-dimensional structure of aquaporin-1. *Nature* **387,** 624–627.

Werten, P., Hasler, L., Koenderinkd, J. B., Klaassen, C. H. W., de Grip, W. J., Engel, A., and Deen, P. M. T. (2001). Large-scale purification of functional recombinant human aquaporin-2. *FEBS Lett.* **504,** 200–205.

Whittembury, G., Gonzalez, E., Gutierrez, A., Echevarria, M., and Hernandez, C. (1997). Length of the selectivity filter of aquaporin-1. *Biol. Cell* **89,** 299–306.

Yang, B., van Hoek, A., and Verkman, A. (1997). Very high single channel water permeability of aquaporin-4 in baculovirus-infected insect cells and liposomes reconstituted with purified aquaporin-4. *Biochemistry* **36,** 7625–7632.

Yang, B., and Verkman, A. S. (1997). Water and glycerol permeabilities of aquaporins 1-5 and MIP determined quantitatively by expression of epitope-tagged constructs in *Xenopus* oocytes. *J. Biol. Chem.* **272,** 16,140–16,146.

Yasui, M., Hazama, A., Kwon, T. H., Nielsen, S., Guggino, W. B., and Agre, P. (1999a). Rapid gating and anion permeability of an intracellular aquaporin. *Nature* **402,** 184–187.

Yasui, M., Kwon, T. H., Knepper, M. A., Nielsen, S., and Agre, P. (1999b). Aquaporin-6: An intracellular vesicle water channel protein in renal epithelia. *Proc. Natl. Acad. Sci. USA* **96,** 5808–5913.

Yool, A., Stamer, W., and Regan, J. (1996). Forskolin stimulation of water and cation permeability in aquaporin 1 water channels. *Science* **273,** 1216–1218.

Zeidel, M., Nielsen, S., Smith, B., Ambudkar, S., Maunsbach, A., and Agre, P. (1994). Ultrastructure, pharmacologic inhibition, and transport selectivity of aquaporin channel-forming integral protein in proteoliposomes. *Biochemisrty* **33,** 1606–1615.

Zeidel, M. L., Ambudkar, S. V., Smith, B. L., and Agre, P. (1992). Reconstitution of functional water channels in liposomes containing purified red cell CHIP28 protein. *Biochemisrty* **31,** 7436–7440.

Zeuthen, T., and Klaerke, D. A. (1999). Transport of water and glycerol in aquaporin 3 is gated by H(+). *J. Biol. Chem.* **274,** 21,631–21,636.

Zhang, R., van Hoek, A. N., Biwersi, J., and Verkman, A. S. (1993). A point mutation at cysteine 189 blocks the water permeability of rat kidney water channel CHIP28. *Biochemisrty* **32,** 2938–2941.

Molecular Physiology of Aquaporins in Plants

C. Maurel,[1] H. Javot, V. Lauvergeat, P. Gerbeau, C. Tournaire, V. Santoni, and J. Heyes[2]
Biochimie et Physiologie Moléculaire des Plantes, Agro-M/CNRS/INRA/UM2, F-34 060 Montpellier Cedex 1, France

In plants, membrane channels of the major intrinsic protein (MIP) super-family exhibit a high diversity with, for instance, 35 homologues in the model species *Arabidopsis thaliana*. As has been found in other organisms, plant MIPs function as membrane channels permeable to water (aquaporins) and in some cases to small nonelectrolytes. The aim of the present article is to integrate into plant physiology what has been recently learned about the molecular and functional properties of aquaporins in plants. Exhaustive compilation of data in the literature shows that the numerous aquaporin isoforms of plants have specific expression patterns throughout plant development and in response to environmental stimuli. The diversity of aquaporin homologues in plants can also be explained in part by their presence in multiple subcellular compartments. In recent years, there have been numerous reports that describe the activity of water channels in purified membrane vesicles, in isolated organelles or protoplasts, and in intact plant cells or even tissues. Altogether, these data suggest that the transport of water and solutes across plant membranes concerns many facets of plant physiology. Because of the high degree of compartmentation of plant cells, aquaporins may play a critical role in cell osmoregulation. Water uptake in roots represents a typical process in which to investigate the role of aquaporins in transcellular water transport, and the mechanisms and regulations involved are discussed.

KEY WORDS: Aquaporin, Major intrinsic protein, Osmoregulation, Permeability, Plant, Plasma membrane, Root, Vacuole, Water transport.

[1]Author to whom correspondence should be addressed.
[2]On leave from NZ Institute for Crop and Food Research, Private Bag 11 600, Palmerston North, New Zealand.

I. Introduction

Terrestrial plants critically depend on a supply of water for their growth and development and have evolved elaborate strategies to achieve water balance in the most adverse environments. This amazing capacity of plants has long been recognized, and there has been extensive physiological and biophysical research on the water relations of plants over the last decades (Boyer, 1985; Steudle, 1994). Despite this interest, the concept of water channels has not always been popular. After being discussed in the late 1950s (reviewed in Maurel, 1997), the strong case made by Dainty (1963) on artifacts due to unstirred layers convinced plant scientists that water channels can scarcely be detected in plant membranes and that they may be marginal for the physiological needs of plants. Thus the concept of water channels has long remained controversial in plant sciences and used only by a few biophysicists who referred to theoretical aqueous pores in model plant membranes (reviewed in Maurel, 1997).

Recent advances in the molecular characterization of membrane proteins, however, have triggered a very rapid evolution in this area of research. Members of the major intrinsic protein (MIP) super-family have now been found in nearly all living organisms and shown to function as channels permeable to water (aquaporins) and/or certain small nonelectrolytes (Agre *et al.,* 1998). In 1993, the first plant aquaporin was functionally characterized (Maurel *et al.,* 1993). The diversity of MIP proteins in plants, their presence in multiple subcellular compartments and their regulated expression throughout plant development and in response to environmental stimuli, have been the object of many reports. Thus, aquaporins have entered many fields of plant biology and revealed that water transport underlies many facets of plant cell functions (Maurel and Chrispeels, 2001; Santoni *et al.,* 2000). While it is widely admitted that aquaporins play an important role in plant water relations (Johansson *et al.,* 2000; Maurel and Chrispeels, 2001; Tyerman *et al.,* 1999), the central challenge remains, however, to understand the real physiological significance of these proteins.

The aim of the present article is to integrate into plant physiology what has been learned recently about the molecular and functional properties of aquaporins in plants. The expression properties of aquaporins and more generally of MIP homologues (Section II) will be used to guide the reader through some of the typical features of plant cell and tissue organization. For instance, the subcellular localization of MIPs highlights the high degree of compartmentation of plant cells, whereas tissue-specific expression of these proteins sets the stage for the intricate water transport processes which occur in plants. Toward the end of the 1990s, it has become possible to connect these expression studies to functional evidence for water channels in organelles, cells, and tissues (Section III). The link is not yet firmly established, but provides a new basis for addressing the integrated function of aquaporins in plants (Section IV).

II. Expression of Plant Aquaporins

A. Cell Level: Aquaporins and the High Degree of Compartmentation of Plant Cells

1. The High Diversity of MIPs in Plants

During the last few years, the number of *MIP* genes reported in the data base steadily increased, and plant genes largely contributed to this inflow. *MIP* genes have been identified in more than 30 plant species, including monocots and dicots. Plant genomes also appear to characteristically encode a high number of MIPs (Tyerman *et al.,* 1999; Weig *et al.,* 1997). For example, 14 full-length or partial MIP cDNAs have been molecularly characterized in the halophytic species, *Mesembryanthemum crystallinum* (Kirch *et al.,* 2000). The genomic sequence of the model species *Arabidopsis thaliana* has now been completed (Arabidopsis Genome Initiative, 2000), thus allowing for the first time a complete count of *mip* genes in a higher eukaryote. Recent data base searches run in our laboratory indicated at least 34 full-length MIP-like sequences in *Arabidopsis*. Their sequence relationship is presented in the dendrogram of Fig. 1. The encoded proteins fall into three main classes, based on their amino acid sequence: tonoplast intrinsic proteins (TIPs), plasma membrane intrinsic proteins (PIPs), and Nodulin-26-like intrinsic proteins (NIPs) (Weig *et al.,* 1997). An additional class, provisionally named small basic integral proteins (SIPs), has been proposed by Johanson and Kjellbom (2000).

Plant MIPs share 27–97% amino acid sequence identity with homologues in the same plant subfamily, but this homology can drop to ~20% when members of two distinct subfamilies are compared. Typical sequence signatures can also be defined to distinguish between plant MIPs of distinct subfamilies. For instance, Schäffner (1998) showed that the N-terminus of PIPs is more extended than that of TIPs, and that amino acid motifs in the N-terminus and the 2nd and 5th extramembrane loops (B and E) allow unambiguous discrimination between the two subfamilies. The PIPs themselves can be subdivided into the PIP1 and PIP2 subclasses (Kammerloher *et al.,* 1994; Schäffner, 1998) (Fig. 1).

The NIP subfamily of plant MIPs is formed by proteins sharing high homologies with soybean Nodulin-26 (NOD26), hence their name. NOD26 is an aquaglyceroporin, which transports water and glycerol in the peribacteroid membrane of N_2-fixing symbiotic root nodules (Dean *et al.,* 1999; Fortin *et al.,* 1987). An increasing number of NIPs has been identified and, for instance, eight homologues are present in the *Arabidopsis* genome (Fig. 1). The function, cellular localization, and expression pattern of NIPs in species that do not differentiate N_2-fixing symbiotic nodules are totally unknown. The SIP sequences have been identified *in silico,* but the genes and their products have not been experimentally characterized. In

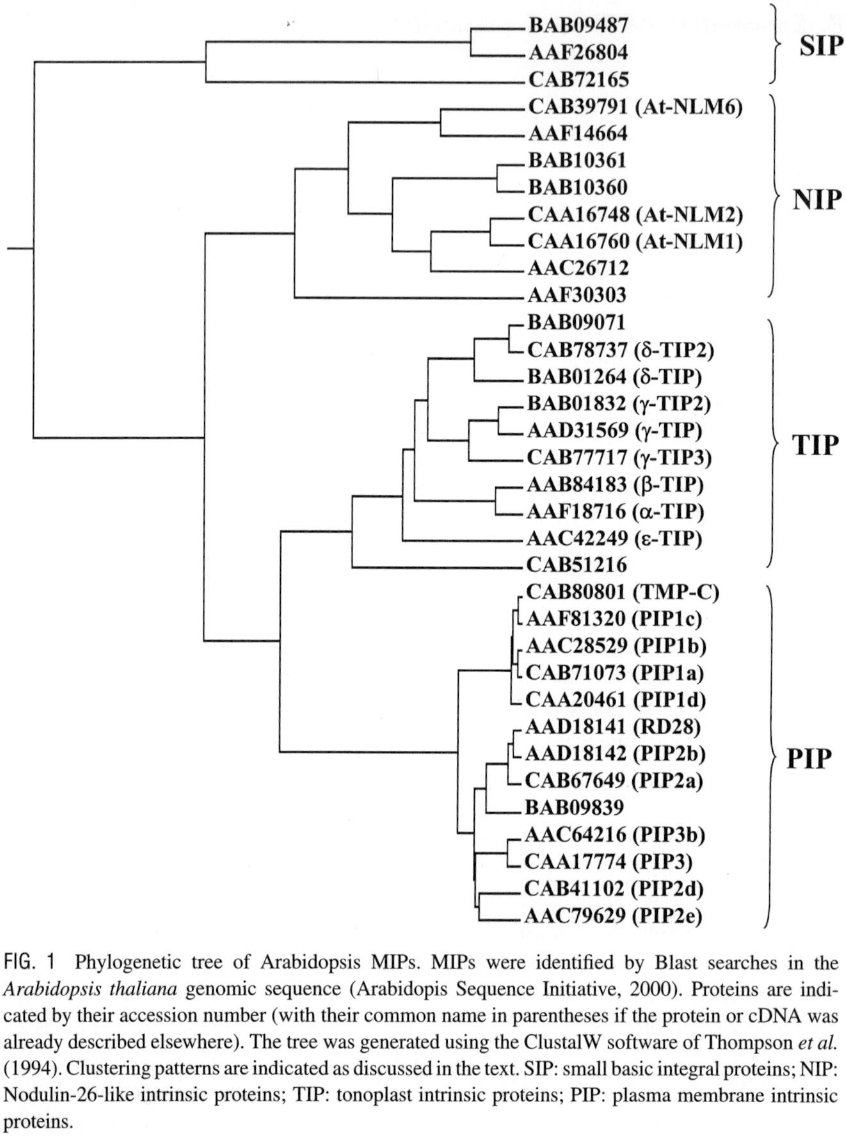

FIG. 1 Phylogenetic tree of Arabidopsis MIPs. MIPs were identified by Blast searches in the *Arabidopsis thaliana* genomic sequence (Arabidopis Sequence Initiative, 2000). Proteins are indicated by their accession number (with their common name in parentheses if the protein or cDNA was already described elsewhere). The tree was generated using the ClustalW software of Thompson *et al.* (1994). Clustering patterns are indicated as discussed in the text. SIP: small basic integral proteins; NIP: Nodulin-26-like intrinsic proteins; TIP: tonoplast intrinsic proteins; PIP: plasma membrane intrinsic proteins.

general, cDNAs of only some of the MIPs identified in *Arabidopsis* (see Table I on pages 113–115.) or other plant species have been described. An even more limited number of them have been characterized as water or solute channels by functional expression in *Xenopus* oocytes or yeast, and the precise allocation of *in planta* functions of the various genes is still awaited.

Plant MIPs possess all typical structural properties of MIPs identified in other organisms with a molecular mass of 25–30 kDa, and an internal homology between the N and C terminal halves, thus defining six putative membrane-spanning domains. The repetition of a Asn-Pro-Ala (NPA) motif in the two halves of the protein has long been taken as an unambiguous signature for MIPs, but exceptions to this rule have now been identified in plant NIPs and SIPs. Despite these variations, the Asn and Pro residues remain perfectly conserved, in agreement with a role proposed for these residues in water permeation and in stabilization of the aqueous pore, respectively (Fu *et al.,* 2000; Murata *et al.,* 2000). Recent molecular structure analyses of two plant aquaporins, bean seed α-TIP and spinach leaf PM28a (Daniels *et al.,* 1999; Fotiadis *et al.,* 2001) confirmed that these proteins have an overall molecular design that resembles that found in MIPs from animals or microorganisms (Fu *et al.,* 2000; Murata *et al.,* 2000; see Engel and Stahlberg, this volume). In particular, projection maps of 2D crystals obtained by cryoelectron microscopy indicated a typical tetrameric structure, with each monomer delineating an individual pore.

2. Differential Expression of TIPs and PIPs in Subcellular Membrane Compartments

a. Vacuolar Localization of TIPs. TIPs were initially identified because of their abundance in the vacuolar membrane, and specific isoforms have been purified and biochemically characterized from bean seeds (Johnson *et al.,* 1989), radish roots (Maeshima, 1992), or spinach leaves (Karlsson *et al.,* 2000). The tonoplast (TP) localization of these and other TIP isoforms in maize, tobacco, *Arabidopsis,* or ice plant has been assessed by immunocytolocalization studies or by immunodetection of the protein in purified subcellular membranes (Daniels *et al.,* 1996; Höfte *et al.,* 1991, 1992; Kirch *et al.,* 2000). Cross-reactivity of an antiserum raised against a TIP-enriched fraction from sugar beet with plasma membranes (PM) of pea cotyledons has been reported, but the specificity of the antibody was not established (Robinson *et al.,* 1996a).

While early studies showed consistent TP localization of all TIPs examined, even more exquisite subcellular localization of these proteins has been recently accomplished. Antibodies that can discriminate between distinct TIP isoforms have been used in pea root tips (Paris *et al.,* 1996), *Mimosa pudica* motor cells (Fleurat-Lessard *et al.,* 1997), or barley aleurone cells (Swanson *et al.,* 1998) to show the existence of vacuolar subtypes which may fulfill distinct functions within the same plant cell. This view was recently refined in work by Jauh *et al.* (1998, 1999). They showed that combinatorial expression of three basic types of TIP isoforms (i.e., α-TIP, δ-TIP, and γ-TIP) participate in the differentiation of a vast array of vacuolar subtypes. Specific expression of γ-TIP in a compartment containing the protease aleuraine would be typical of large lytic vacuoles in vegetative tissues. The expression of δ-TIP would rather be associated with vacuoles accumulating vegetative storage proteins, whereas α-TIP would be expressed in protein storage vacuoles (PSVs), when present in conjunction with δ-TIP or δ-TIP and γ-TIP. Expression of α-TIP alone would be typical of autophagic vacuoles (Jauh *et al.,* 1999). It is not clear, however, whether TIPs can only be taken as markers of intracellular membrane differentiation or whether they determine to a large extent the functional specialization of vacuolar subtypes.

The poor specificity of the anti-aquaporin polyclonal antibodies used in some early studies may, however, question some of the proposed interpretations. Paris *et al.* (1996) used a serum raised against a TP fraction purified from beet roots that recognized not only a major TIP isoform in this membrane but also larger molecular weight (MW) products (Jauh *et al.,* 1999). The specificity of anti-aquaporin antibodies has been improved, and monoclonal antibodies and affinity-purified antibodies raised against oligopeptides were used in more recent studies (Jauh *et al.,* 1998, 1999; Kirch *et al.,* 2000). Although the cross-reactivity of the antibody with respect to other homologues was checked, it remains unclear how these tools can differentiate between very close homologues within the same subfamily and, most importantly, what is the relevance of these antibodies for detecting putative homologues in distant plant species. These homologues may have different cellular and subcellular expression patterns (Bethke and Jones, 2000).

b. Localization of PIPs. Initial studies on the localization of PIPs in *Arabidopsis* using membrane fractions purified by aqueous two-phase partitioning suggested that these proteins were mostly expressed in the PM (Daniels *et al.,* 1994; Kaldenhoff *et al.,* 1995; Kammerloher *et al.,* 1994). The subcellular localization of three PIPs (MIP-A, MIP-B, MIP-C) in *M. crystallinum* has recently been investigated by immunodetection of the proteins in membrane fractions separated by sucrose density gradient centrifugation (Barkla *et al.,* 1999; Kirch *et al.,* 2000). Surprisingly, three distinct patterns were revealed using antibodies raised against each of the proteins. In particular, proteins immunoreactive to anti-MIP-A antibodies were totally absent from the PM fraction but abundant in a fraction of intermediate density, distinct from the TP and PM. MIP-C was weakly detected in the PM but strongly in TP. These experiments questioned the general idea that PIPs are mostly expressed in the PM. The presence of PIPs in intracellular membranes may not be surprising and may simply reflect their routing to the PM throughout the secretory system. Barkla *et al.* (1999) and Kirch *et al.* (2000) have hypothesized, however, the existence of a vesicle shuttling mechanism which, like that described for AQP2 in the kidney (Sasaki *et al.,* 1998), would allow a regulated exposure of PIPs at the cell surface.

In another study (Robinson *et al.,* 1996b), antibodies raised against *Arabidopsis* PIP1 allowed the immunolocalization of PIP homologues in convoluted invaginations of *Arabidopsis* leaf PM. These structures called plasmalemmosomes may favor direct exchanges between vacuole and apoplasm. Alternatively, they may correspond to the cycling of endomembrane vesicles between the cell surface and the cell interior. But here again, dynamic evidence for this model in plants is lacking, and signals that allow a regulated trafficking of aquaporins in plant cells remain to be identified.

In contrast to the use of polyclonal antibodies which possibly cross-react with close aquaporin homologues, expression of aquaporins fused to reporter proteins such as green fluorescent protein (GFP) in transgenic plants provides an unambiguous approach for localization of a well-identified aquaporin isoform. Cutler *et al.* (2000) made random fusions between plant proteins and GFP with the aim of identifying labeling markers of specific subcellular structures. Among proteins that yielded typical labeling of the PM or the TP, a majority (3/5 for the PM; 3/3 for the TP) were isoforms of PIPs and TIPs. GFP fusions have also been used to determine the subcellular localization of two PIP homologues in maize (Chaumont *et al.,* 2000). Expression of the fusion proteins was detected at the cell surface but substantial labeling of a perinuclear compartment, corresponding most probably to the endoplasmic reticulum (ER), was also observed.

c. How Do TIPs and PIPs Reach Their Respective Compartments? Because they form a family of membrane proteins with distinct subcellular localization properties, aquaporins provide interesting models for investigating the molecular bases of membrane protein biogenesis and targeting in plants cells. However, these studies have proved difficult because aquaporins are polytopic membrane proteins, the mutation of which can deeply interfere with insertion and stability within the membrane. Nevertheless, Höfte and Chrispeels (1992) showed that a region containing the sixth transmembrane domain of bean α-TIP with or without the cytoplasmic tail is sufficient to allow localization of a reporter protein in the TP of transformed tobacco cells. Jiang and Rogers (1998) dissected these mechanisms in more detail and studied two TIP isoforms, γ-TIP and αTIP, which are targeted to lytic vacuoles and PSVs, respectively. The cytoplasmic tail of α-TIP was able to redirect a chimeric construct containing the plant vacuolar sorting receptor BP-80 fused to a reporter protein from the lytic vacuole to PSVs, whereas the cytoplasmic tail of γ-TIP was without effect. They concluded the existence of two sorting pathways for plant TP membrane proteins which would contribute to the biogenesis of distinct vacuolar compartment in plant cells. One pathway would direct proteins to lytic vacuoles via the Golgi complex, whereas the second path would be a direct route from the ER to PSVs.

B. Tissue Level

Because they showed a high expression level, a specialized tissue-specific expression pattern, and/or responsiveness to hormonal or environmental factors, many

MIP genes of plants were identified and their expression described, well before the function of their products was understood (Yamaguchi-Shinozaki *et al.,* 1992; Yamamoto *et al.,* 1991). Tables I and II summarize studies performed in *Arabidopsis* and other species, respectively, and show that data on MIP expression in more than 30 plant species are available. It should be stressed, however, that, in most early studies, the existence of an extended multigene *MIP* family had not been suspected and that the specificity of the antibodies and cDNA probes used was not assessed. It is now clear that full-length cDNAs do cross-hybridize with mRNAs of close MIP homologues of the same subfamily. For instance, the five *PIP1* genes in *Arabidopsis* share >75% nucleotidic sequence homology. The 5′ and 3′ untranslated ends of aquaporin genes may not even allow the distinction of very close homologues. The poor specificity of some of the anti-aquaporin antibodies used in plants has been discussed above. Thus, depending on the technique or the probe used for the experiments, the specificity of the expression profile should be considered with caution.

Because of their function in water uptake, roots have been investigated in detail with respect to aquaporin expression. Reporter gene activities (i.e., GFP or GUS) driven by aquaporin promoters, aquaporin mRNAs or proteins have been detected in root tips (TobRB7: Yamamoto *et al.,* 1991), epidermis and root hairs (ZmTIP1: Barrieu *et al.,* 1998a; MipA: Kirch *et al.,* 2000), cortex (MipA: Yamada *et al.,* 1995), endodermis (ZmTIP1: Barrieu *et al.,* 1998a; MipA: Yamada *et al.,* 1995), and in the stele (TobRB7: Yamamoto *et al.,* 1991). The various cell types present in roots have distinct sizes and cytoplasmic density, which makes a comparison of their respective aquaporin expression levels difficult. Yet, studies in *Arabidopsis* and in the ice plant (Schäffner, 1998; Yamada *et al.,* 1995) suggested that PIP homologues would have an expression level higher in the stele than in the cortex. This is consistent with the uptake path of water that converges from the root surface toward the stele for mobilization into the whole plant.

Consistent with this, high expression levels of several PIP and TIP aquaporins in the vascular tissues of roots and shoots have been reported in many plant species. For instance, ZmTIP was expressed in the parenchyma cells that surround the early and late metaxylem of maize roots and stems (Barrieu *et al.,* 1998a), whereas a δ-TIP homologue of sunflower accumulated in root phloem tissues (Sarda *et al.,* 1997, 1999). In *Arabidopsis* leaves, PIP1b (AthH2) was strongly expressed in the parenchyma cells around vascular bundles and within the bundles themselves, in the protoxylem and protophloem (Kaldenhoff *et al.,* 1995). These expression patterns suggest a critical role of transmembrane water transport in loading and unloading phloem and xylem vessels, for continuous circulation between the two types of vessels, and in multidirectional transport of sap throughout the plant.

Although the expression of some aquaporins such as *Arabidopsis* γ-TIP or spinach So-δ-TIP is strictly excluded from dividing cells (Karlsson *et al.,* 2000; Ludevid *et al.,* 1992), specific expression of other isoforms in meristematic cells

TABLE I
Expression Properties of MIP Homologues in *Arabidopsis*

MIP homologue[a]	EST[b]	Expression in tissues	Dependence on stimuli[c]	Technique[d]	Ref.[e]
BAB09487	+				
AAF26804	+				
CAB72165	+				
CAB39791 (At-NLM6)	+				
AAF14664	+				
BAB10361	–				
BAB10360	–				
CAA16748 (At-NLM2)	+				
CAA16760 (At-NLM1)	+	Seedling	*Down:* ABA, NaCl, dark, dehydration; *Stable:* mannitol	QP	(1)
AAC26712	–				
AAF30303	+				
BAB09071	+				
CAB78737 (δ-TIP2)	+	Root		SB (cDNA)	(1)
BAB01264 (δ-TIP)	+	Vascular tissues of shoots		GUS	(2)
		Leaf, bolt	*Down:* mannitol; *Stable:* ABA, NaCl, dark, dehydration	SB (cDNA), QP	(1)
BAB01832 (γ-TIP2)	+	Leaf > root and bolt		SB (cDNA)	(1)
AAD31569 (γ-TIP)	+	Hypocotyl, root, vascular tissues, leaf elongating cells		GUS, IH (cDNA)	(3)
		Root, stem, leaf, flower		SB (cDNA), ID (γ-TIP)	(4)
		Stem, flower > root > leaf	*Up:* GA3	NB (3′ cDNA end)	(5)
		Root, bolt	*Down:* mannitol; *Stable:* ABA, NaCl, dark, dehydration	SB (cDNA), QP	(1)

(continued)

TABLE I (*continued*)

MIP homologue[a]	EST[b]	Expression in tissues	Dependence on stimuli[c]	Technique[d]	Ref.[e]
CAB77717 (γ-TIP3)	–				
AAB84183 (β-TIP)	+				
AAF18713 (α-TIP)	+	Embryo, seed		GUS, IH and SB (cDNA), ID (TP25 bean)	(3, 4)
AAC42249 (ε-TIP)	+				
CAB51216	–				
CAB80801 (TMP-C)	+	Flower and stem > root and leaf		NB (cDNA); SB (cDNA)	(1, 6)
AAF81320 (PIP1c)	+	Leaf, root	*Up:* heat-shock; *Stable:* dehydration	NB (3′ or 5′ cDNA ends), SB (cDNA)	(1, 7)
AAC28529 (PIP1b)	+	Root, leaf	*Up:* heat-shock; *Stable:* dehydration	NB (3′ or 5′ cDNA ends)	(7)
	+	Elongating and differentiating tissues, vascular tissues, guard cells, stamens, young siliques	*Up:* light (blue >white> red) *Up:* ABA, GA; *Stable:* IAA, heat shock, wounding	NB (cDNA) GUS, IH, ID (Nter PIP1b peptide)	(8–10)
		Leaf and root > bolt and flower		SB (cDNA)	(1)
		Pip1b>pip2a>pip1a>pip2b in each organ (light)		RP (3′ end)	(11)
CAB71073 (PIP1a)	+	Root > leaf	Dehydration	NB (3′ or 5′ cDNA ends)	(7)
		Leaf > root		SB (cDNA)	(1)
		Pip1b>pip2a>pip1a>pip2b in each organ (light)		RP (3′ end)	(11)
CAA20461 (PIP1d)	+				
AAD18141 (RD28)	+	Whole plant	*Up:* dehydration; *Stable:* ABA	NB (cDNA)	(12)
		Whole plant except seeds	*Stable:* dehydration	ID (RD28, PIP2A and PIP2B)	(13)
		Pip2b>rd28>pip2a in root	*Down:* mannitol; *Stable:* ABA, NaCl, dark, dehydration	SB (cDNA), QP	(1)

AAD18142 (PIP2b)	+	Root, leaves	*Up:* heat-shock; *Stable:* dehydration	NB (3′ or 5′ cDNA ends)	(7)
		Pip2b>rd28>pip2a in root		SB (cDNA)	(1)
		Pip1b>pip2a>pip1a>pip2b in each organ (light)		RP (3′ end)	(11)
CAB67649 (PIP2a)	+	Root, leaves	*Stable:* heat-shock, dehydration	NB (3′ or 5′ cDNA ends)	(7)
		Pip2b>rd28>pip2a in root		SB (cDNA)	(1)
		Pip1b>pip2a>pip1a>pip2b in each organ (light)		RP (3′ end)	(11)
BAB09839	+				
AAC64216 (PIP3b)	+				
CAA17774 (PIP3)	+	Flower and stem >root and leaf		NB (cDNA)	(6)
		Cell culture, flower and silique > leaf and bolt		SB (cDNA)	(1)
CAB41102 (PIP2d)	+				
AAC79629 (PIP2e)	+	Leaf		SB (cDNA)	(1)

[a]MIP homologues are indicated by their protein accession number (with their common name in parentheses if the protein or cDNA has already been described).

[b]The presence or the absence of EST in data bank is indicated by (+) or (−), respectively.

[c]Stimuli that up-regulate (*Up*), down-regulate (*Down*), or do not alter (*Stable*) MIP expression are indicated.

[d]The techniques used to characterize gene or protein expression were as follows: GUS, promoter: GUS fusion; ID, immunodetection; IH, *in situ* hybridization; NB, northern blot; QP, quantitative PCR; RP, RNAse protection assay; SB, slot-blot. Proteins or peptides used to raise antibodies or DNA probes used in hybridization experiments are indicated in parentheses. Full-length cDNAs are referred to as cDNA and probes with the 3′ or 5′ ends of cDNAs are mentioned.

[e]References: (1) Weig *et al.,* 1997; (2) Daniels *et al.,* 1996; (3) Ludevid *et al.,* 1992; (4) Höfte *et al.,* 1992; (5) Phillips and Huttly, 1994; (6) Utsugi *et al.,* 1996; (7) Kammerloher *et al.,* 1994; (8) Kaldenhoff *et al.,* 1993; (9) Kaldenhoff *et al.,* 1995; (10) Kaldenhoff *et al.,* 1996; (11) Grote *et al.,* 1998; (12) Yamaguchi-Shinozaki *et al.,* 1992; (13) Daniels *et al.,* 1994.

TABLE II
Expression Properties of MIP Homologues in Various Plant Species

Plant species	MIP homologue[a]	Homologue in *Arabidopsis*[b]	Expression in tissues	Dependence on stimuli[c]	Technique[d]	Ref.[e]
Antirrhinum sp.	DIP	δ-TIP2	Cotyledons	*Up:* dark; *Stable:* GA3	QP, IH (cDNA)	(1)
Beta vulgaris	PIP1 and PIP2like	PIP1, PIP2	Storage tissues		ID (Cter PIP *Arabidopsis*)	(2)
	BPM2, BPM3	PIP1	Storage tissues, stems, leaves		NB (cDNA)	(3)
Brassica napus	BnPIP1	PIP2a	Primed seeds, germinating seeds	*Up:* priming with ABA or water	NB (cDNA)	(4)
	BnγTIP2	γ-TIP2	Germinating seeds	*Stable:* priming	NB (cDNA)	(4)
Brassica oleracea	MIPa	PIP1b	Stamen, sepals, petals, carpels	*Up:* drought	NB (cDNA: full length and 3′ end)	(5)
	MIPb	PIP1b	Stamen, sepals, petals, carpels, roots	*Stable:* drought	NB (3′ cDNA end)	(5)
	BobTIP26-1	γ-TIP	Meristems, vascular bundles	*Up:* drought	NB (3′ cDNA end), ID (BobTIP26)	(6, 7)
	BobTIP26-2	γ-TIP	Meristems, vascular bundles, elongating tissues	*Up:* drought	NB and IH (3′ cDNA end), ID (BobTIP26)	(6, 7)
Brassica rapa	MOD	PIP1b	Stigma, leaves, anthers	*Stable:* pollination	NB (partial cDNA)	(8)
Craterostigma sp.	CpPIPa2	PIP1	Leaves, roots, calli	*Up:* drought, ABA	NB (5′ and 3′ cDNA ends)	(9)
	CpPIPa6	PIP1	Leaves, roots	*Up:* drought; *Stable:* ABA	NB (5′ and 3′ cDNA ends)	(9)
	CpPIPa7	PIP1	Roots	*Up:* drought; *Stable:* ABA	NB (5′ and 3′ cDNA ends)	(9)
	CpPIPb	PIP1	Leaves, roots, calli	*Stable:* drought, ABA	NB (5′ and 3′ cDNA ends)	(9)
	CpPIPc	PIP3	Leaves, roots, calli	*Up:* drought; *Stable:* ABA	NB (5′ and 3′ cDNA ends)	(9)
	CpTIP	TIP	Leaves, calli	*Down:* drought, ABA		(9)
Cucurbita sp.	VM23 like	γ-TIP1/2	Hypocotyls, cotyledons		ID (VM23)	(10, 11)
	MP23, MP28	α/β-TIP	Seedlings		ID (MP28 and MP23), NB (cDNA)	(12, 13)

Glycine max	TIP	α-TIP	Cotyledons, pro-vascular cells		ID (TP25)	(14, 15)
	NOD26	NLM1/2	Infected cells of root nodules		IP (MPB proteins), EP (5′ cDNA), GUS	(16, 17)
	SPCP1	γ-TIP	Root elongation zone, vegetative tissues		PE (5′ cDNA end)	(17)
	γ-TIP	γ-TIP1/2	Inner cortex and pericycle of roots, transfer cells		ID (VM23)	(18)
	δ-TIPlike	δ-TIP	Leaves	*Up:* depodding	ID (δ-TIP peptide)	(19)
	γ-TIPlike	γ-TIP	Leaves	*Stable:* depodding	ID (γ-TIP peptide)	(19)
Gossypium sp.	δ-TIP	δ-TIP	Etiolated cotyledons, elongating fiber cells		RT-PCR	(20)
Helianthus sp.	SunTIP7	δ-TIP	Leaves, stems, guard cells, phloem, cortical tissues	*Up:* drought, light	NB, IH (3′ cDNA end)	(21, 22)
	SunTIP18	δ-TIP	Roots and stem phloem	Down: drought	NB, IH (3′ cDNA end)	(22)
	SunTIP20	δ-TIP	Leaves, guard cells	*Up:* drought, light	NB, IH (3′ cDNA end)	(21, 22)
	SunRB7	δ-TIP2	Roots	*Stable:* drought	NB (cDNA)	(22)
	Sunγ-TIP	γ-TIP	Vegetative organs	*Stable:* drought	NB (cDNA)	(22)
Lotus sp.	PIP-like	PIP1,/2	Roots	*Up:* light	NB (AtPIP1a, 2b), ID (AthPIP1a)	(23)
	PIP1-like	PIP1	Roots	Down: nitrate starvation	NB (AtPIP1a)	(24)
	LIMP1	γ-TIP1/2	Nodules, roots		NB (partial cDNA)	(25)
	LIMP2	NLM1/2	Nodules		NB (partial cDNA)	(25)
Lycopersicon sp.	TRAMP1	PIP1	Pericarp, leaves	*Up:* ripening, senescence, ethylene	NB (partial cDNA)	(26)
	TRAMP1	PIP1	Roots, stems, leaves, fruits	*Up:* drought; *Stable:* ABA	NB (partial cDNA)	(27)
	δ-TIPlike	δ-TIP	Leaf epidermis, bundle sheath cells, petals		ID (δ-TIP peptide)	(19)
Medicago truncatula	MtAQP1	γ-TIP1/2	Roots	*Up:* mycorrhiza	NB (partial cDNA)	(28)
Mesembryanthemum crystallinum	MIPA	PIP1	Root epidermis (root hair), root and stem vascular tissues	*Stable:* salt stress	NB, IH (cDNA), ID (MIPA peptide)	(29, 30)

(continued)

TABLE II (*continued*)

Plant species	MIP homologue[a]	Homologue in *Arabidopsis*[b]	Expression in tissues	Dependence on stimuli[c]	Technique[d]	Ref.[e]
	MIPB	PIP1	Root cortex, root hairs, stem vascular tissues	*Stable:* salt stress	NB, IH (cDNA), ID (MIPB peptide)	(29, 30)
	MIPC	PIP2	Root elongation zone	*Up:* salt stress	NB (cDNA), ID (MIP C peptide)	(29, 30)
	MIPF	γ-TIP1/2	Roots, mesophyll cells, vascular tissues	*Down:* salt stress	ID (MIP F peptide)	(30)
Mimosa pudica	VM23like	γ-TIP1/2	Pulvini, petioles		ID (VM23)	(31)
Nicotiana excelsior	NeMIP1	PIP1	Roots, flowers	*Up:* salt, drought	NB (3′cDNA end)	(32)
	NeMIP2 and 3	PIP1	Roots, flowers, leaves	*Up:* salt, drought, light	NB (3′cDNA end)	(32)
Nicotiana tabacum	tobRB7	δ-TIP2	Root meristem, immature vascular cylinder, giant cells, leaf		NB and IH (cDNA), *GUS*	(33–35)
	Nt-TIPa	ε-TIP	Suspension cells		cDNA cloning	(36)
	NtAQP1	PIP1	Root apex, vascular tissues, leaf spongy parenchyma		NB and IH (5′ cDNA), ID (N_{ter} NtAQP1)	(37)
Oryza sativa	rTIP1	γ-TIP	Shoots, roots	*Up:* mannitol, NaCl, ABA	NB (cDNA)	(38)
	rMIP1	NLM1/2	Shoots	*Stable:* drought and salt stress	NB (cDNA)	(38)
	OsPIP2a	PIP3/3b	Meristematic, elongating,and differentiating cells–seedlings, leaves	*Up:* GA3; *Down:* ABA, drought, light	NB (3′ cDNA end)	(39)
	OsPIP1a	PIP1	Meristematic, elongating, and differentiating cells–seedlings, leaves	*Up:* GA3; *Down:* ABA, drought	NB (3′ cDNA end)	(39)
	RWC1	PIP1		*Down:* mannitol, NaCl, chilling	NB (cDNA), CP (3′ cDNA end)	(40)
Petroselinum sp.	PcRB7	δ-TIP	Root arbuscular mycorrhiza	*Up:* mycorrhiza, pathogen	NB (partial cDNA)	(41)
Phaseolus vulgaris	TP25	α-TIP	Cotyledons, embryonic axis		ID (TP25)	(14)
Picea abies	TIP	TIP	Root and hypocotyl of seedlings, somatic and zygotic embryos		ID (TP25)	(42)
Pisum sativum	α-TIP	α-TIP	Cotyledons		ID (TP25)	(43)
	γ-TIP	γ-TIP	Cotyledons		ID (TIP-Ma27)	(43)
	trg31	PIP1	Roots, shoots, young leaves and flowers, trichomes, pollen	*Up:* wilting, mannitol, heat shock, ABA	NB (cDNA), RO, GUS	(44–46)

Raphanus sativus	γ-VM23	γ-TIP1/2	Root, hypocotyl, cotyledon, immature tap root, petiole, leaf	*Down:*light; *Up:* light	ID (VM23), NB (5′ cDNA end)	(10, 47)
	δ-VM23	δ-TIP	Hypocotyls, immature tap root, petiole, leaf (vein)	*Down:* light	NB (5′ cDNA end)	(47)
Spinacia oleracea	PM28a	PIP3	Roots, petiole, leaf veins		NB (partial cDNA)	(48)
	PM28b	PIP1	Leaves		cDNA cloning	(48)
	δ-TIP	δ-TIP	Leaves, petiole, roots		ID (So-δ-TIP peptide)	(49)
Tulipa gesneriana	TIP1	γ-TIP	Vascular tissues, elongationg stalk		QP	(50)
Zea mays	ZmPIP1a	PIP1	Root tip	*Stable:* glucose starvation	NB (cDNA)	(51)
	CHEM8	γ-TIP	Leaves	*Up:* salt stress, heat shock	NB (cDNA)	(52)
	ZmTIP1	γ-TIP	Root meristem, primordia, vascular bundles, epidermis, endodermis, nucellus, pedicels		NB, IH (3′ cDNA end)	(53, 54)
	ZmPIP1a	PIP1	Shoots, roots, leaves, tassels		NB (3′ cDNA end)	(55)
	ZmPIP1b	PIP1	Whole plant, especially developing tassel		NB (3′ cDNA end)	(55)
	ZmPIP2a	PIP2a	Roots		NB (3′ cDNA end)	(55)

[a] Name of MIP homologue as indicated in the original reference.

[b] Closest MIP homologue in *Arabidopsis* or name of the corresponding subfamily.

[c] Stimuli that up-regulate (*Up*), down-regulate (*Down*), or do not alter (*Stable*) MIP expression are indicated.

[d] The techniques used to characterize gene or protein expression were as follows: CP, competitive PCR; GUS, promoter: GUS fusion; ID, immunodetection; IH, *in situ* hybridization; IP, immunoprecipitation; NB, northern blot; PE, primer extension analysis; QP, quantitative PCR; RO, run-on in isolated nuclei. Proteins or peptides used to raise antibodies or DNA probes used in hybridization experiments are indicated in parentheses. Full-length cDNAs are referred to as cDNA and probes with the 3′ or 5′ ends of cDNAs are mentioned.

[e] References: (1) Culianez-Macia and Martin, 1993; (2) Qi *et al.,* 1995; (3) Barone *et al.,* 1998; (4) Gao *et al.,* 1999; (5) Ruiter *et al.,* 1997; (6) Barrieu *et al.,* 1998b; (7) Barrieu *et al.,* 1999; (8) Ikeda *et al.,* 1997; (9) Mariaux *et al.,* 1998; (10) Maeshima, 1992; (11) Maeshima *et al.,* 1994; (12) Inoue *et al.,* 1995a; (13) Inoue *et al.,* 1995b; (14) Johnson *et al.,* 1989; (15) Melroy and Herman, 1991; (16) Fortin *et al.,* 1987; (17) Miao *et al.,* 1992; (18) Serraj *et al.,* 1998; (19) Jauh *et al.,* 1998; (20) Ferguson *et al.,* 1997; (21) Sarda *et al.,* 1997; (22) Sarda *et al.,* 1999; (23) Henzler *et al.,* 1999; (24) Clarkson *et al.,* 2000; (25) Guenther and Roberts, 2000; (26) Davies and Grierson, 1989; (27) Fray *et al.,* 1994; (28) Krajinski *et al.,* 2000; (29) Yamada *et al.,* 1995; (30) Kirch *et al.,* 2000; (31) Fleurat-Lessard *et al.,* 1997; (32) Yamada *et al.,* 1997; (33) Conkling *et al.,* 1990; (34) Yamamoto *et al.,* 1991; (35) Opperman *et al.,* 1994; (36) Gerbeau *et al.,* 1999; (37) Otto and Kaldenhoff, 2000; (38) Liu *et al.,* 1994; (39) Malz and Sauter, 1999; (40) Li *et al.,* 2000; (41) Roussel *et al.,* 1997; (42) Oliviusson and Hakman, 1995; (43) Hoh *et al.,* 1995; (44) Guerrero *et al.,* 1990; (45) Guerrero and Crossland, 1993; (46) Jones and Mullet, 1995; (47) Higuchi *et al.,* 1998; (48) Johansson *et al.,* 1996; (49) Karlsson *et al.,* 2000; (50) Balk and de Boer, 1999; (51) Chevalier *et al.,* 1995; (52) Didierjean *et al.,* 1996; (53) Chaumont *et al.,* 1998; (54) Barrieu *et al.,* 1998a; (55) Chaumont *et al.,* 2000.

has been observed (Barrieu *et al.,* 1998b; Chaumont *et al.,* 1998). In these cells, vacuolar aquaporins may participate in the biogenesis of the vacuolar apparatus and contribute to the adjustments of vacuolar surface and volume that are required for differentiation of a large central vacuole.

Expression of MIPs in most aerial parts of plants, such as hypocotyls (Higuchi *et al.,* 1998), stems and petioles (Otto and Kaldenhoff, 2000), and cotyledons and leaves (Johansson *et al.,* 1996), has also been reported, with a preferential expression in vascular tissues and in cells that undergo rapid elongation and differentiation (Chaumont *et al.,* 1998; Higuchi *et al.,* 1998; Kaldenhoff *et al.,* 1995; Ludevid *et al.,* 1992). Highly specialized cells such as trichomes (Jones and Mullet, 1995), motor cells of *Mimosa pudica* pulvini (Fleurat-Lessard *et al.,* 1997), and guard cells (Kaldenhoff *et al.,* 1995; Sarda *et al.,* 1997) also express specific aquaporin isoforms. Water transport in the two latter cell types plays a crucial role in cell movements.

Flowers exhibit exquisite patterns of aquaporin expression (Jones and Mullet, 1995; Ludevid *et al.,* 1992; Yamada *et al.,* 1997). Plant reproduction requires subtle water exchanges to achieve pollen grain desiccation (maturation) (Ruiter *et al.,* 1997), imbibition on the stigma surface of pollinated flowers, and dramatic growth of the pollen tube. Flower expansion and blooming itself also require accurate control of water relations in sepals and petals. Fruit and seeds also express specific aquaporin isoforms (Fray *et al.,* 1994; Johnson *et al.,* 1989).

C. Whole Plants: Aquaporins in a Changing Environment

Because of their lack of mobility, plants have to adapt constantly to subtle or drastic changes of their environment. For this, they have developed an amazing capacity to perceive the nature and intensity of a large variety of environmental signals. Many of these changes can interfere with plant water relations and, accordingly, with expression of aquaporins.

1. Abiotic Factors

Among all abiotic factors that potentially interfere with aquaporin expression, water and salt stresses are those which have been most investigated. For instance, thorough studies in the model halophytic plant *M. crystallinum* revealed a coordinated and transient decrease in the expression of three aquaporin genes in the 24–48 hr following salt exposure (Yamada *et al.,* 1995). Aquaporin genes that are up- or down-regulated by drought stress and osmotic stress have been identified in numerous species such as *Arabidopsis,* sunflower, cauliflower, and rice (Barrieu *et al.,* 1999; Liu *et al.,* 1994; Sarda *et al.,* 1999; Yamaguchi-Shinozaki *et al.,* 1992). These regulations may allow efficient mobilization of water in tissues whose metabolism is critical for the plant, or they may accompany the deep alteration of plant cell and tissue morphology observed upon prolonged stress (Barrieu *et al.,* 1999; Kirch *et al.,* 2000). The role of aquaporins under extreme desiccation conditions can typically be explored in resurrection plants such as *Craterostigma plantagineum* (Mariaux *et al.,* 1998).

Cold stress results in a water deficit, and a role of a rice PIP1 homologue (RWC1) in chilling tolerance has been proposed (Li *et al.,* 2000), based on the observation that both an osmotic challenge (0.5 M mannitol) and a cold stress (4°C) decreased, independently and synergistically, the amount of RWC1 mRNA. Aquaporins may also be involved in the response of plants to heat shock, since a 38–42°C treatment dramatically enhanced the expression of the CHEM8 aquaporin in maize (Didierjean *et al.,* 1996).

The expression of many plant aquaporins can also be determined by the light conditions during growth. Expression of mRNAs encoding PIP1 homologues in *Lotus japonicus* roots (Henzler *et al.,* 1999) and a δ-TIP homologue in sunflower guard cells (Sarda *et al.,* 1997) showed a marked increase during daytime. These regulations may be related to diurnal variations in root hydraulic conductivity and stomatal aperture in these plants. The *pip1b* gene of *Arabidopsis* is specifically induced by blue light, a signal for floral induction (Kaldenhoff *et al.,* 1993). In contrast, other aquaporins are up-regulated in etiolated tissues, a pattern which may be related to the preferential expression of these aquaporins in elongating cells (Higuchi *et al.,* 1998; Ludevid *et al.,* 1992).

2. Hormones and Biotic Factors

Aquaporin expression is controlled by endogenous hormonal factors. For instance, the stress hormone abscisic acid (ABA), which plays a crucial role in the response of plants to drought or in the maturation of seeds, is known to regulate aquaporins in various species (Gao *et al.,* 1999; Kaldenhoff *et al.,* 1993; Mariaux *et al.,* 1998). Note, however, that some aquaporin genes are induced by drought through an ABA-independent pathway (Fray *et al.,* 1994; Mariaux *et al.,* 1998). Induction of *Arabidopsis* γ-TIP by gibberellic acid, a hormone that promotes cell expansion, is consistent with the expression pattern of this aquaporin, specifically in the elongating zones of roots and stems (Phillips and Huttly, 1994).

Interaction of plants with symbiotic or pathogenic organisms is mediated by biotic signals, many of which remain to be discovered. Arbuscular miccorhiza or N_2-fixing nodules in legumes illustrate some of the deep morphological alterations that plant roots can undergo to accommodate the presence of symbionts. Plant-encoded aquaporins are specifically expressed in these organs and have been localized in the highly compartmented vacuole of arbuscule cells (Roussel *et al.,* 1997) and in the symbiosome membrane that surrounds N_2-fixing bacteroids (Fortin *et al.,* 1987). These aquaporins probably optimize nutrient and water exchange between the two symbiotic partners. They may also permit efficient osmoregulation of the highly compartmented root cells. Plant infection by root-knot nematodes results in a dramatic increase in size for a few root cells that form a feeding site for the parasite. Nutrient and water uptake by the parasite from these reservoirs seems to be facilitated by induction of aquaporin TobRB7 (Opperman *et al.,* 1994). A promoter region of the *tobRB7* gene that is specifically responsive to the pathogen has been identified.

In conclusion, aquaporin expression has been found in virtually all plant tissues examined. Strong aquaporin expression typically occurs in the root cortex, vascular bundles of shoots and roots, and elongating tissues, consistent with the idea that these tissues have special needs for intense water exchange. The diversity of aquaporin expression patterns and regulation in other tissues has shown, however, that very fine adjustments of water transport are needed in more physiological situations than was initially suspected.

Despite the wealth of data collected over recent years, we are far from a complete picture of aquaporin expression properties in plants. This is due mostly to the very high diversity of aquaporin genes in plants. Extensive characterization of all aquaporins expressed in a given plant cell type is lacking, and coexpression of several aquaporin isoforms in the same cell type seems to be common, which makes expression patterns even more difficult to decipher. Basic quantitative questions also remain difficult to answer unambiguously. Which aquaporin isoforms are the most strongly expressed? Do some organs, tissues, and cell types have a higher aquaporin content than others?

The use of a restricted number of models such as *Arabidopsis* or maize will now allow coordinated efforts by several laboratories. For instance, the analysis of ESTs in *Arabidopsis* suggests that certain MIPs such as PIP2d (CAB41102) are specifically expressed in siliques, whereas others such as γ-TIP (BAB01832) or PIP1b (AAF81320) are strongly expressed throughout the plant (Javot *et al.,* unpublished results). The latter results confirm expression studies using RNAse protection assays (Grote *et al.,* 1998) or plants containing a promoter:GUS fusion construct (Ludevid *et al.,* 1992). Because the sequence of all the MIPs in *Arabidopsis* is now accessible, expression studies using macro- or microarrays will become extremely useful, and it will be possible to estimate the cross-reactivity of probes with very close homologues.

However, studies in model plants may be difficult to generalize, since the expression profile of an aquaporin cannot be deduced from data obtained on a close homologue in another plant species. α-TIP homologues in bean, *Arabidopsis,* and Norway spruce all showed abundant expression in seeds (Höfte *et al.,* 1992; Johnson *et al.,* 1989; Ludevid *et al.,* 1992; Oliviusson and Hakman, 1995). In contrast, *Arabidopsis* γ-TIP showed preferential expression in elongating tissues (Ludevid *et al.,* 1992), whereas a close homologue in cauliflower was strongly expressed in meristematic tissues (Barrieu *et al.,* 1998b), and a homologue in maize showed a broader expression pattern, with preferential expression in root vascular tissues (Barrieu *et al.,* 1998a). Thus, expression properties of aquaporins cannot be predicted based on their amino acid sequence identity.

III. Functional Evidence for Plant Aquaporins

As in other organisms, the presence of active water channels in plant membranes can be assessed based on the conjunction of the following parameters:

i. A high osmotic water permeability ($P_f > 50$–$100\ \mu m\ s^{-1}$)
ii. A sensitivity of water transport to mercury inhibition and reversion of mercury effects by reducing agents
iii. A low dependency on temperature, reflecting a low Arrhenius activation energy (E_a)
iv. A ratio of osmotic to diffusional water permeability (P_f/P_d) greater than unity.

Before the molecular identification of plant aquaporins, these parameters had been separately characterized in plant membranes but, with the exception of a study in *Chara* cells by Wayne and Tazawa (1990), had not been assembled as evidence for water channels in plants. This gap has been filled over the last few years, and numerous reports now describe the activity of water channels in plant membranes, in intact cells, and even in tissues. Although interpretations in the two latter systems may not be straightforward, these approaches open new perspectives to address the specific function of aquaporins in plants. The possibility that plant aquaporins transport small neutral solutes has also led several laboratories to revisit the solute transport properties of plant membranes.

Tables III and IV summarize some of the water transport measurements that have been reported in different plant species over the last 4 years, with emphasis on those studies where mercury inhibition of water transport was investigated. The techniques used are also mentioned but will not be detailed here since they have been described in recent reviews (Maurel, 1997; Steudle, 1993; Verkman, 2000).

A. Subcellular Membranes

1. Intracellular Membranes

a. Water Transport Measurements in Isolated Vesicles or Organelles. The discovery of aquaporins in the intracellular membranes of plant cells has led several laboratories to investigate in closer detail the water transport properties of subcellular membrane compartments. Initial attempts were done by cell membrane fractionation and stopped-flow measurements on isolated membrane vesicles (Maurel *et al.,* 1997b; Niemietz and Tyerman, 1997; Rivers *et al.,* 1997). Extremely high water permeability ($P_f \geq 500\ \mu m\ s^{-1}$) and strong inhibition by mercury ($\geq$80%), and/or $P_f/P_d \geq 7$ have been reported both in the peribacteroid membrane of soybean N_2-fixing root nodules (Rivers *et al.,* 1997) and in the TP of tobacco suspension

TABLE III

Characteristics of Water Transport in Isolated Membrane Vesicles, Vacuoles, Protoplasts, and Intact Cells

Plant material[a]	Mean P_F (μm/s)	Technique[b]	E_a (kcal/mol)	Inhibition by $HgCl_2$: Dose (mM)	Treatment (min)	Inhibition (%)	Reversion[c]	References
Membrane vesicles								
Wheat root PM	12.5	SF	11.4					Niemietz and Tyerman (1997)
Tobacco PM	6.1	SF	13.5					Maurel *et al.* (1997b)
Tobacco TP	690	SF	2.5	1	5	80	+	Maurel *et al.* (1997b)
Wheat root IM	86	SF	5.5	0.1		70		Niemietz and Tyerman (1997)
Soybean PBM	500	SF	3.3	0.1		92		Rivers *et al.* (1997)
Isolated vacuoles								
Rape leaf	623	PMM-TC						Morillon and Lassalles (1999)
Petunia leaf	955	PMM-TC						Morillon and Lassalles (1999)
Onion leaf	184	PMM-TC		1	10	70	+	Morillon and Lassalles (1999)
Rape hypocotyl	1100	PMM-TC						Morillon and Lassalles (1999)
Rape root	656	PMM-TC						Morillon and Lassalles (1999)
Red beet root	270	PMM-TC		1	10	85	+	Morillon and Lassalles (1999)
Isolated protoplasts								
Onion leaf	9	PMM-TC	15.5					Ramahaleo *et al.* (1999)
Rape hypocotyl	370	PMM-TC						Ramahaleo *et al.* (1999)
Rape root	305	PMM-TC	0.3	0.2		98		Ramahaleo *et al.* (1999)
Wheat root	2.5	PMM-TC	11.7					Ramahaleo *et al.* (1999)
Melon root	11	PMM						Martinez-Ballesta *et al.* (2000)
Intact cells								
Chara (low CO_2)	353	TCO		1	60	63		Wayne *et al.* (1994)
Chara (high CO_2)	244	TCO		1	60	81		Wayne *et al.* (1994)
Chara	260	TCO		0.1	2	90	+	Tazawa *et al.* (1996)
Onion root	257	CPB		0.05	20	26		Barrowclough *et al.* (2000)
Wheat	76	CPB		0.1	60	75	+	Zhang and Tyerman (1999)

[a] Abbreviations for plant materials are as follows: PM, plasma membrane; TP, tonoplast; IM, intracellular membranes; PBM, peribacteroid membrane.

[b] The P_f value for onion roots is from a root segment with mature endodermis and exodermis. P_f values were measured using the following methods: SF, stopped-flow spectrophotometry; PMM, plasmometric method without or with a transference chamber (TC); TCO, transcellular osmosis; CPP, cell pressure probe.

[c] Reversion of $HgCl_2$ inhibition by a reducing agent (β-mercaptoethanol, DTT) is indicated by (+).

TABLE IV

Characteristics of Water Transport in Root Tissues

Plant species	Hydraulic conductivity (L_p)	Technique[a]	Inhibition by $HgCl_2$				References
			Dose (mM)	Treatment (min)	Inhibition (%)	Reversion[b]	
Tomato	4.6×10^{-7} m s^{-1} MPa^{-1}	HPC	0.5	1	57	+	Maggio and Joly (1995)
Wheat	760 mg g^{-1} h^{-1} MPa^{-1}	EM	0.05	5	64	+	Carvajal *et al.* (1996)
Barley	0.29 pm s^{-1} Pa^{-1}	TRO	0.1	20	85	−	Tazawa *et al.* (1997)
Sugar beet	0.145[c] μL min^{-1} cm^{-2}	HPC	1		80	+	Amodeo *et al.* (1999)
Paprika	800 mg g^{-1} h^{-1} MPa^{-1}	EM	0.05	5	85	+	Carvajal *et al.* (1999)
Sunflower	756 μL g^{-1} h^{-1} MPa^{-1}	EM	1	160	21	n.d.	Quintero *et al.* (1999)
Aspen	9.7×10^{-8} m s^{-1} MPa^{-1}	HPC	0.1	30	47	+	Wan and Zwiazek (1999)
Onion[d]	2×10^{-6} m s^{-1} MPa^{-1}	P	0.05	20	72	+	Barrowclough *et al.* (2000)
Melon	645 mg g^{-1} h^{-1} MPa^{-1}	EM	0.05	5	68	+	Martinez-Ballesta *et al.* (2000)

[a] Water transport measurements were made using the following methods: HPC, hydrostatic pressure chamber; EM, exudate measurement; TRO, trans-root osmosis; P, potometer.

[b] Reversion of $HgCl_2$ inhibition by a reducing agent (β-mercaptoethanol, DTT), or a lack of reversion, is indicated by (+) or (−), respectively. n.d.: not determined.

[c] Value refers to a flux not a hydraulic conductivity.

[d] Measurements on a root segment with mature endodermis and exodermis.

and wheat root cells (Maurel *et al.*, 1997b; Niemietz and Tyerman, 1997). This provided unambiguous evidence for active water channels in these two classes of endomembranes. In complement to measurements on purified TP vesicles, water transport properties have also been characterized in intact vacuoles isolated from protoplasts of various origin (Morillon and Lassalles, 1999). These studies confirmed the idea that the TP of most plant cells exhibits a high P_f ($\geq 200\ \mu\text{m s}^{-1}$). The significance of a high water permeability for plant endomembranes will be discussed further in Section IV of the present review.

b. Solute Transport Properties. TP-enriched vesicles purified from tobacco cells also exhibited a high permeability to small neutral solutes such as glycerol and urea ($P_{\text{urea}} = 74.9 \pm 9.6 \times 10^{-6}$ cm s^{-1}) with features of a facilitated (channel-mediated) transport (Gerbeau *et al.*, 1999). Functional expression of NtTIPa, a novel aquaporin expressed in the TP of these cells, showed this aquaporin to transport urea and to a lesser extent glycerol, thus paralleling the solute transport properties of native membranes (Gerbeau *et al.*, 1999). It was thus proposed that Nt-TIPa can account for most of the high permeability of the tobacco TP to small nonelectrolytes.

The contribution of NOD26 to solute transport in the peribacteroid membrane has been investigated in even greater detail. Early experiments in which NOD26 was reconstituted in artificial membranes revealed an ion channel activity for the protein (Weaver *et al.*, 1994). Accordingly, it has been suggested that NOD26 could mediate the export of organic acids from the bacteroid toward the plant cell (Ouyang *et al.*, 1991). Serious doubt about this hypothesis has been cast by the failure of NOD26 (and of its mammalian homologue MIP/AQP0) to elicit any ion conductance after expression in *Xenopus* oocytes (Dean *et al.*, 1999). In contrast, NOD26 exhibited a good capacity to transport glycerol but not urea, both in oocytes and in artificial membranes (Dean *et al.*, 1999). This agrees with the high and low permeability of the symbiosome membrane to glycerol and urea, respectively (Rivers *et al.*, 1997). Two *Arabidopsis* homologues of NOD26, NLM1 and NLM2, also transport glycerol (Weig and Jakob, 2000), but *Arabidopsis* does not establish any N_2-fixing symbiosis, and the membrane to which NLM1 and NLM2 confer this property is unknown.

NH_3 may be another substrate for plant aquaporins and may be more physiologically relevant than small polyols or urea in the particular case of NOD26. NH_3 entry into peribacteroid membrane vesicles was recently shown to be partially protein-mediated (Niemietz and Tyerman, 2000) based upon the low activation energy of NH_3 uptake and its inhibition by mercury. NOD26 is regulated by phosphorylation (Lee *et al.*, 1995), and the participation of this protein in NH_3 permeation was evaluated by preincubating vesicles with ATP: ATP addition increased water transport on the one hand and inhibited NH_3 transport on the other. This suggested either that two transport mechanisms for water and NH_3 coexist inside NOD26 or that another protein is responsible for NH_3 transport.

2. Plasma Membrane

Whereas a high water and solute permeability has been consistently found in both the TP and the peribacteroid membrane, measurements with isolated PM vesicles have pointed to a much reduced permeability. In two initial studies where tobacco suspension cells or wheat roots were used, a low membrane P_f associated with a high E_a for water transport suggested a lack of active water channel in purified PMs (Maurel *et al.,* 1997b; Niemietz and Tyerman, 1997). PM vesicles isolated from tobacco also showed a reduced permeability to urea and to a series of small linear polyols (Gerbeau *et al.,* 1999), with no facilitated (channel-mediated) transport for these molecules. Recently, Nt-AQP1, a PIP homologue from *Nicotiana tabacum,* was reported to be permeable to water and glycerol upon expression in *Xenopus* oocytes (Biela *et al.,* 1999). This aquaporin is probably targeted to the PM, but the activity of solute transporting aquaporins in this membrane remains to be demonstrated.

The respective water transport properties of the vacuolar and plasma membranes have also been recently investigated in yeast cells (see Hohmann, this volume). Surprisingly, no active water channel activity could be detected in purified membrane vesicles (Coury *et al.,* 1999), and the reason for these results is not yet clear. Because of the restricted number of MIPs in yeast and because of the power of genetic tools developed in this organism, future studies on water transport in yeast will nicely complement studies on the function of aquaporins in plant cells.

3. Posttranscriptional Regulation of Plant Aquaporins

There is to date no definite molecular explanation of the dramatic difference in water channel activity in vesicles isolated from various subcellular compartments. Clearly, this difference cannot be simply accounted for by differences in gene and/or protein expression level, since aquaporins seem to be abundant both in the PM and in the endomembranes of plants. Studies in animals showed that aquaporins can exhibit a large range of unitary water conductance that span up to two orders of magnitude (AQP0: $p_f = 0.25 \times 10^{-14}$ cm^3 s^{-1}; AQP4: $p_f = 24 \times 10^{-14}$ cm^3 s^{-1}) (Yang and Verkman, 1997). These parameters will have to be investigated and compared in plant TP and PM aquaporins.

Observations by several laboratories also point to regulatory mechanisms that may target plant aquaporins. For instance, Niemietz and Tyerman (1997) failed to detect active water channels in PM vesicles purified from wheat roots, whereas cortex cells from the same material had a L_p which was inhibited by mercury (Zhang and Tyerman, 1999). Mercury inhibition appears to take place both at the PM and at the TP of wheat cells (Zhang and Tyerman, 1999), suggesting that inactivation of PM water channels may have occurred during membrane purification. Along these lines, we recently discovered that active water channels can be detected in membrane vesicles purified from *Arabidopsis* suspension cells, provided that chelators of divalent cations are used during cell homogenization to prevent water channel down-regulation by these cations (Gerbeau *et al.,* submitted). The molecular bases of this inhibition are not yet clearly understood. Other clues may arise from the failure of several laboratories to functionally express homologues of the PIP1 family in *Xenopus* oocytes. At least in the case of maize ZmPIP1 (Chaumont *et al.,* 2000), the aquaporin was correctly expressed and targeted to the oocyte PM, suggesting the requirement of posttranslational modifications for water channel activity.

Reversible phosphorylation of plant aquaporins has been described in various subcellular membranes and may represent such a kind of regulatory mechanism. For instance, α-TIP can be phosphorylated in PSVs of bean seeds or in transgenic tobacco by a TP-bound calcium-dependent protein kinase (CDPK) (Johnson and Chrispeels, 1992). CDPKs are also responsible for the phosphorylation of NOD26 and PM28a, in the symbiosome membrane of soybean root nodules and in the PM of spinach leaves, respectively (Johansson *et al.,* 1996; Weaver *et al.,* 1991). Interestingly, phosphorylation of PM28a is also dependent on extracellular water potential and could be reduced by external addition of mannitol to leaf tissues (Johansson *et al.,* 1996). The functional significance of plant aquaporin phosphorylation has been addressed after expression in *Xenopus* oocytes (α-TIP, PM28a) (Johansson *et al.,* 1998; Maurel *et al.,* 1995) or after reconstitution of the protein in artificial membranes (NOD26) (Lee *et al.,* 1995). *Xenopus* oocytes proved to be a very convenient expression system since aquaporin phosphorylation could be modulated by agonists and antagonists of endogenous protein kinases and phosphatases. The water channel activity of both α-TIP and PM28a was enhanced by phosphorylation at sites that are recognized by plant CDPKs, but the role of subsidiary sites, whose occurrence remains to be demonstrated *in planta,* was also uncovered (Johansson *et al.,* 1998; Maurel *et al.,* 1995). In another study where purified NOD26 was reconstituted in planar lipid bilayers, phosphorylation by a recombinant CDPK from *Arabidopsis* altered the ion channel behavior of NOD26 (Lee *et al.,* 1995). From all these reports, it has been hypothesized that, in contrast to the phosphorylation of mammalian AQP2 which provides a signal for AQP2 targeting to the cell surface (Fushimi *et al.,* 1997; Katsura *et al.,* 1997), the phosphorylation of plant aquaporins may provide a means for rapid and reversible *in situ* gating. This property remains, however, to be established in native plant membranes.

B. Cell Level

The observation that plant cells from different species, tissues, or even developmental stages can exhibit a large range of water permeability has been a longstanding one (Maurel, 1997; Steudle, 1989) and can now be taken as an indication of membrane water transport being the object of a strict control in which aquaporins may play a crucial role. For instance, the differential regulation of aquaporins may explain why the mean water permeability of intact wheat root cells was 76 μm s^{-1}, whereas a value of 2.5 μm s^{-1} was measured in protoplasts from the same origin (Ramahaleo *et al.,* 1999; Zhang and Tyerman, 1999). The recent work by Ramahaleo *et al.* (1999) also emphasizes the large variability of P_f values (over more than two orders of magnitude) that can be found in isolated protoplasts from the same preparation. These authors showed that, despite this very high variability, the averaged P_f of root protoplasts from maize, rape, and wheat all increased >10-fold with root age, suggesting a modulation of water channel activity during root development (Ramahaleo *et al.,* 1999).

1. Water Transport Measurements in Intact Cells Have Established the Presence of Active Water Channels in Algae and in Higher Plants

Because of their size, the internodal cells of charophytes are a convenient model for water transport studies and provided the first evidence for water channels in plant cells (Wayne and Tazawa, 1990). A high hydraulic conductivity, low E_a, and inhibition by sulfhydryl reagents have been reported by several authors using either transcellular osmosis or, more recently, pressure probe measurements (Henzler and Steudle, 1995; Schütz and Tyerman, 1997; Tazawa *et al.,* 1996; Wayne and Tazawa, 1990). Evidence for water channels in intact cells of higher plants has proved more difficult to obtain, but recent work by Zhang and Tyerman (1999) also demonstrated inhibition of L_p by mercury in cortex cells of wheat roots.

There have been concerns about the general effects that mercury may exert on cell functions. In *Chara* and wheat cells, for instance, mercury rapidly depolarized the membrane (without altering the membrane electrical conductance in *Chara*) with a dose-dependence that was similar to that of L_p inhibition (Schütz and Tyerman, 1997; Tazawa *et al.,* 1996; Zhang and Tyerman, 1999). It was also found that mercury inhibits cell respiration and blocks cell cyclosis (Schütz and Tyerman, 1997; Zhang and Tyerman, 1999). Thus, the effects of $HgCl_2$ on cell water transport may be mediated by direct blockage of water channels or by a general alteration of cell metabolism that would in turn influence water channel activity.

2. Genetic Evidence for the Activity of Aquaporins in Plant Membranes

Reverse genetics provides more solid evidence for aquaporin function in the plant PM. Transgenic *Arabidopsis* plants that contained a *PIP1B* antisense construct displayed an overall reduction in the level of mRNA and proteins for several PIP1 homologues, consistent with the very high sequence homology that exists between members of the PIP1 subfamily in *Arabidopsis* (Kaldenhoff *et al.,* 1998). The P_f of protoplasts prepared from transgenic leaves was 3- to 4-fold lower than the P_f

of wild-type protoplasts and was reduced to the level of the latter treated with $HgCl_2$. Although the P_f values reported in this work are strikingly low, within a range typical for membranes deprived of water channels, these data demonstrate that PIP aquaporins significantly contribute to water transport at the PM level.

3. Channel-Mediated Solute and Gas Transport in Intact Plant Cells

In *Chara* cells the effects of mercury and temperature on the permeability and reflection coefficient of lipophilic molecules, such as small size alcohols, formamide, or acetone, have been taken to demonstrate the capacity of plant water channels to transport solutes (Henzler and Steudle, 1995; Hertel and Steudle, 1997; Schütz and Tyerman, 1997). The contribution of the water channel path remained minor as compared to transport through lipid membranes. Yet these studies questioned for the first time the notion that plant water channels function as highly selective channels. In particular, extremely low reflection coefficient values of the water channel path suggested that strong friction between water and solutes may occur within the aqueous pore (Schütz and Tyerman, 1997; Steudle and Henzler, 1995).

In a previous study, Wayne *et al.* (1994) established a link between the permeability of *Chara* cells to water and CO_2 availability. Whereas transport of H_2O and HCO_3^- showed distinct pharmacological properties, the correlation between H_2O and CO_2 permeability suggested either that the PM lipid composition was modified by CO_2 in a way affecting water transport or that proteins that are induced by CO_2 mediate the transport of both CO_2 and water. The second idea conforms with recent evidence showing that mammalian AQP1 can transport CO_2 (Nakhoul *et al.*, 1998; Prasad *et al.*, 1998).

4. The Water Permeability of the PM and the TP Can Be Deduced from Measurements Made in Cellular Systems

The water transport properties of wheat roots cells have been investigated both by stopped-flow measurements on purified membrane vesicles (Niemietz and Tyerman, 1997) and by means of a cell pressure probe (Zhang and Tyerman, 1999). Water transport simulations in a model cell revealed, however, that the behavior of intact cells cannot be consistently explained by the water permeability values determined in isolated vesicles (Zhang and Tyerman, 1999). To fit their cell pressure probe data, Zhang and Tyerman (1999) had to assume that the PM water permeability of living cells was higher by a factor of 1.2 to 10 than the value recorded in isolated PM vesicles. To interpret the effects of mercury on the L_p of root cells, they also had to assume that mercury blocked water transport both at the PM and at the TP, suggesting that water channels were active in these two membranes (see above). It was concluded, however, that in all cases, the water transport properties of the PM dominated the whole cell behavior.

Water transport measurements made in protoplasts and in vacuoles isolated from the same materials can also provide hints at the respective permeabilities of the TP and the PM (Morillon and Lassalles, 1999; Ramahaleo *et al.,* 1999; Url, 1971). A rough estimate of PM water permeability can be obtained by assuming that the TP and the PM act as two hydraulic resistances in series ($1/L_{\mathrm{p\ cell}} \sim 1/L_{\mathrm{p\ PM}} + 1/L_{\mathrm{p\ TP}}$). In certain plant materials such as the hypocotyls and roots from rape (Morillon and Lassalles, 1999; Ramahaleo *et al.,* 1999), the PM must be nearly as permeable as the TP. In contrast, most of the hydraulic resistance of onion bulb protoplasts resides in the PM (Url, 1971). In the giant cells of *Chara* internodes, the TP can be disrupted by intracellular perfusion of an EGTA solution. Measurements using this approach suggested that the TP does not significantly contribute to the overall hydraulic resistance of the cell (Kiyosawa and Tazawa, 1977) and that mercury exerts its most significant inhibitory effects at the PM level (Tazawa *et al.,* 1996).

C. Tissue Level

With the exception of one study on sunflower hypocotyls (Hejnowicz and Sievers, 1996), the search for water channel activity in plant tissues has exclusively focused on roots. In this organ, the radial transport of water occurs via two parallel routes, either across cell walls (apoplastic path) or from cell to cell. Cell-to-cell flow is itself a blend of transcellular and symplastic flow (Steudle and Peterson, 1998). Using traditional biophysical analyses, it has not been possible to experimentally distinguish between the symplastic and transcellular contributions to root water transport. The barrier that membranes may present to water flow has thus remained uncertain.

Inhibition of root L_p by mercuric chloride ($HgCl_2$) has been reported in several species including tomato (Maggio and Joly, 1995), wheat (Carvajal *et al.,* 1996), barley (Tazawa *et al.,* 1997), aspen (Wan and Zwiazek, 1999), sugar beet (Amodeo *et al.,* 1999), sunflower (Quintero *et al.,* 1999), paprika (Carvajal *et al.,* 1999), and melon (Martinez-Ballesta *et al.,* 2000) (Table IV). Note, however, that depending on the studies, high (0.5 mM) or much lower (50 μM) doses of $HgCl_2$ were used. Nevertheless, mercury inhibition has been interpreted in all cases as a direct blockage of water channels and supports, for the first time, the idea that cell membranes can represent a major penetration resistance for water in roots. Similar to results obtained in cells, mercury treatment has strong effects on whole-organ physiology and reduced both root respiration and stomatal conductance in aspen seedlings (Wan and Zwiazek, 1999). Because of this general toxicity, the reversal of mercury effects by scavenging agents such as β-mercaptoethanol or dithiothreitol is mandatory when water channel functions are investigated, to ensure that no irreversible damage has been made to roots. Some authors have also checked that ion uptake in roots was not altered, by measuring the ionic composition of the xylem sap (Maggio and Joly, 1995; Wan and Zwiazek, 1999). The dependence on temperature of water transport in controls and in mercury-treated aspen roots has recently been investigated by Wan and Zwiazek (1999). In contrast to what is commonly observed in isolated membranes or cells, E_a was decreased after mercury treatment, and this observation is awaiting explanation.

To dissect mercury effects even further, Barrowclough *et al.* (2000) have adapted mini-potometers to measure water influx into well-defined anatomical zones of onion roots. They probed the effects of mercury in each of these zones and found that water channel contribution is most important in the oldest part of the root where both an exodermis and endodermis were differentiated, and had developed Casparian bands and suberin lamellae. The penetration of mercury in roots was investigated by precipitation by hydrogen sulfide vapor and microscopic observation of the precipitate in root sections. Surprisingly, mercury did not noticeably enter the root beyond the outer cell layers (exodermis) (Barrowclough *et al.,* 2000).

IV. Putative Functions for Plant Aquaporins

A. Cell Osmoregulation

1. What Is the Physiological Significance of Distinct Water Transport Properties at the PM and the TP?

a. Single-Cell Osmoregulation. It is commonly acknowledged that the PM must represent the limiting barrier that regulates water exchange between the cell and its bathing medium. This is consistent with estimations of PM and TP water permeability derived from measurements on purified membrane vesicles or most intact cells (see above). Theoretical analyses of model plant cells have been performed to evaluate the significance of these parameters in single-cell osmotic regulation. Water transport simulations showed that, if the water permeability of the PM remains low, an additional resistance at the TP will marginally determine the rate of cell equilibration (Maurel *et al,* unpublished results; Tyerman *et al.,* 1999). However, additional constraints for cell osmoregulation are imposed by the compartmentation of plant cells into cytosol and vacuole. The cytosol is the most critical compartment for most cell metabolic functions and requires very fine regulation of its volume and osmotic potential. Because it also has to mediate water exchange between the vacuole and the cell exterior, the cytosol is subject to possible volume fluctuations if water mobilization into and from the vacuole becomes limiting. A swelling or shrinking can then be expected in case of a sudden water influx or efflux, respectively (Fig. 2A). Although these processes are only transient, they can represent very rapid and drastic changes well beyond those that can be felt at the whole-cell or vacuole level. For instance, sudden exposure to hypotonicity

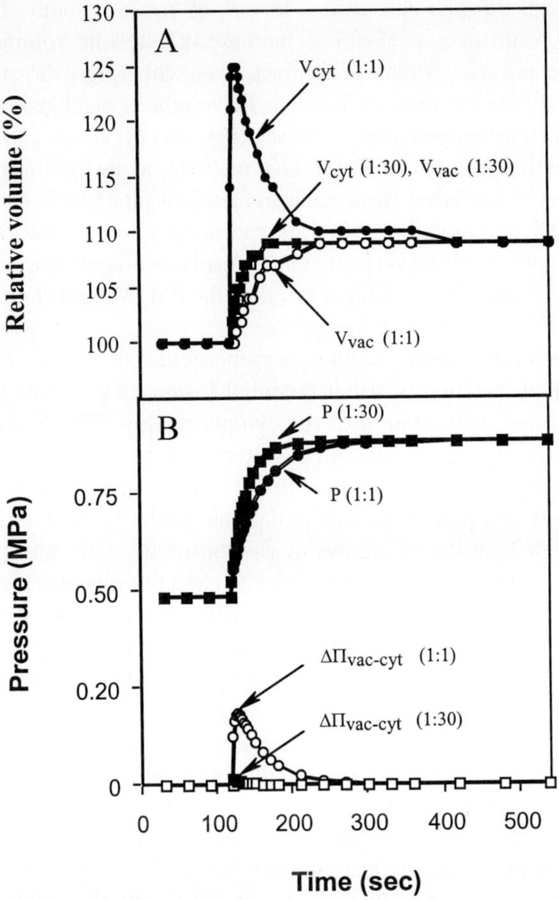

FIG. 2 Response of a model plant cell to hypotonicity. Water transport simulations were done essentially as described in Tyerman *et al.* (1999), considering a spherical cell with a spherical vacuole, and surrounded by a cell wall. The P_f of the PM was set to a constant value ($P_{f-PM} = 6\ \mu m\ sec^{-1}$; see Maurel *et al.,* 1997b) and two distinct P_f values for the TP (i.e., two cell configurations) were investigated: $P_{f-TP} = 6\ \mu m\ sec^{-1}$ (configuration 1:1) and $P_{f-TP} = 180\ \mu m\ sec^{-1}$ (configuration 1:30). For details on the equations used to model the cell, see Fig. 3 of Tyerman *et al.* (1999). (A) Variations of the relative volumes of the cytosol (V_{cyt}) and of the vacuole (V_{vac}) in response to an instantaneous change in external osmotic pressure, at time $t = 120$ sec, from 200 to 0 mOsmol kg^{-1} H_2O. (B) Variations in turgor pressure (P) and osmotic pressure gradient between the vacuole and cytosol ($\Delta\Pi_{vac\text{-}cyt}$) in response to the same hypotonic shock as in (A). Initial conditions were as follows: cell compartment volumes, $V_{cyt} = 0.05 \times V_{cell} = 1.68 \times 10^{-15}\ m^3$ and $V_{vac} = 0.95 \times V_{cell} = 3.18 \times 10^{-14}$; osmotic pressures, $\Pi_{ext} = 0.49$ MPa and $\Pi_{cyt} = \Pi_{vac} = 0.97$ MPa; cell volumetric elastic modulus, $\varepsilon = 4$ MPa; temperature, T = 293 K.

of a hypothetical tobacco cell with a low P_f (6 μm s^{-1}) both at the PM and at the TP can result in a + 15–25 % increase in cytosolic volume in <30 sec (Fig. 2A) (Tyerman *et al.*, 1999). In contrast, a cell with a higher P_f at the TP (180–600 μm s^{-1}) exhibited an increase in cytosolic volume of much reduced amplitude (+ 5–9 %) which developed over >150 sec (Fig. 2A) (Tyerman *et al.*, 1999). The water permeability of the plant PM *in vivo* may not be always minimal as in this model. This can be inferred from measurements in intact cells (see above) and from the possible up-regulation of PM aquaporins by phosphorylation. Simulation studies show, however, that a very efficient cytosol osmoregulation can be achieved as long as the TP has a ≥5-fold higher P_f than the PM (Maurel *et al.*, unpublished results).

Another view of the osmotic regulation properties needed by plant cells involves considering the dissipation of water potential (osmotic) gradients across the TP rather than volume equilibration of the cytoplasm (Fig. 2B). In fact, these two processes are linked since the amount of water that is exchanged between the vacuole and the cytoplasm is proportional to the osmotic gradient across the TP and contributes to its dissipation. Thus, simulations similar to those discussed above show that a high P_f at the TP allows quasi-isoosmotic water transport whatever the rate of water transport in and out of the cell and the vacuole (Fig. 2B). Therefore, physical constraints on the TP of plant cells, and activation of mechano- and osmo-sensitive ion channels, that have been described in this membrane (Alexandre and Lassalles, 1991), could be damped by the water transport properties of the TP.

b. Transcellular Water Transport. Under optimal growing conditions, enormous amounts of water (i.e., the equivalent of the plant fresh weight in a couple of hours) are taken up from the soil, flow across the living tissues of roots, the xylem vessels, the mesophyll, and are lost by transpiration. The differential regulation of water transport at the TP and PM of plant cells can also be physiologically relevant in this context.

Although the contribution of the transcellular path to the overall flow may be critical in certain tissues only, it is still assumed that most of the flow control resides at the PM. Because the plant vacuole can be bypassed by water flow across cytoplasmic strands, it is indeed not clear whether the vacuole can create a significant resistance to transcellular water flow. Thus, the water permeability of vacuoles will hardly determine the overall flow rate of water across tissues. However, it can be critical that the cytosol and vacuole remain in perfect osmotic equilibrium at all times, and this can be achieved if the TP has a high P_f. Trans-tissue water transport possibly fluctuates in response to environmental changes and concomitant modulation of stomatal aperture. As detailed above, the resistance of the vacuolar membranes is critical in determining the kinetic changes in water potential on both sides of the membrane. If the vacuolar membrane poses significant hydraulic resistance, any fluctuation in transcellular water flow would generate a strong osmotic gradient, perturbate vacuolar volume, and result in a potential drag of the vacuole within the cell.

Thus, a unified model would propose that the plant TP be maintained in a state of high water permeability, whereas water transport at the PM can show marked regulation and determine the overall exchange of water between the cell and its exterior. It is assumed that this property can be maintained under various physiological conditions and plays an important role for single cell osmoregulation or under conditions where plant cells mediate an important transcellular water flow.

2. Solute Transport across the PM and Peribacteroid Membrane

The physiological significance of solute transport by aquaporins is uncertain, but may be relevant to their role in plant cell osmoregulation. Gerbeau *et al.* (1999) have proposed that the controlled transport of compatible solutes by TP aquaporins, such as tobacco NtTIPa, may provide a mechanism for long-term regulation of the respective volumes of the cytosol and the vacuole. Soybean NOD26 (Dean *et al.,* 1999) and a close homologue in *Lotus* (Guenther and Roberts, 2000) are other intracellular aquaporins that transport small solutes. They may contribute to dissipating osmotic gradients across the symbiosome membrane. Solute transporting aquaporins at the plant PM, similar to yeast Fps1 (Luyten *et al.,* 1995), may mediate the efflux of compatible solutes, thus allowing the down-regulation of turgor under hypoosmotic conditions. For either membrane, however, the physiological relevance of such transport remains speculative because none of the small molecules involved has clear physiological relevance.

3. Under Which Conditions Do Rapid and Significant Changes in Water Potential Occur?

The models for cell osmoregulation presented above imply that plant cells can experience rapid changes in water potential and/or volume. Under normal conditions, the volume of plant cells can be altered, but the processes involved occur over several tens of minutes (stomatal guard cell movements) or several hours or even days (cell elongation). There are, however, several types of environmental or developmental constraints where the models discussed above may apply.

Firstly, the transpiration regime of plants is determined in part by the water demand exerted onto the aerial parts. Rapid alterations in this demand, due to variations in wind regime, temperature, or light intensity, may occur and surpass the rapidity of stomatal regulations. These alterations are propagated through changes in tension in xylem vessels (Schneider *et al.,* 1997). Even more dramatic changes in tension also occur locally following vessel cavitation. *In situ* measurements using a cell pressure probe have shown that changes in xylem tension result in immediate and large amplitude alterations in the turgor of cells adjacent to the xylem vessels (Wegner and Zimmermann, 1998).

Secondly, root cells at the plant/soil interface can be exposed to drastic changes in water potential, upon sudden flooding of the soil after, for instance, prolonged drought. In certain plant species such as tobacco, root cells have a very low cell wall elastic modulus. This means that significant changes in root cell volume can be observed in response to a sudden decrease in soil solution osmotic potential (Tyerman *et al.,* 1989). A down-regulation of turgor was experimentally measured in this situation, suggesting that a release of ions or neutral solutes was triggered in response to hypotonicity (see above). Resurrection plants can lose more than 90% of their cell water upon exposure to drought (Ingram and Bartels, 1996). Subsequent tissue rehydration imposes severe osmotic constraints that may be overcome by the presence of aquaporins (Mariaux *et al.,* 1998).

The normal developmental cycle of plants also implies drastic desiccation and hydration processes. Seed germination is one such process. It is triggered by the imbibition of seed tissues and involves the hydrolysis of products stored in PSVs, their release in the cytosol, and a complete reorganization of the fragmented vacuolar apparatus to yield a large central vacuole. The water and possibly solute transport activity of seed-specific TP aquaporins may be critical for cytoplasm osmoregulation during these processes (Maurel *et al.,* 1997a). A role for PM aquaporins (PIPs) in the early stages of seed germination has also been proposed (Gao *et al.,* 1999). Priming is a treatment of plant seeds that can substantially improve their subsequent germination efficiency. For this, seeds are hydrated in the presence of polyethyleneglycol (PEG) or ABA and further desiccated prior to their conservation. Gao *et al.* (1999) investigated various priming treatments in rape seeds and found a strong correlation between the subsequent seed germination index and mRNA levels for a PIP homologue. It was suggested that strong expression of this aquaporin may facilitate tissue imbibition and/or counteract osmotic perturbations linked to this process (Gao *et al.,* 1999).

Pollen germination is yet another process which is critically dependent on water transport (Sarker *et al.,* 1988). A role for aquaporins in the self-incompatibility response of *Brassica* has been proposed, based on genetic evidence (Ikeda *et al.,* 1997). However, the finding that loss of self-incompatibility was associated with mutation of a PIP homologue has not received any clear physiological explanation.

B. Control of Transcellular Water Flow: The Example of Root Water Transport

1. General Mechanisms of Water Transport

One of the most interesting generalizations which can be made about roots is that water flow into roots can be modeled as though there were only a single membrane that water must cross on entry. In herbaceous species, the hydraulic conductivity (L_p) of the whole root, as determined by osmotic experiments, is in the range of 10^{-7} m s^{-1} MPa^{-1} and is of the same order as the L_p of a single cell (Steudle, 2000; Steudle and Peterson, 1998). This may be surprising because roots are formed by several radial cell layers which potentially oppose a resistance to water flow between the soil solution and the xylem vessels. Singular properties of the root indicate, however, that this organ cannot be modeled by a single membrane barrier. The reflection coefficient for ionic compounds is far below unity, indicating that the root is not as selective as a single membrane barrier. The hydraulic conductance is dependent on the type of force applied, whether osmotic or hydrostatic, and increases with the amplitude of these forces (Steudle and Peterson, 1998).

When water moves radially into a root, there can be both apoplastic and cell-to-cell flow. In the apoplast, water movement is purely hydraulic; in the cell-to-cell route, which involves passage across cell membranes, there will be an osmotic and hydraulic component to the flow. There is a constant and rapid local water flow between compartments along the path, but the net driving force for water entry is a water potential gradient between the xylem vessels and the medium in which the roots are growing. At high rates of transpiration, there is a strong gradient of water potential, and the apoplastic route will predominate. At lower rates of flow (closed stomata, times of water shortage), interactions between water and solute flow should be more important (Steudle and Peterson, 1998).

Longitudinal (axial) flow along the xylem can be described by the Hagen–Poiseuille equation, which gives great importance to vessel diameter. Axial flow is generally regarded as having a much lower hydraulic resistance than radial flow (1–2 orders of magnitude for maize roots) and therefore does not contribute a significant resistance to total flow (Steudle and Peterson, 1998).

2. Anatomical Dissection of Root Water Transport

In root tips and young roots of certain species such as maize and onion (Frensch *et al.*, 1996; Peterson and Steudle, 1993), water flow can be described by a "uniform resistance" model, that is, the root L_p is constant along the specified region of the root, and cell L_p is not highly variable across the cortex. In more mature root tissue, an alternate "nonuniform resistance" model of water flow applies. Mature roots are characterized by the development of significant barriers to radial apoplastic water flow, such as a mature exodermis, and stelar tissues that surround the phloem and xylem vessels (Steudle and Peterson, 1998).

To understand the cellular mechanisms of root water transport, a major issue has been to map the resistance barriers for radial and axial water flow. Different approaches have been pursued to experimentally explore this idea and to identify which cell layer can be limiting for radial water transport. Zimmermann and

Steudle (1998), using maize plants grown in hydroponic or aeroponic conditions, found that plants grown aeroponically had reduced hydraulic conductivity. This change in root L_p was associated to the differentiation of a root exodermis, with Casparian bands which impose water-tight wall barriers.

Some workers have used apoplastic dyes to directly measure the extent of apoplastic water flow (e.g., rhodamine B; Skinner and Radin, 1994). These data need to be used with caution, since the dyes are generally large and do not move as freely as water through the apoplast. A good example of their use was the approach of Wan and Zwiazek (1999), who showed that the *proportion* of water flowing through the apoplast was increased after mercury treatment; they took this as evidence that the mercury effect genuinely reduced cell-to-cell flow rather than caused, some nonspecific effect on all paths of water movement.

In onion roots, cell hydraulic conductivities were similar in all root regions. Root hydraulic conductivity was, however, greater in the middle region (which had a mature exodermis with both Casparian bands and suberin lamellae), and in older root regions (in which the endodermis also had developed suberin lamellae), as compared with the root tips (which had Casparian bands in the endodermis and an immature exodermis (Barrowclough *et al.,* 2000)). Mercury effectively inhibited cell L_p in the youngest regions of onion roots, but not root L_p. In contrast, it did inhibit L_p in older regions of the root. This suggested that apoplastic flow predominated in the root tips. In older regions of the root, the mature exodermis provided a significant resistance to apoplastic water uptake, leaving only the mercury-sensitive cell-to-cell route across that cell layer (Barrowclough *et al.,* 2000).

3. Regulations with Aquaporins Involved?

Root water transport is regulated by various environmental factors, and the contribution of aquaporins to these regulations has been inferred, based mostly on their sensitivity to mercury inhibition.

In many plant species including maize, melon, tomato, and paprika, salinization with 50–100 mM NaCl reduces root L_p by >65% after 2–4 days (Azaizeh and Steudle, 1991; Carvajal *et al.,* 1999; Martinez-Ballesta *et al.,* 2000; Peyrano *et al.,* 1997) and, in maize and melon, this effect can be counteracted with the addition of 10 mM calcium chloride. More detailed measurements of cellular L_p in maize roots showed the same effects, but these were even more dramatical than on whole roots (Azaizeh *et al.,* 1992). In melon, the P_f of root protoplasts was reduced by 70–90% after 2 days of salt treatment, and these effects were antagonized by calcium (Carvajal *et al.,* 2000; Martinez-Ballesta *et al.,* 2000). These data suggested that a major effect of salinity was on transcellular water transport. The role of aquaporins has been further proposed based on the observations that, under salinity conditions, the reduced L_p of roots and of isolated protoplasts became independent of mercury inhibition. These results are consistent with the transient down-regulation of the water channel message observed in *M. crystallinum* (Yamada *et al.*, 1995), but other regulatory mechanisms such as calcium-dependent phosphorylation of aquaporins may be involved (Carvajal *et al.*, 2000; Johansson *et al.*, 1998; Martinez-Ballesta *et al.*, 2000). More generally, it is assumed that down-regulation of root L_p under water stress conditions may allow the plant to limit a backflow of water into the soil when the driving force for water uptake is reduced because of limited transpiration (Steudle, 2000).

Anaerobiosis is another stress condition experienced by roots. It is associated with flooding or with mechanical compression of the soil and results in a decrease in root L_p (Birner and Steudle, 1993; Zhang and Tyerman, 1991, and references cited therein). Cell pressure probe measurements demonstrated that part of the control occurs at the cell membrane level (Zhang and Tyerman, 1991). These effects could be mimicked by NaN_3 treatment. Both anaerobiosis and NaN_3 induced a decrease in the apparent osmotic volume of the cells, as estimated with a cell pressure probe (Zhang and Tyerman, 1991). From this, it was deduced that the reduction in cell L_p may follow from a closure of plasmodesmata. In more recent studies (Zhang and Tyerman, 1999), it was shown that hypoxic or NaN_3 treated cells became insensitive to mercury inhibition, and the idea that water channels are down-regulated under these conditions seems to be favored at present.

Nutrient deficiency is yet another condition that induces a down-regulation of root L_p (Clarkson *et al.*, 2000). In wheat, deprivation of nitrogen or phosphorus resulted in a reversible 80–85% reduction in root L_p, similar in amplitude to the inhibition observed after mercury treatment (Carvajal *et al.*, 1996). In addition, the L_p of nutrient-deprived roots was insensitive to mercury inhibition, further suggesting that water channel down-regulation was occurring under these conditions. The mechanisms that allow plant roots to sense nutrient deficiency before any metabolic inhibition occurs have been recently discussed by Clarkson *et al.* (2000) but remain unknown.

Diurnal fluctuation of root L_p has also been observed in numerous plant species (Carvajal *et al.*, 1996; Clarkson *et al.*, 2000; Henzler *et al.*, 1999). In *Lotus corniculatus,* these variations were paralleled by a diurnal fluctuation in the expression of mRNAs homologous to an *Arabidopsis pip1a* probe (Clarkson *et al.*, 2000; Henzler *et al.*, 1999). This provides, besides mercury inhibition (Carvajal *et al.*, 1996), a different line of argument to link aquaporin expression and/or activity with the regulation of root water uptake.

Despite the numerous stress conditions that have been supposed to involve water channel regulation, it should be borne in mind that root anatomy is relatively plastic according to the growing conditions of the plant. For instance, water deficit induces the differentiation of an exodermis and promotes root growth as compared to that of shoot (Steudle, 2000). Thus, it is necessary to carefully evaluate to what extent anatomical changes may be affecting water flow, distinct from aquaporin-mediated effects. In general, a time course may help to distinguish between short-term physiological responses and longer-term anatomical changes.

C. Conclusion

Root water transport clearly represents a paradigm with which to investigate the role of plant aquaporins, but the amazing diversity of aquaporins and their expression throughout the plant can allow us to speculate that these proteins play a critical role in many other physiological processes. Functional genomics now provides new tools to address gene functions in large multigene families (Gura, 2000). Original functions for aquaporins, such as facilitating CO_2 diffusion in leaf tissues (Tyerman *et al.,* 1999), or embolism repair in xylem vessels (Holbrook and Zwieniecki, 1999), have also been proposed and will have to be experimentally explored. The role of water uptake in cell expansion has been largely controversial (Cosgrove, 1993), but the intense expression of aquaporins in elongating tissues provides a strong reason to revisit this question. Thus, there is still a critical need to combine molecular biology, genetics, and biophysics to develop an integrated concept of aquaporin function in whole-plant physiology. This effort should test our presumption that aquaporins are relevant targets for plant improvement and biotechnology.

Note added in proof. A description of 35 full-length *MIP* genes in *Arabidopsis* has recently appeared (Johanson *et al.,* 2001, *Plant Physiol.* **126,** 1358–1369). A new nomenclature was proposed and the *Arabidopsis* MIPs mentioned in the present review will have to be renamed accordingly.

References

Agre, P., Bonhivers, M., and Borgnia, M. J. (1998). The aquaporins, blueprints for cellular plumbing systems. *J. Biol. Chem.* **273,** 14,659–14,662.

Alexandre, J., and Lassalles, J.-P. (1991). Hydrostatic and osmotic pressure activated channel in plant vacuole. *Biophys. J.* **60,** 1326–1336.

Amodeo, G., Dorr, R., Vallejo, A., Sutka, M., and Parisi, M. (1999). Radial and axial water transport in the sugar beet storage root. *J. Exp. Bot.* **50,** 509–516.

Arabidopsis Genome Initiative (2000). Analysis of the genome of the flowering plant *Arabidopsis thaliana. Nature* **408,** 796–815.

Azaizeh, H., Gunse, B., and Steudle, E. (1992). Effects of NaCl and $CaCl_2$ on water transport across root cells of maize (*Zea mays* L.) seedlings. *Plant Physiol.* **99,** 886–894.

Azaizeh, H., and Steudle, E. (1991). Effects of salinity on water transport of excised maize (*Zea mays* L.) roots. *Plant Physiol.* **97,** 1136–1145.

Balk, P., and de Boer, A. D. (1999). Rapid stalk elongation in tulip (*Tulipa gesneriana* L. cv Apeldoorn) and the combined action of cold-induced invertase and the water-channel protein γTIP. *Planta* **209,** 646–354.

Barkla, B. J., Vera-Estrella, R., Pantoja, O., Kirch, H.-H., and Bohnert, H. J. (1999). Aquaporin localization — How valid are the TIP and PIP labels. *Trends Plant Sci.* **4,** 86–88.

Barone, L. M., Mu, H. H., Shih, C. J., Kashlan, K. B., and Wasserman, B. P. (1998). Distinct biochemical and topological properties of the 31- and 27-kilodalton plasma membrane intrinsic protein subgroups from red beet. *Plant Physiol.* **118,** 315–322.

Barrieu, F., Chaumont, F., and Chrispeels, M. J. (1998a). High expression of the tonoplast aquaporin ZmTIP1 in epidermal and conducting tissues of maize. *Plant Physiol.* **117,** 1153–1163.

Barrieu, F., Martymazars, D., Thomas, D., *et al.* (1999). Desiccation and osmotic stress increase the abundance of mRNA of the tonoplast aquaporin BobTIP26-1 in cauliflower cells. *Planta* **209,** 77–86.

Barrieu, F., Thomas, D., Martymazars, D., Charbonnier, M., and Marty, F. (1998b). Tonoplast intrinsic proteins from cauliflower (*Brassica oleracea* L. var. botrytis): Immunological analysis, cDNA cloning and evidence for expression in meristematic tissues. *Planta* **204,** 335–344.

Barrowclough, D. E., Peterson, C. A., and Steudle, E. (2000). Radial hydraulic conductivity along developing onion roots. *J. Exp. Bot.* **51,** 547–557.

Bethke, P. C., and Jones, R. L. (2000). Vacuoles and prevacuolar compartments. *Curr. Op. Plant Biol.* **3,** 469–475.

Biela, A., Grote, K., Otto, B., *et al.* (1999). The *Nicotiana tabacum* plasma membrane aquaporin NtAQP1 is mercury-insensitive and permeable for glycerol. *Plant J.* **18,** 565–570.

Birner, T. P., and Steudle, E. (1993). Effects of anaerobic conditions on water and solute relations, and on active transport in roots of maize (*Zea mays* L.). *Planta* **190,** 474–483.

Boyer, J. S. (1985). Water transport. *Annu. Rev. Plant Physiol.* **36,** 473–516.

Carvajal, M., Cerda, A., and Martinez, V. (2000). Does calcium ameliorate the negative effect of NaCl on melon root water transport by regulating aquaporin activity?. *New Phytol.* **145,** 439–447.

Carvajal, M., Cooke, D. T., and Clarkson, D. T. (1996). Responses of wheat plants to nutrient deprivation may involve the regulation of water-channel function. *Planta* **199,** 372–381.

Carvajal, M., Martinez, V., and Alcaraz, C. F. (1999). Physiological function of water channels as affected by salinity in roots of paprika pepper. *Physiol. Plant.* **105,** 95–101.

Chaumont, F., Barrieu, F., Herman, E. M., and Chrispeels, M. J. (1998). Characterization of a maize tonoplast aquaporin expressed in zones of cell division and elongation. *Plant Physiol.* **117,** 1143–1152.

Chaumont, F., Barrieu, F., Jung, R., and Chrispeels, M. J. (2000). Plasma membrane intrinsic proteins from maize cluster in two sequence subgroups with differential aquaporin activity. *Plant Physiol.* **122,** 1025–1034.

Chevalier, C., Bourgeois, E., Pradet, A., and Raymond, P. (1995). Molecular cloning and characterization of six cDNAs expressed during glucose starvation in excised maize (*Zea mays* L.) root tips. *Plant Mol. Biol.* **28,** 473–485.

Clarkson, D. T., Carvajal, M., Henzler, T., *et al.* (2000). Root hydraulic conductance: Diurnal aquaporin expression and the effects of nutrient stress. *J. Exp. Bot.* **51,** 61–70.

Conkling, M. A., Cheng, C.-L., Yamamoto, Y. T., and Goodman, H. M. (1990). Isolation of transcriptionally regulated root-specific genes of tobacco. *Plant Physiol.* **93,** 12003–12011.

Cosgrove, D. J. (1993). Water uptake by growing cells: An assessment of the controlling roles of wall relaxation, solute uptake and hydraulic conductance. *Int. J. Plant Sci.* **154,** 10–21.

Coury, L. A., Hiller, M., Mathai, J. C., *et al.* (1999). Water transport across yeast vacuolar and plasma membrane-targeted secretory vesicles occurs by passive diffusion. *J. Bacteriol.* **181,** 4437–4440.

Culianez-Macia, F. A., and Martin, C. (1993). DIP: a member of the MIP family of membrane proteins that is expressed in mature seeds and dark-grown seedlings of *Antirrhinum majus. Plant J.* **4,** 717–725.

Cutler, S. R., Ehrhardt, D. W., Griffitts, J. S., and Somerville, C. R. (2000). Random GFP::cDNA fusions enable visualization of subcellular structures in cells of *Arabidopsis* at high frequency. *Proc. Natl. Acad. Sci. USA* **97,** 3718–3723.

Dainty, J. (1963). Water relations of plant cells. *Adv. Bot. Res.* **1,** 279–326.

Daniels, M. J., Chaumont, F., Mirkov, T. E., and Chrispeels, M. J. (1996). Characterization of a new vacuolar membrane aquaporin sensitive to mercury at a unique site. *Plant Cell* **8,** 587–599.

Daniels, M. J., Chrispeels, M. J., and Yeager, M. (1999). Projection structure of a plant vacuole membrane aquaporin by electron cryo-crystallography. *J. Mol. Biol.* **294,** 1337–1349.

Daniels, M. J., Mirkov, T. E., and Chrispeels, M. J. (1994). The plasma membrane of *Arabidopsis thaliana* contains a mercury-insensitive aquaporin that is a homolog of the tonoplast water channel protein TIP. *Plant Physiol.* **106,** 1325–1333.

Davies, K. M., and Grierson, D. (1989). Identification of cDNA clones for tomato (*Lycopersicon esculentum* Mill.) mRNAs that accumulate during fruit ripening and leaf senescence in response to ethylene. *Planta* **179,** 73–80.

Dean, R. M., Rivers, R. L., Zeidel, M. L., and Roberts, D. M. (1999). Purification and functional reconstitution of soybean Nodulin 26. An aquaporin with water and glycerol transport properties. *Biochemistry* **38,** 347–353.

Didierjean, L., Frendo, P., Nasser, W., *et al.* (1996). Heavy-metal-responsive genes in maize: Identification and comparison of their expression upon various forms of abiotic stress. *Planta* **199,** 1–8.

Ferguson, D. L., Turley, R. B., and Kloth, R. H. (1997). Identification of a δ-TIP cDNA clone and determination of related A and D genome subfamilies in *Gossypium* species. *Plant Mol. Biol.* **34,** 111–118.

Fleurat-Lessard, P., Frangne, N., Maeshima, M., *et al.* (1997). Increased expression of vacuolar aquaporin and H+-ATPase related to motor cell function in Mimosa pudica L. *Plant Physiol.* **114,** 827–834.

Fortin, M. G., Morrison, N. A., and Verma, D. P. S. (1987). Nodulin-26, a peribacteroid membrane nodulin is expressed independently of the development of the peribacteroid compartment. *Nucleic Acids Res.* **15,** 813–824.

Fotiadis, D., Jeno, P., Mini, T., *et al.* (2001). Structural characterization of two aquaporins isolated from native spinach leaf plasma membranes. *J. Biol. Chem.* **276,** 1707–1714.

Fray, R. G., Wallace, A., Grierson, D., and Lycett, G. W. (1994). Nucleotide sequence and expression of a ripening and water stress-related cDNA from tomato with homology to the MIP class of membrane channel proteins. *Plant Mol. Biol.* **24,** 539–543.

Frensch, J., Hsiao, T. C., and Steudle, E. (1996). Water and solute transport along developing maize roots. *Planta* **198,** 348–355.

Fu, D., Libson, A., Miercke, L. J. W., *et al.* (2000). Structure of a glycerol-conducting channel and the basis for its selectivity. *Science* **290,** 481–486.

Fushimi, K., Sasaki, S., and Marumo, F. (1997). Phosphorylation of serine 256 is required for cAMP-dependent regulatory exocytosis of the aquaporin-2 water channel. *J. Biol. Chem.* **272,** 14,800–14,804.

Gao, Y. P., Young, L., BonhamSmith, P., and Gusta, L. V. (1999). Characterization and expression of plasma and tonoplast membrane aquaporins in primed seed of *Brassica napus* during germination under stress conditions. *Plant Mol. Biol.* **40,** 635–644.

Gerbeau, P., Güclü, J., Ripoche, P., and Maurel, C. (1999). Aquaporin Nt-TIPa can account for the high permeability of tobacco cell vacuolar membrane to small neutral solutes. *Plant J.* **18,** 577–587.

Grote, K., vonTrzebiatovski, P., and Kaldenhoff, R. (1998). RNA levels of plasma membrane aquaporins in *Arabidopsis thaliana*. *Protoplasma* **204,** 139–144.

Guenther, J. F., and Roberts, D. M. (2000). Water-selective and multifunctional aquaporins from *Lotus japonicus* nodules. *Planta* **210,** 741–748.

Guerrero, F. D., and Crossland, L. (1993). Tissue-specific expression of a plant turgor-responsive gene with amino acid sequence homology to transport-facilitating proteins. *Plant Mol. Biol.* **21,** 929–935.

Guerrero, F. D., Jones, J. T., and Mullet, J. E. (1990). Turgor-responsive gene transcription and RNA levels increase rapidly when pea shoots are wilted. Sequence and expression of three inducible genes. *Plant Mol. Biol.* **15,** 11–26.

Gura, T. (2000). Reaping the plant gene harvest. *Science* **287,** 412–414.

Hejnowicz, Z., and Sievers, A. (1996). Reversible closure of water channels in parenchymatic cells of sunflower hypocotyl depends on turgor status of the cells. *J. Plant Physiol.* **147,** 516–520.

Henzler, T., and Steudle, E. (1995). Reversible closing of water channels in *Chara* internodes provides evidence for a composite transport model of the plasma membrane. *J. Exp. Bot.* **46,** 199–209.

Henzler, T., Waterhouse, R. N., Smyth, A. J., *et al.* (1999). Diurnal variations in hydraulic conductivity and root pressure can be correlated with the expression of putative aquaporins in the roots of *Lotus japonicus*. *Planta* **210,** 50–60.

Hertel, A., and Steudle, E. (1997). The function of water channels in *Chara:* The temperature dependence of water and solute flows provides evidence for composite membrane transport and for a slippage of small organic solutes across water channels. *Planta* **202,** 324–335.

Higuchi, T., Suga, S., Tsuchiya, T., *et al.* (1998). Molecular cloning, water channel activity and tissue specific expression of two isoforms of radish vacuolar aquaporin. *Plant Cell Physiol.* **39,** 905–913.

Höfte, H., and Chrispeels, M. J. (1992). Protein sorting to the vacuolar membrane. *Plant Cell* **4,** 995–1004.

Höfte, H., Faye, L., Dickinson, C., Herman, E. M., and Chrispeels, M. J. (1991). The protein-body proteins phytohemagglutinin and tonoplast intrinsic protein are targeted to vacuoles in leaves of transgenic tobacco. *Planta* **184,** 431–437.

Höfte, H., Hubbard, L., Reizer, J., *et al.* (1992). Vegetative and seed-specific forms of tonoplast intrinsic protein in the vacuolar membrane of *Arabidopsis thaliana. Plant Physiol.* **99,** 561–570.

Hoh, B., Hinz, G., Jeong, B.-K., and Robinson, D. G. (1995). Protein storage vacuoles form de novo during pea cotyledon development. *J. Cell Sci.* **108,** 299–310.

Holbrook, N. M., and Zwieniecki, M. A. (1999). Embolism repair and xylem tension: do we need a miracle? *Plant Physiol.* **120,** 7–10.

Ikeda, S., Nasrallah, J. B., Dixit, R., Preiss, S., and Nasrallah, M. E. (1997). An aquaporin-like gene required for the Brassica self-incompatibility response. *Science* **276,** 1564–1566.

Ingram, J., and Bartels, D. (1996). The molecular basis of dehydration tolerance in plants. *Annu. Rev. Plant Physiol. Plant Mol. Biol.* **47,** 377–403.

Inoue, K., Motozaki, A., Takeuchi, Y., Nishimura, M., and Hara-Nishimura, I. (1995a). Molecular characterization of proteins in protein-body membrane that disappear most rapidly during transformation of protein bodies into vacuoles. *Plant J.* **7,** 235–243.

Inoue, K., Takeuchi, Y., Nishimura, M., and Hara-Nishimura, I. (1995b). Characterization of two integral membrane proteins located in the protein bodies of pumpkin seeds. *Plant Mol. Biol.* **28,** 1089–1101.

Jauh, G. Y., Fischer, A. M., Grimes, H. D., Ryan, C. A., and Rogers, J. C. (1998). δ-tonoplast intrinsic protein defines unique plant vacuole functions. *Proc. Natl. Acad. Sci. USA* **95,** 12,995–12,999.

Jauh, G. Y., Phillips, T. E., and Rogers, J. C. (1999). Tonoplast intrinsic protein isoforms as markers for vacuolar functions. *Plant Cell* **11,** 1867–1882.

Jiang, L., and Rogers, J. C. (1998). Integral membrane protein sorting to vacuoles in plant cells: evidence for two pathways. *J. Cell. Biol.* **143,** 1183–1199.

Johanson, U., and Kjellbom, P. (2000). An almost-complete guide to the MIPs in *Arabidopsis. In* "3rd International Conference on Molecular Biology and Physiology of Water and Solute Transport." Abstr. p. 117.

Johansson, I., Karlsson, M., Johanson, U., Larsson, C., and Kjellbom, P. (2000). The role of aquaporins in cellular and whole plant water balance. *Biochim. Biophys. Acta* **1465,** 324–342.

Johansson, I., Karlsson, M., Shukla, V. K., *et al.* (1998). Water transport activity of the plasma membrane aquaporin PM28A is regulated by phosphorylation. *Plant Cell* **10,** 451–459.

Johansson, I., Larsson, C., Ek, B., and Kjellbom, P. (1996). The major integral proteins of spinach leaf plasma membranes are putative aquaporins and are phosphorylated in response to Ca^{2+} and apoplastic water potential. *Plant Cell* **8,** 1181–1191.

Johnson, K. D., and Chrispeels, M. J. (1992). Tonoplast-bound protein kinase phosphorylates tonoplast intrinsic protein. *Plant Physiol.* **100,** 1787–1795.

Johnson, K. D., Herman, E. M., and Chrispeels, M. J. (1989). An abundant, highly conserved tonoplast protein in seeds. *Plant Physiol.* **91,** 1006–1013.

Jones, J. T., and Mullet, J. E. (1995). Developmental expression of a turgor-responsive gene that encodes an intrinsic membrane protein. *Plant Mol. Biol.* **28,** 983–996.

Kaldenhoff, R., Grote, K., Zhu, J.-J., and Zimmermann, U. (1998). Significance of plasmalemma aquaporins for water-transport in *Arabidopsis thaliana. Plant J.* **14,** 121–128.

Kaldenhoff, R., Kölling, A., Meyers, J., *et al.* (1995). The blue light-responsive *AthH2* gene of *Arabidopsis thaliana* is primarily expressed in expanding as well as in differentiating cells and encodes a putative channel protein of the plasmalemma. *Plant J.* **7,** 87–95.

Kaldenhoff, R., Kölling, A., and Richter, G. (1993). A novel blue light- and abscisic acid-inducible gene of *Arabidopsis thaliana* encoding an intrinsic membrane protein. *Plant Mol. Biol.* **23,** 1187–1198.

Kaldenhoff, R., Kölling, A., and Richter, G. (1996). Regulation of the *Arabidopsis thaliana* aquaporin-gene AthH2 (PIP1b). *J. Photobiol. Photobiochem.* **36,** 351–354.

Kammerloher, W., Fischer, U., Piechottka, G. P., and Schäffner, A. R. (1994). Water channels in the plant plasma membrane cloned by immunoselection from a mammalian expression system. *Plant J.* **6,** 187–199.

Karlsson, M., Johansson, I., Bush, M., *et al.* (2000). An abundant TIP expressed in mature highly vacuolated cells. *Plant J.* **21,** 83–90.

Katsura, T., Gustafson, C. E., Ausiello, D. A., and Brown, D. (1997). Protein kinase A phosphorylation is involved in regulated exocytosis of aquaporin-2 in transfected LLC-PK1 cells. *Am. J. Physiol.* **41,** F816–F822.

Kirch, H.-H., Vera-Estrella, R., Golldack, D., *et al.* (2000). Expression of water channel proteins in *Mesembryanthemum crystallinum. Plant Physiol.* **123,** 111–124.

Kiyosawa, K., and Tazawa, M. (1977). Hydraulic conductivity of tonoplast-free *Chara* cells. *J. Membr. Biol.* **37,** 157–166.

Krajinski, F., Biela, A., Schubert, D., *et al.* (2000). Arbuscular mycorrhiza devlopment regulates the mRNA abundance of *Mtaqp1* encoding a mercury-insensitive aquaporin of *Medicago truncatula. Planta* **211,** 85–90.

Lee, J. W., Zhang, Y., Weaver, C. D., *et al.* (1995). Phosphorylation of Nodulin 26 on Serine 262 affects its voltage-sensitive channel activity in planar lipid bilayers. *J. Biol. Chem.* **270,** 27,051–27,057.

Li, L.-G., Li, S.-F., Tao, Y., and Kitagawa, Y. (2000). Molecular cloning of a novel water channel from rice: Its products expression in *Xenopus* oocytes and involvement in chilling tolerance. *Plant Sci.* **154,** 43–51.

Liu, Q., Umeda, M., and Uchimiya, H. (1994). Isolation and expression analysis of two rice genes encoding the major intrinsic protein. *Plant Mol. Biol.* **26,** 2003–2006.

Ludevid, D., Höfte, H., Himelblau, E., and Chrispeels, M. J. (1992). The expression pattern of the tonoplast intrinsic protein γ-TIP in *Arabidopsis thaliana* is correlated with cell enlargement. *Plant Physiol.* **100,** 1633–1639.

Luyten, K., Albertyn, J., Skibbe, W. F., *et al.* (1995). Fps1, a yeast member of the MIP family of channel proteins, is a facilitator for glycerol uptake and efflux and is inactive under osmotic stress. *EMBO J.* **14,** 1360–1371.

Maeshima, M. (1992). Characterization of the major integral protein of vacuolar membrane. *Plant Physiol.* **98,** 1248–1254.

Maeshima, M., Hara-Nishimura, I., Takeuchi, Y., and Nishimura, M. (1994). Accumulation of vacuolar H^+-pyrophosphatase and H^+-ATPase during reformation of the central vacuole in germinating pumpkin seeds. *Plant Physiol.* **106,** 61–69.

Maggio, A., and Joly, R. J. (1995). Effects of mercuric chloride on the hydraulic conductivity of tomato root systems. Evidence for a channel-mediated water pathway. *Plant Physiol.* **109,** 331–335.

Malz, S., and Sauter, M. (1999). Expression of two PIP genes in rapidly growing internodes of rice is not primarily controlled by meristem activity or cell expansion. *Plant Mol. Biol* **40,** 985–995.

Mariaux, J. B., Bockel, C., Salamini, F., and Bartels, D. (1998). Desiccation- and abscisic acid-responsive genes encoding major intrinsic proteins (MIPs) from the resurrection plant *Craterostigma plantagineum. Plant Mol. Biol.* **38,** 1089–1099.

Martinez-Ballesta, M. d. C., Martinez, V., and Carvajal, M. (2000). Regulation of water channel activity in whole roots and in protoplasts from roots of melon plants grown under saline conditions. *Aust. J. Plant Physiol.* **27,** 685–691.

Maurel, C. (1997). Aquaporins and water permeability of plant membranes. *Annu. Rev. Plant Physiol. Plant Mol. Biol.* **48,** 399–429.

Maurel, C., and Chrispeels, M. J. (2001). Aquaporins: a molecular entry into plant water relations. *Plant Physiol.* **125,** 135–138.

Maurel, C., Chrispeels, M. J., Lurin, C., *et al.* (1997a). Function and regulation of seed aquaporins. *J. Exp. Bot.* **48,** 421–430.

Maurel, C., Kado, R. T., Guern, J., and Chrispeels, M. J. (1995). Phosphorylation regulates the water channel activity of the seed-specific aquaporin α-TIP. *EMBO J.* **14,** 3028–3035.

Maurel, C., Reizer, J., Schroeder, J. I., and Chrispeels, M. J. (1993). The vacuolar membrane protein γ-TIP creates water specific channels in *Xenopus* oocytes. *EMBO J.* **12,** 2241–2247.

Maurel, C., Tacnet, F., Güclü, J., Guern, J., and Ripoche, P. (1997b). Purified vesicles of tobacco cell vacuolar and plasma membranes exhibit dramatically different water permeability and water channel activity. *Proc. Natl. Acad. Sci. USA* **94,** 7103–7108.

Melroy, D. L., and Herman, E. M. (1991). TIP, an integral membrane protein of the protein-storage vacuoles of the soybean cotyledon undergoes developmentally regulated membrane accumulation and removal. *Planta* **184,** 113–122.

Miao, G.-H., Hong, Z., and Verma, D. P. S. (1992). Topology and phosphorylation of soybean nodulin-26, an intrinsic protein of the peribacteroid membrane. *J. Cell Biol.* **118,** 481–490.

Morillon, R., and Lassalles, J.-P. (1999). Osmotic water permeability of isolated vacuoles. *Planta* **210,** 80–84.

Murata, K., Mitsuoka, K., Hirai, T., *et al.* (2000). Structural determinants of water permeation through aquaporin-1. *Nature* **407,** 599–605.

Nakhoul, N. L., Davis, B. A., Romero, M. F., and Boron, W. F. (1998). Effect of expressing the water channel aquaporin-1 on the CO_2 permeability of *Xenopus* oocytes. *Am. J. Physiol.* **274,** C543–C548.

Niemietz, C. M., and Tyerman, S. D. (1997). Characterization of water channels in wheat root membrane vesicles. *Plant, Physiol.* **115,** 561–567.

Niemietz, C. M., and Tyerman, S. D. (2000). Channel-mediated permeation of ammonia gas through the peribacteroid membrane of soybean nodules. *FEBS Lett.* **465,** 110–114.

Oliviusson, P., and Hakman, I. (1995). A tonoplast intrinsic protein (TIP) is present in seeds, roots and somatic embryos of Norway spruce (*Picea abies*). *Physiol. Plant.* **95,** 288–295.

Opperman, C. H., Taylor, C. G., and Conkling, M. A. (1994). Root-knot nematode-directed expression of a plant root-specific gene. *Science* **263,** 221–223.

Otto, B., and Kaldenhoff, R. (2000). Cell-specific expression of the mercury-insensitive plasma-membrane aquaporin NtAQP1 from *Nicotiana tabacum. Planta* **211,** 167–172.

Ouyang, L.-J., Whelan, J., Weaver, C. D., Roberts, D. M., and Day, D. A. (1991). Protein phosphorylation stimulates the rate of malate uptake across the peribacteroid membrane of soybean nodules. *FEBS Lett.* **293,** 188–190.

Paris, N., Stanley, C. M., Jones, R. L., and Rogers, J. C. (1996). Plant cells contain two functionally distinct vacuolar compartments. *Cell* **85,** 563–572.

Peterson, C. A., and Steudle, E. (1993). Lateral hydraulic conductivity of early metaxylem vessels in *Zea mays* L. roots. *Planta* **189,** 288–297.

Peyrano, G., Taleisnik, E., Quiroga, M., de Forchetti, S. M., and Tigier, H. (1997). Salinity effects on hydraulic conductance, lignin content and peroxydase activity in tomato roots. *Plant Physiol. Biochem.* **35,** 387–393.

Phillips, A. L., and Huttly, A. K. (1994). Cloning of two gibberellin-regulated cDNAs from *Arabidopsis thaliana* by subtractive hybridization: Expression of the tonoplast water channel, γ-TIP, is increased by GA_3. *Plant Mol. Biol.* **24,** 603–615.

Prasad, G. V. R., Coury, L. A., Finn, F., and Zeidel, M. L. (1998). Reconstituted aquaporin 1 water channels transport CO_2 across membranes. *J. Biol. Chem.* **273,** 33,123–33,126.

Qi, X., Tai, C.-Y., and Wasserman, B. P. (1995). Plasma membrane intrinsic proteins of *Beta vulgaris* L. *Plant Physiol.* **108,** 387–392.

Quintero, J. M., Fournier, J. M., and Benlloch, M. (1999). Water transport in sunflower root systems: Effects of ABA, Ca^{2+} status and $HgCl_2$. *J. Exp. Bot.* **50,** 1607–1612.

Ramahaleo, T., Morillon, R., Alexandre, J., and Lassalles, J.-P. (1999). Osmotic water permeability of isolated protoplasts. Modifications during development. *Plant Physiol.* **119,** 885–896.

Rivers, R. L., Dean, R. M., Chandy, G., *et al.* (1997). Functional analysis of Nodulin 26, an aquaporin in soybean root symbiosomes. *J. Biol. Chem.* **272,** 16,256–16,261.

Robinson, D. G., Haschke, H. P., Hinz, G., *et al.* (1996a). Immunological detection of tonoplast polypeptides in the plasma membrane of pea cotyledons. *Planta* **198,** 95–103.

Robinson, D. G., Sieber, H., Kammerloher, W., and Schäffner, A. R. (1996b). PIP1 aquaporins are concentrated in plasmalemmasomes of *Arabidopsis thaliana* mesophyll. *Plant Physiol.* **111,** 645–649.

Roussel, H., Bruns, S., Gianinazzipearson, V., Hahlbrock, K., and Franken, P. (1997). Induction of a membrane intrinsic protein-encoding mRNA in arbuscular mycorrhiza and elicitor-stimulated cell suspension cultures of parsley. *Plant Sci.* **126,** 203–210.

Ruiter, R. K., Vaneldik, G. J., Vanherpen, M. M. A., Schrauwen, J. A. M., and Wullems, G. J. (1997). Expression in anthers of two genes encoding Brassica oleracea transmembrane channel proteins. *Plant Mol. Biol.* **34,** 163–168.

Santoni, V., Gerbeau, P., Javot, H., and Maurel, C. (2000). The high diversity of aquaporins reveals novel facets of plant membrane functions. *Curr. Opin. Plant. Biol* **3,** 476–481.

Sarda, X., Tousch, D., Ferrare, K., *et al.* (1999). Characterization of closely related delta-TIP genes encoding aquaporins which are differentially expressed in sunflower roots upon water deprivation through exposure to air. *Plant Mol. Biol.* **40,** 179–191.

Sarda, X., Tousch, D., Ferrare, K., *et al.* (1997). Two TIP-like genes encoding aquaporins are expressed in sunflower guard cells. *Plant J.* **12,** 1103–1111.

Sarker, R. H., Elleman, C. J., and Dickinson, H. G. (1988). Control of pollen hydration in *Brassica* requires continued protein synthesis, and glycosylation is necessary for intraspecific incompatibility. *Proc. Natl. Acad. Sci. USA* **85,** 4340–4344.

Sasaki, S., Ishibashi, K., and Marumo, F. (1998). Aquaporin-2 and -3: Representatives of two subgroups of the aquaporin family colocalized in the kidney collecting duct. *Annu. Rev. Physiol.* **60,** 199–220.

Schäffner, A. R. (1998). Aquaporin function, structure, and expression: Are there more surprises to surface in water relations? *Planta* **204,** 131–139.

Schneider, H., Zhu, J. J., and Zimmermann, U. (1997). Xylem and cell turgor pressure probe measurements in intact roots of glycophytes: Transpiration induces a change in the radial and cellular reflection coefficients. *Plant Cell Environ.* **20,** 221–229.

Schütz, K., and Tyerman, S. D. (1997). Water channels in *Chara corallina. J. Exp. Bot.* **48,** 1511–1518.

Serraj, R., Frangne, N., Maeshima, M., Fleuratlessard, P., and Drevon, J. J. (1998). A γ-TIP cross-reacting protein is abundant in the cortex of soybean N_2-fixing nodules. *Planta* **206,** 681–684.

Skinner, R. H., and Radin, J. W. (1994). The effect of phosphorus nutrition on water flow through the apoplastic bypass in cotton roots. *J. Exp. Bot.* **45,** 423–428.

Steudle, E. (1989). Water flow in plants and its coupling to other processes: An overview. *Methods Enzymol.* **174,** 183–225.

Steudle, E. (1993). Pressure probe techniques: Basic principles and application to studies of water and solute relations at the cell, tissue and organ level. *In* Water Deficits: Plant Responses from Cell to Community. (Smith, J. A. C., and Griffiths, H., Eds.), Bios Scientific Publishers Ltd, Oxford, UK, pp. 5–36.

Steudle, E. (1994). The regulation of plant water at the cell, tissue, and organ level: Role of active processes and of compartmentation. *In* Flux Control in Biological Systems. From Enzymes to Populations and Ecosystems. (Schultze, E. D., Ed.), Academic Press, Inc., San Diego, pp. 237–299.

Steudle, E. (2000). Water uptake by roots: Effects of water deficit. *J. Exp. Bot.* **51,** 1531–1542.

Steudle, E., and Henzler, T. (1995). Water channels in plants: Do basic concepts of water transport change? *J. Exp. Bot.* **46,** 1067–1076.

Steudle, E., and Peterson, C. A. (1998). How does water get through roots? *J. Exp. Bot.* **49,** 775–788.

Swanson, S. J., Bethke, P. C., and Jones, R. L. (1998). Barley aleurone cells contain two types of vacuoles: Characterization of lytic organelles by use of fluorescent probes. *Plant Cell* **10,** 685–698.

Tazawa, M., Asai, K., and Iwasaki, N. (1996). Characteristics of Hg- and Zn-sensitive water channels in the plasma membrane of *Chara* cells. *Bot. Acta* **109,** 388–396.

Tazawa, M., Ohkuma, E., Shibasaka, M., and Nakashima, S. (1997). Mercurial-sensitive water transport in barley roots. *J. Plant Res.* **110,** 435–442.

Thompson, J. D., Higgins, D. G., and Gibson, T. J. (1994). CLUSTAL W: Improving the sensitivity of progressive multiple sequence alignment through sequence weighting, positions-specific gap penalties and weight matrix choice. *Nucleic Acids Res.* **22,** 4673–4680.

Tyerman, S. D., Bohnert, H. J., Maurel, C., Steudle, E., and Smith, J. A. (1999). Plant aquaporins: Their molecular biology, biophysics and significance for plant water relations. *J. Exp. Bot.* **50,** 1055–1071.

Tyerman, S. D., Oats, P., Gibbs, J., Dracup, M., and Greenway, H. (1989). Turgor-volume regulation and cellular water relations of *Nicotiana tabacum* roots grown in high salinities. *Austr. J. Plant Physiol.* **16,** 517–531.

Url, W. G. (1971). The site of penetration resistance to water in plant protoplasts. *Protoplasma* **72,** 427–447.

Utsugi, S., Sakamoto, W., Ogura, Y., Murata, M., and Motoyoshi, F. (1996). Isolation and characterization of cDNA clones corresponding to the genes expressed preferentially in floral organs of *Arabidopsis thaliana. Plant Mol. Biol.* **32,** 759–765.

Verkman, A. S. (2000). Water permeability measurement in living cells and complex tissues. *J. Membr. Biol.* **173,** 73–87.

Wan, X. C., and Zwiazek, J. J. (1999). Mercuric chloride effects on root water transport in aspen seedlings. *Plant Physiol.* **121,** 939–946.

Wayne, R., Mimura, T., and Shimmen, T. (1994). The relationship between carbon and water transport in single cells of *Chara corallina. Protoplasma* **180,** 118–135.

Wayne, R., and Tazawa, M. (1990). Nature of the water channels in the internodal cells of *Nitellopsis. J. Membr. Biol.* **116,** 31–39.

Weaver, C. D., Crombie, B., Stacey, G., and Roberts, D. M. (1991). Calcium-dependent phosphorylation of symbiosome membrane proteins from nitrogen-fixing soybean nodules. *Plant Physiol.* **95,** 222–227.

Weaver, C. D., Shomer, N. H., Louis, C. F., and Roberts, D. M. (1994). Nodulin 26, a nodule-specific symbiosome membrane protein from soybean, is an ion channel. *J. Biol. Chem.* **269,** 17,858–17,862.

Wegner, L. H., and Zimmermann, U. (1998). Simultaneous recording of xylem pressure and trans-root potential in roots of intact glycophytes using a novel xylem pressure probe technique. *Plant Cell Environ.* **21,** 849–865.

Weig, A., Deswarte, C., and Chrispeels, M. J. (1997). The major intrinsic protein family of Arabidopsis has 23 members that form three distinct groups with functional aquaporins in each group. *Plant Physiol.* **114,** 1347–1357.

Weig, A. R., and Jakob, C. (2000). Functional identification of the glycerol permease activity of *Arabidopsis thaliana* NLM1 and NLM2 proteins by heterologous expression in *Saccharomyces cerevisiae. FEBS Lett.* **481,** 293–298.

Yamada, S., Katsuhara, M., Kelly, W. B., Michalowski, C. B., and Bohnert, H. J. (1995). A family of transcripts encoding water channel proteins: Tissue-specific expression in the common ice plant. *Plant Cell* **7,** 1129–1142.

Yamada, S., Nelson, D. E., Ley, E., Marquez, S., and Bohnert, H. J. (1997). The expression of an aquaporin promoter from *Mesembryanthemum crystallinum* in tobacco. *Plant Cell Physiol.* **38,** 1326–1332.

Yamaguchi-Shinozaki, K., Koizumi, M., Urao, S., and Shinozaki, K. (1992). Molecular cloning and characterization of 9 cDNAs for genes that are responsive to desiccation in *Arabidopsis thaliana:*

Sequence analysis of one cDNA clone that encodes a putative transmembrane channel protein. *Plant Cell Physiol.* **33,** 217–224.

Yamamoto, Y. T., Taylor, C. G., Acedo, G. H., Cheng, C.-L., and Conkling, M. A. (1991). Characterization of *cis*-acting sequences regulating root-specific gene expression in tobacco. *Plant Cell* **3,** 371–382.

Yang, B. X., and Verkman, A. S. (1997). Water and glycerol permeabilities of aquaporins 1–5 and MIP determined quantitatively by expression of epitope-tagged constructs in Xenopus oocytes. *J. Biol. Chem.* **272,** 16,140–16,146.

Zhang, W. H., and Tyerman, S. D. (1991). Effect of low O_2 concentration and azide on hydraulic conductivity and osmotic volume of the cortical cells of wheat roots. *Aust. J. Plant Physiol.* **18,** 603–613.

Zhang, W.-H., and Tyerman, S. D. (1999). Inhibition of water channels by $HgCl_2$ in intact wheat root cells. *Plant Physiol.* **120,** 849–857.

Zimmermann, H. M., and Steudle, E. (1998). Apoplastic transport across young maize roots: Effect of the exodermis. *Planta* **206,** 7–19.

Osmotic Adaptation in Yeast—Control of the Yeast Osmolyte System

Stefan Hohmann
Department of Cell and Molecular Biology, Microbiology, Göteborg University, S-405 30 Göteborg, Sweden

The yeast *Saccharomyces cerevisiae* (baker's yeast or budding yeast) is an excellent eukaryotic model system for cellular biology with a well-explored, completely sequenced genome. Yeast cells possess robust systems for osmotic adaptation. Central to the response to high osmolarity is the HOG pathway, one of the best-explored MAP kinase pathways. This pathway controls via different transcription factors the expression of more than 150 genes. In addition, osmotic responses are also controlled by protein kinase A via a general stress response pathway and by presently unknown signaling systems. The HOG pathway partially controls expression of genes encoding enzymes in glycerol production. Glycerol is the main yeast osmolyte, and its production is essential for growth in a high osmolarity medium. Upon hypo-osmotic shock, yeast cells transiently stimulate another MAP kinase pathway, the so-called PKC pathway, which appears to orchestrate the assembly of the cell surface and the cell wall. In addition, yeast cells show signs of a regulated volume decrease by rapidly exporting glycerol through Fps1p. This unusual MIP channel is gated by osmotic changes and thereby plays a key role in controlling the intracellular osmolyte content. Yeast cells also possess two aquaporins, Aqy1p and Aqy2p. The production of both proteins is strictly regulated, suggesting that these water channels play very specific roles in yeast physiology. Aqy1p appears to be developmentally regulated. Given the strong yeast research community and the excellent tools of genetics and functional genomics available, we expect yeast to be the best-explored cellular organism for several years ahead, and osmotic responses are a focus of interest for numerous yeast researchers.

KEY WORDS: Yeast, Osmotic shock, MAP kinase signaling, Glycerol, Transport, Gene expression, Adaptation

I. Yeast as a Model System

A. Yeast in Its Natural Environment

Yeasts are ubiquitous in nature and are commonly found on any sugar-containing plant material such as decomposing fruits and in flowers. Yeasts are nonmotile and hence require help from other organisms for transmission, typically insects such as the fruit fly. Hundreds of yeast species are known today, but in molecular biology the term "yeast" is synonymous with mainly two species: the budding yeast *Saccharomyces cerevisiae* and the fission yeast *Schizosaccharomyces pombe*.

The natural environment of yeasts is highly variable with respect to nutrient availability, temperature, oxygen levels, pH, and osmotic pressure. Yeasts adapt to different nitrogen- and carbon-source quality and levels, to temperatures ranging over more than 30 degrees, to oxygen levels between atmospheric concentrations and almost anaerobicity, to pH levels between 3 and 7 (must from the grape has a pH of as low as 3), and to water activities ranging from distilled water to almost saturated sugar solutions (Onishi, 1963). With a reasonable survival rate, yeasts also manage sudden shifts between those extremes, although it is well known that preadaptation to mild stress makes yeasts much more tolerant to extreme stress (Hohmann and Mager, 1997). The need to adapt to dramatic osmotic changes is illustrated by a yeast cell situated on a drying fruit with very high osmotic pressure, while the next moment, the cell may be washed off in rainwater.

The metabolic activity of the yeast itself causes osmotic changes. The metabolism of *Saccharomyces cerevisiae* is programmed to catabolize external sugar's rapidly to ethanol, irrespective of the presence of oxygen (Gancedo and Serrano, 1989). It is thought that this behavior provides a competitive advantage: *S. cerevisiae* is tolerant to up to 20% ethanol, while most competing yeasts and bacteria are killed at more than 5% ethanol. Subsequently, yeast further catabolizes ethanol to carbon dioxide, provided oxygen is available. While high external sugar at the beginning of fermentation is associated with high osmotic pressure, the resulting high-ethanol environment (e. g., wine) is, in terms of turgor pressure, equal to water. This is due to the ability of ethanol to permeate membranes readily. Ethanol, however, causes water stress since it affects hydration and hence the conformation of biomolecules (Hallsworth, 1998). In this article we will, however, focus on turgor rather than on water stress. The two types of osmotic stress, turgor and water stress, invoke different, although partially overlapping, responses in yeast cells.

B. Yeast as a Model Organism in Molecular Biology

The species *Saccharomyces cerevisiae,* budding yeast, is synonymous with baker's yeast and brewer's yeast; the same yeast is used in wine making, the production of spirits, and also for the production of commercial proteins. In addition, budding yeast has become a most important model organism in molecular biology. *S. cerevisiae* is a single-celled eukaryote of 3–7 μm in diameter, with a generation time of around 90 min. It can be grown as stable cultures of both diploid and haploids. Haploids of opposite mating types (a and alpha) can be crossed to form diploids, which in turn can be induced to undergo meiosis to four haploid spores. This feature has been used for genetic analysis since the 1940s (Mortimer and Hawthorne, 1966). Budding yeast can readily be transformed so that foreign DNA is maintained as replicating plasmids of different copy numbers, or is integrated into the genome in a targeted fashion, allowing for reliable and easy gene knockout (Guthrie and Fink, 1991). The genome sequence of *S. cerevisiae* was completed and made public in 1995 (Mewes *et al.,* 1997), and excellent websites provide up-to-date gene and protein annotation (www.proteome.com/databases/YPD/; genome-www.stanford.edu/Saccharomyces/; www.mips.biochem.mpg.de). *S. cerevisiae* has become the paradigm organism for functional genomics. Microarray analyses of global gene expression are commonplace in yeast research. A novel resource in yeast genetics now available to the research community are sets of yeast knockout mutants encompassing each of the 6,200 yeast genes. Hence, for several years ahead, budding yeast will be by far the best-understood eukaryotic cell. Systems with high relevance to the function of eukaryotic cells have been studied by genetic analysis in yeast. They include the cell cycle, vesicular transport, nuclear organization, signal transduction, transcription and translation, metabolic control, and stress responses, to name only a few.

The fission yeast, *Schizosaccharomyces pombe,* has similar advantages for laboratory analysis, and its genome sequence became available in 2001. A significant advantage concerns the apparent evolutionary distance between budding and fission yeast. It appears that these two yeasts are as distant from each other as is each from the human! The fission yeast's cell cycle and its MAP kinase signaling, the areas most intensively studied in this organism, appear to be organized in a way more similar to those of the human than of budding yeast. Still, the use of *S. pombe* in molecular research has so far been much more restricted.

C. Genetic Analysis of Yeast Osmoregulation

The mechanisms of osmoadaption have been studied in yeasts for many years (Brown, 1976; Onishi, 1963; Rose, 1975). A breakthrough came with the analysis of mutants that fail to grow in high-osmolarity medium, i.e., a medium that contains more than 0.8 M NaCl or more than 1.2 M sorbitol (Brewster *et al.,* 1993; Bruning *et al.,* 1998; Larsson *et al.,* 1993). It is now clear that salt-sensitivity or osmosensitivity can be caused by mutations in a large number of different genes (Ferreira *et al.,* 2001; Ross-Macdonald *et al.,* 1999) which do not necessarily encode proteins important for the adaptation mechanisms. However, the first systematic mutant screens revealed genes encoding proteins operating in a MAP kinase pathway and encoding enzymes in glycerol metabolism. Both are now known to be essential for osmotic regulation in yeast.

Salt (NaCl) is very commonly used in research to increase medium osmolarity. However, sodium ions are toxic to yeast cells (Serrano, 1996). Sodium toxicity causes additional responses in yeast cells such as stimulation of expression of the *ENA1/PMR2* gene, which encodes a sodium export pump (Banuelos *et al.,* 1995; Rodriguez-Navarro *et al.,* 1994; Wieland *et al.,* 1995). The response to sodium toxicity is mediated by the calcineurin pathway and hence involves calcium signaling (Danielsson *et al.,* 1996; Marquez and Serrano, 1996; Mendoza *et al.,* 1994; Serrano, 1996). In addition, other specific salt tolerance mechanisms include uptake of potassium and control of the membrane potential, which involves the Hal4p and Hal5p protein kinases (Mulet *et al.,* 1999). The details of sodium tolerance are not a subject of this review.

The adaptation to high osmolarity (hyperosmotic shock) has been studied far more intensively than adaptation to hypo-osmotic shock. The main reason is the large number of yeast mutants that are known not to grow on low-osmolarity medium and rather require 1 M sorbitol as an osmotic stabilizer. Most of these mutants have a defect in cell wall biosynthesis, which is a complex process requiring many proteins (de Nobel *et al.,* 2000; Smits *et al.,* 1999). Hence, isolation of mutants sensitive to low osmolarity usually reveals cell wall mutants and not mutants specifically defective in the adaptation mechanism. After first eliminating all cell wall mutants, we tried to isolate mutants specifically sensitive to hypo-osmotic shock. We found only the single gene we knew already: *FPS1,* which encodes a glycerol export protein to be discussed later. The present review largely concentrates on the adaptation to hyperosmolarity.

II. Yeast Glycerol Metabolism

A. Glycerol Catabolism and Production

Glycerol and its metabolism have several different roles in yeast physiology:

1. As a source for carbon and energy
2. As an osmolyte or osmoprotectant
3. In redox regulation
4. As a precursor for glycerophospholipids

Figure 1 provides an overview of yeast glycerol metabolism and the transport proteins and enzymes involved. For catabolism, glycerol is taken up by passive diffusion through the lipid bilayer and/or with the assistance of two similar transport proteins, Gup1p and Gup2p (Holst *et al.,* 2000) (p denotes the protein encoded by the gene). These proteins, as well as the glycerol kinase, Gut1p (Pavlik *et al.,*

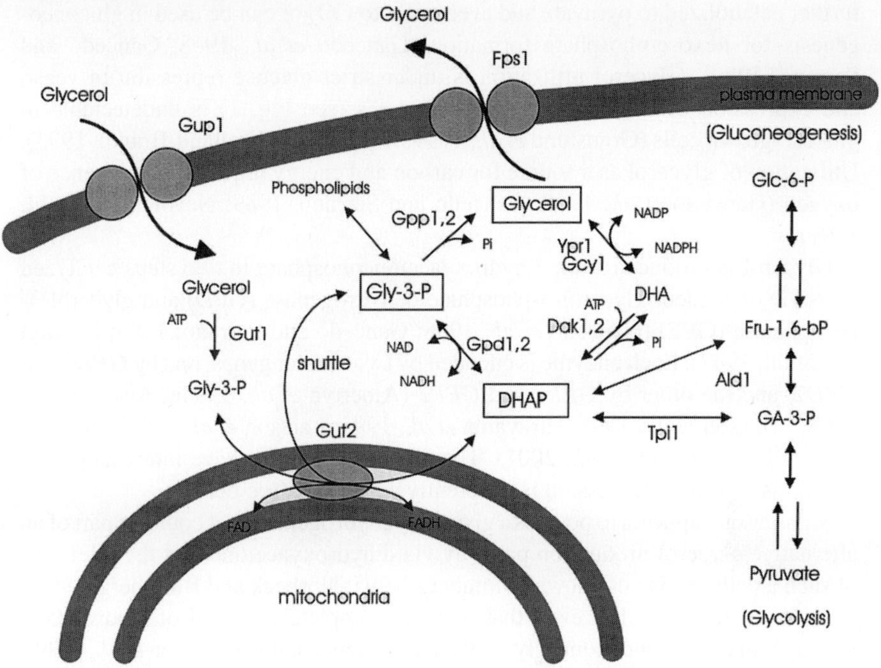

FIG. 1 Yeast glycerol metabolism. Glycerol is taken up via a system that involves the Gup1p/GUP2p membrane protein and the Gut1p glycerol kinase. Glycerol-3-phosphate is oxidized by the Gut2p FAD-dependent glycerol-3-phosphate dehydrogenase. Dihydroxyacetonephosphate is an intermediate of glycolysis and gluconeogenesis (Ald1p: aldolase; Tpi1p: triosephosphate isomerase). Glycerol is produced via NAD-dependent glycerol-3-phosphate dehydrogenase (Gpd1p and Gpd2p) and glycerol-3-phosphatase (Gpp1p and Gpp2p) and exported from the cell via the Fps1p MIP channel. *S. cerevisiae* has enzymes or the genes potentially encoding the enzymes dihydroxyacetone kinase (Dak1p and Dak2p) as well as glycerol dehydrogenase (Gcy1p and Ypr1p), which could form an alternative pathway for glycerol production/degradation. It is not known if this pathway truly exists. The two pathways could together form a futile cycle. Such a cycle could operate as an energy-dependent transhydrogenase due to different (predicted) cofactor specificity of the enzymes. Gut2p and Gpd1p/Gpd2p form a shuttle for transfer of redox equivilants between mitochondria and the cytosol.

1993; Sprague and Cronan, 1977), are together needed for the apparent active glycerol uptake, which has been reported to occur via proton symport (Holst *et al.,* 2000; Lages *et al.,* 1999; Sutherland *et al.,* 1997). It is possible that glycerol uptake is associated with vectorial phosphorylation, as in *E. coli* (Voegele *et al.,* 1993). This mechanism could effectively capture glycerol in the cell, since glycerol-3-phosphate does not permeate the plasma membrane. Glycerol-3-phosphate is then oxidized by a mitochondrial FAD-dependent glycerol-3-phosphate dehydrogenase, Gut2p (Rønnow and Kielland-Brandt, 1993; Sprague and Cronan, 1977); the electrons are directly channeled into the respiratory chain (Larsson *et al.,* 1998). The resulting dihydroxyacetonephosphate is a glycolytic intermediate and can be further catabolized to pyruvate and eventually to CO_2 or can be used in gluconeogenesis for hexose-phosphate formation (Gancedo *et al.,* 1968; Gancedo and Serrano, 1989). Glycerol utilization is under strict glucose repression in yeast, and expression of the *GUT1* and *GUT2* genes (see Fig. 1) is undetectable in glucose-grown cells (Grauslund *et al.,* 1999; Rønnow and Kielland-Brandt, 1993). Utilization of glycerol as a source for carbon and energy requires the presence of oxygen (Gancedo *et al.,* 1968; Gancedo and Serrano, 1989; Nevoigt and Stahl, 1997).

Glycerol is produced from dihydroxyacetonephosphate in two steps catalyzed by NAD-dependent glycerol-3-phosphate dehydrogenase (GPD) and glycerol-3-phosphatase (GPP) (Gancedo *et al.,* 1968; Gancedo and Serrano, 1989; Nevoigt and Stahl, 1997). Each enzyme is encoded by two similar genes, one by *GPD1* and *GPD2,* and the other by *GPP1* and *GPP2* (Albertyn *et al.,* 1994a; Ansell *et al.,* 1997; Eriksson *et al.,* 1995; Hirayama *et al.,* 1995; Larsson *et al.,* 1993; Norbeck *et al.,* 1996; Påhlman *et al.,* 2001). The different isoforms have interchangeable functions, but their expression is differently controlled (see below).

S. cerevisiae appears to possess a glycerol dehydrogenase that could be part of an alternative glycerol production pathway via dihydroxyacetone, but the relevance of such a pathway is not known (Blomberg, 1995; Norbeck and Blomberg, 1995). It has been reported, however, that mutants completely devoid of Gpd or Gpp activity still can produce some glycerol under certain conditions (Shen *et al.,* 1999). Glycerol dehydrogenase could also be part of a glycerol assimilation pathway, and the existence of a dihydroxyacetone kinase encoded by *DAK1* and perhaps *DAK2* has been confirmed (Blomberg, 2000; Norbeck and Blomberg, 1997). Since glycerol dehydrogenase is NADP-dependent, and glycerol-3-phosphate dehydrogenase is NAD-dependent, the Gpd-Gpp and Gcy-Dak pathways together could result in a cycle that generates a transhydrogenase activity (Blomberg, 2000; Norbeck and Blomberg, 1997). *S. cerevisiae* is lacking such an activity. It has, however, not been demonstrated yet that such a "transhydrogenase cycle" could truly operate.

B. Role of Glycerol in Osmoregulation

The production of glycerol under conditions of high osmolarity is essential for yeast cells. This is best illustrated by the observation that mutants unable to produce glycerol, i.e., *gpd1 gpd2* or *gpp1 gpp2* double mutants, are highly sensitive to high osmolarity and do not grow in media containing moderate amounts of NaCl (about 0.4 M) or sorbitol (about 0.8 M) (Ansell *et al.,* 1997; Påhlman *et al.,* 2001). In fact, this observation, although intuitively apparent, is the only clear demonstration so far in a eukaryotic organism that production of an osmolyte is truly necessary to cope with hyperosmolar conditions.

Glycerol is apparently the only osmolyte produced by *S. cerevisiae* in high osmolarity medium (Brown and Edgley, 1980). However, other polyols have been found as osmoprotectants in other yeast species (Jennings, 1984). In *S. cerevisiae,* strains have been engineered to produce xylitol, mannitol, or sorbitol (Krallish *et al.,* 1997; Shen *et al.,* 1999) or engineered to take up erythritol and xylitol (Karlgren and Hohmann, 2001). These compounds can replace glycerol as osmolytes. The function of glycerol as an osmolyte may be dual: to increase the intracellular osmotic pressure and thereby to allow water inflow, and to partially replace water in the cell (André, 1990; Blomberg and Adler, 1989; Gekko and Timasheff, 1981; Hallsworth, 1998; Kenyon *et al.,* 1986; Prior and Hohmann, 1997; Van Laere and Hulsman, 1987; Yagi, 1992).

It is the production of glycerol that is stimulated when yeast cells are shifted to high osmolarity (André, 1990; Blomberg and Adler, 1989). This is due, at least to a certain extent, to the enhanced expression of genes encoding enzymes in glycerol biosynthesis and the concomitant increased amount of the corresponding enzymes (Albertyn *et al.,* 1994a; Ansell *et al.,* 1997; Blomberg, 1995; Blomberg and Adler, 1989; Norbeck and Blomberg, 1996; Norbeck *et al.,* 1996; Rep *et al.,* 1999a, 2000). As outlined above, each of the two enzymes, the dehydrogenase and the phosphatase, is encoded by two isogenes. After osmotic shock, the expression of *GPD1* and *GPP2* is transiently induced up to 50-fold (Rep *et al.,* 1999a,b) and under steady-state growth in high osmolarity, expression of both genes is about 2- to 4-fold enhanced (Ansell *et al.,* 1997; Eriksson *et al.,* 2000; Norbeck and Blomberg, 1996; Påhlman *et al.,* 2001). The enhanced gene expression leads to 5- to 10-fold increased specific enzyme activity and hence enhanced capacity of the glycerol production pathway (Albertyn *et al.,* 1994a,b; Ansell *et al.,* 1997; Blomberg and Adler, 1989; Larsson *et al.,* 1993; Påhlman *et al.,* 2001). In contrast, the expression of *GPD2* is not stimulated after osmotic upshock (Ansell *et al.,* 1997). Expression of *GPP1,* whose basal expression level is about 5-fold higher than that of *GPP2* (Norbeck *et al.,* 1996; Påhlman *et al.,* 2001), is transiently stimulated after osmotic shock by a factor of about 4–5 (Påhlman *et al.,* 2001; Rep *et al.,* 2000). Hence, the genes encoding the different isoenzymes in glycerol biosynthesis are differentially regulated. The expression pattern and the similar effects caused by different regulatory mutants suggest that expression of *GPD1* and *GPP2* is strictly coregulated (Rep *et al.,* 1999a,b).

Increased capacity of the glycerol production pathway alone cannot explain enhanced glycerol biosynthesis, as illustrated by the observation that glycerol production is stimulated even in a *gpd1*Δ mutant (Albertyn *et al.,* 1994a). Hence other control systems redirect metabolic fluxes toward glycerol, but the underlying mechanisms for this have not been systematically investigated.

C. Connection to Redox Metabolism

Glycerol metabolism also plays important roles in the control of yeast redox metabolism. This is best illustrated by the observation that mutants unable to produce glycerol (i.e., *gpd1 gpd2* or *gpp1 gpp2* mutants) cannot grow under anaerobic conditions (Ansell *et al.,* 1997; Björkqvist *et al.,* 1996; Påhlman *et al.,* 2001). Formally, alcoholic fermentation, the production of two molecules each of ethanol and CO_2 from one molecule of glucose, is a redox-neutral process. However, since a fraction of oxidized intermediates from the final steps of glycolysis is used in biosynthetic pathways, fermenting yeast cells always produce a surplus of NADH. This can be reoxidized in the presence of oxygen via the respiratory chain in mitochondria. In the complete absence of oxygen, however, NADH is apparently reoxidized via the glycerol-3-phosphate dehydrogenase reaction leading to enhanced glycerol production. *Gpd1 gpd2* double mutants arrest growth under anaerobic conditions with strongly enhanced NADH levels (Ansell *et al.,* 1997; Björkqvist *et al.,* 1996). The glycerol redox valve constitutes a sophisticated adaptation mechanism, since expression of *GPD2* and *GPP1* is specifically enhanced under strictly anaerobic conditions (Ansell *et al.,* 1997; Påhlman *et al.,* 2001). Expression of *GPD1* and *GPP2* does not respond to anaerobiosis (Ansell *et al.,* 1997; Påhlman *et al.,* 2001).

It is apparent from this discussion that enhanced glycerol production under osmotic stress causes a redox imbalance due to the consumption of NADH and the production of NAD. This has to be compensated for by enhanced production of more oxidized metabolic endproducts. It has been observed that the engineering of glycerol production to achieve higher glycerol yields leads to increased acetate production (Michnick *et al.,* 1997; Remize *et al.,* 1999, 2000a,b).

As will be discussed below, the response to osmotic stress also appears to include the stimulation of mechanisms protecting yeast cells from oxidative damage leading to a substantial reprogramming of the cellular redox metabolism.

D. Detoxification of Byproducts

The production of glycerol may lead to the formation of reactive and hence toxic byproducts, such as dihydroxyacetone and methylglyoxal. The expression of genes encoding enzymes that could metabolize such compounds to nontoxic metabolites is induced upon osmotic shock: the *DAK1* gene for dihydroxyacetone kinase and the *GLO1* gene for glyoxylase (Inoue *et al.,* 1998; Norbeck and Blomberg, 1997).

III. Control of Gene Expression in Yeast Glycerol Metabolism

The control of stress-induced gene expression has attracted major interest in the last couple of years. Stress-induced signal transduction is a field where the tools of yeast genetics and molecular biology have been exploited to the most effective extent, and recently global expression analyses have successfully been used to address novel and interesting aspects.

Central to the osmotic responses in yeast are two interacting mitogen-activated protein (MAP) kinase pathways, the high osmolarity glycerol (HOG), and the protein kinase C (PKC) pathway (Gustin *et al.,* 1998). In addition, protein kinase A (cAMP-dependent protein kinase, PKA) transmits general stress signals and is thereby involved in osmostress responses as well (Ruis and Schüller, 1995).

A. The HOG and PKC MAP Kinase Pathways Mediate Osmotic Responses

1. The HOG Pathway and Hyperosmotic Stress

The HOG pathway (Fig. 2) was first discovered by analysis of mutants sensitive to high osmolarity. Two of the mutants defined the genes for, respectively, the MAP kinase (MAPK) Hog1p and the MAP kinase kinase (MAPKK) Pbs2p (Boguslawski, 1992; Brewster *et al.,* 1993). Independently, Saito and co-workers were studying the protein phosphatase Ptp2p. In an attempt to identify genes encoding proteins that functionally interacted with Ptp2p, they screened for mutations that caused lethality unless Ptp2p was overproduced (Maeda *et al.,* 1994). This screen yielded *SLN1* and *YPD1,* encoding proteins homologous to prokaryotic sensor histidine kinases and response regulators, respectively (Maeda *et al.,* 1994; Posas *et al.,* 1996). The lethality of an *sln1*Δ mutant was then used to identify suppressor mutations (using a construct that allows *SLN1* to be expressed only under certain growth conditions). This approach yielded *SSK1* and *SSK2* and again *PBS2* and *HOG1* (Maeda *et al.,* 1994, 1995; Posas and Saito, 1997). A search for genes that could suppress the lethality of the *sln1*Δ mutant when overexpressed identified the phospho-tyrosine phosphatases *PTP2* and *PTP3* as well as the phospho-serine/threonine phosphatases *PTC1* and *PTC3* (Maeda *et al.,* 1993, 1994). These studies, while identifying numerous pathway components, also suggested that deletion of *SLN1* causes lethality due to inappropriate hyperactivation of the HOG pathway.

Saito and colleagues noted that while deletion of *PBS2* or *HOG1* caused strong osmosensitivity, deletion of *SSK2* did not cause such an effect. They searched for genes with high sequence similarity to *SSK2,* which could encode a redundant function, and found *SSK22* (Posas and Saito, 1997). However, even the double mutant *ssk2*Δ *ssk22*Δ was not sensitive to high osmolarity. Hence, they performed a screen to identify mutants that cause osmosensitivity in a *ssk2*Δ *ssk22*Δ double mutant and identified *SHO1* and *STE11,* thereby defining a second upstream branch of the pathway (Posas and Saito, 1997). The identification of Ste11p as part of the osmosensing HOG pathway was intriguing: this MAP kinase kinase kinase (MAPKKK) had been known for many years to be part of the pathway that

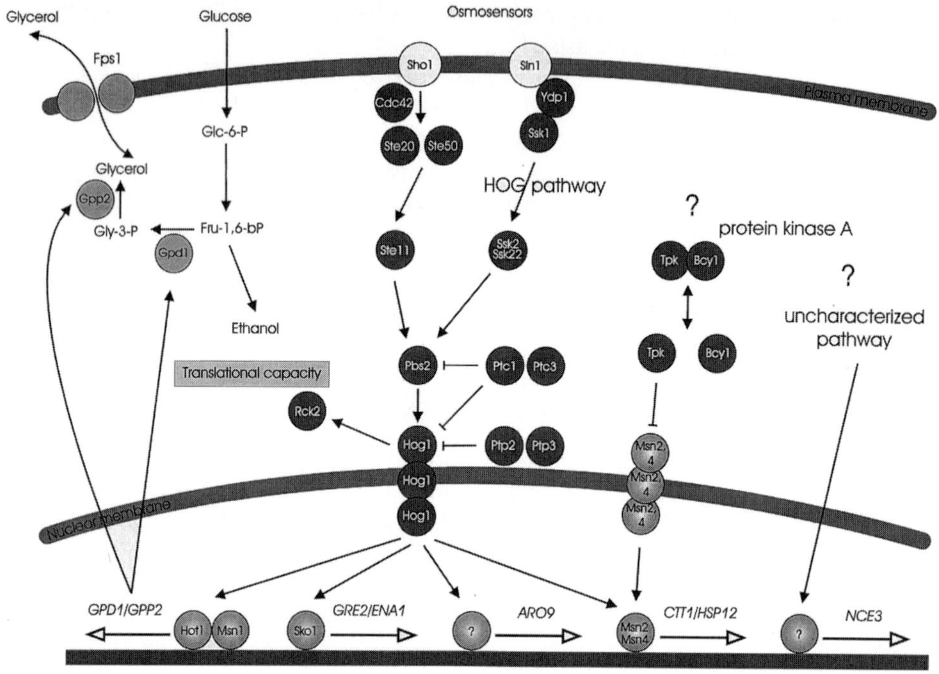

FIG. 2 The yeast HOG pathway and the control of transcriptional responses to hyperosmotic shock. The central cascade of the HOG pathway consists of the MAP kinase (MAPK) Hog1p, the MAPKK Pbs2p and three upstream MAPKKK, the redundant Ssk2p and Ssk22p, as well as Ste11p. The former branch is controlled by the phosphorelay system consisting of the Sln1p sensor kinase and the response regulators Ypd1p and Ssk1p. The latter branch consists of the Sho1p membrane protein (which is not a sensor) and several other proteins. The functional relationship between these proteins is unclear, but all or most of them appear to be part of a complex with the scaffold domain of Pbs2p. The HOG pathway controls transcription via different regulators, Hot1p, Msn1p, Msn2p, Msn4p, and Sko1p. Msn2p and Msn4p are also controlled by protein kinase A, and several genes are controlled by osmotic shock independently of both pathways. Hog1p also has cytosolic targets via the Rck2p kinase. Several phosphoprotein phosphatases, Ptc1p, Ptc3p, Ptp2, and Ptp3p, control the HOG pathway.

mediates the yeast response to mating pheromones and thereby orchestrates the yeast mating process (Gustin *et al.,* 1998; Herskowitz, 1995; Posas *et al.,* 1998b). This finding lead to the identification of further shared components, Ste20p, Ste50p, and Cdc42p (Posas *et al.,* 1998a; Raitt *et al.,* 2000; Reiser *et al.,* 2000). It is now well established that three yeast MAP kinase pathways, the HOG pathway, the mating pheromone response pathway, and also the pathway that stimulates the development of pseudohyphae in response to nutrient limitation, share certain protein kinases (Gustin *et al.,* 1998). Pathway specificity seems to be ensured through different signaling complexes, such as the Ste5p complex of the mating pheromone response pathway and the Pbs2p complex in the HOG pathway (reviewed in Gustin *et al.,* 1998; Herskowitz, 1995; Posas *et al.,* 1998b). Crosstalk can be observed in mutations overactivating or blocking certain pathway components (Davenport *et al.,* 1999; O'Rourke and Herskowitz, 1998).

The operation of the HOG pathway has been studied intensively, making use of mutants that block or overactivate certain components, as well as sensitive monitoring of the phosphorylation state of the Hog1p kinase using anti-phospho-Hog1p specific antibodies. The pathway responds to turgor changes rather than to water stress per se (Tamás *et al.,* 2000). Stimulation is rapid, and phosphorylation of Hog1p on both serine/threonine and tyrosine can be observed within 1 min after upshock (Maeda *et al.,* 1994, 1995). The default state of the sensor kinase Sln1p is "on." Sln1p activity leads to autophosphorylation on His576 and phosphotransfer to Asp1144 in the receiver domain within Sln1p, which is passed further to His64 in Ypd1p and eventually to Asp554 on Ssk1p (Posas *et al.,* 1996). Phosphorylated Ssk1p is inactive. The series of phosphotransfers and its effects have been extremely well documented *in vitro* and *in vivo* (Posas *et al.,* 1996). Upon osmotic shock, Sln1p is transiently inhibited in an unknown way, which leads to a diminished level of phosphorylated Ssk1p. Unphosphorylated Ssk1p can interact with the N-terminal regulatory domain of the MAPKKK Ssk2p/Ssk22p and lift inhibition (Posas and Saito, 1998). Deletion of this regulatory domain renders the protein constitutively active with the same lethal consequences as deletion of the *SLN1* or *YPD1* genes (Posas and Saito, 1998). Ssk2p/Ssk22p phosphorylates Pbs2p on Ser514 and Thr518, which activates this MAPKK (Posas and Saito, 1997). Pbs2p phosphorylates Hog1p on Thr174 and Tyr176 (Posas and Saito, 1998). This event causes Hog1p to translocate to the nucleus (Ferrigno *et al.,* 1998; Reiser *et al.,* 1999).

Sho1p, a transmembrane protein from the alternative activation branch, has been thought until very recently to function as a second osmosensor (Maeda *et al.,* 1995). However, elegant domain swapping experiments have shown that the only function of Sho1p appears to be that of an anchoring protein (Raitt *et al.,* 2000). Upon osmotic shock, Pbs2p, which seems to fulfill the role of a scaffold protein (Maeda *et al.,* 1995; Posas and Saito, 1997), is transiently recruited to the plasma membrane where it interacts through its proline-rich sequence with an SH3-domain of Sho1p (Raitt *et al.,* 2000; Reiser *et al.,* 2000). Ste11p, Ste50, Ste20p, and Cdc42p appear to be part of this complex, at least during certain stages of activation (Raitt *et al.,* 2000; Reiser *et al.,* 2000). The sensor in this branch of the pathway is unknown. Also unknown is why at least two apparently independent sensing systems are involved in Hog1p activation. However, evidence has been presented that the two branches might present different sensitivities (Maeda *et al.,* 1995). In addition, the Sho1p-Pbs2p complex is specifically localized to those parts of the cell surface where active growth occurs (Reiser *et al.,* 2000), indicating either that these domains are specifically sensitive to osmotic changes or that the Sho1p branch monitors osmotic changes associated with cell growth and expansion. Interestingly, under severe osmotic stress, stimulation of Hog1p phosphorylation has been observed even in a *sho1*Δ *ssk2*Δ *ssk22*Δ triple mutant, suggesting the existence of additional mechanisms to stimulate the HOG pathway (Van Wuytswinkel *et al.,* 2000).

Activated Hog1p accumulates in the nucleus, where it is involved in the stimulation of expression of about 150 genes (Posas *et al.,* 2000; Rep *et al.,* 2000). Most of those target genes depend on Hog1p only to a moderate extent, while expression of about 50 genes is more than 80% diminished in a *hog1*Δ mutant (Rep *et al.,* 2000). Several transcription factors have been associated to the HOG pathway, of which only Sko1p has been demonstrated to be a substrate of the Hog1p kinase (Proft *et al.,* 2001; Proft and Serrano, 1999). Sko1p is a cAMP-response element (CRE) binding protein and has been described as a transcriptional repressor but may also participate in gene activation (Pascual-Ahuir *et al.,* 2001; Proft and Serrano, 1999; Rep *et al.,* 2001). Sko1p controls the expression of a small set of genes including the *ENA1* gene (encoding the sodium export pump), *HAL1,* as well as genes encoding proteins that presumably play in role in the defense from oxidative damage (Pascual-Ahuir *et al.,* 2001; Proft and Serrano, 1999; Rep *et al.,* 2001). Hot1p is a DNA binding protein that interacts with Hog1p. Hot1p controls the expression of a small set of genes including *GPD1* and *GPP2* (Rep *et al.,* 1999b). Hot1p has been shown to bind to the *GPD1* promoter, and it appears to associate with Hog1p on the promoter, opening the intriguing possibility that the Hog1p kinase itself controls gene expression (Alepuz *et al.,* 2001). Msn2p and Msn4p are two redundant transcription factors that bind to stress response elements (STREs) upstream of many genes whose expression is stimulated by a variety of stress conditions (Ruis and Schüller, 1995). It has been shown that Hog1p-dependent activation of a STRE-driven reporter gene requires Msn2p/Msn4p (Schüller *et al.,* 1994). In addition, all genes that require Msn2p and Msn4p for induction upon osmotic shock (about 50 genes) are also Hog1p-dependent (Rep *et al.,* 2000). These observations suggest a direct link between Hog1p and Msn2p/Msn4p, but such a connection has not yet been addressed experimentally.

Hog1p phosphorylation and its residence in the nucleus are transient events, controlled by strict feedback regulation (Maeda *et al.,* 1994, 1995). The time course of Hog1p phosphorylation correlates well with the level of mRNAs of Hog1p-dependent genes. The duration depends very much on the osmotic strength: lasting a few minutes at osmolarities less than 0.5 M NaCl and up to 2 hr at 1.2 M NaCl. Genetic evidence suggests the involvement in feedback regulation of the phosphatases Ptp2p, Ptp3 and Ptc1p and Ptc3p, which can suppress a *sln1*Δ deletion as outlined above (Jacoby *et al.,* 1997; Maeda *et al.,* 1994; Wurgler-Murphy *et al.,* 1997). Ptp2p and Ptp3p have been shown to interact with the Hog1p kinase, strongly suggesting that this protein is a substrate for the phosphatases (Jacoby *et al.,* 1997; Mattison and Ota, 2000; Wurgler-Murphy *et al.,* 1997). Furthermore, a *ptp2*Δ *ptc1*Δ double mutant grows very poorly, probably due to an overactive HOG pathway (Mattison and Ota, 2000; Mattison *et al.,* 1999). In addition, expression of *PTP2* and *PTP3* is stimulated after osmotic shock, suggesting the existence of a feedback loop (Jacoby *et al.,* 1997; Wurgler-Murphy *et al.,* 1997).

However, the time course of events seems to exclude increased transcription of the phosphatase genes as the main event in feedback control. Interestingly, the drop in Hog1p phosphorylation (and Hog1p-dependent mRNAs) correlates well with the onset of glycerol accumulation, suggesting that the beginning of cell expansion after the initial osmoshock-induced cell shrinking provides the signal for downregulation of the HOG pathway (unpublished data).

Hog1p has at least one cytosolic target, although upon osmotic shock, the protein appears to accumulate in the nucleus. Rck2p is a protein kinase that interacts with Hog1p, and it has been shown to be a substrate of the Hog1p kinase (Bilsland-Marchesan *et al.,* 2000). Rck2p seems to control translation. Upon osmotic shock, translation efficiency is diminished, and this effect requires Rck2p and Hog1p (M. Teige and G. Ammerer, personal communication).

2. The PKC Pathway and Hypo-osmotic Stress

The role and function of the PKC pathway appears to be even more complex than that of the HOG pathway (for reviews see Gustin *et al.,* 1998; Heinisch *et al.,* 1999; Posas *et al.,* 1998b). The PKC pathway (Fig. 3) can be stimulated rapidly by hypo-osmotic shock, hinting at a role opposite to that of the HOG pathway (Davenport *et al.,* 1995). Indeed, it appears that activation of the HOG and the PKC pathways is mutually exclusive (Davenport *et al.,* 1995). In addition, while mutants in the HOG pathway cause strong sensitivity to osmotic upshock, mutants in the PKC pathway require for growth an osmotic stabilizer, such as 1 M sorbitol, and hence increased osmolarity (Levin and Bartlett-Heubusch, 1992). These observations appear to indicate that the PKC pathway could orchestrate the cellular response to hypo-osmotic stress. However, the sensitivity to low osmolarity of mutants lacking certain PKC pathway components is apparently due to an altered cell wall composition and hence to a weaker cell wall (Klis, 1994; Roemer *et al.,* 1994; Shimizu *et al.,* 1994). In fact, it is very well documented that the PKC pathway controls, in different ways, the assembly of the yeast cell wall (Cabib *et al.,* 1998; de Nobel *et al.,* 2000; Jung and Levin, 1999; Smits *et al.,* 1999).

Central to the PKC pathway is the only yeast homologue to mammalian PKCs, Pkc1p. Although the protein has significant similarity to mammalian PKC isoforms and the deletion mutant can be complemented by PKC-η (Nomoto *et al.,* 1997), it is presently unclear if these biochemical functions are really that similar (Heinisch *et al.,* 1999); for instance, stimulation of the kinase activity by calcium or diacylglycerol could not be demonstrated (Antonsson *et al.,* 1994), although several mutant phenotypes link Pkc1p to calcium signaling (reviewed in Heinisch *et al.,* 1999). In any case, genetic evidence suggests that Pkc1p controls a MAP kinase cascade consisting of Bck1p, Mkk1p/Mkk2p, and Slt2p (Irie *et al.,* 1993; Lee *et al.,* 1993; Lee and Levin, 1992; Levin *et al.,* 1994; Martin *et al.,* 1993; Soler *et al.,* 1995). Slt2p probably controls different transcription factors such as Rlm1p (Watanabe *et al.,* 1995) and thereby regulates expression of genes encoding

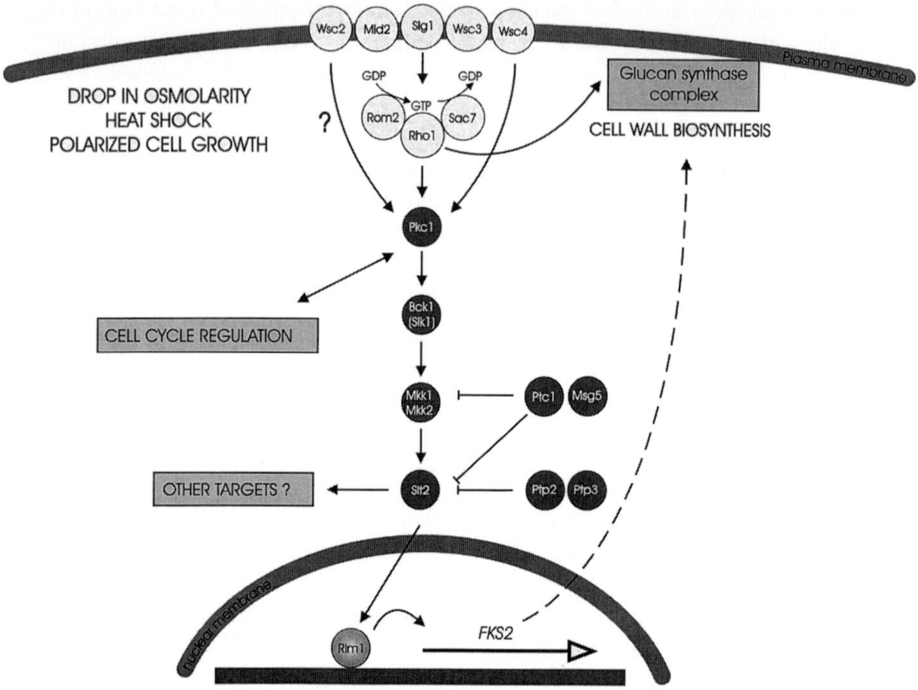

FIG. 3 PKC pathway. The central MAP kinase cascade consists of the MAPK Slt2p, the MAKKs Mkk1p and Mkk2p, as well as the MAPKKK Bck1p. Pkc1p is the single yeast protein kinase C homologue and controls the MAP kinase cascade and additional, unknown, pathways. Upstream of Pkc1p, the Rho1p G-protein and its GDP/GTP exchange factor Rom2p and GTPase Sac7p also link directly to cell wall assembly, which is controlled by the signaling cascade at the transcriptional level. The PKC pathway appears to be stimulated by hypo-osmotic shock, heat stress, and polarized cell growth (e.g., during formation of mating projections) via several potential plasma membrane sensors. Adapted from Heinisch *et al.* (1999).

enzymes in cell wall metabolism (Jung and Levin, 1999). The fact that mutants in the MAP kinase cascade have a much weaker phenotype than *pkc1* mutants suggests that Pkc1p controls additional parallel pathways (reviewed in Heinisch *et al.*, 1999). Upstream components that could control Pkc1p include the putative membrane sensors Slg1p, Mid2p and Wsc2p-4p (Jacoby *et al.*, 1998; Ketela *et al.*, 1999; Lodder *et al.*, 1999; Raboy *et al.*, 1999; Verna *et al.*, 1997) as well as the Rho1p G-protein (Nonaka *et al.*, 1995; Qadota *et al.*, 1996) and its GTP–GDP exchange factor (Ozaki *et al.*, 1996) and GTPases (Peterson *et al.*, 1994; Wang and Bretscher, 1995). The Rho1p G-protein appears to control cell wall assembly by direct interaction with the glucan synthase complex (Qadota *et al.*, 1996).

Genetic evidence, i.e., analysis of suppressor mutations and high-copy suppressors, links the PKC pathway to cell cycle control as well as to nutrient sensing. A functional PKC pathway is also required for cell fusion during mating (reviewed in Heinisch *et al.,* 1999). These observations are consistent with a key role in cell integrity, and one aspect of this function may hence relate to cell turgor control as well.

B. The PKA Pathway Mediates a General Stress Response

PKA plays a central role in yeast (Thevelein, 1992, 1994; Thevelein and de Winde, 1999). It has been estimated that altered PKA activity affects the expression of about one-fourth of all yeast proteins (Boy-Marcotte *et al.,* 1996; Norbeck and Blomberg, 2000) and, in addition, PKA directly controls activity and stability of a large number of enzymes and transport proteins. Yeast PKA, Similar to that of all eukaryotes, consists of a catalytic subunit, which is encoded by the three isogenes *TPK1–3,* and a regulatory, cAMP-binding subunit encoded by *BCY1* (Toda *et al.,* 1987a,b). Mutants lacking *TPK1–3* or *BCY1* grow extremely poorly, demonstrating that both lack of PKA and hyperactive PKA cause deleterious effects. The activity of PKA appears to correlate with cell proliferation: it is generally high in rapidly proliferating cells and low in slowly proliferating or resting cells (Thevelein, 1992, 1994; Thevelein and de Winde, 1999). High PKA activity seems to promote glycolytic metabolism and breakdown of cellular reserves. Low PKA activity seems to promote gluconeogenesis and accumulation of reserve carbohydrates. PKA activity is obviously controlled by the cAMP level, but cAMP-independent mechanisms have been proposed (Crauwels *et al.,* 1997). Much effort has gone into the analysis of the mechanisms that stimulate PKA activity upon a shift from poor to rich substrates. Studies on the glucose-induced cAMP spike indicate the involvement of a G-protein coupled receptor (Colombo *et al.,* 1998; Kraakman *et al.,* 1999; Thevelein *et al.,* 2000; Thevelein and de Winde, 1999). While these observations suggest that PKA activity is controlled in response to external stimuli, it is unclear if altered PKA activity is a response to changes in the proliferation potential of the cell, or if PKA controls that potential.

PKA activity also affects stress tolerance (Fig. 4). High PKA activity, either in rapidly growing cells or in *bcy1* mutants, is associated with a general stress sensitivity, while low PKA in slowly growing/resting cells or in certain *tpk* mutants is associated with a general stress tolerance (de Winde *et al.,* 1997; Thevelein, 1994; Thevelein *et al.,* 2000). One mechanism with which PKA appears to contribute to stress tolerance is control of the transcription factors Msn2p and Msn4p (Estruch and Carlson, 1993; Martinez-Pastor *et al.,* 1996). These two zinc-finger proteins appear to fulfill overlapping functions although their overall sequence identity is only 32%. PKA-mediated phosphorylation of Msn2p and Msn4p causes these proteins to localize in the cytosol. Upon stress, and under conditions of low PKA activity, the proteins localize in the nucleus, where they activate gene expression (Estruch, 2000; Garreau *et al.,* 2000; Görner *et al.,* 1998). It has been observed that,

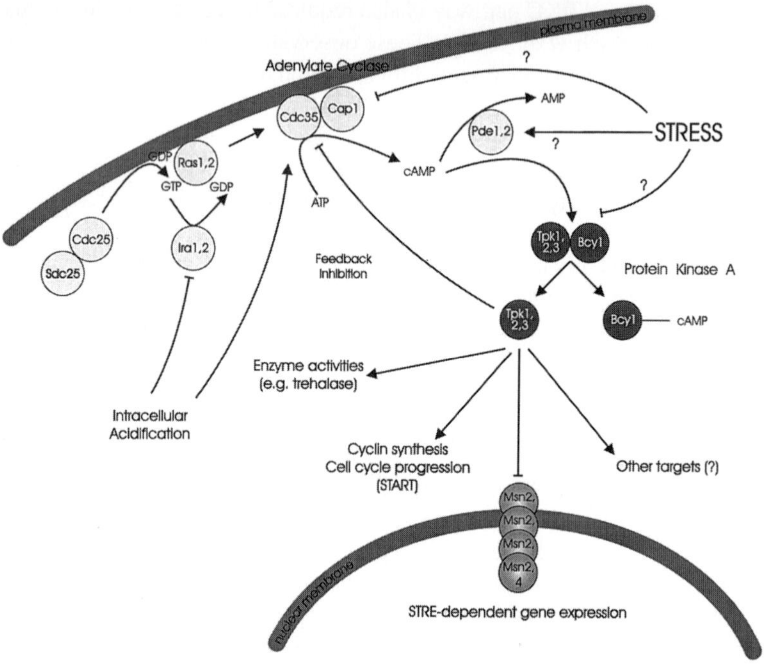

FIG. 4 PKA pathway and Msn2p and Msn4p. The subcellular localization of Msn2p and Msn4p is controlled by protein kinase A. Under optimal conditions, PKA activity is high and Msn2p/Msn4p are cytosolic; under stress, PKA activity is low and Msn2p/Msn4p are nuclear. The enzymes and regulators involved in cAMP metabolism are known, but the mechanisms that control PKA activity under stress have not been elucidated; some possible points of interference are shown. On the other hand, stress leads to intracellular acidification, which is know to stimulate, at least transiently, cAMP production and hence PKA activity.

under stress, Msn2p and Msn4p become hyperphosphorylated in an apparently stress-specific manner, but the kinases responsible for this effect have not been identified (Garreau *et al.,* 2000). Msn2p/Msn4p bind to so-called STREs, which have the canonical sequence CCCCT (or AGGGG) and appear necessary and sufficient for Msn2p/Msn4p mediated stress-induced gene expression (Kobayashi and McEntee, 1993; Marchler *et al.,* 1993; Martinez-Pastor *et al.,* 1996; Schmitt and McEntee, 1996; Schüller *et al.,* 1994; Treger *et al.,* 1998). Typically, the promoters of STRE-controlled genes contain several such elements, but presence or absence of the STRE sequence in a promoter is not sufficient as a diagnostic test for STRE control of a gene (Moskvina *et al.,* 1998; Rep *et al.,* 2000). In any case, Msn2p/Msn4p and STRE elements appear to mediate induced gene expression by different stress conditions, including starvation, heat shock, oxidative stress, acidity stress and osmotic stress (i.e., any conditions that cause lower PKA). A variant of the STRE is the post-diauxic shift element PDSE, which mediates stimulated gene expression when yeast cells have consumed glucose and start ethanol consumption (the diauxic shift). In this case, the binding protein is Gis1p (Pedruzzi *et al.,* 2000).

STRE-controlled genes encode proteins involved in a general stress response, encompassing for instance heat shock proteins, catalase, and enzymes in trehalose metabolism (trehalose is a stress protectant/storage sugar) (Moskvina *et al.,* 1998). As indicated above, induction of these genes by osmotic shock requires Msn2p and Msn4p (Rep *et al.,* 2000). Inappropriately high PKA activity (such as in *bcy1* mutants) strongly affects the induction of these genes, most probably because it prevents nuclear translocation of Msn2p/Msn4p upon osmotic shock (Norbeck and Blomberg, 2000). However, nuclear translocation of Msn2p and Msn4p does not seem to be sufficient for normal induction of STRE controlled genes by osmotic shock. It appears that Msn2p/Msn4p require an additional Hog1p-dependent activation to stimulate gene expression under osmotic stress (Rep *et al.,* 2000; Schüller *et al.,* 1994). This activation could be due to Hog1p-dependent phosphorylation or interaction with Hog1p or a Hog1p-dependent protein. This question has not as yet been investigated.

C. Glycerol Biosynthesis Genes Are Controlled in a Complex Way

Global gene expression analysis has revealed only very few genes whose induction is completely dependent on the HOG pathway. One such example is *STL1,* which is more than 100-fold induced after osmotic shock but remains unexpressed in *hog1*Δ and *hot1*Δ mutants (Rep *et al.,* 2000). Expression of the genes *GPD1* and *GPP2* appears to be controlled in a complex way. As with many other stress-controlled genes, expression of *GPD1* and *GPP2* is rapidly and strongly stimulated after osmotic shock (Fig. 5), and the time of peak induction depends on the severity of the shock (the stronger the shock the later the peak; no peak at all at very severe shock) (Rep *et al.,* 1999a,b). Subsequently, the mRNA level falls back to about the same level as before stress and then increases again to reach a steady state about 2- to 4-fold above basal level. The HOG signaling pathway contributes to both basal and induced levels; however, even in a *hog1*Δ mutant, significant and rapid induction of *GPD1* and *GPP2* still occurs (Rep *et al.,* 1999a,b). This residual induction does not seem to be dependent on PKA and Msn2p/Msn4p, since mutations that alter PKA activity, deletion of the genes *MSN2* and *MSN4,* as well as deletion of three putative STREs in the *GPD1* promoter only marginally affect expression (Eriksson *et al.,* 2000; Norbeck and Blomberg, 2000; Rep *et al.,* 1999a,b; Alipour and Blomberg, personal communication). In fact, it is completely unknown which signaling pathways and transcription factors mediate the HOG-independent response of the glycerol biosynthesis genes.

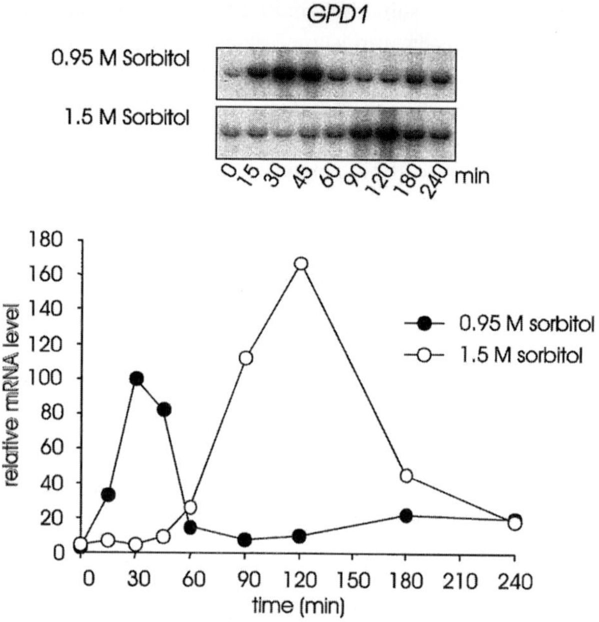

FIG. 5 *GPD1* induction profile. Typical profile of induction of HOG-dependent genes after osmotic shock. Expression is transiently very strongly induced before a moderately stimulated steady-state level is attained. Amplitude and timing of the response depend of severity of the shock: the stronger the shock, the higher and later the peak induction. Relative mRNA levels have been calculated with reference to the level of the *IPP1* mRNA, which does not change during this experiment. (Adapted from Rep *et al.,* 1999a.)

Analysis of the *GPD1* promoter revealed that binding of the Rap1p general chromatin remodeling factor is essential for *GPD1* expression (Eriksson *et al.,* 2000). In addition, it has been shown that Hot1p and Msn1p are both present on the *GPD1* promoter, but their exact binding sites are unknown (Alepuz *et al.,* 2001). Since Rap1p associates in the control of expression of genes encoding glycolytic enzymes with Gcr1p (Tornow *et al.,* 1993), and Gcr1p is distantly related to Hot1p and Msn1p, one might speculate that Rap1p facilitates Hot1p and Msn1p binding to the *GPD1* promoter. Deletion of *HOT1* and of *HOT1* plus *MSN1* diminishes *GPD1* expression, albeit to a lesser extent than deletion of *HOG1*. Remarkably, it has been shown that the Hog1p kinase is also present on the *GPD1* promoter (Alepuz *et al.,* 2001). In an attractive, although still highly speculative model, Hot1p, perhaps with the help of Rap1p, recruits the Hog1p kinase, the promoter, and the kinase itself, then functions as the activator of RNA polymerase II. Certainly, the situation is complex, and intensive studies are ongoing in several laboratories to better understand how the Hog1p MAP kinase stimulates gene expression.

D. Osmotic Responses in Yeast Comprise More Than 300 Genes

Global expression analysis using microarrays has shown that the expression of 200 genes is rapidly stimulated after osmotic shock, and the expression of about 100 genes is rapidly diminished (Posas *et al.,* 2000; Rep *et al.,* 2000). This analysis suggests that osmotic shock affects numerous cellular processes (Fig. 6). Expression changes indicate a reprogramming of carbohydrate metabolism and its isoenzyme pattern, of redox metabolism, amino acid metabolism, and lipid metabolism. In addition, vacuolar function, the cytoskeleton, and the cell wall appear to be affected. Several of the genes, such as those encoding heat shock proteins or catalase, are induced under all stress conditions (Ruis and Schüller, 1995). A significant proportion of the repressed genes encode ribosomal proteins and translation factors, indicating diminished translational capacity. This effect is

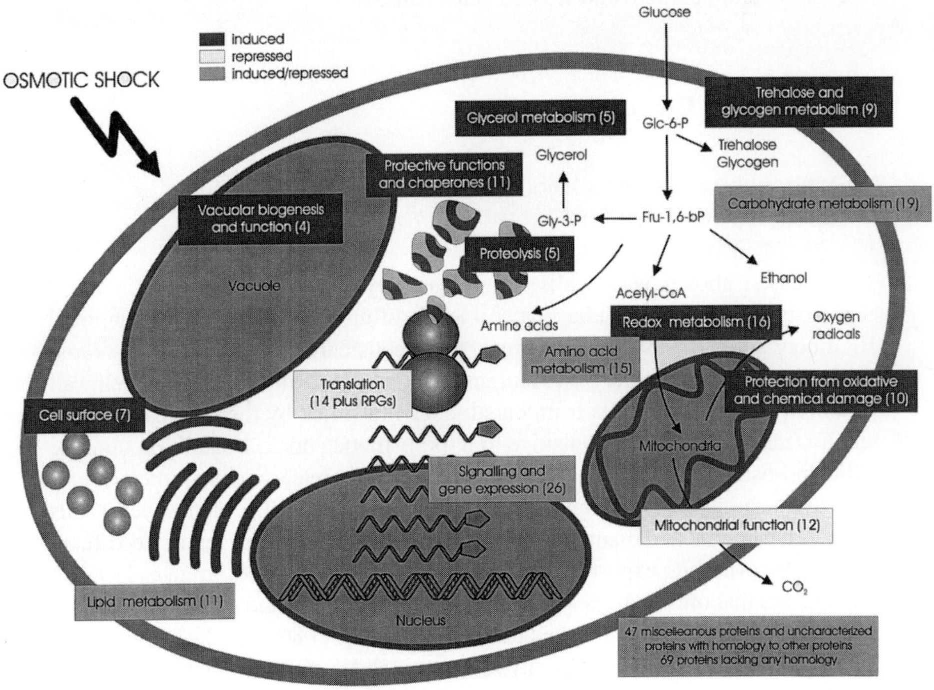

FIG. 6 Scheme for global responses. A presentation of cellular processes that seem to adjust to an osmotic shock on the basis of an analysis on changes in gene expression. The number of genes in each category is indicated, and the different gray shadings indicate categories with exclusively induced or repressed genes or where both induced and repressed genes occur. Within carbohydrate, amino acid, and lipid metabolism, the changes suggest that certain pathways are stimulated and others repressed.

probably due to the transient proliferation arrest upon osmotic shock and is controlled, at least in part, by PKA (Griffioen *et al.,* 1994). While the physiological role or at least the biochemical function of most of the strongly osmoregulated genes is known, about one-third of the genes whose expression is affected by osmotic shock is uncharacterized (Rep *et al.,* 2000).

As indicated above, there is a large overlap in the expression pattern caused by different stress conditions, and a significant part of this general stress response is mediated via PKA and Msn2p and Msn4p. In addition, there are a number of genes encoding functions known or thought to be involved in oxidative stress response whose expression is stimulated also under heat shock or upon osmotic shock. We have recently studied a class of those genes, and it appears that their promoters are controlled by separate elements mediating osmotic and oxidative responses (Rep *et al.,* 2001). It is possible that different stress conditions stimulate oxidative metabolism and hence also the potential production of reactive oxygen species, invoking an oxidative stress response. This, together with an altered NAD/NADH balance caused by glycerol production could explain the changes in redox metabolism in osmostressed yeast cells.

IV. Control of Glycerol Export

A. Response of Yeast to Hypo-osmotic Shock: Glycerol Export through Fps1p

As discussed above, yeast cells accumulate glycerol when growing in high-osmolarity media. Intracellular glycerol levels of up to 1 M have been monitored (Blomberg and Adler, 1992). In nature, yeast cells can be exposed to very rapid and dramatic hypo-osmotic shock, for instance, during a rain shower. The cell wall certainly protects yeast cells from bursting, as illustrated by the fact that mutants with a weaker cell wall are sensitive to hypo-osmotic shock. In addition, to cope with hypo-osmotic shock, yeast cells can very rapidly release most of the glycerol they have accumulated (Luyten *et al.,* 1995; Tamás *et al.,* 1999). Hence, yeast cells share with bacterial and mammalian cells the ability to perform a regulated volume decrease by osmolyte export (Booth and Louis, 1999; Kirk and Strange, 1998).

It appears that only a single protein, Fps1p, is necessary and sufficient for rapid glycerol export. While wild-type cells dispose of up to 80% of the accumulated glycerol within the first 3 min after hypo-osmotic shock, *fps1*Δ mutants need about 1 to 2 hr for this process (Luyten *et al.,* 1995; Tamás *et al.,* 1999). This suggests that passive diffusion of glycerol through the lipid bilayer is a fairly inefficient process in yeast. Mutants lacking Fps1p are sensitive to hypo-osmotic shock and survive at a 10- to 100-fold lower proportion than wild type; surviving cells resume growth much more slowly (Tamás *et al.,* 1999). Double mutants lacking Fps1p and bearing a mutation that causes a slightly weaker cell wall show an even stronger hypo-osmosensitivity. To our knowledge, the phenotype of the *fps1*Δ mutant is the only genetic evidence that osmolyte export is essential for adaptation to low external osmolarity in eukaryotic cells.

We have attempted to identify further mutants with a phenotype similar to that of *fps1*Δ mutants, but we have failed (Ferreira *et al.*, 2001), thus suggesting that glycerol export is the only process needed for yeast cells to cope with an osmotic downshock. On the other hand, *fps1* mutants have been identified in different genetic screens and show different phenotypes, which illustrates the diverse roles of glycerol.

Fps1 mutant yeast cells mate at a much lower efficiency (Philips and Herskowitz, 1997). When yeast cells of the opposite sex mate, they grow toward each other and then locally degrade the cell wall so that the plasma membranes can fuse. In this situation, yeast cells are likely to be very sensitive to osmotic changes. It is also conceivable that turgor pressure must be diminished to prevent cell bursting. Fps1p might play a role in this process. This idea is supported by the observation that the cell fusion defect of *fps1* mutants can be suppressed by the addition of an osmotic stabilizer such as 1 M sorbitol or by deletion of the *GPD1* gene, which leads to lower glycerol production (Philips and Herskowitz, 1997). On the other hand, cell fusion is usually tested on normal growth medium where the cellular glycerol content is very low. Therefore, glycerol production may be enhanced in mating cells, perhaps during the cell expansion leading to formation of mating projections. Alternatively, the cell might be able to generate higher glycerol levels locally. These options have not been systematically studied.

FPS1 has also been identified in a screen for mutants affecting the activity of an osmosensitive reporter gene and has subsequently been found to affect signaling through the HOG pathway (Tamás *et al.*, 2000; Tao *et al.*, 1999). This is not surprising. We have observed that mutants lacking Fps1p show a diminished and shorter activation of the HOG pathway after osmotic shock, presumably because they can accumulate glycerol somewhat faster (Tamás *et al.*, 2000). Mutants with overactive Fps1p, on the other hand, show a stronger and more sustained activation of the HOG pathway because they cannot accumulate glycerol as rapidly as the wild type. So far, there is no evidence that Fps1p directly affects osmosignaling; all observations can be explained by the role of Fps1p as a glycerol exporter and its effects on intracellular glycerol levels.

An additional observation confirms the role of Fps1p as a glycerol export channel. The protein is needed for normal growth under strictly anaerobic conditions, when yeast cells produce glycerol for redox regulation (Tamás *et al.*, 1999). *Fps1* mutants hyperaccumulate glycerol under these conditions and grow poorly, probably due to inappropriate high turgor pressure. Related to this observation is that Fps1p was originally identified as a suppressor of a mutant that causes an imbalance in central metabolism associated with enhanced glycerol production (Luyten *et al.*, 1995; Van Aelst *et al.*, 1991).

B. Fps1p Is an Unusual Member of the MIP Family

Fps1p is clearly a member of the MIP family of channel proteins, and its closest homologues are mammalian AQP3, 7, 9 and the bacterial glycerol facilitators, with sequence identities around 35%. On the other hand, Fps1p also has a number of unusual features (Fig. 7), and phylogenetic analyses place it in some distance from other MIP channels (Hohmann *et al.*, 2000; see Engels and Stahlberg, this volume). First of all, the canonical NPA boxes in loops B and E, which line the channel and are important for transport function and specificity (Fu *et al.*, 2000; Murata *et al.*, 2000), are altered to NPS and NLA, respectively. We have recently studied the functional significance of these motifs (Bill *et al.*, 2001). Interestingly, when NPS in loop B is altered to NPA, we observed a reduction in glycerol transport, while changing NLA to NPA in loop E did not have any effect. In fact, when both, NPS and NLA were changed to NPA, the protein behaved like wild-type Fps1p, indicating that NPS in loop B functions better when combined with an NLA in loop E. When altering the NPA motifs in the homologous *E. coli* GlpF, we observed interesting effects: The double mutant (NPS and NLA in loops B and E, respectively), as well

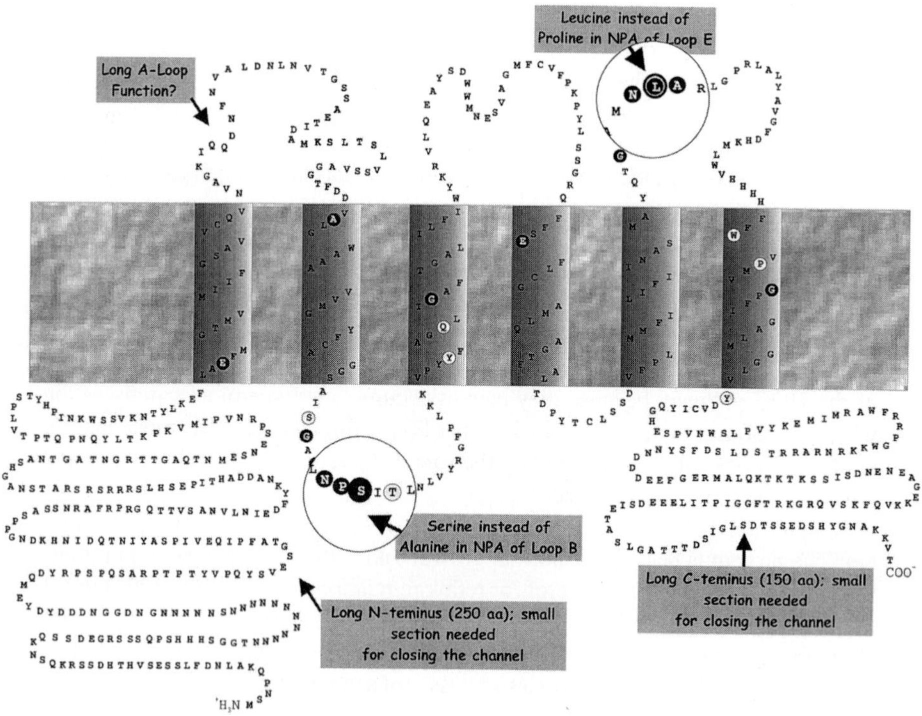

FIG. 7 Unusual features of Fps1p. Adapted from an illustration kindly provided by K. Hedfalk.

as the single NLA mutation in loop E resulted in nonfunctional (albeit correctly localized) channels. The single NPS mutation in loop B caused a diminished transport activity only for uptake, but not efflux, of glycerol. These data indicate that additional domains contribute to the functionality of the NLA/NPS boxes in Fps1p. Those are probably located within the transmembrane domains, because swapping the entire loops B and E from Fps1p to GlpF did not result in an active channel. The data also indicate that certain mutations can have more pronounced effects when transport is measured in the one, rather than in the other, direction. This may be due to the way the substrate interacts with the first NPA on its way through the channel.

Another unusual feature of Fps1p is its size: 669 amino acids, that is, more than twice that of a "conventional" MIP channel (Van Aelst *et al.,* 1991). Loop A has an additional ~30 amino acids that are not found in other MIP channels. The role, if any, of this domain is not known. Fps1p also has a 250-amino acid N-terminal and a 150-amino acid C-terminal extension. Both extensions are highly hydrophilic and not homologous to any protein in the database. As discussed below, certain parts of these extensions have regulatory functions.

Saccharomyces cerevisiae has an additional, unusual, MIP channel. We have tried to identify a role for this protein but have failed so far to associate a phenotype to the deletion mutant. Hence the gene still has its systematic name *YFL054c*. While Yfl054p has NPA boxes, it is also an unusually large MIP channel: 646 amino acids. The N-terminal extension is even longer than that of Fps1p (about 300 amino acids as compared to 250 amino acids), but the two extensions do not share any homology. In fact, while no homologues for Fps1p have so far been found in any other organism, proteins clearly related to Yfl054p have been found by systematic sequencing in *Schizosaccharomyces pombe* and *Botrytis cinerea* (Hohmann *et al.,* 2000). This suggests that Yfl054p is likely to have a function relevant for fungal biology.

C. Fps1p Is a Regulated Glycerol Export Channel

The fact that Fps1p exports the osmolyte glycerol from yeast cells suggests that its function is regulated: under conditions of high external osmolarity, the yeast cell needs to accumulate the glycerol it produces, but upon a hypo-osmotic shock, glycerol is rapidly exported. We have monitored Fps1p activity by determining the uptake of radiolabeled glycerol into yeast cells, and these data indicate that Fps1p function is indeed very rapidly regulated (Luyten *et al.,* 1995; Tamás *et al.,* 1999). It appears that the channel is closed upon hyperosmotic shock within less than 15 sec. Furthermore, the very rapid, Fps1p-dependent release of glycerol upon hypo-osmotic shock indicates that the channel is rapidly reopened again. This rapid regulation, together with data indicating that the production of Fps1p is not regulated by osmotic changes, suggests that the control occurs at the level of channel activity (Luyten *et al.,* 1995; Tamás *et al.,* 1999). This notion is further supported by certain mutations in Fps1p that cause an apparent inability of the channel to close upon hyperosmotic shock. Such mutants have severely delayed glycerol accumulation. They overproduce and excrete large amounts of glycerol and are sensitive to high osmolarity. Interestingly, yeast cells expressing the bacterial glycerol facilitator GlpF show the same behavior (Luyten *et al.,* 1995; Tamás *et al.,* 1999). GlpF, whose function in *E. coli* is in glycerol uptake for catabolism, is not regulated by osmotic shock when expressed in yeast.

Although the mechanism of Fps1p regulation is not understood, several interesting pieces of information have been collected. First of all, we have failed to associate any of the known, osmosensing signaling pathways with the control of Fps1p (Luyten *et al.,* 1995; Tamás *et al.,* 1999). In addition, among the many osmosensitive yeast mutants known none seems to be defective in Fps1p control. Finally, we have failed to identify proteins in yeast that interact with Fps1p in the two-hybrid system, and no such protein has shown up in a genome-wide search for interacting proteins (Uetz *et al.,* 2000). Although these negative results are, of course, no strong argument, they suggest that Fps1p might not need any other proteins for its regulation.

We have recently demonstrated that the unusual N- and C-terminal extensions contain domains that are needed for the control of Fps1p. Remarkably, however, the bulk of the extensions does not seem to be involved in regulation: up to 216 amino acids can be deleted from the N-terminus without affecting the closing upon hyperosmotic shock (Tamás *et al.,* 1999). However, deletion of amino acids 216–231 (the first transmembrane domain starts at about amino acid 254) results in a channel that can no longer be closed. Mutational analysis within this segment further defines amino acids 218–240 as apparently the most important (Tamás *et al.,* 2001). An interesting characteristic of this region is that several proline residues present in an almost regular spacing (if one adds a leucine in place of one proline): P_{217}xxxxPxxLxxxPxxPxxxPxxxxP_{241}. It seems that mutations of the inner proline/leucine residues are not tolerated, leading to a fully open channel. But other residues in this region may also contribute. The significance of these observations is unclear, but such a domain could serve as an interface for interactions with other proteins or other parts of the same proteins. Alternatively, the regulatory domain could be involved in orienting transmembrane domains in a certain conformation, thereby enhancing the sensitivity of the channel to osmotic changes. It has been observed in dynamic models of AQP1 that this protein, and its ability to transport water, maybe sensitive to lateral pressure, providing a possible means of channel regulation by membrane stretching (de Groot and Grubmüller, personal communication). Extensions and loops certainly contribute to orienting transmembrane domains, and certain conformations could well be more susceptible than others to membrane stretching.

In the C-terminal extension, we have recently identified amino acids 535–577 (last transmembrane domain predicted to end at 526) as important for channel closure (Hedfalk *et al.,* 2001). A remarkable feature of this sequence is the presence of 10 positively charged residues within 22 amino acids. Again, while the significance of this observation is unclear, this domain may interact with other proteins, with other parts of the same protein, or it may determine the orientation of neighboring domains in space.

D. Approaches to Studying Fps1p Regulation

Further work is needed to better understand the regulation of Fps1p. Of specific interest is now the construction of a derivative of Fps1p containing the shortest possible extensions that still allow normal channel regulation. The very long hydrophilic extensions have so far prevented overproduction and purification of the protein, which is obviously a condition for *in vitro* studies as well as structural analysis. Especially, projections maps generated by the atomic force microscope are expected to provide interesting information on possible mechanisms of regulation. Complementary to these approaches, we continue to employ genetic analysis to better understand regulation of Fps1p. Further targeted mutational analysis of the regulatory domains identified in the N- and C-terminal extension should lead to determination of the residues involved in regulation. In addition, we have recently devised a genetic screen, with which we expect to identify random mutations in Fps1p that block channel closure under hyperosmotic conditions. Finally we are making use of the fact that GlpF, when expressed in yeast, is not regulated by osmolarity, but rather confers constitutive glycerol export. By swapping domains between Fps1p and GlpF, we hope to identify those parts of Fps1p that are needed to convert GlpF into a regulated channel.

E. Transport of Toxic Metal Ions through Fps1p

GlpF has been shown to transport antimonite ions (Sanders *et al.,* 1997). It has been suggested that trivalent antimonite ions in their hydrated form resemble glycerol and are transported for that reason. Based on the structure of GlpF, it has also been suggested that ions could be transported through a potential single central cavity formed by the four subunits of a GlpF tetramer (Fu *et al.,* 2000). This interesting possibility, which could lead to a much wider definition of the physiological roles of MIP channels, needs to be investigated.

Recently, transport of trivalent antimonite and arsenite through yeast Fps1p has been studied by analyzing sensitivity of yeast cells to these toxic metal ions, as well as by direct transport assays in cell lacking Fps1p or expressing the truncated, constitutively active version of Fps1p (Wysocki *et al.,* 2001). The data strongly suggest that Fps1p mediates transport of antimonite and arsenite. Most interestingly, cells lacking Fps1p are more tolerant. These findings indicate that MIP channels might be important entry pathways for antimonite and arsenite into cells, with apparent important clinical and pharmacological implications. Even more surprisingly, yeast cells seem to be "aware" of Fps1p transporting arsenite into the cell. Expression of the *FPS1* gene is rapidly and strongly repressed when arsenite is added. Hence, the cell appears to have developed defense mechanisms, including reduction of possible entry pathways. The transport of antimonite and arsenite ions via MIP channels appears to be a most interesting area for future research, with respect to the mechanisms, the physiological implications, and the responses of cells to these compounds (Wysocki *et al.,* 2001).

V. Water Channels in Yeast

The yeast *Saccharomyces cerevisiae* has successfully been used to express mammalian and plant aquaporins for functional studies (Laizé *et al.,* 1997; Shinbo *et al.,* 1999; Sullivan *et al.,* 2000; Weig and Jakob, 2000). What has attracted much less attention, however, is the fact that the genome of this organism itself encodes two aquaporins. Aquaporins have also been found in other eukaryotic microorganisms, such as the slime mold *Dictyostelium* and the pathogenic yeast *Candida albicans* (Calamita, 2000; Hohmann *et al.,* 2000).

A. Yeast Has Two Water Channels

Curiously, it has turned out that the majority of laboratory yeast strains, including the one whose genome sequence has been determined (S288C), carry mutations in both *Saccharomyces cerevisiae* aquaporin genes *AQY1* and *AQY2* (André 1995; Bonhivers *et al.,* 1998; Laizé *et al.,* 2000). This is a most unusual situation. Sequence differences between yeast strains are certainly not uncommon, but that two homologous genes are both mutated has not been reported previously. At present, only one single strain with the name Σ1278b seems to have wild-type versions of both aquaporin genes (Bonhivers *et al.,* 1998; Laizé *et al.,* 2000). This strain has apparently retained a number of properties that have been lost in laboratory strains, such as the ability to switch to a pseudohyphal mode of growth under nutrient starvation.

Aqy1p from S288C and Σ1278b differ in three amino acid positions (Fig. 8) and in addition, the C-terminus is entirely different due to a frameshift (Bonhivers *et al.,* 1998). It has been demonstrated that the alterations in position 121 and 255 both contribute to the inability of Aqy1p from S288C to transport water in the *Xenopus* oocyte system (Bonhivers *et al.,* 1998). Aqy1p from Σ1278b transports water in oocytes and hence has a demonstrated function as an aquaporin (Bonhivers *et al.,* 1998).

AQY2 does not appear as one, but rather as two overlapping open reading frames in S288C, while there is one open reading frame in Σ1278b. This is due to the lack of 11 bp in the sequence of the S288C gene in the center of *AQY2* (Laizé

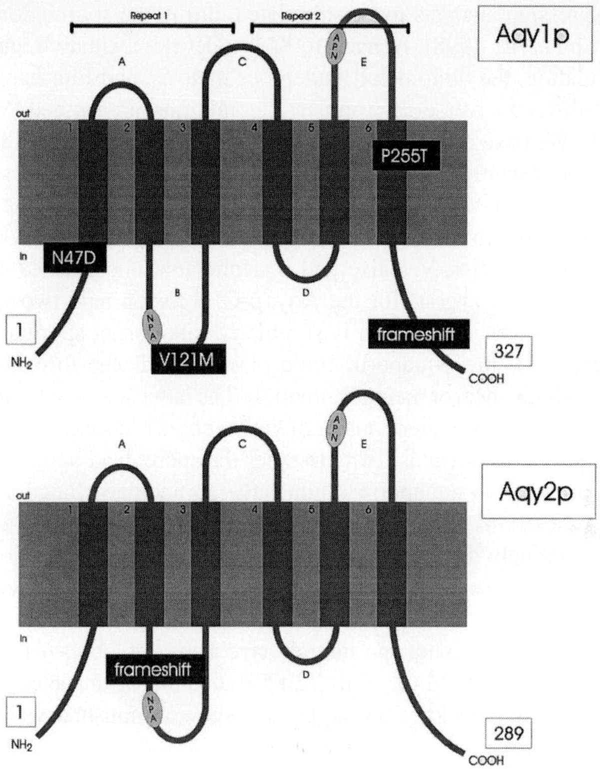

FIG. 8 Aqy1p and Aqy2p. The sketch indicates the sequence differences between the two proteins in the Σ1278b strain, where both proteins appear to be active, and the common laboratory strain, where both proteins appear to be inactive.

et al., 2000). However, even the Aqy2p from strain Σ1278b does not transport water in the oocyte system (Meyrial *et al.,* 2001). This negative result of course does not exclude its functionality in yeast: Fps1p also does not function in oocytes.

Aqy1p and Aqy2p from strain Σ1278b are 87% identical and 93% similar, and hence derive from gene duplication in yeast. They are most closely related to plant aquaporins (Bonhivers *et al.,* 1998).

B. Aqy1p

The physiological role of neither of the two yeast aquaporins is understood at present. However, gene expression and localization data suggest that the two proteins fulfill very different functions. The expression of *AQY1* has been observed by global expression analysis to be stimulated during the sporulation of diploid yeast cells (Chu *et al.,* 1998). Hence, *AQY1* is a developmentally regulated gene. During sporulation, the diploid cell undergoes meiosis, and four haploid spores, which can withstand adverse environmental conditions, are formed (Vershon and Pierce, 2000). We have confirmed this gene expression pattern, but the protein is undetectable in vegetative yeast cells (Meyrial *et al.,* 2001). Interestingly, however, we observed a boost of expression of the Aqy1p-GFP fusion in the developing yeast spore. The protein localizes to both the endoplasmic reticulum and the cell surface (F. Sidoux-Walter, V. Laizé and S. Hohmann, unpublished data). Since in a diploid cell heterozygous for the Aqy1p-GFP fusion only two spores show this expression pattern, expression is stimulated late during sporulation. Therefore, it appears as if this aquaporin could play a role in the formation of yeast spores, their maintenance, or their germination. The latter seems to be an especially attractive possibility. The water content of yeast spores is diminished. Hence, during spore germination, water is likely to enter the spore, leading to cell swelling. During this process, an aquaporin might have an ancillary function. However, since most yeast strains used in the laboratory, where sporulation and spore germination are routinely done in genetic crosses, do not have a functional Aqy1p, this gene cannot be essential for spore germination, at least not under laboratory conditions.

Aquaporins associated with sporulation have also been observed in *Dictyostelium* (Flick *et al.,* 1997; Mitra *et al.,* 2000) and, in one instance, it has been observed that mutants lacking the aquaporin show diminished viability during extended spore dormancy.

C. Aqy2p

Expression of *AQY2* has not been observed to any significant extent in standard laboratory yeast strains, and this is probably due to the deletion of 11 bp in the center of the gene. This frameshift most likely causes translation to stop, which in turn leads to rapid degradation of the mRNA. Hence, the global gene expression data in current databases do not correctly reflect the expression of *AQY2* correctly.

In strain Σ1278b, we have observed that *AQY2* is a fairly strongly expressed gene. The Aqy2 protein is readily detectable in Western blot analysis (Meyrial *et al.,* 2001). In addition, electron microscopy with immunogold labeling shows that the protein is localized to the plasma membrane (Meyrial *et al.,* 2001), a result that is also confirmed by light microscopy using an Aqy2p-GFP fusion (Hohmann *et al.,* unpublished data). While no Aqy2p-dependent water transport has been monitored in *Xenopus* oocytes, microsomal vesicles from strain Σ1278b show an Aqy2p-dependent water transport activity (Meyrial *et al.,* 2001). In conclusion, in strain Σ1278b, Aqy2p is an abundantly expressed plasma membrane water channel.

Expression of the *AQY2* gene is strictly controlled. We have observed strong expression of *AQY2* only under optimal growth conditions, which allow for rapid cell proliferation (Meyrial *et al.,* 2001, and our unpublished results). Consistent with this observation, it has been reported recently that deletion of the *TPK2* gene, which encodes one of three catalytic subunits of protein kinase A, causes a three-fold lower *AQY2* expression in Σ1278b (Robertson *et al.,* 2000). As discussed above, a high PKA activity is usually associated with high proliferation potential. Finally, we have recently observed that hyperosmotic shock leads to diminished, and hypo-osmotic shock to stronger, expression of *AQY2*. Although this analysis is still preliminary, it points to Aqy2p being more abundant in rapidly growing, and hence rapidly expanding, cells. As has been discussed for AqpZ from *Escherichia coli* (Calamita, 2000; Calamita *et al.,* 1997, 1998; Delamarche *et al.,* 1999), Aqy2p might have an ancillary role in cell expansion, especially during rapid growth. If this notion is confirmed, an interesting question would be whether Aqy2p facilitates water uptake and cell swelling, or works to counteract swelling and hence facilitates water efflux. Whatever the role of Aqy2p, the preliminary results suggest that analysis of Aqy2p function and the regulatory connections to the osmosensing signaling PKC and HOG pathways and the proliferation controlling PKA pathway will be a most interesting subject for future studies.

VI. Conclusions and Perspectives

A. In How Much Is Yeast a Model and What Can We Learn from Yeast in General?

Yeasts (*Saccharomyces cerevisiae* and *Schizosaccharomyces pombe*) are excellent study objects and indeed have model character for many aspects of eukaryotic cell function. With respect to osmoregulation, as far as we know at present, this is especially true for signal transduction through MAP kinase pathways. In this respect, it appears that *S. pombe* might even be the better model because details of sensing, signaling, and physiological response may be more similar to that of mammalian cells. The overall strategy of controlling osmolyte accumulation appears to be similar in yeast and in mammalian cells. Production of osmolytes is controlled at the level of gene expression, but probably also by redirection of metabolic fluxes via unknown mechanisms (at least in yeast cells). Hypo-osmotic shock triggers rapid export of osmolytes in yeast and in mammalian cells. As shown for yeast glycerol export, a MIP channel (Fps1p) is involved in regulated osmolyte export, but it is not known if mammalian MIP channels are also involved in the process of regulated volume decrease.

MIP channels are ubiquitous, but it is apparent that their specific role and regulation is subject to the evolutionary pressure imposed by the natural environment.

Therefore, the role, function, and regulation of MIP channels in yeast may serve to only a limited extent as a model for higher eukaryotes. As described in this chapter, Aqy1p appears to play a role in sporulation, and Aqy2p is expressed only in rapidly proliferating cells. Hence, the physiological roles of these MIP channels appear to be specific for (fungal) microorganisms. Still, these examples illustrate that MIP channels seem to constitute a tool employed by evolution in many different physiological and evolutionary contexts, which makes studies on different biological systems such as fungi, plants, and mammalian cells most interesting.

Acknowledgments

Work in this laboratory has been supported by the European Commission through Contracts FMRX-CT96-0007, FMRX-CT97-0128, and BIO4-CT98-0024. S.H. is a special researcher supported by the Swedish Natural Science Research Council.

References

Albertyn, J., Hohmann, S., Thevelein, J. M., and Prior, B. A. (1994a). *GPD1,* which encodes glycerol-3-phosphate dehydrogenase, is essential for growth under osmotic stress in *Saccharomyces cerevisiae* and its expression is regulated by the high-osmolarity glycerol response pathway. *Mol. Cell. Biol.* **14,** 4135–4144.

Albertyn, J., Hohmann, S., and Prior, B. A. (1994b). Characterization of the osmotic-stress response in Saccharomyces cerevisiae: Osmotic stress and glucose repression regulate glycerol-3-phosphate dehydrogenase independently. *Curr. Genet.* **25,** 12–18.

Alepuz, P. M., Jovanovic, A., Reiser, V., and Ammerer, G. (2001). Stress induced MAP kinase Hog1 is part of transcription activation complexes. *Mol. Cell* **1,** 767–777.

André, B. (1995). An overview of membrane transport proteins in *Saccharomyces cerevisiae. Yeast* **11,** 1575–1611.

André, L. (1990). "On Glycerol Accumulation and Its Role in Osmoregulation in Yeast." University of Göteborg, Sweden.

Ansell, R., Granath, K., Hohmann, S., Thevelein, J. M., and Adler, L. (1997). The two isoenzymes for yeast NAD+-dependent glycerol 3-phosphate dehydrogenase encoded by GPD1 and GPD2 have distinct roles in osmoadaptation and redox regulation. *EMBO J.* **16,** 2179–2187.

Antonsson, B., Montessuit, S., Friedli, L., Payton, M. A., and Paravicini, G. (1994). Protein kinase C in yeast. Characteristics of the *Saccharomyces cerevisiae PKC1* gene product. *J. Biol. Chem.* **269,** 16,821–16,828.

Banuelos, M. A., Quintero, F. J., and Rodriguez-Navarro, A. (1995). Functional expression of the ENA1(PMR2)-ATPase of Saccharomyces cerevisiae in Schizosaccharomyces pombe. *Biochim. Biophys. Acta* **1229,** 233–238.

Bill, R. M., Hedfalk, K., Karlgren, S., Mullins, J. E., Rydström, J., and Hohmann, S. (2001). Analysis of the pore of the unusual MIP channel, yeast Fps1p. *J. Biol. Chem.* **276,** 36543–36549.

Bilsland-Marchesan, E., Arino, J., Saito, H., Sunnerhagen, P., and Posas, F. (2000). Rck2 kinase is a substrate for the osmotic stress-activated mitogen-activated protein kinase Hog1. *Mol. Cell. Biol.* **20,** 3887–3895.

Björkqvist, S., Ansell, R., Adler, L., and Lidén, G. (1996). Glycerol-3-phosphate dehydrogenase mutants of Saccharomyces cerevisiae grown under aerobic and anaerobic conditions. *Appl. Environ. Microbiol.* **63,** 128–132.

Blomberg, A. (1995). Global changes in protein synthesis during adaptation of the yeast Saccharomyces cerevisiae to 0.7 M NaCl. *J. Bacteriol.* **177,** 3563–3572.

Blomberg, A. (2000). Metabolic surprises in *Saccharomyces cerevisiae* during adaptation to saline conditions: Questions, some answers and a model. *FEMS Microbiol. Lett.* **182,** 1–8.

Blomberg, A., and Adler, L. (1992). Physiology of osmotolerance in fungi. *Adv. Microb. Physiol.* **33,** 145–212.

Blomberg, A., and Adler, L. (1989). Roles of glycerol and glycerol-3-phosphate dehydrogenase (NAD^+) in acquired osmotolerance of *Saccharomyces cerevisiae. J. Bacteriol.* **171,** 1087–1092.

Boguslawski, G. (1992). *PBS2,* a yeast gene encoding a putative protein kinase, interacts with the *RAS2* pathway and affects osmotic sensitivity of *Saccharomyces cerevisiae. J. Gen. Microbiol.* **138,** 2425–2432.

Bonhivers, M., Carbrey, J. M., Gould, S. J., and Agre, P. (1998). Aquaporins in Saccharomyces. Genetic and functional distinctions between laboratory and wild-type strains. *J. Biol. Chem.* **273,** 27,565–27,572.

Booth, I. R., and Louis, P. (1999). Managing hypoosmotic stress: Aquaporins and mechanosensitive channels in Escherichia coli. *Curr. Opin. Microbiol.* **2,** 166–169.

Boy-Marcotte, E., Tadi, D., Perrot, M., Boucherie, H., and Jacquet, M. (1996). High cAMP levels antagonize the reprogramming of gene expression that occurs at the diauxic shift in Saccharomyces cerevisiae. *Microbiology* **142,** 459–467.

Brewster, J. L., de Valoir, T., Dwyer, N. D., Winter, E., and Gustin, M. C. (1993). An osmosensing signal transduction pathway in yeast. *Science* **259,** 1760–1763.

Brown, A. D. (1976). Microbial water stress. *Bacteriol. Rev.* **40,** 803–846.

Brown, A. D., and Edgley, M. (1980). Osmoregulation in yeast. *In* "Symposium on Genetic Engineering of Osmoregulation" (D. W. Rains, R. C. Valentine, and A. Hollander, Eds.), pp. 75–90. New York, Plenum Press.

Bruning, A. R., Bauer, J., Krems, B., Entian, K. D., and Prior, B. A. (1998). Physiological and genetic characterisation of osmosensitive mutants of Saccharomyes cerevisiae. *Arch. Microbiol.* **170,** 99–105.

Cabib, E., Drgonova, J., and Drgon, T. (1998). Role of small G proteins in yeast cell polarization and wall biosynthesis. *Annu. Rev. Biochem.* **67,** 307–333.

Calamita, G. (2000). The Escherichia coli aquaporin-Z water channel. *Mol. Microbiol.* **37,** 254–262.

Calamita, G., Kempf, B., Bonhivers, M., Bishai, W. R., Bremer, E., and Agre, P. (1998). Regulation of the Escherichia coli water channel gene aqpZ. *Proc. Natl. Acad. Sci. USA* **95,** 3627–3631.

Calamita, G., Kempf, B., Rudd, K. E., Bonhivers, M., Kneip, S., Bishai, W. R., Bremer, E., and Agre, P. (1997). The aquaporin-Z water channel gene of Escherichia coli: Structure, organization and phylogeny. *Biol. Cell* **89,** 321–329.

Chu, S., DeRisi, J., Eisen, M., Mulholland, J., Botstein, D., Brown, P. O., and Herskowitz, I. (1998). The transcriptional program of sporulation in budding yeast. *Science* **282,** 699–705.

Colombo, S., Ma, P., Cauwenberg, L., Winderickx, J., Crauwels, M., Teunissen, A., Nauwelaers, D., de Winde, J. H., Gorwa, M. F., Colavizza, D., and Thevelein, J. M. (1998). Involvement of distinct G-proteins, Gpa2 and Ras, in glucose- and intracellular acidification-induced cAMP signalling in the yeast *Saccharomyces cerevisiae. EMBO J.* **17,** 3326–3341.

Crauwels, M., Donaton, M. C., Pernambuco, M. B., Winderickx, J., de Winde, J. H., and Thevelein, J. M. (1997). The Sch9 protein kinase in the yeast Saccharomyces cerevisiae controls cAPK activity and is required for nitrogen activation of the fermentable-growth-medium-induced (FGM) pathway. *Microbiology* **143,** 2627–2637.

Danielsson, A., Larsson, C., Larsson, K., Gustafsson, G., and Adler, L. (1996). A genetic analysis of the calcineurin and calmodulin in Ca^{++}-dependent improvement of NaCl tolerance of *Saccharomyces cerevisiae. Curr. Genet.* **30,** 476–484.

Davenport, K. D., Williams, K. E., Ullmann, B. D., and Gustin, M. C. (1999). Activation of the Saccharomyces cerevisiae filamentation/invasion pathway by osmotic stress in high-osmolarity glycogen pathway mutants. *Genetics* **153,** 1091–1103.

Davenport, K. R., Sohaskey, M., Kamada, Y., Levin, D. E., and Gustin, M. C. (1995). A second osmosensing signal transduction pathway in yeast—hypotonic shock activates the PKC1 protein kinase-regulated cell integrity pathway. *J. Biol. Chem.* **270,** 30,157–30,161.

Delamarche, C., Thomas, D., Rolland, J. P., Froger, A., Gouranton, J., Svelto, M., Agre, P., and Calamita, G. (1999). Visualization of AqpZ-mediated water permeability in *Escherichia coli* by cryoelectron microscopy. *J. Bacteriol.* **181,** 4193–4197.

de Nobel, H., van Den Ende, H., and Klis, F. M. (2000). Cell wall maintenance in fungi. *Trends Microbiol.* **8,** 344–345.

de Winde, J. H., Thevelein, J. M., and Winderickx, J. (1997). From feast to famine: Adaptation to nutrient depletion in yeast. *In* "Yeast Stress Responses" (S. Hohmann and W. H. Mager, Eds.), pp. 7–52. Austin, TX, R. G. Landes.

Eriksson, P., Alipour, H., Adler, L., and Blomberg, A. (2000). Rap1p-binding sites in the *Saccharomyces cerevisiae GPD1* promoter are involved in its response to NaCl. *J. Biol. Chem.* **275,** 29,368–29,376.

Eriksson, P., Andre, L., Ansell, R., Blomberg, A., and Adler, L. (1995). Molecular cloning of *GPD2,* a second gene encoding *sn*-glycerol 3-phosphate dehydrogenase (NAD^+) in *Saccharomyces cerevisiae,* and its comparison to *GPD1. Mol. Microbiol.* **17,** 95–107.

Estruch, F. (2000). Stress-controlled transcription factors, stress-induced genes and stress tolerance in budding yeast. *FEMS Microbiol. Rev.* **24,** 469–486.

Estruch, F., and Carlson, M. (1993). Two homologous zinc finger genes identified by multicopy suppression in a SNF1 protein kinase mutant of *Saccharomyces cerevisiae. Mol. Cell. Biol.* **13,** 3872–3881.

Ferreira, M. C., Bao, X.-M., Laizé, V., and Hohmann, S. (2001). Transposon mutagenesis reveals novel loci affecting tolerance to salt stress and growth at low temperature. *Curr. Genet.* **40,** 27–39.

Ferrigno, P., Posas, F., Koepp, D., Saito, H., and Silver, P. A. (1998). Regulated nucleo/cytoplasmic exchange of HOG1 MAPK requires the importin beta homologs NMD5 and XPO1. *EMBO J.* **17,** 5606–5614.

Flick, K. M., Shaulsky, G., and Loomis, W. F. (1997). The wacA gene of Dictyostelium discoideum is a developmentally regulated member of the MIP family. *Gene* **195,** 127–130.

Fu, D., Libson, A., Miercke, L. J., Weitzman, C., Nollert, P., Krucinski, J., and Stroud, R. M. (2000). Structure of a glycerol- conducting channel and the basis for its selectivity. *Science* **290,** 481–486.

Gancedo, C., Gancedo, J. M., and Sols, A. (1968). Glycerol metabolism in yeasts. Pathways of utilization and production. *Eur. J. Biochem.* **5,** 165–172.

Gancedo, C., and Serrano, R. (1989). Energy-yielding metabolism. *In* "The Yeasts" (A. H. Rose and J. S. Harrison, Eds.), pp. 205–257. New York. Academic Press.

Garreau, H., Hasan, R. N., Renault, G., Estruch, F., Boy-Marcotte, E., and Jacquet, M. (2000). Hyperphosphorylation of Msn2p and Msn4p in response to heat shock and the diauxic shift is inhibited by cAMP in *Saccharomyces cerevisiae. Microbiology* **146,** 2113–2120.

Gekko, K., and Timasheff, S. N. (1981). Mechanism of protein stabilization by glycerol: Preferential hydration in glycerol-water mixtures. *Biochemistry* **20,** 4667.

Görner, W., Durchschlag, E., Martinez-Pastor, M. T., Estruch, F., Ammerer, G., Hamilton, B., Ruis, H., and Schüller, C. (1998). Nuclear localization of the C2H2 zinc finger protein Msn2p is regulated by stress and protein kinase A activity. *Genes Dev.* **12,** 586–597.

Grauslund, M., Lopes, J. M., and Ronnow, B. (1999). Expression of GUT1, which encodes glycerol kinase in Saccharomyces cerevisiae, is controlled by the positive regulators Adr1p, Ino2p and Ino4p and the negative regulator Opi1p in a carbon source-dependent fashion. *Nucleic Acids Res.* **27,** 4391–4398.

Griffioen, G., Mager, W. H., and Planta, R. J. (1994). Nutritional upshift response of ribosomal protein gene transcription in Saccharomyces cerevisiae. *FEMS Microbiol. Lett.* **123,** 137–144.

Gustin, M. C., Albertyn, J., Alexander, M., and Davenport, K. (1998). MAP kinase pathways in the yeast *Saccharomyces cerevisiae. Microbiol. Mol. Biol. Rev.* **62,** 1264–1300.

Guthrie, C., and Fink, G. R. (1991). "Guide to Yeast Genetics and Molecular Biology," Volume 194. New York, Academic Press.

Hallsworth, J. E. (1998). Ethanol-induced water stress in yeast. *J. Ferm. Bioeng.* **85,** 125–137.

Hedfalk, K., Bill, R., Rydström, J., and Hohmann, S. (2001). The C-terminal extension of the *Saccharomyces cerevisiae* osmosensitive glycerol channel Fps1p contributes to channel regulation. *in preparation.*

Heinisch, J. J., Lorberg, A., Schmitz, H. P., and Jacoby, J. J. (1999). The protein kinase C-mediated MAP kinase pathway involved in the maintenance of cellular integrity in *Saccharomyces cerevisiae. Mol. Microbiol.* **32,** 671–680.

Herskowitz, I. (1995). MAP kinase pathways in yeast: for mating and more. *Cell* **80,** 187–197.

Hirayama, T., Maeda, T., Saito, H., and Shinozaki, K. (1995). Cloning and characterization of seven cDNAs for hyperosmolarity-responsive (HOR) genes of *Saccharomyces cerevisiae Mol. Gen. Genet.* **249,** 127–138.

Hohmann, S., Kayingo, G., Bill, R. M., and Prior, B. A. (2000). Microbial MIP channels. *Trends Microbiol.* **8,** 33–38.

Hohmann, S., and Mager, W. H. (1997). "Yeast Stress Responses." Austin, TX, R. G. Landes Company.

Holst, B., Lunde, C., Lages, F., Oliveira, R., Lucas, C., and Kielland-Brandt, M. C. (2000). *GUP1* and its close homologue *GUP2,* encoding multimembrane-spanning proteins involved in active glycerol uptake in *Saccharomyces cerevisiae. Mol. Microbiol.* **37,** 108–124.

Inoue, Y., Tsujimoto, Y., and Kimura, A. (1998). Expression of the glyoxalase I gene of *Saccharomyces cerevisiae* is regulated by high osmolarity glycerol mitogen-activated protein kinase pathway in osmotic stress response. *J. Biol. Chem.* **273,** 2977–2983.

Irie, K., Takase, M., Lee, K. S., Levin, D. E., Araki, H., Matsumoto, K., and Oshima, Y. (1993). *MKK1* and *MKK2,* which encode *Saccharomyces cerevisiae* mitogen-activated protein kinase–kinase homologs, function in the pathway mediated by protein kinase C. *Mol. Cell. Biol.* **13,** 3076–3083.

Jacoby, J. J., Nilius, S. M., and Heinisch, J. J. (1998). A screen for upstream components of the yeast protein kinase C signal transduction pathway identifies the product of the *SLG1* gene. *Mol. Gen. Genet.* **258,** 148–155.

Jacoby, T., Flanagan, H., Faykin, A., Seto, A. G., Mattison, C., and Ota, I. (1997). Two protein-tyrosine phosphatases inactivate the osmotic stress response pathway in yeast by targeting the mitogen-activated protein kinase, Hog1. *J. Biol. Chem.* **272,** 17,749–17,755.

Jennings, D. H. (1984). Polyol metabolism in fungi. *Adv. Microb. Physiol.* **25,** 149.

Jung, U. S., and Levin, D. E. (1999). Genome-wide analysis of gene expression regulated by the yeast cell wall integrity signalling pathway. *Mol. Microbiol.* **34,** 1049–1057.

Karlgren, S., and Hohmann, S. (2001). A system to study polyol uptake in yeast by simple growth tests. *in preparation.*

Kenyon, C. P., Prior, B. A., and van Vuuren, H. J. J. (1986). Water relations of ethanol fermentation by *Saccharomyces cerevisiae:* Glycerol production and solute stress. *Enzyme Microb. Technol.* **8,** 461–464.

Ketela, T., Green, R., and Bussey, H. (1999). *Saccharomyces cerevisiae* Mid2p is a potential cell wall stress sensor and upstream activator of the PKC1-MPK1 cell integrity pathway. *J. Bacteriol.* **181,** 3330–3340.

Kirk, K., and Strange, K. (1998). Functional properties and physiological roles of organic solute channels. *Annu. Rev. Physiol.* **60,** 719–739.

Klis, F. M. (1994). Review: Cell wall assembly in yeast. *Yeast* **10,** 851–869.

Kobayashi, N., and McEntee, K. (1993). Identification of *cis* and *trans* components of a novel heat shock stress regulatory pathway in *Saccharomyces cerevisiae. Mol. Cell. Biol.* **13,** 248–256.

Kraakman, L., Lemaire, K., Ma, P., Teunissen, A. W., Donaton, M. C., Van Dijck, P., Winderickx, J., de Winde, J. H., and Thevelein, J. M. (1999). A *Saccharomyces cerevisiae* G-protein coupled receptor, Gpr1, is specifically required for glucose activation of the cAMP pathway during the transition to growth on glucose. *Mol. Microbiol.* **32,** 1002–1012.

Krallish, I., Jeppsson, H., Rapoport, A., and Hahn-Hagerdal, B. (1997). Effect of xylitol and trehalose on dry resistance of yeasts. *Appl. Microbiol. Biotechnol.* **47,** 447–451.

Lages, F., Silva-Graca, M., and Lucas, C. (1999). Active glycerol uptake is a mechanism underlying halotolerance in yeasts: A study of 42 species. *Microbiology* **145,** 2577–2585.

Laizé, V., Ripoche, P., and Tacnet, F. (1997). Purification and functional reconstitution of the human CHIP28 water channel expressed in Saccharomyces cerevisiae. *Protein Expr. Purif.* **11,** 284–288.

Laizé, V., Tacnet, F., Ripoche, P., and Hohmann, S. (2000). Polymorphism of Saccharomyces cerevisiae aquaporins. *Yeast* **16,** 897–903.

Larsson, C., Påhlman, I. L., Ansell, R., Rigoulet, M., Adler, L., and Gustafsson, L. (1998). The importance of the glycerol 3-phosphate shuttle during aerobic growth of *Saccharomyces cerevisiae. Yeast* **14,** 347–357.

Larsson, K., Eriksson, P., Ansell, R., and Adler, L. (1993). A gene encoding *sn*-glycerol 3-phosphate dehydrogenase (NAD^+) complements an osmosensitive mutant of *Saccharomyces cerevisiae. Mol. Microbiol.* **10,** 1101–1111.

Lee, K. S., Hines, L. K., and Levin, D. E. (1993). A pair of functionally redundant yeast genes (PPZ1 and PPZ2) encoding type 1-related protein phosphatases function within the PKC1-mediated pathway. *Mol. Cell. Biol.* **13,** 5843–5853.

Lee, K. S., Irie, K., Gotoh, Y., Watanabe, Y., Araki, H., Nishida, E., Matsumoto, K., and Levin, D. E. (1993). A yeast mitogen-activated protein kinase homolog (Mpk1p) mediates signalling by protein kinase C. *Mol. Cell. Biol.* **13,** 3067–3075.

Lee, K. S., and Levin, D. E. (1992). Dominant mutations in a gene encoding a putative protein kinase (*BCK1*) bypass the requirement for a *Saccharomyces cerevisiae* protein kinase C homolog. *Mol. Cell. Biol.* **12,** 172–182.

Levin, D. E., and Bartlett-Heubusch, E. (1992). Mutants in the *S. cerevisiae PKC1* gene display a cell cycle-specific osmotic stability defect. *J. Cell. Biol.* **116,** 1221–1229.

Levin, D. E., Bowers, B., Chen, C. Y., Kamada, Y., and Watanabe, M. (1994). Dissecting the protein kinase C/MAP kinase signalling pathway of *Saccharomyces cerevisiae. Cell. Mol. Biol. Res.* **40,** 229–239.

Lodder, A. L., Lee, T. K., and Ballester, R. (1999). Characterization of the Wsc1 protein, a putative receptor in the stress response of *Saccharomyces cerevisiae. Genetics* **152,** 1487–1499.

Luyten, K., Albertyn, J., Skibbe, W. F., Prior, B. A., Ramos, J., Thevelein, J. M., and Hohmann, S. (1995). Fps1, a yeast member of the MIP family of channel proteins, is a facilitator for glycerol uptake and efflux and is inactive under osmotic stress. *EMBO J.* **14,** 1360–1371.

Maeda, T., Takekawa, M., and Saito, H. (1995). Activation of yeast PBS2 MAPKK by MAPKKKs or by binding of an SH3-containing osmosensor. *Science* **269,** 554–558.

Maeda, T., Tsai, A. Y. M., and Saito, H. (1993). Mutations in a protein tyrosin phosphatase gene (*PTP2*) and a protein serine/threonine phosphatase gene (*PTC1*) cause a synthetic growth defect in *Saccharomyces cerevisiae. Mol. Cell. Biol.* **13,** 5408–5417.

Maeda, T., Wurgler-Murphy, S. M., and Saito, H. (1994). A two-component system that regulates an osmosensing MAP kinase cascade in yeast. *Nature* **369,** 242–245.

Marchler, G., Schüller, C., Adam, G., and Ruis, H. (1993). A *Saccharomyces cerevisiae* UAS element controlled by protein kinase A activates transcription in response to a variety of stress conditions. *EMBO J.* **12,** 1997–2003.

Marquez, J. A., and Serrano, R. (1996). Multiple transduction pathways regulate the sodium-extrusion gene *PMR2/ENA1* during salt stress in yeast. *FEBS Lett.* **382,** 89–92.

Martin, H., Arroyo, J., Sanchez, M., Molina, M., and Nombela, C. (1993). Activity of the yeast MAP kinase homologue Slt2 is critically required for cell integrity at 37 degrees C. *Mol. Gen. Genet.* **241,** 177–184.

Martinez-Pastor, M. T., Marchler, G., Schüller, C., Marchler-Bauer, A., Ruis, H., and Estruch, F. (1996). The *Saccharomyces cerevisiae* zinc finger proteins Msn2p and Msn4p are required for transcriptional induction through the stress-response element (STRE). *EMBO J.* **15,** 2227–2235.

Mattison, C. P., and Ota, I. M. (2000). Two protein tyrosine phosphatases, Ptp2 and Ptp3, modulate the subcellular localization of the Hog1 MAP kinase in yeast. *Genes Dev.* **14,** 1229–1235.

Mattison, C. P., Spencer, S. S., Kresge, K. A., Lee, J., and Ota, I. M. (1999). Differential regulation of the cell wall integrity mitogen-activated protein kinase pathway in budding yeast by the protein tyrosine phosphatases Ptp2 and Ptp3. *Mol. Cell. Biol.* **19,** 7651–7660.

Mendoza, I., Rubio, F., Rodriguez-Navarro, A., and Pardo, J. M. (1994). The protein phosphatase calcineurin is essential for NaCl tolerance of *Saccharomyces cerevisiae. J. Biol. Chem.* **269,** 8792–8796.

Mewes, H. W., Albermann, K., Bahr, M., Frishman, D., Gleissner, A., Hani, J., Heumann, K., Kleine, K., Maierl, A., Oliver, S. G., Pfeiffer, F., and Zollner, A. (1997). Overview of the yeast genome. *Nature* **387(6632 Suppl.),** 7–65.

Meyrial, V., Laizé, V., Gobin, R., Ripoche, P., Hohmann, S., and Tacnet, F. (2001). Existence of a tightly regulated water channel in *Saccharomyces cerevisiae. Eur. J. Biochem.* **268,** 334–343.

Michnick, S., Roustan, J. L., Remize, F., Barre, P., and Dequin, S. (1997). Modulation of glycerol and ethanol yields during alcoholic fermentation in Saccharomyces cerevisiae strains overexpressed or disrupted for *GPD1* encoding glycerol 3-phosphate dehydrogenase. *Yeast* **13,** 783–793.

Mitra, B. N., Yoshino, R., Morio, T., Yokoyama, M., Maeda, M., Urushihara, H., and Tanaka, Y. (2000). Loss of a member of the aquaporin gene family, aqpA affects spore dormancy in *Dictyostelium. Gene* **251,** 131–139.

Mortimer, R. K., and Hawthorne, D. C. (1966). Yeast genetics. *Annu. Rev. Microbiol.* **20,** 151–168.

Moskvina, E., Schuller, C., Maurer, C. T. C., Mager, W. H., and Ruis, H. (1998). A search in the genome of *Saccharomyces cerevisiae* for genes regulated via stress response elements. *Yeast* **14,** 1041–1050.

Mulet, J. M., Leube, M. P., Kron, S. J., Rios, G., Fink, G. R., and Serrano, R. (1999). A novel mechanism of ion homeostasis and salt tolerance in yeast: The Hal4 and Hal5 protein kinases modulate the Trk1-Trk2 potassium transporter. *Mol. Cell. Biol.* **19,** 3328–3337.

Murata, K., Mitsuoka, K., Hirai, T., Walz, T., Agre, P., Heymann, J. B., Engel, A., and Fujiyoshi, Y. (2000). Structural determinants of water permeation through aquaporin-1. *Nature* **407,** 599–605.

Nevoigt, E., and Stahl, U. (1997). Osmoregulation and glycerol metabolism in the yeast *Saccharomyces cerevisiae. FEMS Microbiol. Rev.* **21,** 231–241.

Nomoto, S., Watanabe, Y., Ninomiya-Tsuji, J., Yang, L. X., Nagai, Y., Kiuchi, K., Hagiwara, M., Hidaka, H., Matsumoto, K., and Irie, K. (1997). Functional analyses of mammalian protein kinase C isozymes in budding yeast and mammalian fibroblasts. *Genes Cells* **2,** 601–614.

Nonaka, H., Tanaka, K., Hirano, H., Fujiwara, T., Kohno, H., Umikawa, M., Mino, A., and Takai, Y. (1995). A downstream target of *RHO1* small GTP-binding protein is *PKC1,* a homolog of protein kinase C, which leads to activation of the MAP kinase cascade in *Saccharomyces cerevisiae. EMBO J.* **14,** 5931–5938.

Norbeck, J., and Blomberg, A. (1995). Gene linkage of two-dimensional polyacrylamide gel electrophoresis resolved proteins from isogene families in *Saccharomyces cerevisiae* by microsequencing of in-gel trypsin generated peptides. *Electrophoresis* **16,** 149–156.

Norbeck, J., and Blomberg, A. (2000). The level of cAMP-dependent protein kinase A activity strongly affects osmotolerance and osmo-instigated gene expression changes in *Saccharomyces cerevisiae. Yeast* **16,** 121–37.

Norbeck, J., and Blomberg, A. (1997). Metabolic and regulatory changes associated with growth of *Saccharomyces cerevisiae* in 1.4 M NaCl. Evidence for osmotic induction of glycerol dissimilation via the dihydroxyacetone pathway. *J. Biol. Chem.* **272,** 5544–5554.

Norbeck, J., and Blomberg, A. (1996). Protein expression during exponential growth in 0.7 M NaCl medium of *Saccharomyces cerevisiae. FEMS Microbiol. Lett.* **137,** 1–8.

Norbeck, J., Påhlman, A. K., Akhtar, N., Blomberg, A., and Adler, L. (1996). Purification and characterization of two isoenzymes of DL-glycerol-3- phosphatase from Saccharomyces cerevisiae. Identification of the corresponding *GPP1* and *GPP2* genes and evidence for osmotic regulation of Gpp2p expression by the osmosensing mitogen-activated protein kinase signal transduction pathway. *J. Biol. Chem.* **271,** 13,875–13,881.

Onishi, H. (1963). Osmophilic yeasts. *Adv. Food Res.* **12,** 53–90.

O'Rourke, S. M., and Herskowitz, I. (1998). The Hog1 MAPK prevents cross talk between the HOG and pheromone response MAPK pathways in *Saccharomyces cerevisiae. Genes Dev.* **12,** 2874–2886.

Ozaki, K., Tanaka, K., Imamura, H., Hihara, T., Kameyama, T., Nonaka, H., Hirano, H., Matsuura, Y., and Takai, Y. (1996). Rom1p and Rom2p are GDP/GTP exchange proteins (GEPs) for the Rho1p small GTP binding protein in *Saccharomyces cerevisiae. EMBO J.* **15,** 2196–2207.

Påhlman, A.-K., Granatk, K., Ansell, R., Hohmann, S., and Adler, L. (2001). The yeast glycerol 3-phosphatases Gpp1p and Gpp2p are required for glycerol biosynthesis and differentially involved in the cellular responses to osmotic, anaerobic, and oxidative stress. *J. Biol. Chem.* **276,** 3555–3563.

Pascual-Ahuir, A., Serrano, R., and Proft, M. (2001). Sko1p-repressor and Gcn4p-activator antagonistically modulate stress-regulated transcription in *Saccharomyces cerevisiae. Mol. Cell. Biol.* **21,** 16–25.

Pavlik, P., Simon, M., Schuster, T., and Ruis, H. (1993). The glycerol kinase (*GUT1*) gene of *Saccharomyces cerevisiae:* cloning and characterization. *Curr. Genet.* **24,** 21–25.

Pedruzzi, I., Burckert, N., Egger, P., and De Virgilio, C. (2000). *Saccharomyces cerevisiae* Ras/cAMP pathway controls post-diauxic shift element-dependent transcription through the zinc finger protein Gis1. *EMBO J.* **19,** 2569–2579.

Peterson, J., Zheng, Y., Bender, L., Myers, A., Cerione, R., and Bender, A. (1994). Interactions between the bud emergence proteins Bem1p and Bem2p and Rho- type GTPases in yeast. *J. Cell Biol.* **127,** 1395–1406.

Philips, J., and Herskowitz, I. (1997). Osmotic balance regulates cell fusion during mating in *Saccharomyces cerevisiae. J. Cell Biol.* **138,** 961–974.

Posas, F., Chambers, J. R., Heyman, J. A., Hoeffler, J. P., de Nadal, E., and Arino, J. (2000). The transcriptional response of yeast to saline stress. *J. Biol. Chem.* **275,** 17249–17255.

Posas, F., and Saito, H. (1998). Activation of the yeast SSK2 MAP kinase kinase kinase by the SSK1 two-component response regulator. *EMBO J.* **17,** 1385–1394.

Posas, F., and Saito, H. (1997). Osmotic activation of the HOG MAPK pathway via Ste11p MAPKKK: Scaffold role of Pbs2p MAPKK. *Science* **276,** 1702–1705.

Posas, F., Witten, E. A., and Saito, H. (1998a). Requirement of STE50 for osmostress-induced activation of the STE11 mitogen-activated protein kinase kinase kinase in the high-osmolarity glycerol response pathway. *Mol. Cell. Biol.* **18,** 5788–5796.

Posas, F., Takekawa, M., and Saito, H. (1998b). Signal transduction by MAP kinase cascades in budding yeast. *Curr. Opin. Microbiol.* **1,** 175–182.

Posas, F., Wurgler-Murphy, S. M., Maeda, T., Witten, E. A., Thai, T. C., and Saito, H. (1996). Yeast HOG1 MAP kinase cascade is regulated by a multi-step phosphorelay mechanism in the SLN1-YPD1-SSK1 "two-component" osmosensor. *Cell* **86,** 865–875.

Prior, B. A., and Hohmann, S. (1997). Glycerol production and osmoregulation. *In* "Yeast Sugar Metabolism" (F. K. Zimmermann and K.-D. Entian, Eds.), pp. 313–337. Lancaster, Technomic Publishing Co. Inc.

Proft, M., Pascual-Ahuir, A., de Nadal, E., Arino, J., Serrano, R., and Posas, F. (2001). Regulation of the Sko1 transcriptional repressor by the Hog1 MAP kinase in response to osmotic stress. *EMBO J.* **20,** 1123–1133.

Proft, M., and Serrano, R. (1999). Repressors and upstream repressing sequences of the stress-regulated *ENA1* gene in *Saccharomyces cerevisiae:* bZIP protein Sko1p confers HOG-dependent osmotic regulation. *Mol. Cell. Biol.* **19,** 537–546.

Qadota, H., Python, C. P., Inoue, S. B., Arisawa, M., Anraku, Y., Zheng, Y., Watanabe, T., Levin, D. E., and Ohya, Y. (1996). Identification of yeast Rho1p GTPase as a regulatory subunit of 1,3-beta-glucan synthase. *Science* **272,** 279–281.

Raboy, B., Marom, A., Dor, Y., and Kulka, R. G. (1999). Heat-induced cell cycle arrest of Saccharomyces cerevisiae: Involvement of the RAD6/UBC2 and WSC2 genes in its reversal. *Mol. Microbiol.* **32,** 729–739.

Raitt, D. C., Posas, F., and Saito, H. (2000). Yeast Cdc42 GTPase and Ste20 PAK-like kinase regulate Sho1-dependent activation of the Hog1 MAPK pathway. *EMBO J.* **19,** 4623–4631.

Reiser, V., Ruis, H., and Ammerer, G. (1999). Kinase activity-dependent nuclear export opposes stress-induced nuclear accumulation and retention of Hog1 mitogen-activated protein kinase in the budding yeast *Saccharomyces cerevisiae. Mol. Biol. Cell* **10,** 1147–1161.

Reiser, V., Salah, S. M., and Ammerer, G. (2000). Polarized localization of yeast Pbs2 depends on osmostress, the membrane protein Sho1 and Cdc42. *Nat. Cell Biol.* **2,** 620–627.

Remize, F., Andrieu, E., and Dequin, S. (2000a). Engineering of the pyruvate dehydrogenase bypass in *Saccharomyces cerevisiae:* Role of the cytosolic Mg^{2+} and mitochondrial K^{+} acetaldehyde dehydrogenases Ald6p and Ald4p in acetate formation during alcoholic fermentation. *Appl. Environ. Microbiol.* **66,** 3151–3159.

Remize, F., Roustan, J. L., Sablayrolles, J. M., Barre, P., and Dequin, S. (1999). Glycerol overproduction by engineered *Saccharomyces cerevisiae* wine yeast strains leads to substantial changes in by-product formation and to a stimulation of fermentation rate in stationary phase. *Appl. Environ. Microbiol.* **65,** 143–149.

Remize, F., Sablayrolles, J. M., and Dequin, S. (2000b). Re-assessment of the influence of yeast strain and environmental factors on glycerol production in wine. *J. Appl. Microbiol.* **88,** 371–378.

Rep, M., Albertyn, J., Thevelein, J. M., Prior, B. A., and Hohmann, S. (1999a). Different signalling pathways contribute to the control of *GPD1* expression by osmotic stress in *Saccharomyces cerevisiae. Microbiology* **145,** 715–727.

Rep, M., Krantz, M., Thevelein, J. M., and Hohmann, S. (2000). The transcriptional response of *Saccharomyces cerevisiae* to osmotic shock. Hot1p and Msn2p/Msn4p are required for the induction of subsets of high osmolarity glycerol pathway-dependent genes. *J. Biol. Chem.* **275,** 8290–8300.

Rep, M., Proft, M., Remize, F., Tamás, M., Serrano, R., Thevelein, J. M., and Hohmann, S. (2001). The *Saccharomyces cerevisiae* Sko1p transcription factor mediates HOG pathway-dependent osmotic regulation of a set of genes encoding enzymes implicated in protection from oxidative damage. *Mol. Microbiol.* **40,** 1067–1083.

Rep, M., Reiser, V., Holzmüller, U., Thevelein, J. M., Hohmann, S., Ammerer, G., and Ruis, H. (1999b). Osmotic stress-induced gene expression in *Saccharomyces cerevisiae* requires Msn1p and the novel nuclear factor Hot1p. *Mol. Cell. Biol.* **19,** 5474–5485.

Robertson, L. S., Causton, H. C., Young, R. A., and Fink, G. R. (2000). The yeast A kinases differentially regulate iron uptake and respiratory function. *Proc. Natl. Acad. Sci. USA* **97,** 5984–5988.

Rodriguez-Navarro, A., Quintero, F. J., and Garciadeblas, B. (1994). Na^{+}-ATPases and Na^{+}/H^{+} antiporters in fungi. *Biochim. Biophys. Acta* **1187,** 203–205.

Roemer, T., Paravicini, G., Payton, M. A., and Bussey, H. (1994). Characterization of the yeast (1– >6)-beta-glucan biosynthetic components, Kre6p and Skn1p, and genetic interactions between the PKC1 pathway and extracellular matrix assembly. *J. Cell Biol.* **127,** 567–579.

Rønnow, B., and Kielland-Brandt, M. (1993). *GUT2,* a gene for mitochondrial glycerol 3-phosphate dehydrogenase of *Saccharomyces cerevisiae. Yeast* **9,** 1121–1130.

Rose, D. (1975). Physical responses of yeast cells to osmotic shock. *J. Appl. Bacteriol.* **38,** 169–175.

Ross-Macdonald, P., Coelho, P. S., Roemer, T., Agarwal, S., Kumar, A., Jansen, R., Cheung, K. H., Sheehan, A., Symoniatis, D., Umansky, L., Heidtman, M., Nelson, F. K., Iwasaki, H., Hager, K., Gerstein, M., Miller, P., Roeder, G. S., and Snyder, M. (1999). Large-scale analysis of the yeast genome by transposon tagging and gene disruption. *Nature* **402,** 413–418.

Ruis, H., and Schüller, C. (1995). Stress signaling in yeast. *Bioessays* **17,** 959–965.

Sanders, O. I., Rensing, C., Kuroda, M., Mitra, B., and Rosen, B. P. (1997). Antimonite is accumulated by the glycerol facilitator GlpF in *Escherichia coli. J. Bacteriol.* **179,** 3365–3367.

Schmitt, A. P., and McEntee, K. (1996). Msn2p, a zinc finger DNA-binding protein, is the transcriptional activator of the multistress response in *Saccharomyces cerevisiae. Proc. Natl. Acad. Sci. USA* **93,** 5777–5782.

Schüller, G., Brewster, J. L., Alexander, M. R., Gustin, M. C., and Ruis, H. (1994). The HOG pathway controls osmotic regulation of transcription via the stress response element (STRE) of the *Saccharomyces cerevisiae CTT1* gene. *EMBO J.* **13,** 4382–4389.

Serrano, R. (1996). Salt tolerance in plants and microorganisms: toxicity targets and defense responses. *Int. Rev. Cytol.* **165,** 1–52.

Shen, B., Hohmann, S., Jensen, R. G., and Bohnert, H. (1999). Roles of sugar alcohols in osmotic stress adaptation. Replacement of glycerol by mannitol and sorbitol in yeast. *Plant Physiol.* **121,** 45–52.

Shimizu, J., Yoda, K., and Yamasaki, M. (1994). The hypo-osmolarity-sensitive phenotype of the Saccharomyces cerevisiae hpo2 mutant is due to a mutation in PKC1, which regulates expression of beta-glucanase. *Mol. Gen. Genet.* **242,** 641–648.

Shinbo, I., Fushimi, K., Kasahara, M., Yamauchi, K., Sasaki, S., and Marumo, F. (1999). Functional analysis of aquaporin-2 mutants associated with nephrogenic diabetes insipidus by yeast expression. *Am. J. Physiol.* **277,** F734–F741.

Smits, G. J., Kapteyn, J. C., van den Ende, H., and Klis, F. M. (1999). Cell wall dynamics in yeast. *Curr. Opin. Microbiol.* **2,** 348–352.

Soler, M., Plovins, A., Martin, H., Molina, M., and Nombela, C. (1995). Characterization of domains in the yeast MAP kinase Slt2 (Mpk1) required for functional activity and in vivo interaction with protein kinases Mkk1 and Mkk2. *Mol. Microbiol.* **17,** 833–842.

Sprague, G. F., and Cronan, J. E. (1977). Isolation and characterization of *Saccharomyces cerevisiae* mutants defective in glycerol catabolism. *J. Bacteriol.* **129,** 1335–1342.

Sullivan, B. M., Harrison-Lavoie, K. J., Marshansky, V., Lin, H. Y., Kehrl, J. H., Ausiello, D. A., Brown, D., and Druey, K. M. (2000). RGS4 and RGS2 bind coatomer and inhibit COPI association with golgi membranes and intracellular transport. *Mol. Biol. Cell* **11,** 3155–3168.

Sutherland, F. C. W., Lages, F., Lucas, C., Luyten, K., Albertyn, J., Hohmann, S., Prior, B. A., and Kilian, S. G. (1997). Characteristics of Fps1-dependent and -independent glycerol transport in *Saccharomyces cerevisiae. J. Bacteriol.* **179,** 7790–7795.

Tamás, M. J., Karlgren, S., Thevelein, J. M., and Hohmann, S. (2001). Characterisation of a regulatory domain in the osmosensitive glycerol channel Fps1p in the yeast *Saccharomyces cerevisiae. in preparation.*

Tamás, M. J., Luyten, K., Sutherland, F. C. W., Hernandez, A., Albertyn, J., Valadi, H., Li, H., Prior, B. A., Kilian, S. G., Ramos, J., Gustafsson, L., Thevelein, J. M., and Hohmann, S. (1999). Fps1p controls the accumulation and release of the compatible solute glycerol in yeast osmoregulation. *Mol. Microbiol.* **31,** 1087–1104.

Tamás, M. J., Rep, M., Thevelein, J. M., and Hohmann, S. (2000). Stimulation of the yeast high osmolarity glycerol (HOG) pathway: Evidence for a signal generated by a change in turgor rather than by water stress. *FEBS Lett.* **472,** 159–165.

Tao, W., Deschenes, R. J., and Fassler, J. S. (1999). Intracellular glycerol levels modulate the activity of Sln1p, a Saccharomyces cerevisiae two-component regulator. *J. Biol. Chem.* **274,** 360–367.

Thevelein, J. M. (1992). The RAS-adenylate cyclase pathway and cell cycle control in Saccharomyces cerevisiae. *Antonie Van Leeuwenhoek* **62,** 109–130.

Thevelein, J. M. (1994). Signal transduction in yeast. *Yeast* **10,** 1753–1790.

Thevelein, J. M., Cauwenberg, L., Colombo, S., De Winde, J. H., Donation, M., Dumortier, F., Kraakman, L., Lemaire, K., Ma, P., Nauwelaers, D., Rolland, F., Teunissen, A., Van Dijck, P., Versele, M., Wera, S., and Winderickx, J. (2000). Nutrient-induced signal transduction through the protein kinase A pathway and its role in the control of metabolism, stress resistance, and growth in yeast. *Enzyme Microb. Technol.* **26,** 819–825.

Thevelein, J. M., and de Winde, J. H. (1999). Novel sensing mechanisms and targets for the cAMP-protein kinase A pathway in the yeast *Saccharomyces cerevisiae. Mol. Microbiol.* **33,** 904–918.

Toda, T., Cameron, S., Sass, P., Zoller, M., Scott, J. D., McMullen, B., Hurwitz, M., Krebs, E. G., and Wigler, M. (1987a). Cloning and characterization of *BCY1,* a locus encoding a regulatory subunit of the cyclic AMP-dependent protein kinase in *Saccharomyces cerevisiae. Mol. Cell. Biol.* **7,** 1371–1377.

Toda, T., Cameron, S., Sass, P., Zoller, M., and Wigler, M. (1987b). Three different genes in S. cerevisiae encode the catalytic subunits of the cAMP-dependent protein kinase. *Cell* **50,** 277–287.

Tornow, J., Zeng, X., Gao, W., and Santangelo, G. M. (1993). GCR1, a transcriptional activator in *Saccharomyces cerevisiae,* complexes with RAP1 and can function without its DNA binding domain. *EMBO J.* **12,** 2431–2437.

Treger, J. M., Magee, T. R., and McEntee, K. (1998). Functional analysis of the stress response element and its role in the multistress response of *Saccharomyces cerevisiae*. *Biochem. Biophys. Res. Commun.* **243,** 13–19.

Uetz, P., Giot, L., Cagney, G., Mansfield, T. A., Judson, R. S., Knight, J. R., Lockshon, D., Narayan, V., Srinivasan, M., Pochart, P., Qureshi-Emili, A., Li, Y., Godwin, B., Conover, D., Kalbfleisch, T., Vijayadamodar, G., Yang, M., Johnston, M., Fields, S., and Rothberg, J. M. (2000). A comprehensive analysis of protein–protein interactions in *Saccharomyces cerevisiae*. *Nature* **403,** 623–627.

Van Aelst, L., Hohmann, S., Zimmermann, F. K., Jans, A. W., and Thevelein, J. M. (1991). A yeast homologue of the bovine lens fibre MIP gene family complements the growth defect of a *Saccharomyces cerevisiae* mutant on fermentable sugars but not its defect in glucose-induced RAS-mediated cAMP signalling. *EMBO J.* **10,** 2095–2104.

Van Laere, A. J., and Hulsman, E. (1987). Water potential, glycerol synthesis, and water content of germinating *Phycomyces* spores. *Arch. Microbiol.* **147,** 257–262.

Van Wuytswinkel, O., Reiser, V., Siderius, M., Kelders, M. C., Ammerer, G., Ruis, H., and Mager, W. H. (2000). Response of *Saccharomyces cerevisiae* to severe osmotic stress: Evidence for a novel activation mechanism of the HOG MAP kinase pathway. *Mol. Microbiol.* **37,** 382–397.

Wang, T., and Bretscher, A. (1995). The rho-GAP encoded by BEM2 regulates cytoskeletal structure in budding yeast. *Mol. Biol. Cell* **6,** 1011–1024.

Verna, J., Lodder, A., Lee, K., Vagts, A., and Ballester, R. (1997). A family of genes required for maintenance of cell wall integrity and for the stress response in Saccharomyces cerevisiae. *Proc. Natl. Acad. Sci. USA* **94,** 13804–13809.

Vershon, A. K., and Pierce, M. (2000). Transcriptional regulation of meiosis in yeast. *Curr. Opin. Cell Biol.* **12,** 334–339.

Voegele, R. T., Sweet, G. D., and Boos, W. (1993). Glycerol kinase of *Escherichia coli* is activated by interaction with the glycerol facilitator. *J. Bacteriol.* **175,** 1087–1094.

Watanabe, Y., Irie, K., and Matsumoto, K. (1995). Yeast *RML1* encodes a serum response factor-like protein that may function downstream of the Mpk1 (Slt2) mitogen-activated protein kinase pathway. *Mol. Cell. Biol.* **15,** 5740–5749.

Weig, A. R., and Jakob, C. (2000). Functional identification of the glycerol permease activity of Arabidopsis thaliana NLM1 and NLM2 proteins by heterologous expression in Saccharomyces cerevisiae. *FEBS Lett.* **481,** 293–298.

Wieland, J., Nitsche, A. M., Strayle, J., Steiner, H., and Rudolph, H. K. (1995). The *PMR2* gene cluster encodes functionally distinct isoforms of a putative Na^+ pump in the yeast plasma membrane. *EMBO J.* **14,** 3870–3882.

Wurgler-Murphy, S. M., Maeda, T., Witten, E. A., and Saito, H. (1997). Regulation of the *Saccharomyces cerevisiae* Hog1 mitogen activated protein kinase by the Ptp2 and Ptp3 protein tyrosine phosphatases. *Mol. Cell. Biol.* **17,** 1289–1297.

Wysocki, R., Chéry, C. C., Wawrzycka, D., Van Hulle, M., Cornelis, R., Thevelein, J. M., and Tamás, M. J. (2001). The glycerol channel Fps1p mediates the uptake of arsenite and antimonite in *Saccharomyces cerevisiae*. *Mol. Microbiol.* **40,** 1391–1401.

Yagi, T. (1992). Regulation of intracellular osmotic pressure during the initial stages of salt stress in a salt-tolerant yeast, Zygosaccharomyces rouxii. *Microbios* **70,** 93–102.

Adaptation of Kidney Medulla to Hypertonicity: Role of the Transcription Factor TonEBP

Seung Kyoon Woo and H. Moo Kwon
Division of Nephrology, Johns Hopkins University, Baltimore, Maryland 21205

The osmolality of the mammalian kidney medulla is very high. The high osmolality provides the driving force for water reabsorption and urinary concentration, key functions of the kidney for maintaining proper body fluid volume and blood pressure. Salt and urea are the major solutes in the renal medullary interstitium. Unfortunately, high salt (hypertonicity) causes DNA damage and cell death. In response, the renal medullary cells adapt to the hypertonicity by accumulating compatible osmolytes. A regulatory protein, tonicity-responsive enhancer binding protein (TonEBP), plays a central role in the accumulation of these compatible osmolytes by stimulating genes whose products either actively transport or synthesize the appropriate osmolytes. TonEBP is active under isotonic conditions. It responds to both an increase and a decrease in ambient tonicity, in opposite directions, which involves changes in its abundance and nucleocytoplasmic distribution. In the kidney medulla, however, nucleocytoplasmic distribution is the major site of control, under normal conditions of diuresis and antidiuresis.

KEY WORDS: TonEBP, NFAT5, Osmolytes, Kidney medulla.

One of the defining features of the mammalian kidney is that osmolality of the medulla is very high. The cortex is isotonic, but osmolality of the interstitial fluid increases gradually down the outer and inner medulla, reaching the highest level at the papilla. Because the basolateral plasma membrane that faces the interstitial fluid is highly permeable to water, the interior of the cells is in osmotic equilibrium with the interstitium. Beck and his colleagues (1984) accurately measured the solute compositions of the renal medulla for the first time. The interstitial fluid of the

0074-7696/02 $35.00

renal papilla from mildly dehydrated rats contains nearly 1000 mOsm NaCl and 1500 mOsm urea. Osmolality of the interstitial fluid varies greatly depending on the diuretic status of the animal. These variations are greatest in the inner medulla where changes in urea concentration account for the majority of changes, while NaCl also contributes significantly.

Although urea and NaCl are the major solutes in the interstitium of renal medulla, their effects on cells are different. Urea permeates the plasma membrane readily, and therefore it does not contribute to effective osmolality (tonicity). On the other hand, NaCl has low membrane permeability and it is osmotically effective. Thus hyperosmotic NaCl is hypertonic (i.e., cells shrink in it) while hyperosmotic urea is not. This chapter describes the adaptation of cells in the kidney medulla to hypertonicity, concentrating on the role of the transcription factor TonEBP (tonicity-responsive enhancer (TonE) binding protein). As described below, hypertonicity imposes a great of deal of stress on cells, and TonEBP is a key molecule in overcoming the stress. For recent literature on cellular effects of hyperosmotic urea, refer to Zhang *et al.* (2000) and references therein.

I. Why Is Hypertonicity Stressful?

When cells are exposed to a hypertonic fluid, they shrink within seconds, and the ionic strength inside the cells increases. Over the course of several hours, cells accumulate organic solutes called organic osmolytes or compatible osmolytes. The concentration of compatible osmolytes rises high enough to lower the ionic strength inside the renal medullary cells to isotonic level (Sone *et al.,* 1995). Thus, the role of compatible osmolytes is to restore the ionic strength inside the cell. When the accumulation of compatible osmolytes is blocked, cells do not survive (Uchida *et al.,* 1991), attesting to the importance of maintaining cellular ionic strength low. The universality of compatible osmolytes throughout evolution in adaptation to hypertonicity (see Hohmann, this volume) suggests that a high ionic strength inside the cell imposes a fundamental and general stress on the cell.

Why is high ionic strength inside the cell stressful? Yancey and his colleagues (1982) proposed that perturbation of protein function is the source of stress. This was based on the observation that the activity of many enzymes is inhibited by increased concentrations of ions, but not by increased concentrations of organic osmolytes. Hence the name "compatible" osmolytes. Although this theory has been widely accepted, it is not known whether particular proteins are the critical targets for the stress that leads to cell death.

Data emerging in the last few years demonstrate that DNA damage is another form of stress caused by hypertonicity. Kültz *et al.* (1998) studied mIMCD3 cells, a cell line established from the renal medulla of transgenic mice expressing the SV40 large T antigen. They found growth arrest and induction of GADD45 (growth arrest- and DNA damage-inducible gene) and GADD153 when mIMCD3 cells were switched to hypertonic medium. The growth arrest is due to delays in all phases of the cell cycle and can be induced also by high concentrations of urea (Michea *et al.,* 2000). However, hypertonicity, but not hyperosmotic urea, induces the expression of the tumor suppressor protein p53 and activates p53 by phosphorylation of serine 15 (Dmitrieva *et al.,* 2000). Therefore, the induction and phosphorylation of p53 are specific for hypertonicity. Forced down-regulation of p53 results in apoptosis, indicating that p53 activation contributes to cell survival. These reactions to hypertonicity are very similar to those induced by ionizing radiation that causes double-stranded breaks in DNA (Zhou and Elledge, 2000). Indeed, hypertonicity, but not hyperosmolar urea, causes double-stranded breaks in genomic DNA of mIMCD3 cells (Kültz and Chakravarty, 2001). These data are consistent with the scheme of events shown in Fig. 1. Depending on the degree of hypertonicity and, presumably, the degree of DNA breaks, cells either die or activate ATM kinases in response to the DNA breaks. The activated ATM kinase, and possibly other kinases, activate p53 induction and phosphorylation. Active p53 prevents apoptosis and stops cell cycle progression to allow DNA repair to proceed.

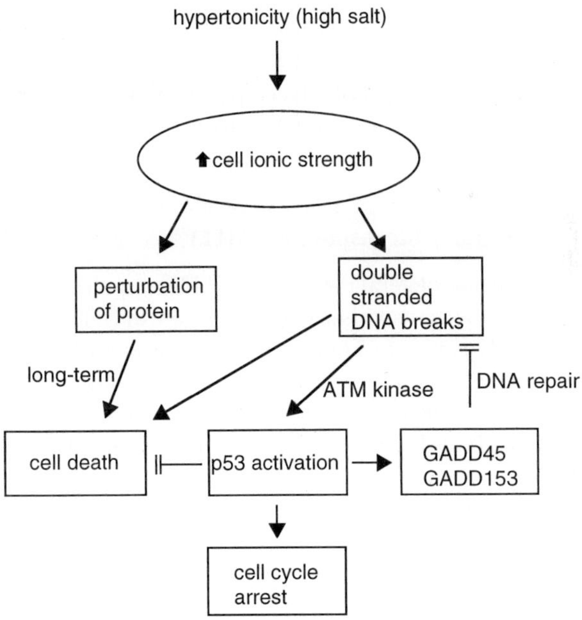

FIG. 1 Nature of stress induced by hypertonicity. Exposure to hypertonic environment results in an immediate rise in the cellular ionic strength, which perturbs protein function and causes double-stranded DNA breaks. Arrows represent activating events, and perpendicular ends represents inhibitory events. See text for details.

GADD45 and GADD153 induction are partially mediated by p53. GADD153 may play a role in DNA repair.

The cell cycle arrest in mIMCD3 cells is relieved after 18 hr of exposure to hypertonic medium (Kültz *et al.,* 1998). The cells then go on to proliferate like control cells. The relief of cell cycle arrest coincides temporally with substantial accumulation of compatible osmolytes (see below) and, therefore, with restoration of cellular ionic strength to near the isotonic level. In summary, hypertonicity is stressful to cells because it increases intracellular ionic strength, which in turn perturbs protein function and damages the genome. Cellular accumulation of compatible osmolytes lowers the ionic strength inside the cells and thereby removes the source of stress.

II. Role of Transcription in Cellular Accumulation of Compatible Osmolytes—A TonE Connection

A. Transporters and an Enzyme That Catalyze Cellular Accumulation of Compatible Osmolytes

A number of compatible osmolytes has been identified in the renal medulla (reviewed in Garcia-Perez and Burg, 1991). The most abundant osmolytes include *myo*-inositol, betaine, sorbitol, glycerophosphorylcholine, and taurine. The proteins that catalyze the cellular accumulation of compatible osmolytes will now be described.

1. Sodium/myo-Inositol Cotransporter (SMIT)

SMIT is located in the basolateral membrane of kidney epithelial cells (Yamauchi *et al.,* 1991). Thus, SMIT transports *myo*-inositol from the interstitial fluid into the cells. It is a single polypeptide of 718 amino acids (Kwon *et al.,* 1992). The K_m for *myo*-inositol is $\sim 50\ \mu$M, close to its concentration in plasma and in the interstitial fluid. SMIT can concentrate *myo*-inositol inside the cell more than 1000-fold over the outside, to over 50 mM, because it couples transport of two sodium ions to each molecule of *myo*-inositol (Hager *et al.,* 1995). In the kidney medulla, SMIT is abundantly expressed in the thick ascending limb of Henle's loop and in the inner medullary collecting ducts (Yamauchi *et al.,* 1995).

2. Sodium/Chloride/Betaine Cotransporter (BGT1)

BGT1 is a polypeptide of 614 amino acids (Yamauchi *et al.,* 1992) that couples three sodium ions and one or two chloride ions for each betaine that is transported (Matskevitch *et al.,* 1999). BGT1 is also located in the basolateral membrane of the medullary thick ascending limbs (Yamauchi *et al.*, 1991; Perego *et al.*, 1999) and inner medullary collecting ducts (Miyai *et al.*, 1996; Moeckel *et al.*, 1997). Although the plasma concentration of betaine is near 50 μM, the betaine concentration in the renal medullary interstitium may be a good deal higher, perhaps close to the K_m of BGT1 of 500 μM, due to a combination of betaine reabsorption in the thin limbs of Henle's loop and recycling between the tubules and the vasa recta (Pummer *et al.*, 2000).

3. Sodium- and Chloride-Coupled Taurine Transporter and Aldose Reductase (AR)

This transporter is expressed in the kidney and has been identified (Uchida *et al.*, 1992), but its localization in the renal medulla and its role in cellular accumulation of taurine are not clear. Sorbitol is synthesized from glucose by a reaction catalyzed by AR. AR is expressed in the ascending thin limb of Henle's loop and the innner medullary collecting ducts (Jung *et al.*, 2000).

B. Role of Transcription in Hypertonicity-Induced Cellular Accumulation of Compatible Osmolytes: Monopoly of TonE

Cell lines derived from kidney have been used to study the mechanism of cellular accumulation of compatible osmolytes in response to hypertonicity. After MDCK cells are switched to hypertonic medium, the cell content of *myo*-inositol and betaine rises slowly and reaches a steady state in 24 hr (Yamauchi *et al.*, 1991). This accumulation is preceded by a few hours by the increased activity of SMIT and BGT1. Northern and nuclear run-off assays clearly demonstrated that transcription plays the key role in the hypertonicity-induced stimulation of SMIT (Yamauchi *et al.*, 1993) and BGT1 (Uchida *et al.*, 1993). Another kidney-derived cell line, PAP-HT25, was used to demonstrate that hypertonicity-induced stimulation of the AR gene is responsible for sorbitol accumulation (Smardo *et al.*, 1992; Moriyama *et al.*, 1989).

Raising the tonicity of the renal medulla by water deprivation results in increased mRNA abundance of SMIT (Yamauchi *et al.*, 1994), BGT1 (Moeckel *et al.*, 1997), and AR (Martial *et al.*, 1995). The increase in mRNA abundance leads to increased activity of BGT1 (Moeckel *et al.*, 1997) and AR (Cowley *et al.*, 1990). These data demonstrate that hypertonicity-induced stimulation of transcription of the SMIT, BGT1, and AR genes also takes place in the renal medulla (Fig. 2).

Intense efforts have led to identification of a genomic regulatory element that mediates hypertonicity stimulation of transcription. It turns out that all the three genes (SMIT, BGT1, and AR) are regulated by the same element TonE (tonicity-responsive enhancer), also known as ORE (osmotic response element). TonE

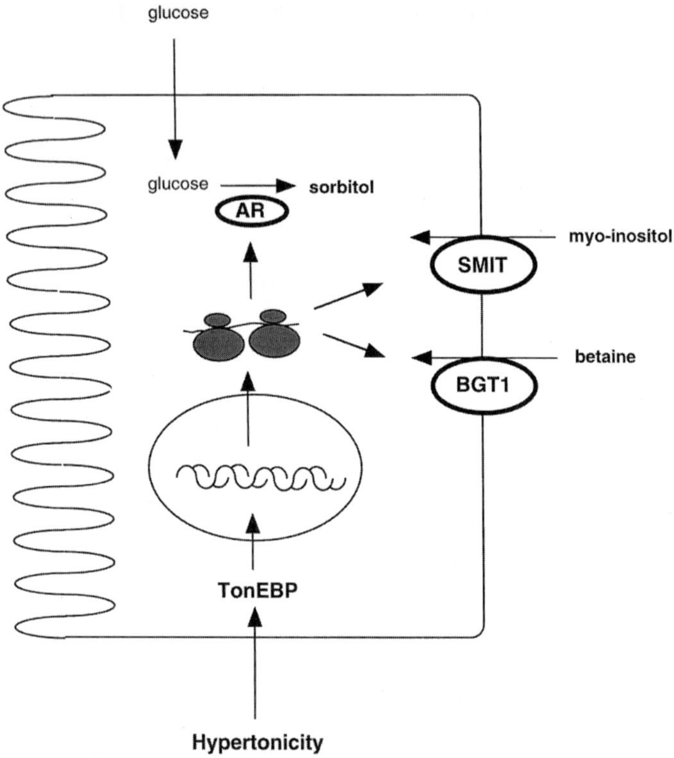

FIG. 2 Role of transcription in cellular accumulation of compatible osmolytes. Upon exposure to a hypertonic environment, the activity of TonEBP is stimulated, leading to increased transcription of SMIT BGT1, and AR genes. Increased synthesis and activity of the transporters and enzyme ensues over the course of 20 hr resulting in accumulation of inositol, betaine, and sorbitol.

is 11bp long and has a consensus sequence TGGAAANNYNY (Y is C or T) (Takenaka *et al.,* 1994; Miyakawa *et al.,* 1998). The BGT1 promoter is regulated by two TonEs within 150 bp upstream (Miyakawa *et al.,* 1999b). The two TonEs act in synergy as do most enhancers. Likewise, the AR promoter is regulated by three TonEs clustered about 1 kb upstream (Ko *et al.,* 1997). It should be noted that TonE:TonEBP binding has a very low affinity, corresponding to a dissociation constant of ~10 nM (Miyakawa *et al.,* 1998) while most transcription factors bind their cognate DNA sites with a high affinity, a dissociation constant of 0.1nM. It is possible that multiple TonEs, closely spaced, increase the binding affinity as a result of interaction between neighboring TonEBP molecules. The synergy of multiple TonEs supports this idea.

The relationship between TonE and the SMIT promoter is more complex. The SMIT promoter *in situ* is clearly regulated by TonE because expression of a dominant negative form of TonEBP (see below) down-regulates SMIT mRNA expression (unpublished observations). Given this, it is surprising that there is only one TonE within 12 kb upstream of the SMIT promoter. This proximal TonE alone stimulates the promoter less than 2-fold in response to hypertonicity (Rim *et al.,* 1997; Mallee *et al.,* 1997), a good deal lower than the 10-fold stimulation of transcription observed. Three more TonEs were identified 35 kb farther upstream (Rim *et al.,* 1998). In other words, there are four TonEs scattered in a span of ~55 kb upstream, and all of them appear to be active, based on *in vivo* footprinting analysis (see below). It is possible that the four TonEs upstream are physically close to the SMIT promoter within the chromatin.

III. Regulation of TonEBP

A. Molecular Features of TonEBP

TonE functions by providing a binding site for the transcription factor TonEBP. TonEBP was cloned using the yeast one-hybrid approach, based on specific binding to TonE (Miyakawa *et al.,* 1999a). Alternative splicing near the 5′ end of mRNA predicts several polypeptides sharing the same 1455 amino acids, but having different lengths of amino acids (up to 94 residues) at the N-terminus (unpublished results). Figure 3 shows the shortest form of TonEBP. Of interest is that the DNA

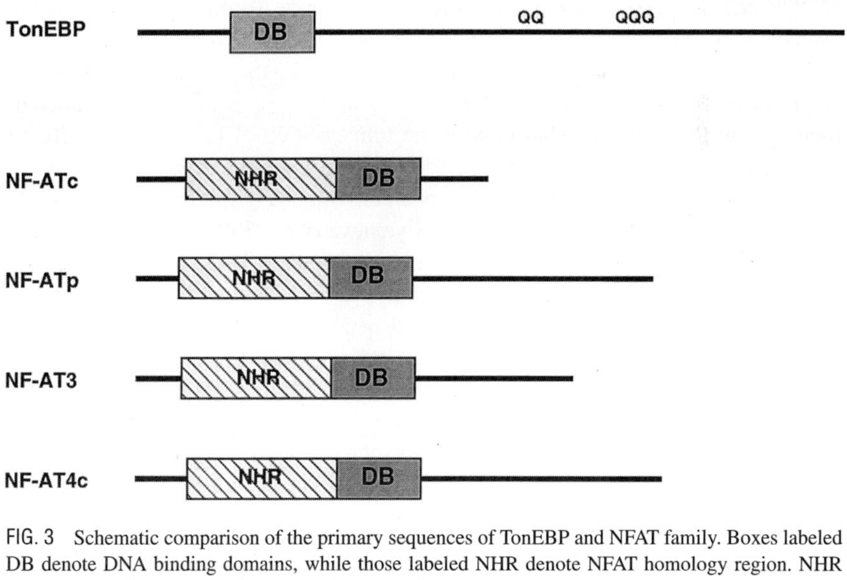

FIG. 3 Schematic comparison of the primary sequences of TonEBP and NFAT family. Boxes labeled DB denote DNA binding domains, while those labeled NHR denote NFAT homology region. NHR binds calcineurin and is dephosphosphorylated when calcineurin is activated. Note that NHR is not found in TonEBP. QQ and QQQ denote stretches of glutamines.

binding domain of TonEBP is 45% identical with those of the nuclear factor of activated T cells (NFAT) family. Members of the NFAT family share 88% of amino acids in the DNA binding domain.

It should be emphasized that the similarity of TonEBP to the NFAT family is limited to the DNA-binding domain. Two highly conserved regions of NFATs are not found in TonEBP. First, those amino acid residues involved in interaction with AP-1 are not found in TonEBP. Consistent with this, there is no evidence for interaction of TonEBP with AP-1 (Lopez-Rodriguez *et al.,* 1999). Second, all members of the NFAT family contain the NFAT homology region (NHR) of ~300 amino acids N-terminal to the DNA-binding domain. Calcineurin binding and its substrate phospho-amino amino acid residues are in the NHR. Accordingly, TonEBP is not regulated by calcineurin (Lopez-Rodriguez *et al.,* 1999).

The region in TonEBP that is C-terminal to the DNA binding domain is rich in glutamines (18% of amino acids are glutamines) and includes two stretches of glutamines (Fig. 3). A truncated TonEBP lacking this glutamine-rich region functions as a dominant negative TonEBP (Miyakawa *et al.,* 1999a) consistent with the view that this region stimulates transcription (Gill *et al.,* 1994). There is a "bipartite" nuclear localization signal (Nakielny and Dreyfuss, 1999) near the N-terminus, although its role has not yet been investigated.

B. Activation of TonEBP in Response to Hypertonicity

MDCK cells have been used extensively to characterize TonEBP regulation. Studies have focused on the regulation of TonEBP in response to hypertonicity. Earlier work, using the electrophoretic mobility shift assay (EMSA) of TonEBP from nuclear extracts, revealed that the DNA binding activity of TonEBP in the nucleus increases in temporal correlation with the transcription of the SMIT and BGT1 gene (Miyakawa *et al.,* 1998). *In vivo* footprinting analysis demonstrated that the increased activity of TonEBP results in a corresponding increase in binding to the TonE sites upstream of the BGT1 (Miyakawa *et al.,* 1998, 1999b) and SMIT (Rim *et al.,* 1998) genes *in situ* (i.e., in the chromatin). Thus, stimulation of the DNA binding activity is the key step in the hypertonic stimulation of transcription by TonEBP (Fig. 4).

Direct measurement of phosphorylation revealed that the TonEBP of MDCK cells is phosphorylated under isotonic conditions. After the cells are switched to hypertonic medium, an increase in phosphorylation is detectable at 30 min, and the level of phosphorylation continues to increase over the next several hours (Dahl *et al.,* 2001). A number of protein kinases, including all the major MAP kinase isoforms and the protein tyrosine kinases, are activated when mammalian cells are exposed to hypertonicity. TonEBP does not appear to be a substrate of these kinases because none of their inhibitors affects phosphorylation of TonEBP. Removal of phosphates from immunoprecipitated TonEBP does not affect TonE

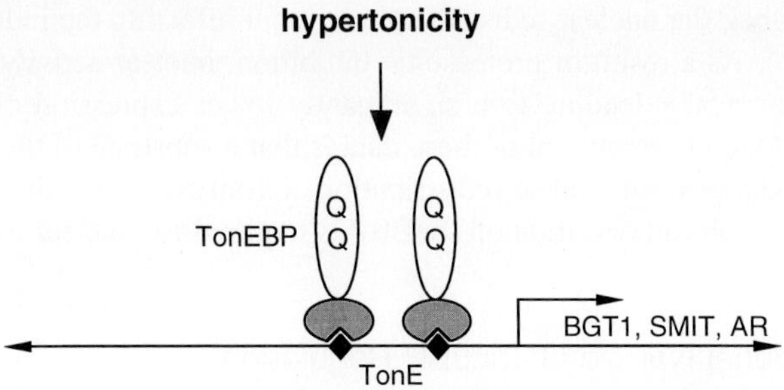

FIG. 4 Stimulation of TonEBP activity in the nucleus is the key step in hypertonicity induction of its downstream genes. TonEBP activity in the nucleus as measured by EMSA increases for 15 hr after exposure to hypertonicity. The rise in TonEBP activity temporally correlates with binding to the TonE sites upstream of the BGT1 and SMIT genes and their transcription.

binding *in vitro,* indicating that phosphorylation is not required for DNA binding. On the other hand, the time course of increased phosphorylation correlates with nuclear redistribution of TonEBP. At this point, it is not clear if phosphorylation is required for nuclear import or for nuclear retention.

The abundance of TonEBP rises slowly after MDCK cells are switched to hypertonic medium, doubling in 6 hr and reaching a steady state of a 4-fold increase at 12 hr. This is due to increased synthesis of TonEBP protein as a result of an increase in mRNA abundance (Woo *et al.,* 2000a). A combination of induction and redistribution to the nucleus results in an approximately 8-fold increase in the abundance of TonEBP in the nucleus (Table I) in parallel with 8-fold increase in DNA binding activity (Miyakawa *et al.,* 1998).

The induction and nuclear redistribution (or localization) of TonEBP are separate pathways; inhibitors of the proteasome, such as MG132 (Table I) and lactacystin,

TABLE I

Summary of Subcellular Distribution of TonEBP in MDCK Cells and the Thin Limbs of Henle's Loop of the Renal Medulla

		Relative amount of TonEBP		
Model	Conditions	Cytoplasm	Nucleus	Total
	Isotonicity	1	1	2
MDCK cells				
16 hr after treatment	Hypertonicity	0.5	7.5	8
	Hypertonicity + MG132	4	4	8
Kidney medulla	Diuresis	9	1	10
Thin limb of Henle's loop	Antidiuresis	1	9	10

selectively block the nuclear redistribution without affecting the induction (Woo *et al.,* 2000b). As a result of proteasome inhibition, nuclear activity of TonEBP is reduced by ~50% leading to a significantly lower expression of SMIT and AR mRNA. One interpretation of these data is that a substrate of the proteasome needs to be degraded for nuclear redistribution of TonEBP. This substrate can be a phosphotase, if phosphorylation of TonEBP is required for nuclear translocation.

C. Bidirectional Nature of TonEBP Regulation

A substantial amount of TonEBP is present in the nuclei of isotonic MDCK cells (relative amounts are given in Table I). Expression of the dominant negative form of TonEBP reduces TonE-mediated luciferase expression under isotonic conditions as well as under hypertonic conditions (Miyakawa *et al.,* 1999a). Stable expression of the dominant negative form of TonEBP reduces the abundance of SMIT and BGT1 mRNA under isotonic conditions (unpublished observations). In addition, switching cells to hypotonic (low osmolality) conditions leads to decreased abundance of TonEBP and the cytoplasmic redistribution, the exact opposite of what happens when cells are switched to hypertonicity (Woo *et al.,* 2000a). Thus, TonEBP is active under isotonic conditions. It responds to both increase and decrease in tonicity in opposite directions, in terms of both nuclear distribution and abundance (Fig. 5).

TonEBP is widely expressed throughout embryogenesis and is expressed in adult tissues including brain and heart (Maouyo *et al.,* 2000). As such, TonEBP is likely to be important in the expression of certain genes (probably housekeeping genes) in nonrenal tissues that are isotonic. Recently, it was reported that TonEBP expression is induced in activated T cells (Trama *et al.,* 2000), further supporting the view that TonEBP has a general role, other than the adaptation to hypertonicity.

D. TonEBP Regulation in the Kidney

In the rat kidney, the abundance of TonEBP mRNA is three to four times higher in the outer and inner medulla compared to the cortex (Cha *et al.,* 2001). Immunohistochemistry, using specific TonEBP antibody, revealed that every cell in the inner medulla and the inner stripe of the outer medulla expresses TonEBP strongly (Cha *et al.,* 2001). In the cortex and outer strip of the outer medulla, the cortical collecting ducts and the thick ascending limbs express TonEBP. Thus, as expected, expression of TonEBP generally follows the degree of hypertonicity. Hydration and dehydration lead to changes in the nucleocytoplasmic distribution of TonEBP, especially in the thin limbs of Henle's loop, without affecting the overall abundance of TonEBP in the renal medulla (Table I). These data suggest that TonEBP plays an important role in the regulation of SMIT, BGT1, and AR in the renal

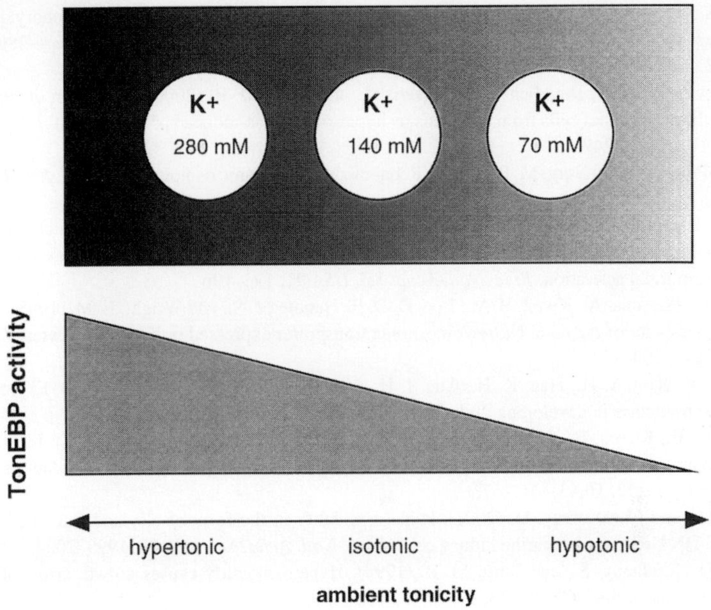

FIG. 5 Bidirectional nature of TonEBP. Three circles in the top depict cellular ionic strength (represented by potassium concentration) under hypertonic, isotonic, and isotonic conditions before compensation by compatible osmolytes takes place. TonEBP is active under isotonic conditions. It is stimulated by an increase in tonicity and inhibited by a decrease in tonicity.

medulla *in vivo*. Studies are under way to address the role and regulation of TonEBP in the kidney.

Acknowledgment

This work was supported by NIH Grants (DK42479 and DK44484) and a fellowship to S. K. W from the Juvenile Diabetes Foundation International.

References

Beck, F., Dörge, A., Rick, R., and Thurau, K. (1984). Intra- and extracellular elements of rat renal pipilla in antidiuresis. *Kidney Int.* **25,** 397–403.

Cha, J. H., Woo, S. K., Han, K. H., Kim, Y. H., Handler, J. S., Kim, J., and Kwon, H. M. (2001). Hydration status affects nuclear distribution of tonicity-responsive enhancer binding protein in rat kidney medulla. *J. Am. Soc. Nephrol.* **12,** 2221–2230.

Cowley, B. D., Ferraris, J. D., Carper, D., and Burg, M. B. (1990). In vivo osmoregulation of aldose reductase mRNA, protein, and sorbitol in renal medulla. *Am. J. Physiol.* **258,** F154–F161.

Dahl, S. C., Handler, J. S., and Kwon, H. M. (2001). Hypertonicity-induced phosphorylation and nuclear localization of the transcription factor TonEBP. *Am. J. Physiol. Cell Physiol.* **280,** C248–C253.

Dmitrieva, N., Kültz, D., Michea, L., Ferraris, J., and Burg, M. B. (2000). Protection of renal inner medullary epithelial cells from apoptosis by hypertonic stress-induced p53 activation. *J. Biol. Chem.* **275,** 18,243–18,247.

Garcia-Perez, A., and Burg, M. B. (1991). Renal medullary organic osmolytes. *Physiol. Rev.* **71,** 1081–1115.

Gill, G., Pascal, E., Tseng, Z. H., and Tjian, R. (1994). A glutamine-rich hydrophobic patch in transcription factor Sp1 contacts the dTAFII110 component of the Drosophila TFIID complex and mediates transcriptional activation. *Proc. Natl. Acad. Sci. USA* **91,** 192–196.

Hager, K., Hazama, A., Kwon, H. M., Loo, D. D. F., Handler, J. S., and Wright, E. M. (1995). Kinetics and specificity of the renal Na/*myo*-inositol cotransporter expressed in *Xenopus* oocytes. *J. Membr. Biol.* **143,** 103–113.

Jung, J. Y., Kim, Y. H., Han, K. H., Cha, J. H., Madsen, K. M., and Kim, J. (2000). Expression of aldose reductase in developing rat kidney. *J. Am. Soc. Nephrol.* **11,** 44A (abstract).

Ko, B. C. B., Ruepp, B., Bohren, K. M., Gabbay, K. H., and Chung, S. S. M. (1997). Identification and characterization of multiple osmotic response sequences in the human aldose reductase gene. *J. Biol. Chem.* **272,** 16,431–16,437.

Kültz, D., and Chakravarty, D. (2001). Hyperosmolality in the form of elevated NaCl but not urea cause DNA damage in murine kidney cells. *Proc. Natl. Acad. Sci. USA* **98,** 1999–2004.

Kültz, D., Madhany, S., and Burg, M. B. (1998). Hyperosmolality causes growth arrest of murine kidney cells. *J. Biol. Chem.* **273,** 13,645–13,651.

Kwon, H. M., Yamauchi, A., Uchida, S., Preston, A. S., Garcia-Perez, A., Burg, M. B., and Handler, J. S. (1992). Cloning of the cDNA for a Na/*myo*-inositol cotransporter, a hypertonicity stress protein. *J. Biol. Chem.* **267,** 6297–6301.

Lopez-Rodgriguez, C., Aramburu, J., Rakeman, A. S., and Rao, A. (1999). NFAT5, a constitutively nuclear NFAT protein that does not cooperate with Fos and Jun. *Proc. Natl. Acad. Sci. USA* **96,** 7214–7219.

Mallee, J. J., Atta, M. G., Lorica, V., Rim, J. S., Kwon, H. M., Lucente, A. D., Wang, Y., and Berry, G. T. (1997). The structural organization of the human Na/*myo*-inositol cotransporter (SLC5A3) gene and characterization of the promoter. *Genomics* **46,** 459–465.

Maouyo, D., Kim, J. Y., Woo, S. K., Handler, J. S., and Kwon, H. M. (2000). Expression of TonEBP during embryogenesis and in multiple diverse adult tissues suggests that TonEBP plays an important undiscovered role. *J. Am. Soc. Nephrol.* **11,** 45A (abstract).

Martial, S., Price, S. R., and Sands, J. M. (1995). Regulation of aldose reductase, sorbitol dehydrogenase, and taurine contransporter mRNA in rat medulla. *J. Am. Soc. Nephrol.* **5,** 1971–1978.

Matskevitch, I., Wagner, C. A., Stegen, C., Bröer, S., Noll, B., Risler, T., Kwon, H. M., Handler, J. S., Waldegger, S., Busch, A. E., and Lang, F. (1999). Functional characterization of the betaine/GABA transporter BGT1 expressed in *Xenopus* oocytes. *J. Biol. Chem.* **274,** 16,709–16,716.

Michea, L., Ferguson, D. R., Peters, E. M., Andrews, P. M., Kirby, M. R., and Burg, M. B. (2000). Cell cycle delay and apoptosis are induced by high salt and urea in renal medullary cells. *Am. J. Physiol.* **278,** F209–F218.

Miyai, A., Yamauchi, A., Moriyama, T., Kaneko, T., Takenaka, M., Sugiura, T., Kitamura, H., Ando A., Tohyama, M., Shimada, S., Imai, E., and Kamada, T. (1996). Expresion of betaine transport mRNA: Its unique localization and rapid regulation in rat kidney. *Kidney Int.* **50,** 819–827.

Miyakawa, H., Woo, S. K., Chen, C. P., Dahl, S. C., Handler, J. S., and Kwon, H. M. (1998). Cis- and trans-acting factors regulating transcription of the BGT1 gene in response to hypertonicity. *Am. J. Physiol.* **274,** F753–F761.

Miyakawa, H., Woo, S. K., Dahl, S. C., Handler, J. S., and Kwon, H. M. (1999a). Tonicity-responsive enhancer binding protein, a Rel-like protein that stimulates transcription in response to hypertonicity. *Proc. Natl. Acad. Sci. USA* **96,** 2538–2542.

Miyakawa, H., Rim, J. S., Handler, J. S., and Kwon, H. M. (1999b). Identification of the second tonicity-responsive enhancer for the BGT1 gene. *Biochim. Biophys. Acta* **1446,** 359–364.

Moeckel, G. W., Lai, L. W., Guder, W. G., Kwon, H. M., and Lien, Y. H. H. (1997). Kinetics and osmoregulation of Na- and Cl-dependent betaine transporter in rat renal medulla. *Am. J. Physiol.* **272,** F100–F106.

Moriyama, T., Garcia-Perez, A., and Burg, M. B. (1989). Osmotic regulation of aldose reductase protein synthesis in renal medullary cells. *J. Biol. Chem.* **264,** 16,810–16,814.

Nakielny, S., and Dreyfuss, G. (1999). Transport of proteins and RNAs in and out of the nucleus. *Cell* **99,** 677–690.

Perego, C., Vanoni, C., Villa, A., Longhi, R., Kaech, S. M., Frohli, E., Hajnal, A., Kim, S. K., and Pietrini G., (1999). PDZ-mediated interactions retain the epithelial GABA transporter on the basolateral surface of polarized epithelial cells. *EMBO J.* **18,** 2394–2393.

Pummer, S., Dantzler, W. H., Lien, Y. H. H., Moeckel, G. W., Völker, K., and Silbernagl, S. (2000). Reabsorption of betaine in Henle's loops of rat kidney in vivo. *Am. J. Physiol.* **278,** F434–F439.

Rim, J. S., Tana, S., Takenaka, M., Handler, J. S., and Kwon, H. M. (1997). The canine sodium/*myo*-inositol cotransporter gene: Structural organization and characterization of the promoter. *Arch. Biochem. Biophys.* **341,** 193–199.

Rim, J. S., Atta, M. G., Dahl, S. C., Berry, G. T., Handler, J. S., and Kwon, H. M. (1998). Transcription of the sodium/myo-inositol cotransporter gene is regulated by multiple tonicity-responsive enhancers spread over 50 kilobase pairs in the 5′ flanking region. *J. Biol. Chem.* **273,** 20,615–20,621.

Smardo, F. L., Burg, M. B., and Garcia-Perez, A. (1992). Kidney aldose reductase gene transcription is osmotically regulated. *Am. J. Physiol.* **262,** C776–C782.

Sone, M., Ohno, A., Abrecht, G. J., Thurau, K., and Beck, F-X. (1995). Restoration of urine concentrating ability and accumulation of medullary osmolytes after chronic diuresis. *Am. J. Physiol.* **269,** F480–F490.

Takenaka, M., Preston, A. S., Kwon, H. M., and Handler, J. S. (1994). The tonicity-sensitive element that mediates increased transcription of the betaine transporter gene in response to hypertonic stress. *J. Biol. Chem.* **269,** 29,379–29,381.

Trama, J., Lu, Q., Hawley, R. G., and Ho, S. N. (2000). The NFAT-related protein NFATL1 (TonEBP/NFAT5) is induced upon T-cell activation in a calcineurin-dependent manner. *J. Immunol.* **165,** 4884–4894.

Uchida, S., Nakanishi, T., Kwon, H. M., Preston, A. S., and Handler, J. S. (1991). Taurine behaves as an osmolyte in Madin-Darby canine kidney cells. *J. Clin. Invest.* **88,** 656–662.

Uchida, S., Kwon, H. M., Yamauchi, A., Preston, A. S., Marumo, F., and Handler, J. S. (1992). Molecular cloning of the cDNA for an MDCK cell Na- and Cl-dependent taurine transporter that is regulated by hypertonicity. *Proc. Natl. Acad. Sci. USA* **89,** 8230–8234.

Uchida, S., Yamuchi, A., Preston, A. S., Kwon, H. M., and Handler, J. S. (1993). Medium tonicity regulates expression of the Na- and Cl-dependent betaine transporter in MDCK cells by increasing transcription of the transporter gene. *J. Clin. Invest.* **91,** 1604–1607.

Woo, S. K., Dahl, S. C., Handler, J. S., and Kwon, H. M. (2000a). Bidirectional regulation of tonicity-responsive enhancer binding protein in response to changes in tonicity. *Am. J. Physiol. Renal Physiol.* **278,** F1006–F1012.

Woo, S. K., Maouyo, D., Handler, J. S., and Kwon, H. M. (2000b). Nuclear redistribution of tonicity-responsive enhancer binding protein requires proteasome activity. *Am. J. Physiol. Cell Physiol.* **278,** C323–C330.

Yamauchi, A., Kwon, H. M., Uchida, S., Preston, A. S., and Handler, J. S. (1991). *Myo*-inositol and betaine transporters regulated by tonicity are basolateral in MDCK cells. *Am. J. Physiol.* **261,** F197–F202.

Yamauchi, A., Uchida, S., Kwon, H. M., Preston, A. S., Robey, R. B., Garcia-Perez, A., Burg, M. B., and Handler, J. S. (1992). Cloning of a Na- and Cl-dependent betaine transporter that is regulated by hypertonicity. *J. Biol. Chem.* **267,** 649–652.

Yamauchi, A., Uchida, S., Preston, A. S., Kwon, H. M., and Handler, J. S. (1993). Hypertonicity stimulates transcription of gene for Na- *myo*-inositol contransporter in MDCK cells. *Am. J. Physiol.* **264,** F20–F23.

Yamauchi, A., Nakanishi, T., Takamitsu, Y., Sugita, M., Imai, E., Noguchi, T., Fujiwara, Y., Kamada T., and Ueda, N. (1994). In vivo osmoregulation of Na/myo-inositol cotransporter mRNA in rat kidney medulla. *J. Am. Soc. Nephrol.* **5,** 62–67.

Yamauchi, A., Miyai, A., Shimada, S., Minami, Y., Tohyama, M., Imai, E., Kamada, T., and Ueda N. (1995). Localization and rapid regulation of Na/*myo*-inositol cotransporter in rat kidney. *J. Clin. Invest.* **96,** 1195–1201.

Yancey, P. H., Clark, M. E., Hand, S. C., Bowlus, R. D., and Somero, G. N. (1982). Living with water stress: Evolution of osmolyte systems. *Science* **217,** 1214–1222.

Zhang, Z., Yang, X-Y., Soltoff, S., and Cohen, D. M. (2000). PI3K signaling in the murine kidney inner medullary cell response to urea. *Am. J. Physiol. Renal Physiol.* **278,** F155–F164.

Zhou, B-B., and Elledge, S. J. (2000). The DNA damage response: Putting checkpoints in perspective. *Nature* **408,** 433–439.

Passive Water Transport in Biological Pores

Thomas Zeuthen and Nanna MacAulay
Institute of Medical Physiology, The Panum Institute, University of Copenhagen, DK 2200 Copenhagen N, Denmark

Three kinds of membrane proteins have been shown to have water channels properties: the aquaporins, the cotransporters, and the uniports. A molecular–kinetic description of water transport in pores is compared to analytical models based on macroscopic parameters such as pore diameter and length. The use and limitations of irreversible thermodynamics is discussed. Experimental data on water and solute permeability in aquaporins are reviewed. No unifying transport model based on macroscopic parameters can be set up; for example, there is no correlation between solute diameter and permeability. Instead, the influence of hydrogen bonds between solute and pore, and the pH dependence of permeability, point toward a model based upon chemical interactions. The atomic model for AQP1 based on electron crystallographic data defines the dimensions and chemical nature of the aqueous pore. These structural data combined with quantum mechanical modeling and computer simulation might result in a realistic description of water transport. Data on water and solute permeability in cotransporters and uniports are reviewed. The function of these proteins as substrate transporters involves a series of conformational changes. The role of conformational equilibria on the water permeability will be discussed.

KEY WORDS: Aquaporins, Cotransporters, Uniports, Water permeability, Reflection coefficient.

I. Introduction

Water is the most abundant molecule in organisms and one of the smallest. Passive transport of water is therefore of vital importance for all aspects of cellular volume homeostasis. Historically, passive water transport across cell membranes was considered an osmotic transport problem in the lipid membrane only. Over the last few years, however, osmotic water transport in membrane proteins has become well established. This concerns the specific water channels (aquaporins) as well as proteins usually thought of in connection with other functions such as cotransporters and uniports.

It has proved difficult to set up an analytical physical model for water permeation in biological pores. For a wide pore, macroscopic concepts such as pore diameter, pore length, and hydrostatic pressures are sufficient to give an analytical description of the transport. Unfortunately, these macroscopic concepts have meaning only for dimensions much larger than the water molecule. For the more biologically relevant narrow pore with diameters similar to those of the water molecule, models would require a molecular–kinetic description, which is unavailable in an analytical version. To put it crudely, there is a good physical theory for pores that are biologically irrelevant; but there is no adequate theory for pores that are relevant. The first part of this chapter summarizes the state of the art in regard to physical descriptions of permeation in water channels.

A comparison among the permeability data obtained for aquaporins shows clearly the difficulties of setting up a unifying physical transport model. There is no correlation between the diameter of the osmolyte and the measured water permeability. Furthermore, some aquaporins are permeable to both water and other substances, say glycerol, and transport parameters may depend on external pH and the hydrogen bonding properties of the solutes. Such data point toward models in which the chemical bonds between water and the pore are of central importance. It is therefore a major incentive to obtain the amino acid composition and tertiary structure of the channel protein. What amino acid residues are responsible for the selectivity filter, what residues line the access pathways, and how is the channel protein stabilized and anchored in the membrane? Answers to such questions are now emerging for the aquaporin water channel AQP1 (Murata *et al.,* 2000; see Engel and Stahlberg, this volume) and the related glycerol channel GlpF (Fu *et al.,* 2000; see Engel and Stahlberg, this volume) based on electron crystallographic data combined with amino acid sequence data and model building (see Engel and Stahlberg, this volume).

Cotransporters and uniports are usually associated with substrate transport of, for example, sugars and amino acids. It is now established that these proteins have a well-defined passive water permeability and may contribute significantly to passive water transport across cell membranes. The understanding of the underlying mechanisms is subject to the same difficulties as those outlined for the aquaporins. Furthermore, the proteins occupy various conformational states during substrate transport, and the passive water permeability may not be the same among the various states. This is the subject of the final part of the chapter.

II. Physical Models

A. Molecular–Kinetic Description

In aqueous solutions, both water and solute molecules perform thermal motions in which particles exchange kinetic energy in a random fashion. As a result, each particle continuously changes position and velocity. Given a physiological salt solution of concentration 110 mmol L^{-1}, there will be about 500 water molecules per molecule of salt. In such dilute solutions, two solute molecules rarely interact, and the properties of the solution can be defined by the interaction between an individual solute molecule and the water molecules. To understand how water permeates a membrane, it is necessary to have a physical model of the molecular–kinetic events in the solution, at the interphase between membrane and solution, and in the membrane itself. Unfortunately, there is no molecular–kinetic model of diffusion in aqueous solutions, let alone of interactions between water solute and a membrane. Dainty (1965) attempted a molecular–kinetic treatment but concluded that this would require knowledge of "the statistical mechanics of liquid flow, in which the transfer of momentum to individual molecules arising from the unbalanced jumping process at the pore mouth is considered in details." Accordingly, researchers have been forced to apply macroscopic or continuum theories where parameters are averages, both temporarily and spatially, of molecular–kinetic events.

B. Diffusion and Osmosis

If the solute concentration is nonuniform, diffusion will take place. More particles will move from regions of high concentration into dilute ones and fewer in the other direction. Since the movement of each particle is entirely random, there is no force associated with diffusion (Fig. 1). The so-called diffusion force "is not a ponderomotive force that can accelerate or impart net velocities. It is a statistical or virtual force describing the increase of 'randomness' due to increasing entropy

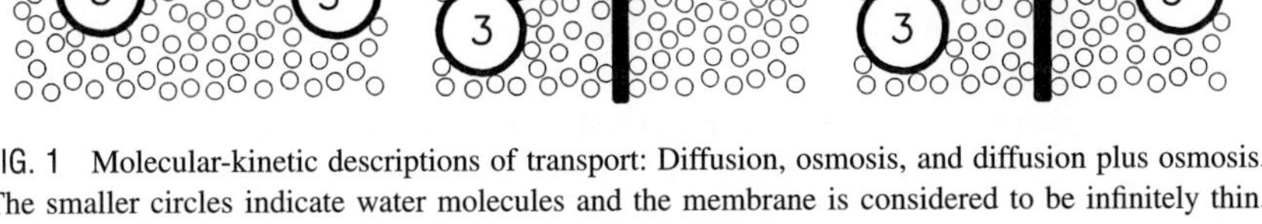

FIG. 1 Molecular-kinetic descriptions of transport: Diffusion, osmosis, and diffusion plus osmosis. The smaller circles indicate water molecules and the membrane is considered to be infinitely thin. Diffusing solute molecules move randomly; each jump may take place either to the left (from 2 to 3) or to the right (from 2 to 3′). There is no net transfer of momentum between solute and water. In osmosis solutes cannot cross the membrane. Compared with diffusion, the jumps from 2 to 3 are not balanced by jumps from 2 to 3′. Consequently there must be a net transfer of momentum from the solutes to the water in the direction towards the concentrated bath. In case the membrane is partly permeable, the situation is intermediate between diffusion and osmosis. The solute molecule can either be reflected giving rise to an osmotic event (3) or jump into the dilute solution which gives rise to a diffusive event (3′). It follows that the water permeability measured by means of a permeable solute is smaller than that recorded by means of an impermeable solute. Redrawn from Zeuthen (1995).

of dilution" (Hille, 1992, p. 263; see also Einstein, 1956; Hartley and Crank, 1949; Sten-Knudsen, 1978). At present, there is no model that combines the microscopic interactions between solutes and solvent during diffusion with macroscopic or measurable properties. This has forced investigators to apply macroscopic or continuum theories. Einstein (1956), for example, used Stokes' relation to arrive at an expression for the diffusion coefficient.

In osmosis, a semipermeable membrane separates two solutions of different concentrations; the membrane is permeable to water but restricts solute movements. Two cases are usually distinguished: one where the membrane is purely lipid and one where it contains pores. In the lipid case, water crosses by a diffusional process charaterized by a relatively high Arrhenius activation energy, about 12 to 20 kcal mol^{-1}. The process probably requires formation of sites for water molecules within the hydrocarbon part of the bilayer (Deamer and Nicols, 1989; Marrink *et al.,* 1996; Nagle, 2000). The rate of transport is independent of the nature of the driving force for water, whether this be diffusive, osmotic, or hydraulic (for a review, see Finkelstein, 1987).

In the case of membrane pores, it is useful first to compare the case where the pore is infinitely short (i.e., the membrane infinitely thin) with the case of simple diffusion (see Fig. 1). In case the membrane is perfectly semipermeable, a solute molecule that has arrived at a position in front of the opening (position 2) can only move away if it returns toward the concentrated solution (position 3). In case of diffusion, however, the solute molecule might equally well have proceeded toward the dilute solution (from position 2 to 3′). In diffusion, therefore, water molecules experience an equal number of solute molecules moving to one side and to the other; consequently, there is no net transfer of momentum from the solute to the water molecules. In osmosis, the water molecules situated in the hole experience more solute molecules moving toward the bath with the high solute concentration. As a consequence, there is a net transfer of momentum to the water molecules toward this side. The mechanism for the osmotic transport must be sought in this asymmetry of momentum transfer.

In case the membrane is not perfectly semipermeable, the diameter of the pore is large enough to allow solute molecules to pass. A solute molecule that permeates the membrane (position 2 to 3′) will behave entirely diffusively, and the event will not lead to net movement of water. If, however, the solute molecule is reflected from the edge of the hole, osmotic forces will arise. This suggests that a molecular–kinetic description of the permeable pore must contain both diffusive and osmotic events and that macroscopic or measured parameters must be continuous functions of solute radius.

C. Pores with a Length

Pores that span plasma membranes have lengths much larger than the diameter of the water molecule and of the solute molecule. For an impermeable pore, all solute molecules are reflected from the solute-facing end of the pore. Following the treatment above, a momentum will be transferred to the water molecules residing inside the pore with direction into the solute-containing compartment, and water transport will ensue. The situation is described in thermodynamic terms, chemical potentials, and hydrostatic pressures in the so-called Vegard-Mauro model (Vegard, 1908; Mauro, 1957; Hill, 1982; Finkelstein, 1987; see also Manning, 1968; Mauro, 1979; Hammel, 1979; Soodak and Iberall, 1979; Tomicki, 1985). The model postulates a gradient of hydrostatic pressure through the pore; at one end, the pressure equals the pressure of the solvent-free bath while it describes a sharp jump at the interface with the solute-containing bath. The unphysical abrupt change in the hydrostatic pressure must be seen as the temporal and spatial average of the stochastic processes. The hydrodynamic nature of the driving force is supported by experiments (Mauro, 1957).

If, on the other hand, the pore is permeable to the solute, there is no agreement with regard to a model for permeation. One kind of model simply combines the mechanisms described for the impermeable pore with the assumption that the pore wall rejects the permeable solute, thus decreasing its concentration (Garby, 1957; Anderson and Malone, 1974; Tomicki, 1985; Finkelstein, 1987). This strategy has been criticized (Hill, 1982, 1989a,b) on the view that if a solute can enter the pore, the activity of the solute, at equilibrium, will be the same outside and inside the pore. As a consequence, there should be no difference in osmotic pressure and therefore no build up of hydrostatic pressure gradients. With no hydrostatic pressure differences, flows should be entirely diffusive and driven by the difference in the activities at the two sides of the membrane.

This conflict does not arise in a molecular–kinetic description. If a solute particle collides with the orifice of the pore and is reflected, forces are set up that move water from the pore into the bath. Such events are driven by a gradient of hydrostatic pressure as described above. If, on the other hand, the solute particle enters the pore, water moves from the pore into the bath by interdiffusion. In short, the permeable pore will be characterized by a mixture of hydrostatic and diffusive events. For the very wide pore, most solute molecules will enter the pore, and an entirely diffusive description will be adequate. In the other limit, the narrow impermeable pore, all solute molecules will be reflected, and the hydrodynamic description is applicable.

In conclusion: the molecular–kinetic description gives an intuitively correct description of osmosis in membrane pores, but it gives no analytical formulae. A thermodynamic model based on macroscopic concepts leads to analytical formulae but not to a unified physical picture.

D. Irreversible Thermodynamics

Irreversible thermodynamics is a useful framework in which to interpret water transport experiments. It defines a set of parameters and defines the conditions under which they should be measured (see review by Katchalsky and Curran, 1965). Unfortunately, the physical meaning of the parameters is less clear and has been subject to various interpretations. Consider, for example, a membrane with water channels that are permeable to glycerol. The equations for the volume and solute fluxes and the coupling between them are:

$$J_V = -RTL_p(\Delta C_i + \sigma_{s,Gl}\Delta C_{Gl}) \tag{1}$$

$$J_{Gl} = RTP_{Gl}\Delta C_{Gl} + J_v C_{Gl}'(1 - \sigma_{f,Gl}) \tag{2}$$

where J_V is the volume flow, J_{Gl} is the flux of glycerol, i.e., determined from tracer uptake, R is the gas constant, and T the absolute temperature. ΔC_i is the transmembrane concentration difference of impermeable solutes such as mannitol; ΔC_{Gl} is the difference in glycerol concentration. C_{Gl}' is the average concentration of glycerol in the aqueous pore. The coupling between the solute and volume fluxes in the pore can be characterized by the reflection coefficient σ. The L_p in Eq. (1) is determined from osmotic challenges by impermeable osmolytes (i.e., mannitol), while gradients implemented by glycerol determine the reflection coefficient $\sigma_{s,Gl}$. The parameters of Eq. (2) are determined from ultrafiltration experiments. In these, J_{Gl} arises from two mechanisms: by diffusion (first term) and by the water flow in the pore itself (second term). This second term is a product of the volume flow J_V, the glycerol concentration C_{Gl}', and a term $(1 - \sigma_{f,Gl})$, which contains a reflection coefficient. Traditionally, the two reflection coefficients of Eqs. (1) and (2) $\sigma_{s,Gl}$ and $\sigma_{f,Gl}$, are assumed to be equal, as would be the case if Onsager symmetry applied. In that case, σ_{Gl} $(= \sigma_{f,Gl} = \sigma_{s,Gl})$ and C_{Gl}' can be determined from numerical solution of Eqs. (1) and (2) (see Macey and Karan, 1993).

It has been argued, however, that for a leaky pore, σ_f and σ_s are in fact different since they arise from two different physical mechanisms for which Onsager symmetry does not apply (Hill, 1982). Only in the limits of either very narrow or wide pores will σ_f and σ_s be equal. Different σ_f and σ_s would make a precise determination of the parameters in Eqs. (1) and (2) difficult. Certainty about glycerol permeation in the aqueous pathway could only be obtained by testing whether osmotic flows increased or decreased the glycerol fluxes, above or below that given by the permeability alone. In such a case, the second term of Eq. (2) would be different from zero. Experimentally, this could be difficult. It might require gradients of several hundred mosm L^{-1} to induce significant solute fluxes, gradients which may not be tolerated by cells. In addition, there is no way to estimate the glycerol concentration C_{Gl}' inside the pore.

So far, published experimental data have been interpreted under the assumption that σ_f and σ_s are equal. This means that L_p and the reflection coefficient for a permeable solute, say, glycerol, σ_{Gl}, can be determined in osmotic experiments using mannitol and glycerol as osmolytes. A σ_{Gl} smaller than 1 will be indicative of glycerol permeation:

$$\sigma_{Gl} = 1 - RTP_{Gl}(\Delta x/\varphi_w)f_{sw} \tag{3}$$

where P_{Gl} is the glycerol permeability, $\Delta x/\varphi_w$ is a constant that gives the ratio of the membrane thickness to the volume fraction of water, and f_{sw} is the friction between the solute (glycerol) and water in the pore. Equation (3) is adapted from Eq. (10–56) in Katchalsky and Curran (1965) with partial molar volume effects ignored. In our discussion of experimental data, we employ Eq. (3) bearing in mind the caveats outlined above. It appears that a smaller σ is associated with a larger permeability P.

E. Pores with Variable Diameters

Consider an aqueous pore with a wide and a narrow section in series (Fig. 2). If the L_p is measured by means of an osmolyte with a diameter larger than the wide part of the pore, then the L_p will represent the water permeability of the whole pore (Fig. 2A). The L_p may also be measured by an osmolyte that is small enough to enter the wide part of the pore in an osmotic active form (Fig. 2B). In that case, the measured L_p will be larger, and given by the narrow part of the pore only. In the language of irreversible thermodynamics (Eq. 1), it might appear as if σ_s for the smaller osmolyte were larger than 1. In case the osmolyte is small enough to permeate throughout the pore (Fig. 2C), the interpretation becomes very

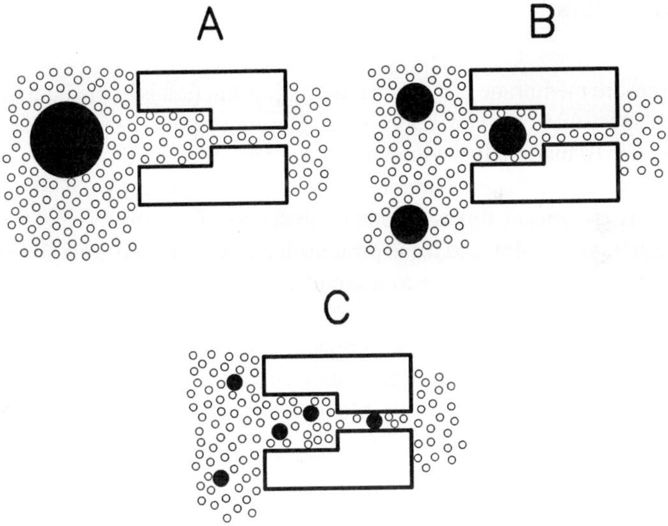

FIG. 2 Pores with a variable diameter. Consider a pore that has a wide and a narrow section in series. In A the water permeability L_p is measured by an osmolyte (filled circles) which is too large to enter the pore at all. In B L_p is measured by an osmolyte which is small enough to enter the wide part but not the narrow part. Under the assumption that the osmolyte remains osmotically active within the wide cavity of the pore, the L_p recorded will be larger than in A. If the osmolyte is small enough to enter both the narrow and the wide part of the protein (C) the water permeability can be recorded to be both smaller, larger or equal to that found in A.

complicated. The σ will appear to increase on account of the partial permeation into the pore as described above; on the other hand, σ will tend to be smaller than 1 on account of the permeation through the narrow section of the pore. All in all, the measured σ may be larger, equal to, or smaller than 1 depending on the size and permeability properties of the probe molecule. A measured, σ equal to 1 might arise as a combination of these various mechanisms. Measurements of σ in filtration experiments (σ_f, Eq. 2) would alleviate the problem.

Recent data suggest that water pores have diameters that vary along the length of the aqueous pathway. This applies to aquaporins (Murata *et al.,* 2000) (see Section III.A and Engel and Stahlberg, this volume), and to some cotransporters where larger L_ps are obtained with smaller osmotic probes (see Section IV.B). It should be emphasized that the concept of the wideness of a pore should be understood in a broad sense. It may include other phenomena such as the properties of surface water in the protein and whether the osmotic probe has access to this layer. Structural information is required to select among the various possibilities for physical mechanisms. This may be achieved in studies in pores of known structure such as gramicidin or from high-resolution atomic models of biological pores such as that described below.

III. Aquaporins

Aquaporins are membrane proteins whose major function is to mediate transmembrane transport of water (for reviews see Borgnia *et al.,* 1999; Ishibashi *et al.,* 2000). So far, 10 mammalian aquaporins have been identified: AQP0 to AQP9. Of these, AQP0, 1, 2, 4, 5, and 8 are predominately water selective, while the others may support significant fluxes of other substances. Glycerol permeates through AQP3, AQP7, and AQP9, and these proteins have been named aquaglyceroporins. In addition, AQP9 is permeable to a broad spectrum of substances, among them urea and mannitol (Tsukaguchi *et al.,* 1998, 1999). Some investigations, however, find that AQP9 is permeable only to urea and water (Ko *et al.,* 1998; Ishibashi *et al.,* 1998). AQP6 exhibits anion conductance (Yasui *et al.,* 1999). Plants express numerous types of aquaporins (see Maurel, this volume) and so do yeasts (see Hohman, this volume).

A. Molecular Architecture of AQP1

AQP1 was initially identified in red blood cells and renal proximal tubules and was shown to be an integral membrane protein (Denker *et al.,* 1988; Preston and Agre, 1991). Expression in *Xenopus laevis* oocytes revealed that the protein functioned as a water channel (Preston *et al.,* 1992). Subsequent sequence analysis and expression studies applying functional epitope-scanning mutants have shown

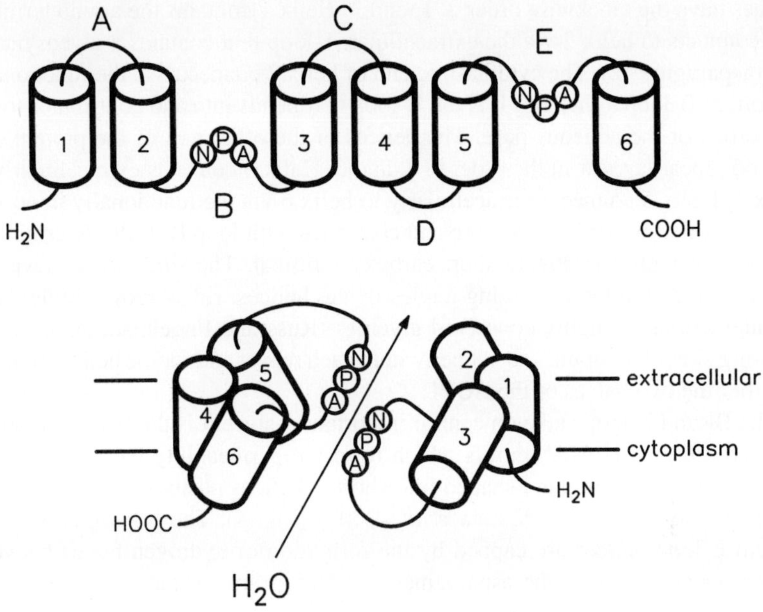

FIG. 3 Structural organization of the AQP1 The AQP1 is made from six membrane spanning helices connected by five loops (top panel). In the functional monomer (lower panel), the hydrophilic loop B, which connects helix 2 and 3, and loop E, which connects helix 5 and 6, are bent back into the hole formed by the helices. The two loops meet in the middle to form the putative water-selective gate that contains two NPA motifs (Asn-Pro-Ala). For clarity, the helices 3, 1 and 2 are drawn separated from helices 5, 4 and 6. The hydrogen-bonding properties of the side-groups of the two asparagine residues are thought to constitute the permeation barrier. In the membrane the six α-helices form a right-handed twisted arrangement. Based on Murata *et al.* (2000), redrawn from Zeuthen (2001).

that AQP1 consists of 269 amino acid residues which form two tandem repeats, each of which has three membrane-spanning α-helices. The carboxy- and amino termini are located to the cytoplasmic side. Two of the loops that connect the α-helices, the B and E loops, fold back and connect in what appears to be the aqueous pathway through the protein. This unique structure is known as the hourglass model (Jung *et al.,* 1994) (Fig. 3). The two loops each contain an NPA motif (asparagine, proline, alanine, or Asn-Pro-Ala), which is well conserved among the aquaporins and believed to play a pivotal role for water selectivity.

Preparation of crystals from highly purified samples of human red blood cell AQP1 has recently permitted the determination of the atomic structure at a high level of resolution (Walz *et al.,* 1997; Cheng *et al.,* 1997; Li and Jap, 1997; Murata *et al.,* 2000). In the latter work, the order and spatial orientation of the constituent α-helices have been determined. The six membrane-spanning α-helices form a right-handed twisted arrangement (Fig. 3B). Seen from the extracellular side, the helices have the clockwise order 2, 1, and 3. Helix 1 contains the amino terminal and connects to helix 2 via the extracellular A-loop that contains a glycosylation site (asparagine 42). The cytoplasmic end of helix 2 connects via the functionally important B-loop to helix 3. It is the B-loop that bends into and contributes to the properties of the aqueous pore. The helices of the other part of the protein (the second repeat) appear in the order 5, 4, and 6. Helix 3 connects extracellularly to helix 4. Helix 5 connects extracellularly to helix 6 via the functionally important E-loop, which bends into the pore and combines with loop B. Helix 6 terminates the protein with a relatively short carboxy terminal. The structure achieves its stability from the large crossing angles of the helices, ridge–grove fittings, and by interactions of highly conserved glycines (Russ and Engelman, 2000) at the crossing sites. This applies to helices within each repeat and for the helices holding together the two halves of the AQP1.

The B- and E-loops take on central importance, since it is the two asparagines (76 and 192) of the NPA motifs which form the permeability barrier for water. The B-loop and E-loop are shaped into short α-helices positioned at an angel of about 90° to each other (Murata *et al.,* 2000) (Fig. 4). The N-terminals of the B- and E-loop helices are capped by the formation of hydrogen bonds between the carbonyl groups of the asparagines and NH groups of the main chain. The

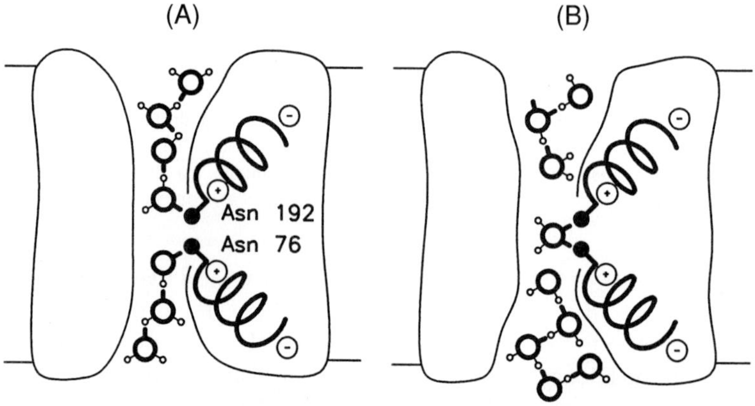

FIG. 4 The atomic model for water permeation in AQP1. The B and E loops which dip into the interior of the AQP1 monomer form short α-helices. The N-termini of these short helices point into the aqueous pathway and create a positive electrostatic field which will orientate the water oxygen towards the two asparagines. These residues are located in a narrow portion of the pore (3 Å diameter) which prevent larger passage of larger molecules. The water molecules are able to form a hydrogen bonds with these polar amino acid residues. In A there is a water molecule connected with each of the two asparagines. In B one water molecule occupies both asparagine sites. As a consequence water molecules pass through the aqueous pathway by making and breaking two hydrogen bonds. Importantly, the string of hydrogen bonded water molecules extending through the pore is broken preventing transport of H^+. The rest of the channel is mainly made up from hydrophobic residues allowing for fast transport rates. Based on Murata *et al.* (2000), Fig. 5.

two loops are locked to each other by van der Waals interactions between the two prolines. In addition, the two helices are tied up with the rest of the protein via a number of ion pairs and hydrogen bonds.

AQP1 exists in the membrane as a homotetramer (Van Hoek *et al.,* 1991; Smith and Agre, 1991). The diameter of the hole formed between the four AQP1s is about 3.5 Å at its narrowest, and 8.5 Å at its widest. It has been discussed whether this center, which is not occupied by the AQPs, could form a pathway for water permeation. Previous experimental studies have discredited the notion, and Murata *et al.* (2000) give a structural rationale for this. The hole is too small to accommodate a lipid molecule, and since the protein exposes hydrophobic side groups toward this cavity, it is considered unlikely that water is present.

B. Atomic Model for Water Permeation in AQP1

What is the molecular architecture of the aqueous pore that provides the high water permeability and selectivity? More than 10^9 water molecules can pass through the channel per second (Zeidel *et al.,* 1992b), while the protein has been found to be impermeable to molecules and ions, in particular H^+. The molecular architecture shows that the pore has a narrow centrally located constriction of about 3 Å, which is only slightly larger than the 2.8 Å usually given for the diameter of the water molecule (Murata *et al.,* 2000). Size alone then would block the passage of larger solutes. The selectivity of the constriction is determined by the two asparagines (76 and 192) which extend their amido groups into the lumen of the pathway. It is by forming hydrogen bonds with these groups that water, and water alone, slips through (Fig. 4). The portions of the pore which connect the constriction with the cellular and extracellular solutions are mainly lined with hydrophobic residues that ensure rapid transit of water. During transport, a water molecule is initially connected to neighboring water molecules by hydrogen bonds. As the water molecule approaches the constriction formed by the two asparagines, the oxygen atom of the water molecule orients toward the two residues. This is because the E and B helices are orientated with their N-termini pointing toward the lumen, giving a positive electrostatic field. The water molecule then breaks its hydrogen bonds with the adjacent water molecules and forms instead, first one, then two hydrogen bonds with the asparagines. The deserted water molecules cannot form other hydrogen bonds with the hydrophobic pore wall. As a consequence, water molecules are predicted to permeate with an energy barrier equivalent to making and breaking one hydrogen bond, about 3–4 kcal mol^{-1}, in good agreement with experimental data. Most importantly, the string of hydrogen bonds that would otherwise extend through the pore and constitute an efficient pathway for H^+ transport is broken (Pomès and Roux, 1996, 1998).

The atomic structure of the aquaporin homolog, the glycerol facilitator GlpF from *Escherichia coli,* has been determined to a resolution of 2.2 Å (Fu *et al.,* 2000). This protein is impermeable to water and ions but allows high rates of diffusive transport of glycerol (Maurel *et al.,* 1993; see Maurel, and Engel and Stahlberg, this volume). The tertiary structure is similar to that of AQP1 with six major α-helices and two shorter α-helices in the pore. The selectivity filter of GlpF is different from that of AQP1 and consists of a 28-Å long, 3.4- to 3.8-Å wide amphiphatic channel. During permeation, the alkyl backbone of the glycerol molecule is wedged against the hydrophobic part of the channel, while the —OH groups of the glycerol form successive hydrogen bonds with side groups from the hydrophilic part.

Several questions arise from the comparison between AQP1 and GlpF. First, in bulk solution, water molecules share hydrogen bonds with up to four others. To permeate a channel in single file, a water molecule would have to reduce this number to two. Unless these bonds are replaced by other bonds to the pore wall, the activation energy would increase significantly, since 3 to 4 kcal mol^{-1} are involved per bond. Fu *et al.* (2000) suggest that the energy barrier for this dehydration suffices to explain the impermeability of GlpF to water. This differs from the mechanism suggested for AQP1 which involved the making and breaking of an additional hydrogen bond inside the pore (Fig. 4) (Murata *et al.,* 2000). The activation energy for water transport in AQP1 is in the range 3 to 7 kcal mol^{-1} (Zeidel *et al.,* 1992; Meinild *et al.,* 1998). Clearly a precise evaluation of the energetics of hydrogen bond formation for water in both AQP1 and GlpF is required. Second, the mechanisms suggested for the impermeability of AQP1 and GlpF to H^+ ions are different. In neither pore can a hydrated ion pass the narrow section, and the hydrophobic part of the wall prevents the costly dehydration of the ion. Fu *et al.* (2000) consider this sufficient to explain the impermeability of GlpF to H^+. They do not invoke any breaking of the string of hydrogen bonds, as suggested for AQP1 (Murata *et al.,* 2000).

C. Transport Parameters of Aquaporins

Most data on transport parameters of aquaporins have been obtained from expression studies in *Xenopus* oocytes. The water permeability L_p has been found to be the same whether obtained from shrinkage or swelling experiments (Meinild *et al.,* 1998b). The unit channel permeabilities have been obtained from counting the number of expressed channels by tagging (Yang and Verkman, 1997) or by freeze-fracture (Zampighi *et al.,* 1995; Chandy *et al.,* 1997). Unit permeabilities were (in units of cm^3 sec^{-1} 10^{-14}) for AQP0: 0.25 (Yang and Verkman, 1997), 0.015 (Zampighi *et al.,* 1995), 0.028 (Chandy *et al.,* 1997); AQP1: 4.6 (Zeidel *et al.,* 1992b), 1.4 (Zampighi *et al.,* 1995), 1.2 (Chandy *et al.,* 1997), and 6 (Yang and Verkman, 1997); AQP 2: 3.3 (Yang and Verkman, 1997); AQP3: 3.1 (Yang and Verkman, 1997); AQP4: 24 (Yang and Verkman, 1997), and for AQP5: 5.0 (Yang and Verkman, 1997). The variability among the data calls for further experiments (see Engel and Stahlberg, this volume).

TABLE I
Reflection Coefficients σ for Aquaporins 0 to 5

	Mannitol	Urea	Glycerol	Acetamide	Formamide
Mole weight	182	60	92	63	45
AQP0	1	1	1	0.8	0.6
AQP1	1	1	0.8	1	1
AQP2	1	1	0.8	1	1
AQP3 23°C	1	1	0.2	0.7	0.4
AQP3 13°C	1	1	0.4	1	0.5
AQP4	1	1	0.9	1	1
AQP5	1	1	1	1	0.8

From Meinild *et al.* (1998b).

The reflection coefficients for various test solutes (σ, Eq.1) have been measured by a highly accurate assay for aquaporins 0 to 5 expressed in oocytes (Table I). The values were obtained as the ratio of the cell shrinkage obtained by the test solute divided by the shrinkage obtained using mannitol. A σ smaller than 1 is taken as an indication of permeation of the test solute in the aqueous channel; the smaller the σ, the larger the permeability (see Eq. 3 and accompanying discussion). A comparison of L_p and σ shows the inadequacy of a purely hydraulic model based on size and length of the pore and the van der Waals dimensions of the solute. Consider the dimensions of the solutes employed (in Å): mannitol (7.4 × 8.2 × 12), urea (3.6 × 5.2 × 5.4), glycerol (4.8 × 5.1 × 7.8), acetamide (3.8 × 5.2 × 5.4), and formamide (3.4 × 4.4 × 5.4). For AQP1, AQP2, and AQP4 only for glycerol was σ lower than 1. Accordingly, glycerol should permeate, yet the glycerol molecule is not the smallest one tested. In AQP3, σ for glycerol is smaller than σ for formamide which is smaller than σ for acetamide. Yet, glycerol is the largest among the three. Furthermore, both glycerol and acetamide are larger than urea, which does not permeate. Thus a straightforward application of van der Waals dimensions is insufficient to explain the sequences of observed σs. Part of the explanation could be solute–pore interactions. If, for example, the $-NH_2$ and the $-OH$ of the solutes formed hydrogen bonds with the pore wall, the effective size of the solute would be smaller.

A similar conclusion is reached from a discussion of pore lengths. The σ for glycerol is close to 1 for all the aquaporins, except AQP3. If this is taken to mean that the cross section of the pores in AQPs 1, 2, 4, and 5 are equal, it follows from the unit permeability that AQP1 and AQP2 have the same length while AQP4 is 6 times shorter. From the σs, it seems that AQP3 has the largest pore cross section, but since its single channel permeability is similar to AQP1 and AQP2, the length of its pore must be greater. These postulated differences in cross section and length are unlikely when the similarity in secondary structure among the aquaporins is considered.

D. Polyol Permeation in AQP3, Role of Hydrogen Bonds and Backbone Length

AQP3 acts as a channel for both water and glycerol. The fluxes are linearly increasing functions of their respective chemical driving forces (Echevarría *et al.,* 1994, 1996; Ishibashi *et al.,* 1994; Zampighi *et al.,* 1995), and the activation energies for the fluxes are low (Ishibashi *et al.,* 1994; Echevarría *et al.,* 1996). Furthermore, the two fluxes share the pathway in the protein, as indicated by the low σs (Echevarría *et al.,* 1996; Meinild *et al.,* 1998b).

The mechanism of glycerol permeation can be investigated by comparing the transport parameters for a spectrum of polyols. The importance of hydrogen bonding between —OH groups of the solute and the pore is supported by the wide range of reflection coefficients, σ (Eq. 3); see Fig. 5. This is particularly clear when σs of the butandiols (1,2-butandiol, 1,3-butandiol, 1,4-butandiol, and 2,3-butandiol) were compared. The location and intramolecular interactions of the two —OH groups had significant effects on the σs. The σs were larger if the two —OH groups were located next to each other and engaged in intramolecular bonding ($\sigma_{B23} > \sigma_{B12} > \sigma_{B13} \cong \sigma_{B14}$). The extent of intramolecular bonding was mirrored by the boiling points, which were lower for 1,2-butandiol and 2,3-butandiol than for 1,3-butandiol and 1,4-butandiol. Accordingly, the —OH groups in 1,2-butandiol and 2,3-butandiol might not be available for interaction with the sites in the pore to the same degree as the —OH groups of 1,3-butandiol and 1,4-butandiol. This would result in smaller permeabilities and therefore larger σs for 2,3-butandiol and 1,2-butandiol. The effects of the locations of the —OH groups on σ were absent for the pentanols. Most likely, the longer carbon chain mitigates the strength of intramolecular bonding between —OH groups, as witnessed by the small variations between the boiling points among this group.

In general, σ increased (permeability decreased) with the number of —OH groups as well as with the number of carbons in the backbone (Fig. 5). A comparison between the σs for glycerol and propanediol shows that σ was larger when the number of —OH groups was increased by one. A comparison between σ for pentandiol and for ethylene glycol (1,2-ethandiol) showed that a further increase in σ was obtained when the number of carbons in the backbone was increased by one. Compare also 1,3-butandiol with 2,4-pentandiol, and 1,4-butandiol with 1,4-pentandiol.

The picture that emerges is one where the test molecules, viewed as cylinders of different lengths and roughly similar diameters, cross the pore of AQP3 with their axis parallel to the pore. During permeation, the —OH groups of the solute

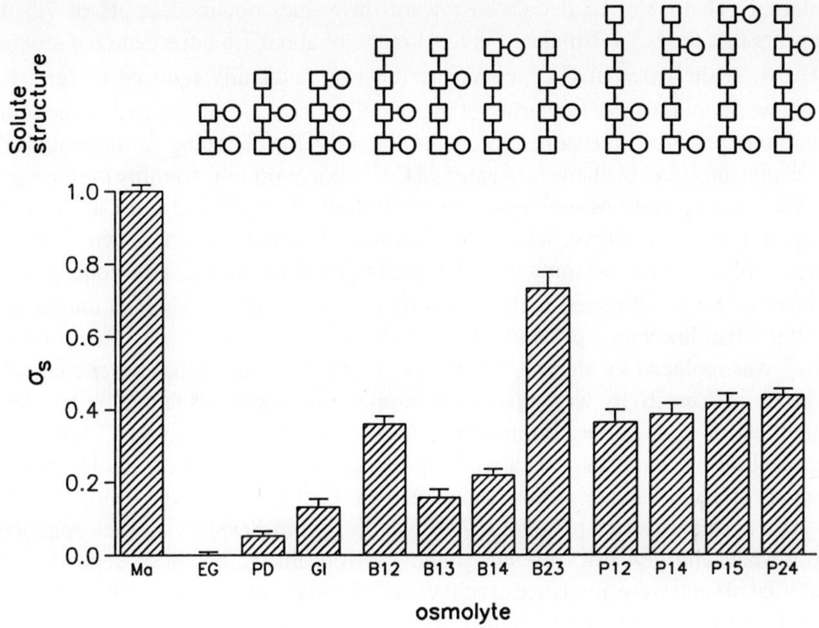

FIG. 5 Reflection coefficients of AQP3 for different straight chain polyols. The aquaporin was expressed in *Xenopus* oocytes and the rate of oocyte shrinkage was measured by exposing the oocyte to solutions made 20 mOsm hyperosmolar by means of addition of various straight chain polyols: Ethylen glycol (EG), 1,2-propanediol (PD), glycerol (Gl), 1,2-, 1,3-, 1,4-, and 2,3-Butandiol (B12, B13, B14, B23), and 1,2-, 1,3-, 1,4-, and 2,3-Pentandiol (P12, P14, P15, P24). The shrinkages produced were compared to those produced by mannitol, the ratio presented as σ (σ_s of Eq. 1). The structures of the test solutes are given above. An open circle indicates an —OH group, and squares indicate backbone carbons. Each carbon has four bonds, but single hydrogen atoms and associated bonds are not shown. In general the data suggest that the shorter the backbone of the polyol and the fewer —OH bonds, the easier the polyol penetrates (small σ, Eq. 3). Interestingly, the two polyols B12 and B23 deviate from this trend having larger σs (low permeabilities). This could be a result of intramolecular bonds between the two abutting —OH groups which are then not available to make contact with transport sites in the pore. Redrawn from (Zeuthen and Klaerke, 1999).

form a succession of single hydrogen bonds with the pore wall as indicated by the low activation energies of around 5 kcal mol^{-1} observed for permeation (Meinild *et al.,* 1998b; Ishibashi *et al.,* 1994; Echevarría *et al.,* 1996).

E. Effects of pH on AQP0, AQP3, and AQP6

The water permeability L_p of AQP0 is regulated by pH and Ca^{2+} (Németh-Cahalan and Hall, 2000). AQP0 was expressed in the *Xenopus* oocytes; when external pH was reduced from 7.5 to 6.4, L_p increased 3- to 4-fold. At further reductions in external pH, to 5.5, L_p decreased toward the values obtained at pH of 7.5. It appears that the L_p is stimulated in a pH range of about 1.5 units centered around pH 6.4. If the external Ca^{2+} concentration was nominally reduced to zero, L_p increased about 4-fold. Lowering of internal Ca^{2+} concentrations and calmodulin inhibitors produced the same effects. Importantly, the clamping of internal Ca^{2+} overrode the effect of changes in external Ca^{2+} concentration; possibly the changes in Ca^{2+} concentrations are sensed intracellularly. The pH and the Ca^{2+} effects were not strictly additive, which suggests that H^+ and Ca^{2+} act at two different sites, which affects the same common pathway regulating L_p. This concept was supported by site-directed mutagenesis of a histidine, His^{40}, which is unique for AQP0. Histidine has a pK around 7, similar to physiological pH values. When His^{40} was replaced by alanine, aspartic acid, or lysine, functional water channels with no pH sensitivity were produced. Similar effects were obtained by covalent modification of histidines. From the predicted membrane topology, His^{40} is located extracellularly. This suggests a working model of the AQP0 with the H^+ sensor on the extracellular face of the channel and the Ca^{2+} sensor on the inside.

At a normal external pH of 7.4, AQP3 expressed in *Xenopus* oocytes conducts both water and glycerol (references above). Both fluxes, however, depended on external pH and were abolished at acid values (Zeuthen and Klaerke, 1999) (Fig. 6). The L_p was characterized by a pK of 6.4 and a Hill coefficient of 3, the glycerol permeability P_{Gl} by a pK of 6.1 and a Hill coefficient of 6. The difference in pH

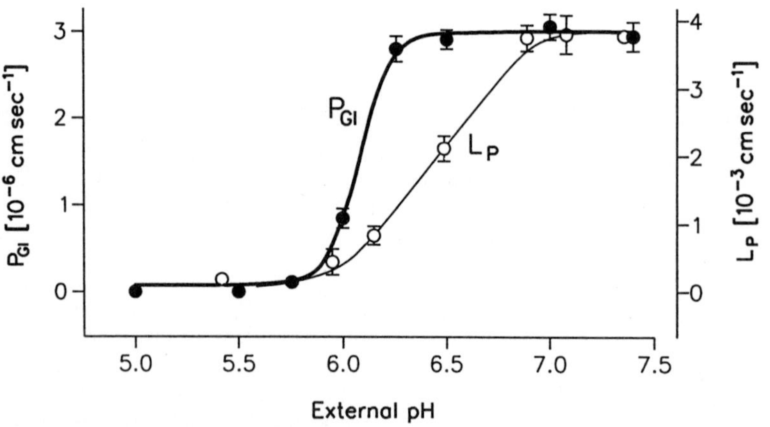

FIG. 6 pH sensitivity of water and glycerol permeability of AQP3 expressed in *Xenopus* oocytes. The water permeability L_p decreased with external pH with a pK of 6.4 and a Hill coefficient of 2.7, it was completely abolished at pH below 5.6. The glycerol permeability P_{Gl} was measured from uptake of [^{14}C]glycerol. It had a pK of 6.1 and a Hill coefficient of 6.2. It was completely abolished for pH below 5.7. The findings are discussed in the text on the basis of a model in which water and glycerol permeates via the same pathway. Redrawn from (Zeuthen and Klaerke, 1999).

sensitivities for L_p and P_{Gl} shows that AQP3 is a water and glycerol channel at neutral pH, while it is predominantly a glycerol channel in the pH range 5.8 to 6.2. It follows that the reduction of L_p by H^+ does not arise from a physical closure of the putative pore, since the aquaporin is still open to glycerol transport while closed for water transport. The pH effects were reversible.

One way to view the data from AQP3 is to interpret them by means of an Eyring energy barrier model (Glasstone *et al.*, 1941). In this model, the molecule permeates by a series of jumps, the energy barriers of which are determined by the chemical bonds between the molecule and specific sites in the pathway. For AQP3, the Arrhenius activation energy (E_a) for L_p is low, around 5 kcal mol^{-1} (Meinild, *et al.*, 1998b; Ishibashi *et al.*, 1994; Echevarría *et al.*, 1996), which suggests that the water molecule at neutral pH permeates by forming a succession of single H^+-bonds. Titration of the sites at acid pH would abolish their H^+-bonding capacity and render them effectively hydrophobic, thereby abolishing the L_p. In analogy to the L_p, P_{Gl} also had a low E_a and a marked dependence on external pH. This would suggest that glycerol also permeates by forming successive H^+-bonds.

Despite the differences in pH dependence, it is possible that the pathways for H_2O and glycerol share some titratable groups. The relatively low pK for glycerol transport could result from a competitive interaction between H^+ and glycerol at the titratable groups. Such competition has been described in intact human red blood cells (Carlsen and Wieth, 1976), where the glycerol transport has been suggested to be mediated by AQP3 (Roudier *et al.*, 1998). For an opposing view on the role of AQP3 in red blood cell, see Yang *et al.* (2001) and Section V below. In these cells, pK for P_{Gl} was about 6.0 at external glycerol concentrations of 1 mmol L^{-1} and 5.5 at external glycerol concentrations of 2 mol L^{-1}, which was taken as evidence for competition between H^+ and glycerol. In addition, the Hill coefficients were estimated to be larger than 2 (see also Stein, 1962). The values are in agreement with those of the present study where glycerol concentrations of 20 mmol L^{-1} were employed. The finding suggests that glycerol, when close to the titratable site, to a certain extent displaces water molecules, an effect that would be enhanced by the confinement of the pore. The resulting lower molar fraction of water near the site would result in a lower local H^+ concentration and consequently in a decrease of the effective pK.

AQP1, 2, 4, and 5 did not exhibit any pH sensitivity (Zeuthen and Klaerke, 1999; Németh-Cahalan and Hall, 2000). AQP1 has a small but significant permeability to glycerol (Meinild *et al.*, 1998b; Abrami *et al.*, 1995, 1996) and the σs for the smaller polyols, ethylene glycol (1,2-ethandiol) and 1,2-pentanediol, were similar to that of glycerol (Zeuthen and Klaerke, 1999). This shows that although these polyols may interact with water in AQP1, some structural incompatibility, not found in AQP3, prevents them from permeating at any higher rate.

AQP6 is functionally distinct from other known aquaporins (Yasui *et al.*, 1999). Expressed in *Xenopus* oocytes, AQP6 exhibits a relatively low water permeability L_p. Application of Hg^{2+} *increases* L_p up to 10-fold and lowers the activation energy to values expected from osmotic water transport. This is surprising, since Hg^{2+} is a well-known inhibitor of most other aquaporins (AQP4 is an exception). Hg^{2+} also stimulated Cl^- conductance in AQP6. The L_p increased about 3-fold when extracellular pH was decreased from control values (7.5) to 4.0. The unique properties of AQP6 might be understood in conjunction with its localization to the membranes of intracellular vesicles of acid-secreting cells of renal collecting duct, where it is colocalized with H^+-ATPase.

IV. Cotransporters

In addition to their primary function, many membrane proteins mediate a passive water flux, given a transmembrane difference in water chemical potential. They act as water channels with a well-defined water permeability L_p. This dual function is found among the cotransporters such as those coupling Na^+ with sugar or amino acid transport. The aquaglyceroporins described above, i.e., AQP3, also transport two substrates, glycerol and water. There is one fundamental difference, however, between cotransporters and aquaglyceroporins. The cotransporters undergo conformational changes during transport of the nonaqueous substrate. Each conformational state can in principle be described by its own L_p and reflection coefficient (Eqs. 1 to 3). Interpretation of data will therefore require additional information about the temporal occupancy of the various states.

To complicate matters even further, water transport by cotransporters is bimodal. In addition to the passive water transport, the proteins also sustain an active mode of transport. In this mode, water is cotransported along with the nonaqueous substrates in a fixed ratio determined by the properties of the protein. (This mode of transport is dealt with in the chapter by Zeuthen and MacAulay, Cotransporters as Molecular Water Pumps, this volume.) In the present article, we restrict ourselves to the water channel properties where transport is induced by transmembrane osmotic gradients.

The hypothesis that a cotransporter could serve as a passive water channel was first suggested in connection with the Na^+-coupled glucose transporter SGLT1 (Fischbarg *et al.,* 1993) and was confirmed by its expression in *Xenopus laevis* oocytes (Zampighi *et al.,* 1995; Loike *et al.,* 1996; Loo *et al.,* 1996, 1999). The finding has been extended to other cotransporters: the Na^+/GABA transporter GAT1 (Loo *et al.,* 1999), the Na^+/dicarboxylate transporter NaDC1 (Meinild *et al.,* 1999), and the Na^+/glutamate transporter EAAT1 (MacAulay *et al.,* 2001). The L_p is characterized by a small unit water permeability. For the Na^+/glucose transporter (SGLT1), it could be estimated to be about two orders of magnitude lower than that of AQP1 (Zampighi *et al.,* 1995). Even so, the contribution of cotransporters to the water permeability of a given cell may be significant due to the large number of cotransporters present; this would apply in particular to cell membranes without aquaporins.

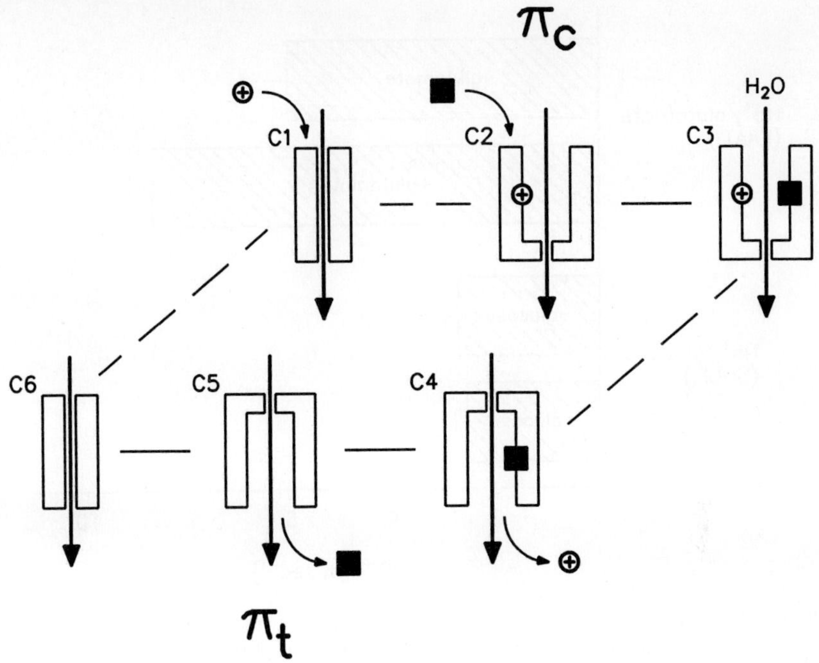

FIG. 7 Relation between passive water permeability and conformational states in a cotransport protein. This model cotransporter is considered to have six conformational states, C1 to C6, each being characterized by a passive water permeability as indicated by the vertical arrow. In this example, state C1 binds to a cationic substrate (+) and the protein changes conformation in order to allow the binding of an organic substrate (■), state 2 and 3. With both substrates bound, the protein undergoes a conformational change that exposes the substrates to the other side of the membrane, state 4 and 5. When the substrates have left the transporter, state C6, the protein can return to state 1. In the presence of a transmembrane osmotic gradient, $\pi_c - \pi_t$, there will be an osmotic flow of water given by the weighed average of L_ps of the individual conformational states: $L_p = p_1 L_{p,1} + p_2 L_{p,2} + \cdots\cdots + p_6 L_{p,6}$. Here p indicate the relative occupancy of the given state under the given conditions, see text.

A. Water Permeability and Conformational Changes

During cotransport, the protein undergoes a cycle of conformational changes (Fig. 7). The passive water permeability of the cotransporter will be a sum of the water permeability of each of these conformational states multiplied by its fractional occupacy. If the cotransporter has six states (C1 to C6 in Fig. 7) and state no. i is occupied for p_i% of the cycle and has a water permeability of $L_{p,i}$, then the total water permeability would be the weighed average:

$$L_p = p_1 L_{p,1} + p_2 L_{p,2} + \cdots\cdots + p_6 L_{p,6}$$

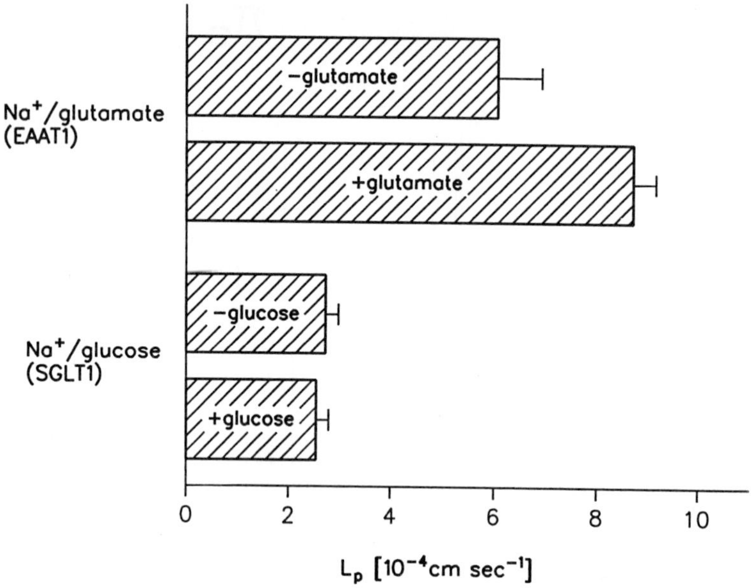

FIG. 8 Water permeability of cotransporters in the presence and absence of the organic substrates. The human Na^+/glutamate cotransporter EAAT1 (MacAulay *et al.*, 2001) and the rabbit Na^+/glucose cotransporter were expressed in *Xenopus* oocytes. The L_p was measured by means of osmotic gradients implemented by mannitol in the absence (−) and presence (+) of the organic substrate. The water permeability of the native oocyte membrane has been subtracted. It is seen that EAAT1 has a higher L_p, and that the L_p of the SGLT1 is unaffected by the presence of the substrate (Loo *et al.*, 1999). The two transporters are electrogenic and the finding applies to both the voltage clamped and unclamped situation.

Since the states may have different water permeabilities, the total L_p would be expected to be a complicated function of the external parameters, such as clamp voltage and substrate concentrations. This is exemplified by the effect of substrate on the L_p of the Na^+/glutamate cotransporter which *increased* its total L_p in the presence of glutamate (MacAulay *et al.*, 2001) (Fig. 8). In contrast, the L_p of the Na^+/glucose transporter was unaffected by the presence of sugar.

The conformational states and their fractional occupancy have been studied in detail for the SGLT1 (see Fig. 7) (Parent *et al.*, 1992a,b; Loo *et al.*, 1999). Under clamped conditions without Na^+, 60% of the cotransporters are thought to be in state C1 and 40% in state C6; under clamped conditions in the presence of Na^+, 80% are in C2, 10% in C1, and 10% in C3; in the presence of Na^+ and sugar, 80% are in state C5 and 20% in C6. Thus C1, C2, C5, and C6 may each contribute to the L_p. In the presence of the inhibitor phlorizin the L_p was abolished (Loo *et al.*, 1999). With phlorizin, SGLT1 is locked in conformational state C3 (Fig. 7). One interpretation of the transport data is that this state has a low passive water permeability.

The activation energy E_a of the L_p of cotransporters was found to be around 5 kcal mol^{-1}, as would be expected from transport via an aqueous pathway (Loo *et al.*, 1996, 1999; Meinild *et al.*, 1998a; MacAulay *et al.*, 2001).

B. Reflection Coefficients of Cotransporters

In order to define the reflection coefficient of a given solute (σ_s, Eq. 1), the shrinkage produced by equimolar addition of the test solute is determined. The reflection coefficient is then obtained as the ratio between this shrinkage and the one obtained with an inert large osmolyte, say, mannitol. Such experiments have been performed for the Na^+/glutamate cotransporter (EAAT1, MacAulay *et al.* unpublished) and the Na^+/glucose cotransporter (SGLT1, Zeuthen unpublished). Five osmolytes were employed: mannitol, urea, glycerol, acetamide, and formamide. The cotransporters were expressed in *Xenopus* oocytes under clamped conditions but with no organic substrates present (glutamate or sugar). The ratio between the shrinkage (or water permeability) measured with a given test osmolyte ($L_{p,x}$) and that obtained with mannitol ($L_{p,m}$) is shown in Fig. 9. Surprisingly, the shrinkages obtained with urea and glycerol for the EAAT1 were larger than those obtained with mannitol. A straightforward use of Eq. (1) would imply that σ was larger than 1, which has no direct interpretation within the framework of irreversible thermodynamics set up by Eqs. (1) and (2). It would require the applications of the equations developed for composite membranes which involve a large number of partly inaccessible transport parameters (Kedem and Katchalsky, 1962, 1963). An intuitive interpretation of the data would be that urea and glycerol can enter into a vestibule of the aqueous pore in an osmotically active form. In this way, the transmembrane osmotic gradient would be applied across a shorter section of the pore, which would result in a larger L_p being measured (see Fig. 2 and related text). The smallest test solutes, acetamide and formamide, may enter the whole channel, yielding a reduced $L_{p,x}/L_{p,m}$ due to the diffusive process in the aqueous pathway. In other words, the reflection coefficients of these substances are probably small for both sections of the pore. The SGLT1 behaved differently from the EAAT1 with respect to urea and glycerol. The L_p recorded with these osmolytes was nearly the same as that obtained with mannitol; the smaller osmolytes behaved similarly for the two cotransporters. The dimensions of the pore of the SGLT1 are dependent on conformational changes in the protein. This is reflected by the temperature dependence of the L_p measured by means of acetate and formamide. At 14°C, the L_ps obtained with these smaller osmolytes were similar to those obtained with mannitol. One interpretation is that the pore narrows at low temperatures. The data obtained with formamide, for example, indicate a doubling of L_p for a 10°C change in temperature. This is equivalent to an E_a of about 11 kcal mol^{-1}.

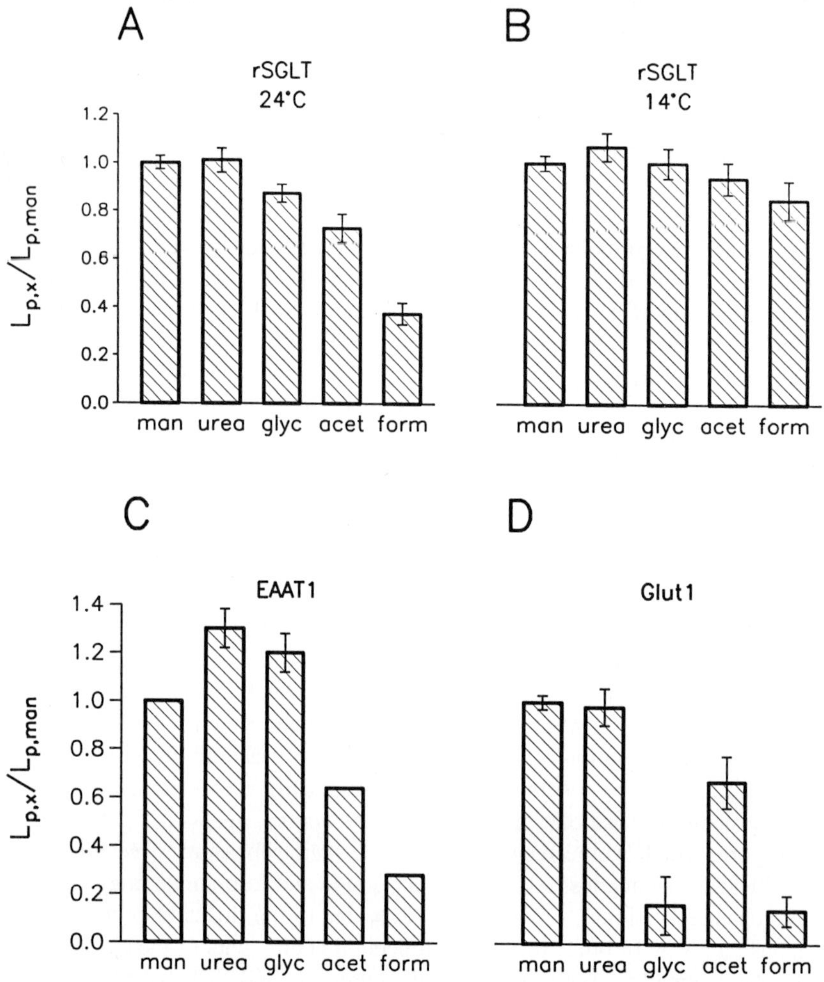

FIG. 9 Water permeabilities L_p obtained with different sized solutes in cotransporters and uniports. The L_ps obtained with urea (mw 60), glycerol (mw 92), acetate (mw 63), and formamide (mw 45) are given relative to that determined with mannitol (mw 182), that is as $L_{p,x}/L_{p,m}$. Different transporters of organic substrates were tested, (A) the rabbit Na^+/glucose transporter SGLT1 at 24° C and (B) at 14° C, Zeuthen unpublished; (C) the human Na^+/glutamate transporter EAAT1, MacAulay and Zeuthen, unpublished; and (D) the glucose monoport GLUT1, Zeuthen, unpublished. The transporters were expressed in *Xenopus* oocytes and L_p was measured by a sensitive optical system (Zeuthen *et al.*, 1997).

V. Uniports

A. Glucose Transporters (GLUT)

Many animal cells posses membrane proteins specifically designed for facilitated glucose transport. The proteins, which belong to the GLUT family, have been identified in the brain, skeletal muscle, hepatocytes, adipocytes, and fetal muscle. The transport of glucose is passive, driven by the transmembrane concentration difference for glucose and involves conformational changes of the protein. The GLUTs also support passive water transport. This was suggested from studies in macrophage cell lines J774 (Fischbarg *et al.,* 1989) and later confirmed by expression studies in *Xenopus* oocytes (Fischbarg *et al.,* 1990; Zhang *et al.,* 1991). The water permeability per protein is low compared to that of aquaporins, estimates of 6 to 60% have been presented (Fischbarg *et al.,* 1993). The Arrhenius activation energy E_a is higher than that of passive water diffusion, between 10 and 13 kcal mol^{-1} (Zeidel *et al.,* 1992a). This high E_a may not reflect solely the mode of water transport through the pore, but could also reflect temperature-dependent conformational changes in the protein leading to a narrowing of the aqueous pathway.

Measurements of the reflection coefficients (σ_s, Eq. 1) of GLUT1 for urea, glycerol, acetamide, and formamide disclose a very low σ for glycerol (Fig. 9; Zeuthen, unpublished). This would indicate a high permeability for glycerol (Eq. 3). The protein was expressed in *Xenopus* oocytes and the L_p and σ were measured by challenging the oocyte with solutions made hyperosmolar with the respective test molecules (Meinild *et al.,* 1998a). When compared to the σs for SGLT1 (Fig. 9), it is seen that acetamide and formamide penetrate through both transporters, but the high glycerol permeability is exclusive for the GLUT. It would be interesting to test if GLUT constituted a significant pathway for glycerol in erythrocytes. Recent studies using double-knockout mice lacking AQP1 and AQP3 showed no reduction in the phloritin sensitivity of the glycerol permeability (Yang *et al.,* 2001). This may contradict transport models for erythrocytes in which AQP3 constitutes the pathway for glycerol. But the glycerol could well permeate through the GLUT1 transporter of these cells (see Fig. 9).

B. Urea Transporters

Uniports for urea have been identified in particular in the kidney. One of these, the UT3, has been shown to be an efficient water channel (Yang and Verkman, 1998). The unit water permeability was 0.14 10^{-14} cm^3 sec^{-1}, which is 40 times less than the L_p for AQP1 and about 20 times less than that of AQP3, but is of the order of the passive water permeability of the Na^+/glucose transporters SGLT1. The activation energy was low, less than 4 kcal mol^{-1}, and the reflection coefficient for urea was

0.3. These are good indications that water and urea share an aqueous pathway in the protein. Urea and water permeation are also combined in AQP 9 (Tsukaguchi *et al.,* 1998), in the Na^+/glucose transporter SGLT1 (Leung *et al.,* 2000), and in the Na^+/glutamate cotransporter EAAT1 (MacAulay *et al.,* unpublished). The question of how water and urea may share an aqueous pathway is an interesting biophysical question that deserves more experiments.

VI. Conclusions

With the discovery of water channel proteins (aquaporins) and constitutive passive water permeability in proteins with other functions (cotransporters and uniports) the amount of data for passive water permeation has increased dramatically and so has the need for a physicochemical model of transport. The permeation of water through aqueous pores is determined by a wide spectrum of parameters. First, there are those that define the water molecule, the pore, and, in the case of osmosis, the solute molecule used for creating the driving force. Secondly, there are the parameters that define the interactions between the participants. Due to this complexity, a realistic physicochemical model has so far been an elusive goal. Simple models based on macroscopic concepts such as pore length, pore diameter, and the physical dimensions of the test solutes do not, in general, comply with the experimental data. Experiments on permeability have hinted at the importance of hydrogen bonding as an important mechanism of interaction between solutes and pore wall.

The advent of the atomic structures for AQP1 (Murata *et al.,* 2000) and for GlpF (Fu *et al.,* 2000) is an important turning point for the description of transport in biological pores. Questions like those presented above can now be approached by quantum mechanical modeling and computer simulation, similar to the approaches to gramicidin A (Pomès and Roux 1996, 1998). The atomic model for AQP1 will provide a useful framework for questions regarding other aquaporins, some of which have unique properties. What makes AQP3, 7, and 9 permeable to both water and glycerol (Borgnia *et al.,* 1999)? Which sites in AQP0 are responsible for its Ca^{2+} sensitivity (Németh-Cahalan and Hall, 2000), and which sites are responsible for the pH sensitivity of the AQP3 (Zeuthen and Klaerke, 1999)? Which structures in AQP6 give rise to its anion conductance (Yasui *et al.,* 1999)? How can AQP9 be permeable to a variety of neutral solutes (Tsukaguchi *et al.,* 1998)?

Models for water transport in cotransporters and uniports face the same problems as those for the aquaporins. In addition, these proteins undergo a series of conformational changes in order to fulfill their other transport functions. Each of these conformational states is characterized by a water permeability, so the total water permeability is a complicated function of the functional state of the protein. This, in turn, depends on external parameters such as substrate availability and electrochemical driving forces.

Note added in proof: In a recent paper, Curry *et al.* (2001) show that single water channels of AQP-1 do not obey the Kedem-Katchalsky equations: Methylurea is impermeable, yet its reflection coefficient is smaller than one. [Curry, M. R., Shachar-Hill, B., and Hill, A. E. (2001). "Single water channels of aquaporin-1 do not obey the Kedem-Katchalsky equations." *J. Membr. Biol.* **181,** 115–123.]

Acknowledgments

Useful discussions with Professor W. D. Stein, Dr. D. A. Klaerke, and Dr. A. E. Hill are gratefully acknowledged. Svend Christoffersen is thanked for the professional artwork.

References

Abrami, L., Berthonaud, V., Deen, P. M. T., Rousselte, G., Tacnet, F., and Riproche, P. (1996). Glycerol permeability of mutant aquaporin 1 and other AQP-MIP proteins: Inhibitions studies. *Pflügers Arch.* **431,** 408–414.

Abrami, L., Tacnet, F., and Riproche, P. (1995). Evidence for a glycerol pathway through aquaporin 1 (CHIP28) channels. *Pflügers Arch.* **430,** 447–458.

Anderson, J. L., and Malone, D. M. (1974). Mechanism of osmotic flow in porous membranes. *Biophys. J.* **14,** 957–982.

Borgnia, M., Nielsen, S., Engel, A., and Agre, P. (1999). Cellular and molecular biology of the aquaporin water channels. *Annu. Rev. Biochem.* **68,** 425–458.

Carlsen, A., and Wieth, J. O. (1976). Glycerol transport in human red cells. *Acta Physiol. Scand.* **97,** 501–513.

Chandy, G., Zampighi, G. G. A., Kreman, M., and Hall, J. E. (1997). Comparison of the water transporting properties of MIP and AQP1. *J. Membr. Biol.* **159,** 29–39.

Cheng, A., Van Hoek, A. N., Yeager, M., Verkman, A. S., and Mitra, A. K. (1997). Three-dimensional organization of a human water channel. *Nature* **387,** 627–630.

Dainty, J. (1965). Osmotic flow. *Symp. Soc. Exp. Bio.* **19,** 75–85.

Deamer, D. W., and Nichols, J. W. (1989). Proton flux mechanisms in model and biological membranes. *J. Membr. Biol.* **107,** 91–103.

Denker, B. M., Smith, B. L., Kuhajda, F. P., and Agre, P. (1988). Identification, purification, and partial characterization of a novel M_r 28,000 integral membrane protein from erythrocytes and renal tubules. *J. Biol. Chem.* **263,** 15,634–15,642.

Echevarría, M., Windhager, E. E., and Frindt, G. (1996). Selectivity of the renal collecting duct water channel Aquaporin-3. *J. Biol. Chem.* **271,** 25,079–25,082.

Echevarría, M., Windhager, E. E., Tate, S. S., and Frindt, G. (1994). Cloning and expression of AQP3, a water channel from the medullary collecting duct of rat kidney. *Proc. Natl. Acad. Sci. USA* **91,** 10,997–11,001.

Einstein, A. (1956). "Investigations on the Theory of the Brownian Movements." Dover, New York.

Finkelstein, A. (1987). "Water Movement through Lipid Bilayers, Pores, and Plasma Membranes." John Wiley & Sons, New York.

Fischbarg, J., Kuang, K., Hirsch, J., Lecuona, S., Rogozinski, L., Silverstein, S. C., and Loike, J. (1989). Evidence that the glucose transporter serves as a water channel in J774 macrophages. *Proc. Natl. Acad. Sci. USA* **86,** 8397–8401.

Fischbarg, J., Kuang, K., Li, J., Arant-Hickman, S., Vera, J. C., Silverstein, S. C., and Loike, J. D. (1993). Facilitative and sodium-dependent glucose transporters behave as water channels. *In*

"Isotonic Transport in Leaky Epithelia, Alfred Benzon Symposium 34" (H. H. Ussing, J. Fischbarg, O. Sten-Knudsen,E. H. Larsen, and N. J. Willumsen, Eds.), pp. 432–446. Munksgaard, Copenhagen.

Fischbarg, J., Kuang, K., Vera, J. C., Arant, S., Silverstein, S. C., Loike, J., and Rosen, O. M. (1990). Glucose transporters serve as water channels. *Proc. Natl. Acad. Sci. USA* **87,** 3244–3247.

Fu, D., Libson, A., Miercke, L. J. W., Weitzman, C., Nollert, P., Krucinski, J., and Stroud, R. M. (2000). Structure of a glycerol-conducting channel and the basis for its selectivity. *Science* **290,** 481–486.

Garby, L. (1957). Studies on transfer of matter across membranes with special reference to the isolated human amniotic membrane and the exchange of amniotic fluid. *Acta Physiol. Scand.* **40,** 1–84.

Glasstone, S., Laidler, K. J., and Eyring, H. (1941). "The Theory of Rate Processes." McGraw-Hill Book Company, Inc., New York, London.

Hammel, H. T. (1979). Forum on osmosis. I. Osmosis: Diminished solvent activity or enhanced solvent tension? *Am. J. Physiol.* **237,** R95–R107.

Hartley, G. S., and Crank, J. (1949). Some fundamental definitions and concepts in diffusion processes. *Trans. Faraday Soc.* **45,** 801–819.

Hill, A. (1982). Osmosis: A bimodal theory with implications for symmetry. *Proc. R. Soc. Lond.* **215,** 155–174.

Hill, A. E. (1989a). Osmosis in leaky pores: The role of pressure. *Proc. R. Soc. Lond.* **(B) 237,** 363–367.

Hill, A. E. (1989b). Osmotic flow equations for leaky porous membranes. *Proc. R. Soc. Lond.* **237,** 369–377.

Hill, A. E. (1994). Osmotic flow in membrane pores of molecular size. *J. Membr. Biol.* **137,** 197–203.

Hille, B. (1992). "Ionic Channels of Excitable Membranes." Sinauer Associates Inc., Sunderland, Massachusetts.

Ishibashi, K., Kuwahara, M., Gu, Y., Tanaka, Y., Marumo, F., and Sasaki, S. (1998). Cloning and functional expression of a new aquaporin (AQP9) abundantly expressed in the peripheral leukocytes permeable to water and urea, but not to glycerol. *Biochem. Biophys. Res. Commun.* **244,** 268–274.

Ishibashi, K., Sasaki, S., Fushimi, K., Uchida, S., Kuwahara, M., Saito, H., Furukawa, T., Nakajima, K., Yamaguchi, M., Gojobori, T., and Marumo, F. (1994). Molecular cloning and expression of a member of the aquaporin family with permeabililty to glycerol and urea in addition to water expressed at the basolateral membrane of kidney collecting duct cells. *Proc. Natl. Acad. Sci. USA* **91,** 6369–6273.

Ishibashi, K., Kuwahara, M., and Sasaki, S. (2000). Molecular biology of aquaporins. *Rev. Physiol. Biochem. Pharmacol.* **141,** 1–32.

Jung, J. S., Preston, G. M., Smith, B. L., Guggino, W. B., and Agre, P. (1994). Molecular structure of the water channel through aquaporin CHIP: The hourglass model. *J. Biol. Chem.* **269,** 14,648–14,654.

Katchalsky, A., and Curran, P. F. (1965). "Nonequilibrium Thermodynamics in Biophysics." Harvard Univ. Press, Cambridge, Massachusetts.

Kedem, O., and Katchalsky, A. (1962). Permeability of composite membranes. *Trans. Faraday Soc.* **59,** 1931–1940.

Kedem, O., and Katchalsky, A. (1963). Permeability of composite membranes. *Trans. Faraday Soc.* **59,** 1941–1953.

Ko, S. B., Uchida, S., Naruse, S., Kuwahara, M., Ishibashi, K., Marumo, F., Hayakawa, T., and Sasaki, S. (1998). Cloning and functional expression of rAQP9L a new member of the aquaporin family from rat liver. *Biochem. Mol. Biol. Int.* **47,** 309–318.

Leung, D. W., Loo, D. F., Hirayama, B. A., Zeuthen, T., and Wright, E. M. (2000). Urea transport by cotransporters. *J. Physiol.* **528(2),** 251–257.

Li, H., and Jap, B. K. (1997). Molecular design of aquaporin-1 water channel as revealed by electron crystallography. *Nat. Struct. Biol.* **4,** 263–265.

Loike, J., Hickman, S., Kuang, K., Xu, M., Cao, L., Vera, J. C., Silverstein, S. C., and Fischbarg, J. (1996). Sodium-glucose cotransporters display sodium- and phlorizin-dependent water permeability. *Am. J. Physiol.* **271,** C1774–C1779.

Loo, D. D. F., Zeuthen, T., Chandy, G., and Wright, E. M. (1996). Cotransport of water by the Na^+/glucose cotransporter. *Proc. Natl. Acad. Sci. USA* **93,** 13,367–13,370.

Loo, D. F., Hirayama, B. A., Meinild, A.-K., Chandy, G., Zeuthen, T., and Wright, E. (1999). Passive water and ion transport by cotransporters. *J. Physiol.* 195–202.

MacAulay, N., Gether, U., Klaerke, D. A., and Zeuthen, T. (2001). Water Transport by the Na^+-coupled glutamate cotransporter. *J. Physiol.* **530,** 367–378.

Macey, R. I., and Karan, D. M. (1993). Independence of water and solute pathways in human red blood cells. *J. Membr. Biol.* **134,** 241–250.

Manning, G. S. (1968). Binary diffusion and bulk flow through a potential-energy profile: A kinetic basis for the thermodynamic equations of flow through membranes. *J. Chem. Phys.* **49,** 2668–2675.

Marrink, S. J., Jähnig, F., and Berendsen, H. J. C. (1996). Proton transport across transient single-file water pores in a lipid membrane studied by molecular dynamics simulation. *Biophys. J.* **71,** 632–647.

Maurel, C., Reizer, J., Schroeder, J. I., and Chrispeels, M. J. (1993). The vacuolar membrane protein TIP creates water specific channels in Xenopus oocytes. *EMBO J.* **12,** 2241–2247.

Mauro, A. (1957). Nature of solvent transfer in osmosis. *Science* **126,** 252–253.

Mauro, A. (1979). Forum on osmosis. III. Comments on Hammel and Scholander's solvent tension theory and its application to the phenomenon of osmotic flow. *Am. J. Physiol.* **237,** R110–R113.

Meinild, A.-K., Klaerke, D. A., Loo, D. D. F., Wright, E. M., and Zeuthen, T. (1998a). The human Na^+/Glucose cotransporter is a molecular water pump. *J. Physiol.* **508(1),** 15–21.

Meinild, A.-K., Klaerke, D. A., and Zeuthen, T. (1998b). Bidirectional water fluxes and specificity for small hydrophilic molecules in aquaporins 0 to 5. *J. Biol. Chem.* **273,** 32,446–32,451.

Meinild, A.-K., Loo, D. F. F., Pajor, A., Zeuthen, T., and Wright, E. M. (2000). Water transport by the renal Na^+/dicarboxylate cotransporter. *Am. J. Physiol.* **278,** F777–F783.

Murata, K., Mitsouka, K., Hirai, T., Walz, T., Agre, P., Heymann, J. B., Engel, A., and Fujiyoshi, Y. (2000). Structural determinants of water permeation through aquaporin-1. *Nature* **407,** 599–605.

Nagle, J. F. (2000). Theory of passive proton conductance in lipid bilayers. *J. Bioenerg. Biomembr.* **19,** 413–426.

Németh-Cahalan, K. L., and Hall, J. E. (2000). pH and calcium regulate the water permeability of aquaporin 0. *J. Biol. Biochem.* **275,** 6777–6782.

Parent, L., Supplisson, S., Loo, D. D. F., and Wright, E. M. (1992a). Electrogenic properties of the cloned Na^+/glucose cotransporter: I. Voltage-clamp studies. *J. Membr. Biol.* **125,** 49–62.

Parent, L., Supplisson, S., Loo, D. D. F., and Wright, E. M. (1992b). Electrogenic properties of the cloned Na^+/glucose cotransporter: II. A transport model under nonrapid equilibrium conditions. *J. Membr. Biol.* **125,** 63–79.

Pomès, R., and Roux, B. (1996). Structure and dynamics of a proton wire: A theoretical study of H? Translocation along the single-file water chain in the gramicidin A channel. *Biophys. J.* **71,** 19–39.

Pomès, R., and Roux, B. (1998). Free energy profiles for H^+ conduction along hydrogen-bonded chains of water molecules. *Biophys. J.* **75,** 33–40.

Preston, G. M., and Agre, P. (1991). Isolation of the cDNA for erythrocyte integral membrane protein of 28 kilodaltons: Member of an ancient channel family. *Proc. Natl. Acad. Sci. USA* **88,** 11,110–11,114.

Preston, G. M., Carroll, T. P., Guggino, W. B., and Agre, P. (1992). Appearance of water channels in Xenopus oocytes expressing red cell CHIP28 protein. *Science* **256,** 385–389.

Roudier, N., Verbavatz, J.-M., Maurel, C., Riproche, P., and Tacnet, F. (1998). Evidence for the presence of aquaporin-3 in human red blood cells. *J. Biol. Chem.* **273,** 8407–8412.

Russ, W. P., and Engelman, M. (2000). The GxxxG motif: A framework for transmembrane helix–helix association. *J. Mol. Biol.* **296,** 911–919.

Smith, B. L., and Agre, P. (1991). Erythrocyte Mr 28,000 transmembrane protein exists as a multisubunit oligomer similar to channel proteins. *J. Biol. Chem.* **266,** 6407–6415.

Soodak, H., and Iberall, A. (1979). Forum on osmosis. IV. More on osmosis and diffusion. *Am. J. Physiol.* **237,** R114–R122.

Stein, W. D. (1962). Spontaneous and enzyme-induced dimer formation and its role in membrane permeability. II. The mechanism of movement of glycerol across the human erythrocyte membrane. *Biochim. Biophys. Acta* **59,** 47–65.

Sten-Knudsen, O. (1978). Passive transport processes. *In* "Membrane Transport in Biology. Concepts and Models" (G. Giebish, D. C. Tosteson, and H. H. Ussing, Eds.), Vol I, pp. 1–113. Springer-Verlag, Berlin.

Tomicki, B. (1985). The role of pressure in osmotic flow. *J. Colloid Interface Sci.* **108,** 484–494.

Tsukaguchi, H., Shayakul, C., Berfer, U. V., Mackenzie, B., Devidas, S., Guggino, W. B., VanHoek, A. N., and Hediger, M. A. (1998). Molecular characterization of a broad selectivity neutral solute channel. *J. Biol. Biochem.* **273,** 24,737–24,743.

Tsukaguchi, H., Weremowicz, S., Morton, C. C., and Hediger, M. A. (1999). Functional and molecular characterization of the human neutral solute channel aquaporin-9. *Am. J. Physiol.* **277,** F685–F696.

Van Hoek, A. N., *et al.* (1991). Functional unit of 30 kDa for proximal tubule water channels as revealed by radiation inactivation. *J. Biol. Chem.* **266,** 16,633–16,635.

Vegard, L. (1908). On the free pressure in osmosis. *Proc. Camb. Phil. Soc. Math. Phys. Sci.* **15,** 13–23.

Verbavatz, J.-M., Brown, D., Sabolic, I., Valenti, G., Ausiello, D. A., Van Hoek, A. N., and Verkman, A. S. (1993). Tetrameric assembly of CHIP28 water channels in liposomes and cell membranes: A freeze-fracture study. *J. Cell Biol.* **123,** 605–618.

Walz, T., Hirai, T., Murata, K., Heyman, J. B., Mitsuoka, K., Fujiyoshi, Y., Smith, B. L., Agre, P., and Engel, A. (1997). The three-dimensional structure of aquaporin-1. *Nature* **387,** 624–626.

Yang, B., and Verkman, A. S. (1997). Water and glycerol permeabilities of aquaporin 1–5 and MIP determined quantitatively by expression of epitope-tagged constructs in *Xenopus* oocytes. *J. Biol. Chem.* **272,** 16,140–14,146.

Yang, B., and Verkman, A. S. (1998). Urea transporter UT3 functions as an efficient water channel. *J. Biol. Chem.* **273,** 9369–9372.

Yang, B., Ma, T., and Verkman, A. S. (2001). Erythrocyte water permeability and renal function in double knockout mice lacking aquaporin-1 and aquaporin-3. *J. Biol. Chem.* **276,** 624–628.

Yasui, M., Hazama, A., Kwon, T.-H., Nielsen, S., Guggino, W. B., and Agre, P. (1999). Rapid gating and anion permeability of an intracellular aquaporin. *Nature* **402,** 184–187.

Zampighi, G. A., Kreman, M., Boorer, K. J., Loo, D. D. F., Bezanilla, F., Chandy, G., Hall, J. E., and Wright, E. M. (1995). A method for determining the unitary functional capacity of cloned channels and transporters expressed in *Xenopus laevis* oocytes. *J. Membr. Biol.* **148,** 65–78.

Zeidel, M. L., Albalak, A., Grossman, E., and Carruthers, A. (1992a). Role of glucose carrier in human erythrocyte water permeability. *Biochemistry* **31,** 589–596.

Zeidel, M. L., Ambudkar, S. V., Smith, B. L., and Agre, P. (1992b). Reconstitution of functional water channels in liposomes containing purified red cell CHIP28 protein. *Biochemistry* **31,** 7436–7440.

Zeuthen, T. (1995). Molecular mechanisms for passive and active transport of water. *Int. Rev. Cyt.* **160,** 99–161.

Zeuthen, T. (2001). How water pass through aquaporins. *Trends Biochem. Sci.* **26,** 77–79.

Zeuthen, T., and Klaerke, D. A. (1999). Transport of water and glycerol in aquaporin 3 is gated by H^+. *J. Biol. Chem.* **274,** 21,631–21,636.

Zeuthen, T., Meinild, A.-K., Klaerke, D. A., Loo, D. D. F., Wright, E. M., Belhage, B., and Litman, T. (1997). Water transport by the Na^+/glucose cotransporter under isotonic conditions. *Biol. Cell* **89,** 307–312.

Zhang, R., Alper, S. L., Thorens, B., and Verkman, A. S. (1991). Evidence from oocyte expression that the erythrocyte water channel is distinct from band 3 and the glucose transporter. *J. Clin. Invest.* **88,** 1553–1558.

Cell Volume Homeostasis: Ionic and Nonionic Mechanisms

The Sodium Pump in the Emergence of Animal Cells

Wilfred D. Stein
Department of Biological Chemistry, Silberman Institute of Life Sciences, Hebrew University, Jerusalem 91904, Israel

Plant cells and bacterial cells are surrounded by a massive polysaccharide wall, which constrains their high internal osmotic pressure (tens of atmospheres). Animal cells, in contrast, are in osmotic equilibrium with their environment, have no restraining surround, and can take on a variety of shapes and can change these from moment to moment. This osmotic balance is achieved, in the first place, by the action of the energy-consuming sodium pump, one of the P-type ATPase transport protein family, members of which are found also in bacteria. The pump's action brings about a transmembranal electrochemical gradient of sodium ions, harnessed in a range of transport systems which couple the dissipation of this gradient to establishing a gradient of the coupled substrate. These transport systems include many which are responsible for short-term regulation of the cell's volume in response to acute changes of their osmotic balance. Thus, the primary role of the sodium pump as a regulator of cell volume has been built upon to provide the basis for an enormous variety of physiological functions.

KEY WORDS: Volume regulation, Sodium pump, Comparative physiology, Evolution, Water balance.

I. The Osmotic Problem of a Cell in a Marine Environment

Plants are sessile; animals move. Plants trap the sun's energy to run their life processes; animals consume these plants and hence their stored energy, either directly or at some distance up the food chain. Plant cells are surrounded by a massive cellulose wall; animal cells are not. Animal cell membranes possess the sodium/potassium pump; plant cells membranes do not. Is there some connection between these facts? Clearly, animals move from plant to plant (or from preyed object to preyed object) to harvest stored energy. It seems reasonable that animals are able to move because their flexible cells lack the burden of the cellulose walls that constrain the bacterial or plant cells. What then is the connection between the presence of the sodium pump and the absence of the cellulose wall?

August Krogh first pointed out this connection in a paper in the *Proceedings of the Royal Society* (Krogh, 1946), an intuition later extended by Montague Maizels (1954), Tom Wilson (1954), and Alexander Leaf (1956). Krogh saw that *all* cells in an aqueous environment are under permanent threat of flooding because they contain nonpermeable matter (nucleic acids, proteins, intermediary metabolites, energy stores), yet are immersed in a medium containing ions to which their membranes are, at least to some extent, permeable. In the absence of countervailing forces, the ions will tend to reach an equal concentration within and without the cell, leaving an osmotic burden within it. Bacterial and plant cells parry this threat of death-by-flooding by enclosing each cell in a restraining cellulose wall, able to withstand the large osmotic pressures that develop. Such cells are then condemned to a largely sessile life. Animal cells have evolved along another path. They possess a membrane-located pump which pumps one of the major ions, sodium, out of the cell. The sodium pump, in essence, immobilizes sodium externally and, as a result, the cell can be in osmotic equilibrium, and the osmotic effect of its impermeable metabolites is compensated for by the absence of intracellular sodium. How is this achieved and what adaptive mechanisms in animal physiology have derived from this?

II. The Evolution of the Sodium Pump

The sodium pump is a membrane-bound ATPase (Glynn, 1993). It splits the energy-storing adenosine triphosphate (ATP), to yield adenosine diphosphate (ADP), releasing inorganic phosphate, and, in the process, moves three sodium ions out of the cell in exchange for two potassium ions. Sodium pumps, from many animal cells, have been cloned and sequenced (Fig. 1). The molecule is composed of three chains (Therien and Blostein, 2000), the largest possessing many hydrophobic sequences presumed to be transmembranal; the smallest, a single such sequence; while the third is involved in the pump's regulation (Therien and Blostein, 2000). Much is known about the details of its action. But how ATP splitting is coupled to ion movements is not yet clear (Glynn, 1993), although progress in this field must now be rapid after the elucidation of the 3-D structure of the related calcium pump (Toyoshima *et al.,* 2000; Zhang *et al.,* 2000). What *is* known is that in the presence of sodium ions, phosphate is transferred from ATP to the pump-enzyme itself, before being liberated into the cell under the influence of potassium ions.

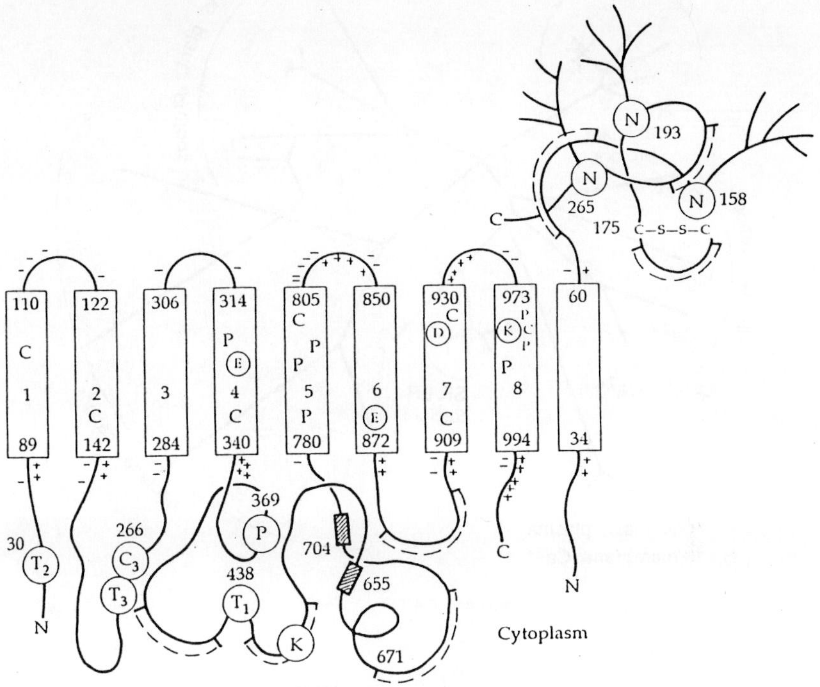

FIG. 1 Schematic diagram of the sodium pump. The rectangles represent putative transmembrane sequences. Sections labeled I, II, and III represent the N-terminal, cytoplasmic, and C-terminal thirds of the molecule that were used for building the phylogenetic trees of Figs. 2A and 2B.

There is a super-family of such membrane-bound ATPases that are phosphorylated during the course of their action. Some require sodium and potassium ions as cofactors. In others, the sodium ion is replaced by hydrogen ions; in still others by calcium ions, while the potassium ions can be replaced by protons. Many of these so-called P-type ATPases have been cloned and sequenced. With one trivial exception, no protein other than a P-type ATPase exhibits significant sequence similarity with this super-family. Fagan and Saier (1994) have analyzed 47 of these sequences and constructed the pertinent phylogenetic trees. Figures 2A and B depict two possible phylogenetic trees. Figure 2A is based on the sequences of the first section of the molecule, containing the N-terminal transmembrane sequences; Fig. 2B on the cytoplasmic loop of the pump which contains the ATP-binding (and phosphorylated) sequences.

The two trees are concordant, although not identical. In both, four major families exist. One includes four of the bacterial P-ATPases, pumping potassium, cadmium, copper ions, or an as yet unknown cation, respectively. More or less equally closely rooted to this family are two large families. One includes the calcium-stimulated ATPases of yeast and other protozoa and of the endoplasmic

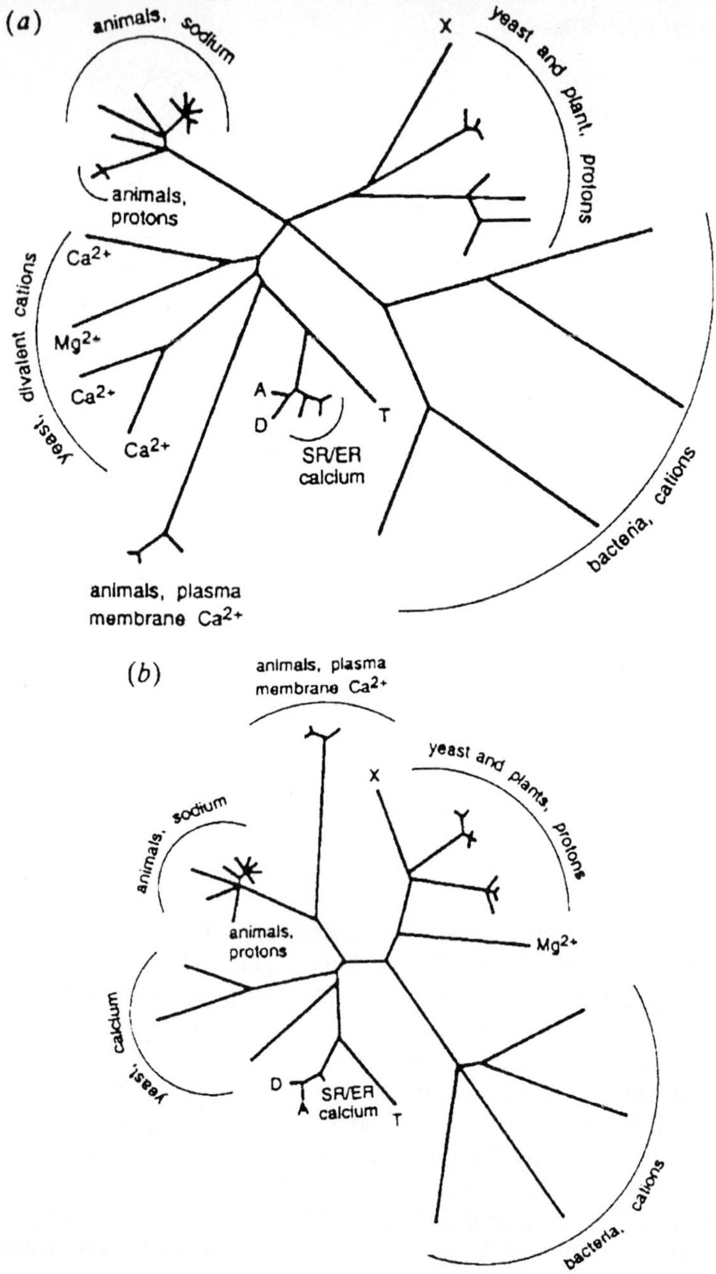

FIG. 2 Unrooted phylogenetic trees for the P-type ATPases. (A) Based on sequences in the N-terminal third of the molecules, as labeled I of Fig. 1. (B) Based on the cytoplasmic loop, labeled II in Fig. 1. Mg^{2+} represents a Mg-activated ATPase from *Salmonella typhimurium*. A, D and T are the calcium pumps from *Artemia salina, Drosophila melanogaster* and *Trypanosoma brucei* respectively while X is a cation pump of unknown specificity from *Leishmania donovani*. (Redrawn from Fagan and Saier, 1994.)

reticulum, and the sarcoplasmic reticulum of many animals. (In the tree of Fig. 2A, the calcium ATPases from the plasma membrane of the animal cells cluster together with the remaining calcium ATPases, but cluster separately in the tree of Fig. 2B.) The second large family contains the proton pumps of plants, yeast, and fungi. A fifth bacterial P-ATPase, a Mg^{2+}-dependent ATPase from *Salmonella typhimurium,* clusters with the eukaryotic calcium ATPases in Fig. 2A but with their proton ATPases in Fig. 2B. This suggests that the split between the remaining four bacterial ATPases and all other members of the P-ATase family occurred *before* the appearance of the eukaryotes. Well-separated from the proton and the calcium pumps is the fourth family, containing the sodium pumps (including those from hydra, drosophila, frog, mouse, and man) and also the gastric and colon proton pumps. The so-called unrooted phylogenetic trees depicted in Fig. 2 cannot be used unequivocally to uncover evolutionary relationships, but they allow one at least to hazard that the bacterial cation P-type ATPase pumps were primitive. After a duplication event, of which the *S. typhimurium* Mg^{2+} ATPase is a survivor, and from this branch, evolved the calcium pumps of the intracellular vesicles and the plant plasma membrane proton pumps, early and separately in eukaryotic evolution. From such vesicular calcium pumps evolved the plasma membrane calcium pumps and (perhaps from them) the sodium pumps, and from these, the gastric and colon proton pumps.

III. Donnan and Double Donnan Distributions

The data in Fig. 2 confirm that the sodium pump may have evolved when the animals split off as a separate eukaryotic kingdom. I want now to show how the possession of a sodium pump enables the animal cell to control its volume. First, notice that many of those intracellular components that constitute the osmotic load are negatively charged. (Consider, for example, the nucleic acids and the intermediates of the tricarboxylic cycle and of glycolysis.) Now, there can be no substantial excess of charge within a closed phase such as a cell. The impermeability of the fixed intracellular anions thus brings about a Donnan effect, so that mobile anions are excluded from the cell and mobile cations are concentrated within it. Thus, a cell immersed in sea water, with no sodium pump present, would accumulate intracellular sodium ions far above that present extracellularly, to an extent determined by the ratio of extracellular anion to intracellular fixed anion.

Figure 3A depicts this situation at the steady state with no excess of charge within the cell. The cell develops an electric potential across the membrane, negative inside, due to the immobile intracellular anions. The distribution of mobile ions, given by this Donnan potential is $Na_{in}/Na_{out} = K_{in}/K_{out} = Cl_{out}/Cl_{in} = \exp(\Delta\psi F/RT)$, where $\Delta\psi$ is the potential across the membrane, F the Faraday, R the gas constant and T, absolute temperature. The cell is in charge balance but not

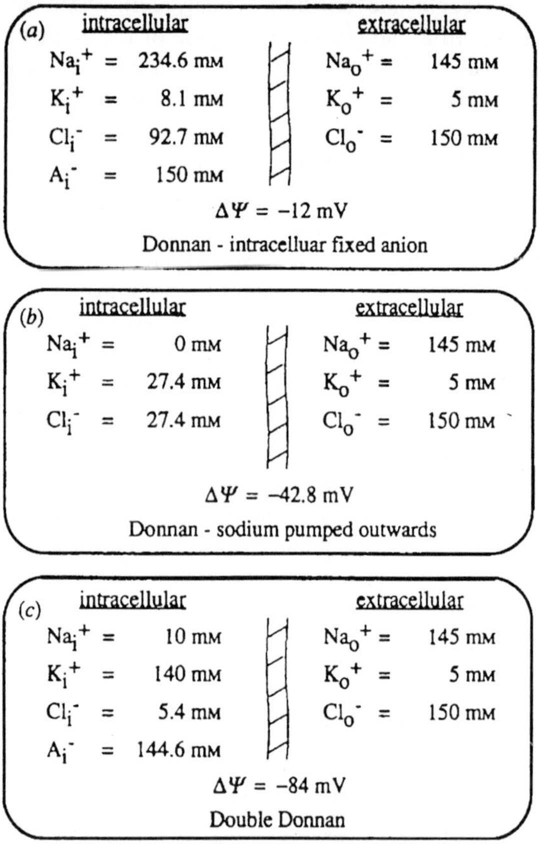

FIG. 3 Ionic distributions across cell membranes at charge equality. (A) With a fixed anion present intracellularly. (B) With the sodium pump maintaining zero sodium within the cell. (C) With both a fixed intracellular anion and an active sodium pump present, the Double Donnan distribution.

in osmotic equilibrium, due to the excess of total ions within the cell. The cell will swell indefinitely. Figure 3B depicts the opposite case. There is now no fixed anion, but a pump ensures that sodium ions are kept out of the cell, bringing about an excess of fixed cations outside the cell, and hence also a Donnan potential. Again, at steady state, the mobile ions distribute themselves in accordance with the Donnan potential. Again there is charge balance, but now an excess of osmotic material is kept outside of the cell. The cell will thus shrink indefinitely.

Finally, Fig. 3C depicts what might be found in a typical animal cell. Here, a fixed anion imposes an osmotic stress inward; the sodium pump balances this with an osmotic stress outward. This "Double Donnan" situation allows the cell to be in osmotic equilibrium as well as in charge equality, at the expense of energy

TABLE I

Measured and Predicted Ion Distributions in Some Animal Cells

	Na (mM)		Ψ (mV)	K (mM)		Ψ (mV)	Cl (mM)		Ψ (mV)	Ψ (mV)	
Cell type	Out	In	Equilbr'm	Out	In	Equilbr'm	Out	In	Equilbr'm	Measured	Ref.
Hydra body	46	17	+25	0.2	51	−140	6.5	12	+15	−40	(a)
Anodonta mantle	21	14	+10	0.7	28	−93	15	10	−10	−31	(b)
Sepia axon	486	77	+46	10	267	−83	560	41	−66	−60	(c)
Rana axon	120	10	+63	2.5	130	−100	121	3.5	−89	−93	(d)
Guinea pig glial cell	118	25	+41	3	66	−82	121	3	−98	−85	(i)
Mammal skeletal muscle	145	12	+67	4	155	−98	123	4.2	−90	−90	(e)
Ehrlich ascites	160	26	+49	4	134	−93	157	51	−30	−15	(d)
L-cells	137	9	+74	5.7	167	−90	147	70	−20	−15	(d)
Amphiuma Rbc	100	16	+49	3.5	155	−101	120	80	−11	−19	(d)
Human rbc	164	13	+69	5	136	−85	154	83	−17	−9	(d)
Lymphocyte	140	37	+36	5	165	−98	145	80	−16	−55	(f)
Monocyte	140	21	+51	5	122	−85	145	104	−9	−35	(g)
Neutrophil	140	25	+49	5	120	−85	145	103	−10	−59	(h)

References: a (Zeuthen, 1992); b (Coimbra *et al.,* 1988); c (Sjodin, 1984); d (Altman and Katz, 1985); e (Wilson, 1979); f (Garcia-Soto and Grinstein, 1990); g(Ince *et al.,* 1987); h (Simchowitz *et al.,* 1982); i (Ballanyi *et al.,* 1987).

consumed by the sodium pump's ATPase. Mobile ions distribute themselves according to this Double Donnan, the size of the Donnan potential being given by the combined effect of the fixed anion and the sodium pump. As an example, with internal anion fixed at 144.6 mM, charge and osmotic equality are achieved if internal sodium ion is set at 10 mM, when the Donnan potential becomes 84 mV, the Donnan ratio 28, internal chloride ion 5.4 mM, and internal potassium ion, 140 mM.

These computed values should be compared with those found for different animal cells, listed in Tablc I. Also listed are the measured resting membrane potentials for each cell and, for each ion, its equilibrium potential; the ions′ distribution to be determined only by this membrane potential. Sodium ion concentrations are everywhere low, far from those expected for a passive distribution—not surprisingly, since all these cells contain the energy-consuming sodium pump. For the excitable cells, the measured intracellular potassium and chloride ion concentrations are not far from the computed values and the Double Donnan model gives an excellent accounting of the ion distributions. We should note that nerve cells are, of course, highly specialized to maintain (and then manipulate) their membrane potential. In contrast, for some cell types commonly cultivated in the laboratory, and also for erythrocytes, the potassium ion concentrations are very much higher (the calculated equilibrium potential far more negative) than predicted from a passive distribution, based only on balancing the fixed anion. For leukocytes (the lymphocytes, monocytes, and neutrophils of Table I), it is the chloride ions whose internal concentrations are higher than expected from passive diffusion. Can we account for these discrepancies?

IV. The Post–Jolly Equation Relating Cell Volume, Cell Content, and the Pump–Leak Ratio

The Post–Jolly equation links the transport properties of cells with their equilibrium volume (Post and Jolly, 1957). Consider first a simple case with only two solutes present, neither being charged (Fig. 4, left). The cell, suspended in a medium containing only a permeant molecule at concentration $[P]_e$, contains an impermeant molecule at concentration $[A]_i$ and the permeant molecule at concentration $[P]_i$. At osmotic equilibrium:

$$[A]_i + [P]_i = [P]_e \tag{1}$$

A pump expels P across the cell membrane with rate constant k_p, while P leaks in and out with leak permeability constant, k_l. At equilibrium, efflux of P(pump + leak), equals influx (leak only), so that

$$k_l[P]_e = k_l[P]_i + k_p[P]_i \tag{2}$$

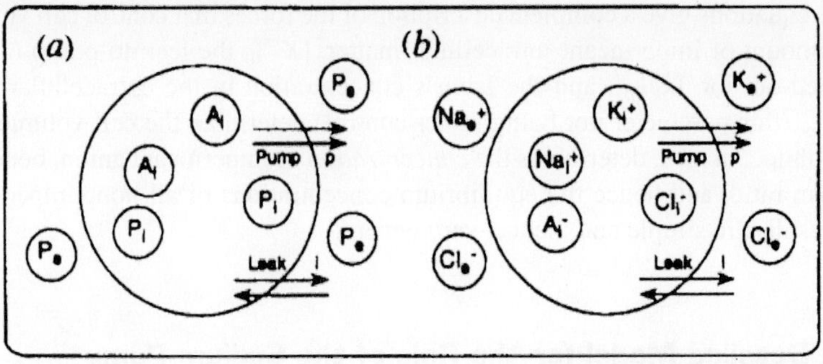

FIG. 4 Pump-leak relations in schematic cells. (A) The scheme for uncharged species. P is permeable, A nonpermeable. (B) The scheme for charged species. A is an impermeable anion. Subscripts i and e denote intracellular and extracellular components, respectively. p is the pump's rate constant, l the leak rate constant.

Solving for $[A]_i$:

$$[A]_i = [P]_e/(1 + k_l/k_p) \tag{3}$$

Thus, $[A]_i$ depends on $[P]_e$ and upon the leak-to-pump ratio, k_l/k_p. Now, $[A]_i$ is simply the amount of internal impermeant material, X_i, divided by the equilibrium cell volume, V. Thus, $[A]_i = X_i/V$. Substituting in Eq. (3), and rearranging, gives the Post–Jolly equation

$$V = (X_i/[P]_e)(1 + k_l/k_p) \tag{4}$$

Thus, the volume at equilibrium of an osmotically responsive cell is determined by the amount of impermeant material (X_i), the concentration of extracellular material ($[P]_e$, and the leak-to-pump ratio (k_l/k_p). Of these values, $[P]_e$ is a given datum of the natural world and is outside the control of the cell. X_i is fixed by the biological requirements of the cell, that is, how much impermeant matter (enzymes, metabolites, etc.) it contains. The leak/pump ratio must then be selected for, to determine an appropriate cellular volume.

In a more realistic model (Fig. 4, right), the cell contains two cations (Na^+ and K^+, only one of which is pumped out of the cell, an impermeant monovalent anion ($[A]$), and a permeant anion (Cl^-). Assuming osmotic equilibrium, a Donnan relation for the permeant species, and chemical equilibrium, as Na^+ is pumped out and leaks back, solution of the relevant equations gives

$$V = (X_i/[Na^+]_e)(1 + k_l/k_p) \tag{5}$$

equivalent to Eq.(4), with the extracellular concentration of pumped ion replacing that of the pumped molecule. The Donnan condition gives

$$[Cl^-]_i/[Cl^-]_e = [K^+]_e/[K^+] = r \tag{6}$$

These equations give a complete description of the forces that control cell volume. The amount of impermeant intracellular matter, $[X^-]_i$, the leak-to-pump ratio of pumped-out ion (k_l/k_p), and the latter's concentration in the extracellular fluid, $[Na^+]_e$, (this parameter not being under control) determine the cell volume. The cell volume, in turn, determines the *concentration* of impermeant anion, hence the Donnan ratio, and hence the equilibrium concentrations of all nonpumped ions. The results are simple and, hence, very general.

V. A Detailed Model for the Role of the Sodium Pump in Cell Volume Regulation

Lew and Bookchin, (1986) and Hernandez and his colleagues (Hernandez and Cristina, 1998; Hernandez and Chifflet, 2000) have greatly extended the Post–Jolly analysis to provide us with full models of the role of the sodium pump in volume regulation. The Lew–Bookchin model is directed toward understanding the volume regulation mechanisms in the erythrocyte. Their treatment derives all kinetic parameters from a single kinetic parameter, the value of the sodium pump flux, taking into account the experimental values of the steady-state parameters of ion concentrations and cell volume, and the known inhibitory effects of internal potassium ions and external sodium ions on the rate of sodium pumping. They provide a set of three specialized treatments based on the full model, of which the first treatment considers the change in volume of the red cell, and of its Na^+ and K^+ contents, when a resting cell is suddenly subjected to nystatin (which increases the fluxes of both Na^+ and K^+). The ionic concentrations change rapidly. The drug is then washed out, and incubation continues, with reversal of this change in ion content. The ion contents are restored within 15 hr. The volume change, which is a result of anion loss following the 3 Na^+ to 2 K^+ exchange of cations, reaches its maximum by that same 15 hr, but the volume is only fully restored over weeks. The activation of a $Na^+ : K^+ : 2Cl^-$ cotransporter speeds up both the rate of loss of volume and the rate of its restoration. Physiologically, volume changes in red cells can arise during their deformation while they flow through the microcirculation. Pathological volume changes also occur during the deoxygenation-induced sickling of homozygotic sickle-cell (SS) red cells. However, these sickling-related volume changes must have an additional component, since the loss of K^+ is in marked excess over the gain of Na^+. Lew and Bookchin (1986) used their model to simulate the effects of replacing the external ion by an impermeant ion, and to simulate the effects of a sudden increase in the diffusional permeability of potassium ions.

The Hernandez model is based on an explicit kinetic scheme for the pump, explicit formulations of the ionic "leak" pathways, and realistic values of the rate

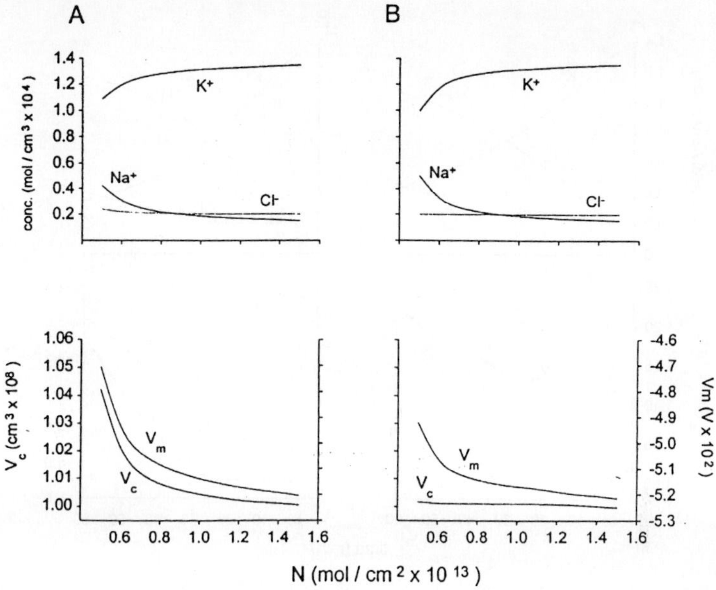

FIG. 5 Theoretical predictions of a pump-leak model for the ion concentrations in (A), a cell that does not regulate or (B), one that can regulate its volume. The upper two figures depict the concentrations of Na^+, K^+, and Cl^- as indicated, while the lower curves depict the cellular volume, V_c, and the electrical potential across the membrane, V_m. In all the figures, the abscissa is the density of the sodium pumps in units as given. (Taken with permission from Hernandez and Cristina, 1998.)

processes involved, for both the pump and the leak pathways. In addition to the assumptions used above in the Post–Jolly analysis, the Hernandez group assumes that fine-tuning of the regulation of cell volume is mediated by coupled fluxes of K^+ and Cl^-, and of Na^+ and Cl^- (where each of these paths may indeed be composed of a more complex set of ion fluxes, with recycling of some components of these fluxes to yield the required net fluxes of Na^+, K^+ and Cl^-). Their model allows them to determine the properties of the system either with these regulatory pathways available to the cell or with their not being available. Figure 5 is taken from the paper by Hernandez and Cristina (1998) and shows plots of the steady-state values of the intracellular Na^+, K^+, and Cl^- (upper figures) and of cell volume V_c, and membrane potential, V_m, (lower figures). In the left-hand panels, these predictions are drawn for the case where the regulatory pathways are not switched on; for the right-hand panels where they are. The values of the ionic and cellular parameters are plotted against N, the *density* of the sodium pumps in the cell membrane (given as the number of pumps per unit cell membrane area). The lines drawn are *not* here extrapolated to low pump densities but it is obvious that, were they so, the cell volume, V_c, would reach very large (and unsupportable) values. In these

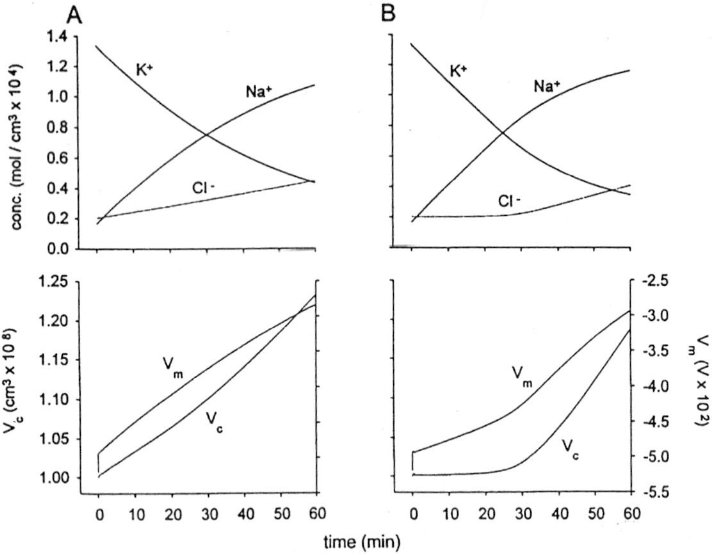

FIG. 6 Effect of inhibiting the sodium pump as predicted on a pump-leak model in (A), a cell that does not regulate or (B), one that can regulate its volume. The upper two figures depict the concentrations of Na^+, K^+, and Cl^- as indicated, while the lower curves depict the cellular volume, V_c, and the electrical potential across the membrane, V_m. In all the figures, the abscissa is the time after the pump is instantly inhibited. The density of the sodium pump is fixed at 1.25 mol/cm^2 $\times$ 10^{13} (compare Fig. 5). (Taken with permission from Hernandez and Cristina, 1998.)

steady-state predictions, it is, of course, of little significance whether or not the regulatory pathways are acting.

Look, however, at the illuminating results depicted in Fig. 6 which again shows panels depicting these four predictions but this time as a function of time, at a particular value of the membrane density of sodium pumps, with these pumps being completely inhibited at time zero. Note how rapidly the volume rises with the increase in cellular sodium ions in the absence of the regulatory pathways (left-hand panels), after the pump is inhibited. However, with the regulatory systems activated (right-hand panels), the cell volume is buffered and is maintained constant for some time, with the ionic sodium inflow being balanced by a potassium outflow. The state of balance continues until the intracellular potassium ion concentration falls so low that its gradient becomes directed inward, so that the regulatory ion and water flows reverse. (A similar phenomenon, having the same basis in the loss of intracellular potassium ions, is found in epithelia; see Zeuthen, this volume.) Note, also, the small, but real, rise in the membrane potential, V_m, at time zero after the sodium pump is switched off. This represents the pump's direct, electrogenic contribution to the membrane potential. The subsequent, large decrease in membrane potential arises from the ionic shifts and emphasizes, if this were still needed, that the resting membrane potential is determined by the transmembranal ionic concentration ratios, which are themselves totally determined by the sodium pump. In the subsequent paper from the Hernandez group (Hernandez and Chifflet, 2000), they point out that the predictions of Fig. 5 are valid only if the *rates* of the ion leaks (the cotransport systems) are high compared with the rate of the sodium pump. If, in contrast, the sodium pump has the faster rate, the cell requires an electroneutral Na^+/K^+ exchange to be active, if conventional values of the membrane potential and the cell volume are to be maintained.

But what determines actual value of the leak/pump ratio in the resting cell or in a cell subjected to acute changes in cell volume? Any mechanism for the regulation of cell volume requires that the cell measure its own volume. We do not know how this is done (Parker, 1993; Strange, 1993). It has been suggested that it is the mere concentration of intramolecular molecules that is sensed by the cell, that what is sensed is the "crowding" of the cell's internal molecules (Minton *et al.,* 1992; Parker, 1993; Garner and Burg, 1994). There is good evidence, however, for the direct involvement of swelling-induced potassium ion channels in cell volume regulation. Vanoye and Reuss (1999), using the whole-cell patch-clamp technique, applied to isolated epithelial cells, were able to show the reversible opening of potassium ion channels either when the cells were exposed to a hyposmotic medium or when the cell membranes were physically stretched by applying a pressure at the patch clamp pipette. These channels had quite the same properties in these two states, suggesting that it was the mere stretching of the membranes that induced the channel opening. There is unlikely to be a change in the macromolecular crowding when the pressure in the pipette is altered. A very direct demonstration of swelling-induced changes that are independent of the intracellular concentration comes from the work of Dubinsky *et al.* (1999) who succeeded in obtaining a membrane vesicle preparation from *Necturus* enterocytes which were quite free from any intracellular molecular material. These vesicles took up rubidium ions, the congener and marker for potassium ions, when exposed to an isosmotic medium; took up more rubidium ions when the osmolarity was lowered and far less when they were placed in a hyperosmotic medium. This uptake was blocked by glibenclamide, an inhibitor of ATP-activated potassium channels. The uptake was also blocked by cytochalasin D, an agent known to interfere with actin polymerization, suggesting strongly that the cytoskeleton of the cell is a mediator of the stretch response that regulates the cell volume.

It is the balance of volume and leak/pump ratios that determine *long-term* volume regulation. But the detailed knowledge of mechanism, that we have described above, does not tell us how evolutionary selection, working on the cell, might have achieved this balance. We know much, however, about how the leak/pump ratio is controlled in *short-term* responses, and it is to this aspect of volume regulation that we will now direct our attention.

VI. The Three Classes of Ion Distribution Ratios

Inspection of Table I reveals a major fact not accounted for by the simple Post–Jolly sodium pump models of Figs. 3 and 4: For all these cells, the potassium ion concentrations are high, often far higher than expected from the simple Double Donnan model. Indeed, most cells (the red blood cells of cats and dogs constituting some of the exceptions) contain potassium ions as their major cation. This is true for animal, plant, and bacterial cells, so the requirement for high intracellular potassium ions seems as old as the RNA world itself. This is an intriguing fact with no accepted explanation (but see Herzberg's paper, 1974, in which he argues that the protein synthesis mechanism itself requires the presence of potassium ions). Recall that the sodium pump of animal cells is not merely a pump but also a sodium/potassium exchanger, bringing potassium ions into the cell in exchange for the sodium ions that are pumped out. This produces the accumulation of intracellular potassium ions, a function that bacterial and plant cells perform by potassium/proton exchange. If a high intracellular concentration of potassium ion is, indeed, a necessity for a nucleotide-based life, animals have linked this requirement to the *sodium* pump's maintenance of cell volume. The low sodium/high potassium intracellular ion ratio appears to be a property of the pump itself. At the intracellular face, the pump binds either sodium ions or potassium ions, the former with high affinity, the latter with low. Pumping continues until sufficient potassium ions have entered to compete with the reduced intracellular sodium ion concetration. This regulation of the rate of sodium ion efflux enables the inward leak to be balanced, determining the ratio k_l/k_p in Eq. (5).

The cells listed in Table I seem to fall into three main classes concerning their transmembrane ion distribution ratios. The three classes are distinguished by the ratio of their potassium to chloride ion permeabilities. There are only three possibilities: their ratio may be roughly equal, the potassium ion may be of far lower permeability than chloride, or the reverse may be true. In the first class are the excitable cells, nerve and muscle. Direct measurements show that the membranes of this class of cells possess channels for both potassium and chloride ions, although the potassium ions cross the resting membrane of the nerve some five times more rapidly than does the chloride ion. For such cells, a pump that exchanges potassium ions for the pumped sodium ions will generate an outward flow of KCl, limited by the rate of chloride flow, but still rapid. The cell will lose its internal osmoticum and will shrink, unless some counter-measure is developed. Indeed, nerve cells possess, as a major contributor to the internal fixed anion, an unusual cell constituent, isethionic acid. This is an impermeable anion, whose role appears to be to ensure that the cell's volume can be maintained, in the face of a high negative resting membrane potential and a high permeability to both potassium and chloride ions. For this first cell class, the ionic and volume interrelationships are well described by Eq. (5), with volume being maintained in the long-term and short-term by manipulating the concentration of the internal fixed anions (Law, 1994).

The second class includes those cells (red blood cells and tumor cells) for which the chloride ion distribution ratio is still close to that determined by the membrane potential, itself computable from the simple Double Donnan model and Eq. (5). For these cells, however, the intracellular potassium ion concentration is not that given by the membrane potential. Potassium ions permeate with difficulty, and the potassium ion concentration is kept high internally by the action of the sodium pump, satisfying a requirement which, as we discussed above, seems even more basic than that of volume regulation itself. Exchange of potassium for sodium ions does not alter the osmotic relations across the cell nor the charge distribution and, hence, neither does it alter the Double Donnan potential. Thus, these cells, where the membrane's resting permeability to potassium ions is low (while that for chloride ions is higher), maintain a high intracellular potassium ion concentration economically in the face of an electrochemical potential driving potassium ions outward. As we will discuss below, the ionic relations of this class of cells demand, however, that they use a system for short-term volume regulation which is different from that used by the former class of cells.

The third class includes the mammalian white blood cells. Here, the cell membrane possesses a high permeability for potassium ions, but a low permeability for chloride ions. The high intracellular potassium ion concentration, brought about by the sodium/potassium pump, is compatible with the transmembrane potential and, indeed, largely determines this potential. The intracellular chloride ion concentration, however, is far higher than that expected from the transmembrane potential, being maintained high by a set of coupled pumps linked, in the final analysis, to the sodium ion gradient (Macknight *et al.,* 1994). Chloride ion is thus fixed intracellularly and contributes to the fixed anion of Eq. (5), actively maintaining the cell's volume, as isethionic acid does in the nerve cell. Again, this class of cells uses a different set of mechanisms for short-term volume regulation.

VII. Volume Regulatory Increase and Volume Regulatory Decrease

A. Ionic Mechanisms

Using this classification of cells, which we see to be based on the cells' ratio of potassium to chloride ion permeabilities, we can approach the regulation of cell volume. Consider again Eq. (5). Here, the cell volume is fixed in part by the ratio of leak/pump rate constants, k_l/k_p, for the pumped sodium ion. Of these rates, the regulation of K_p, the pump's rate constant, seems to play a minor role in the short term in most animal cells. Thus, the effect of inhibition of the sodium pump by pump-specific drugs such as ouabain gives variable, generally small, effects which can take a long time to appear (as is predicted by the theoretical considerations depicted in Fig. 6). The membrane's calcium pump is, in contrast, tightly regulated in the short term (Carafoli, 1991). The reverse effect, that experimentally manipulated changes in cellular volume can affect the properties of the sodium pump, has been clearly established (Whalley *et al,* 1993). Figure 7, taken from Whalley *et al.* (1993), depicts how the rate of sodium pumping across rabbit cardiac myocyte membranes, measured as a sodium current in a whole-cell patch clamp preparation, depends on the sodium ion concentration at what is the intracellular face of the cell membrane. Figure 7 (top) shows how the current depends on $[Na^+]$ in cells maintained under isosmolar (open circles) and hyposmolar (filled circles) conditions, while Fig. 7 (bottom) shows similarly a comparison between isosmolar and hyperosmolar conditions. The apparent affinity of the pump for sodium ions is increased when the cells face a hyposmolar environment and decreased when this environment is hyperosmolar. Such a change in affinity should drive more of the sodium ions out of the cell when the cell's volume is increased, with the reverse happening when it is decreased. This behavior would be in the direction required for volume homeostasis, but the quantitative significance of these affinity changes in the living cell is not clear.

Long-term regulation of the pump is achieved by altering the *number* of sodium pumps per unit surface area of the cell, i.e., by synthesis of more pumps (Gick and Ismail-Beigi, 1990), a slow process generally taking hours, or by increased degradation of existing pump molecules. The leak of sodium ions occurs through membrane channels or through membrane carriers (Stein, 1990). Excitable cells regulate their sodium ion fluxes through specialized channels (Hille, 1992), but more typical cells regulate their sodium fluxes through cotransport or symport (Geck and Heinz, 1986; Hoffmann and Ussing, 1992). Here, a membrane-bound carrier moves sodium ions across the membrane together with chloride ions in a 1:1 stoichiometry, or together with an additional potassium ion and *two* chloride ions (for a comprehensive review of these systems see Russell, 2000). In both cases, the product of the existing gradients of sodium, chloride, and potassium ions (if transported by the carrier) at each face of the membrane constitutes a force driving sodium ions into the cell, opposed to the outwardly directed pump. This condition is fulfilled for all the cells in Table I. The cotransporter for sodium and chloride ions (with or without potassium) *is* tightly regulated, being activated when the cell's volume is *reduced* from normal, i.e., with the cell in a hyperosmotic milieu. On activating the cotransporter, sodium ions enter the cell, and water follows it osmotically, correcting the volume deficit. This is *volume regulatory increase,* or RVI. At the resting cell volume, the cotransporter is deactivated and the basal leak/pump ratio for sodium again determines the cell's volume. Sodium/chloride cotransport (with or without potassium) is found in many animal cells, being the major mechanism for regulation of cell volume in the direction of its increase (Russell, 2000).

If the cell is to regulate its volume against an *increase* (perhaps brought about by a temporary cessation of sodium pumping), down-regulation of the sodium ion leak will not suffice. The leak is already at a basal value. If the internal concentrations

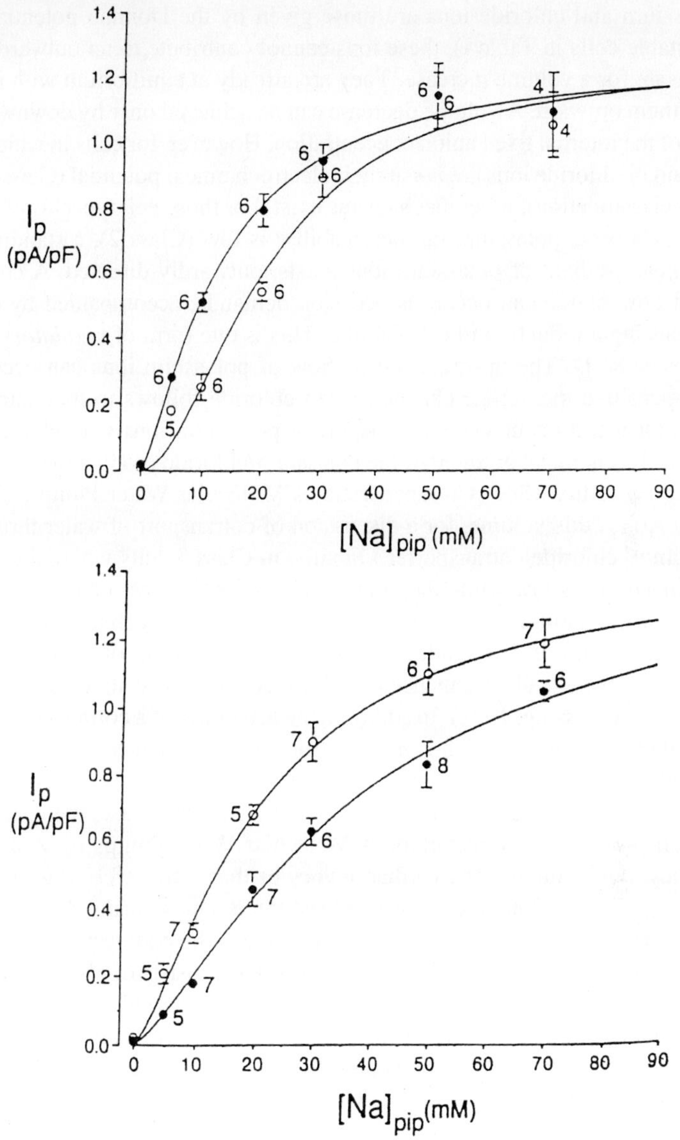

FIG. 7 Top figure: Dependence of transmembrane current, I_p, on the concentration of sodium in the patch-clamp pipette, [Na]pip, in isolated rabbit cardiac myocytes, in isosmolar (○) and hyposmolar (●) solutions. The numbers in parentheses are the numbers of cells studied at each sodium concentration. The concentrations of sodium that gave half-maximal current were 12.8 and 21.4 mM for hyposmolar and isosmolar conditions, respectively. Bottom figure: As on the top, except that now the symbols refer to isosmolar (○) and hyperosmolar (●) solutions. The concentrations of sodium that gave half-maximal current were 39.1 and 21.8 mM for hyperosmolar and isosmolar conditions, respectively. (Taken with permission from Whalley *et al.*, 1993.)

of potassium and chloride ions are those given by the Donnan potential (as in the excitable cells in Table I), these ions cannot contribute to an outward leak to compensate for a volume increase. They are already at equilibrium with no force driving them outward. A volume decrease can be achieved only by downward regulation of the internal fixed anion concentration. However, for cells in which either potassium or chloride ions are *not* at their electrochemical potential (Classes 2 and 3 of our classification), other mechanisms exist. For those cell types for which the membrane's basal potassium ion permeability is low (Class 2), a standing electrochemical gradient of potassium ions exists, outwardly directed. A controlled outward flow of K^+ can occur, as occasion demands, accompanied by osmotic water, causing a reduction in cell volume. This is one form of *regulatory volume decrease* or RVD. The up-regulated outflow of potassium ions can occur via a volume-sensitive membrane channel, when chloride follows to maintain charge balance. Or it can occur via a cotransport of potassium ions with chloride ions, when it is the chemical gradient of the potassium/chloride product that is the driving force for efflux. (See "Cotransporters as Molecular Water Pumps," Zeuthen and MacAuley, this volume, for a discussion of cotransport of water through the potassium + chloride cotransporter.) Finally, in Class 3 cells where the chloride ion permeability is low while the potassium ions are in electrochemical equilibrium across the membrane, an up-regulation of a potassium channel cannot bring about a potassium-driven conductive efflux. Up-regulation of a chloride channel can, however, lead to enhanced KCl efflux and hence to volume decrease (Hoffmann and Ussing, 1992). In addition, the activation of a cotransport of potassium and chloride ions, which is *not* electrogenic and is hence not dependent on the membrane potential, can, here again, lead to enhanced ion efflux and a volume decrease (Garcia-Soto and Grinstein, 1990). In such a case, the water moves by cotransport (see Cotransporters as Molecular Water Pumps by Zeuthen and MacAuley, this volume). The cardiac myocytes that were referred to in the discussion around Fig. 7 apparently use at least three routes during the process that allows potassium ions to leave the cell. One route is the cotransport of potassium and chloride ions just described. The second route consists of channels for these ions. The third route is the cotransport of potassium and chloride ions, but now together with sodium ions. The evidence for this is a decreased ability of these cells to regulate their volume when bumetanide is added, under conditions where this drug should be specific for the Na/K/Cl cotransporter. This is a surprising result since the net driving force for sodium ions is generally inward and this, it might appear, should lead to an *increase* of the cell volume, rather than the reduction that the volume response of the cell requires.

These studies of the short-term regulation of volume, involving RVI and RVD, enable us to identify at least two different mechanisms used to allow a sodium ion leak, the Na/Cl cotransporter and the Na/K/2Cl cotransporter. Similarly, several mechanisms exist for the potassium ion leak. This can be a K/Cl cotransporter or a K channel accompanied by a separate chloride pathway, or a chloride channel with potassium in a separate pathway. The data of Fig. 2 suggest that the sodium pump appeared only once in evolution, implying, if the above arguments are correct, that the pump is the primary mechanism of volume regulation. In contrast, volume regulatory decrease and increase mechanisms have appeared several times, and are apparently fine-tuning mechanisms, superimposed on the basic sodium pump system. A very fine review on these systems, with 1400 citations, is given by Lang *et al.,* 1998. They discuss in a comprehensive manner how RVD and RVI are signaled and list the many systems in the cell that respond to these signals to bring about the regulation of cell volume.

B. Nonionic Mechanisms

Many additional cotransport and counter-transport systems link the animal cell's transmembrane sodium (and/or potassium and chloride) ion gradients to the driving of a range of metabolites into and out of cells. These include the transporters for various amino acids, many sugars, protons, calcium ions, iodide, vitamins, and other metabolites. These transporters form numerous evolutionarily related families (Saier, 1994a,b)and are found in all phyla (Lang *et al.,* 1998). Their action underlies much of cell physiology and of general physiology. In animal cells, most such systems are linked to the gradient of that ion, sodium, whose transmembrane concentrations are determined by the need to fix and to regulate cell volume. By freeing the animal cell from the constraint of an external wall, the evolution of the sodium pump allowed the cell subsequently to evolve a wealth of coupled transport systems, opening up new modes of existence compatible with an increasingly complex organization.

Consider, for example, the interesting data presented by Biagini and his colleagues (2000) on the osmoregulation of the free-living protozoon *Hexamita inflata*. This creature seems to have no mechanism for RVI. Perhaps it seldom finds itself in such an environment, or if it does, it merely remains shrunken and unharmed until the danger is past. But it does have an excellent down-regulation of its volume (RVD), presumably since cell swelling can lead to a catastrophic bursting of the cell if it is not countered. What leaves the cell, under conditions where RVD is turned on, are the neutral amino acids. The amounts of the amino acids that leave the cell accord pretty well with their cellular content, alanine being by far the major contributor at 78% of the total. The sum of the content of the amino acids is just greater than the amount of intracellular potassium, so these organic osmolytes make a major contribution to the resting osmolarity of the cell and, in accordance with this, to the outflow of osmotic material during RVD. The route by which the amino acids leave the cell is via the well-studied transporters for these compounds (Haussinger and Lang, 1991). The technique for establishing this was to show that a nonnatural amino acid, amino isobutyric acid or (AIB, often used as a probe for one of the amino acid transport systems), enters the cells when RVD is initiated. This is presumably by exchange with amino acids that leave the cell, these latter riding out on the transport system that is switched on to bring about RVD.

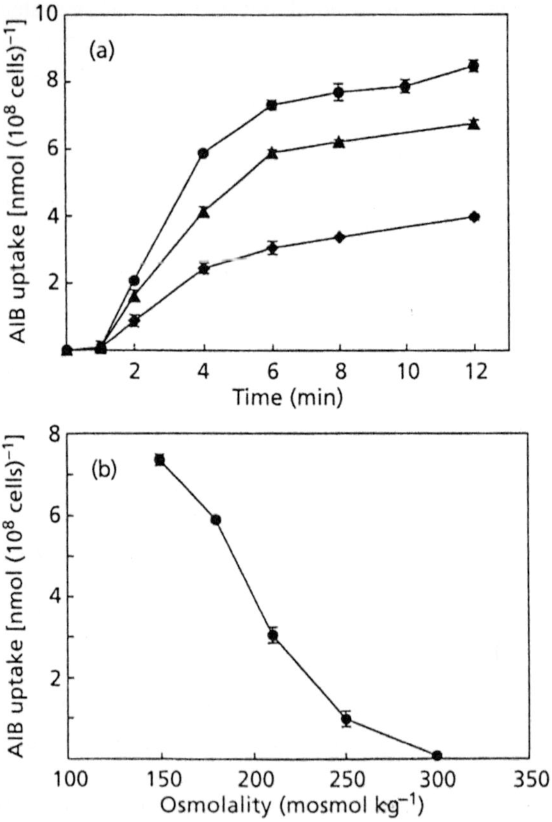

FIG. 8 Uptake of amino-isobutyric acid (AIB) into the protozoon *Hexamita inflata*. In (a) AIB uptake is depicted as a function of time after suspension of the cells in buffers of osmolalities of 210 (●), 180 (▲) or 150 (◆) mosmol/kg. (The isosmolar molality is 300 msom/kg). In (b), the uptake of AIB was measured 6 min after suspending the cells in solutions of the osmolalites recorded on the abscissa. (Taken with permission from Biagini *et al.,* 2000.)

Figure 8a shows how rapidly AIB transport is initiated when RVD is switched on, while Fig. 8b shows how the rate of this transport varies with the osmolarity of the medium in which the protozoa find themselves. It is not only the amino acids that leave the cell during RVD. Potassium ions, accompanied by chloride ions, also leave the cell by a volume-regulated mechanism that has not been fully characterized. It is not blocked by conventional channel and transporter inhibitors such as barium, DIDS, quinine, and tamoxifen. The protozoon *Leishmania major,* the pathogen of the human disease schistosomiasis, also controls its RVD response by manipulating an efflux route for amino acids (Vieira *et al.,* 1996).

Of course, it is not only in the protozoa that amino acids are used in volume regulation, Thus, a full study using human vascular endothelial cells taken from coronary artery bypass patients (Dall'Asta *et al.*, 1999) showed dramatic changes in the amino acid pools of these cells during RVI. A doubling of the total amino acid content occurred when the medium's hypertonicity was increased from 290 to 370 mosmol/kg. The concentration of the amino acid proline went up 7-fold, but the most significant increase in absolute terms occurred in the case of glutamine, which increased from 103 to 260 mosmol/kg of protein, these values being equivalent to 13.8 and 36.1 mM. Overall, the total concentration of the amino acids went up from 72 to 139 mM, sufficient to compensate for the increase in extracellular osmolarity. In the absence of external glutamine, RVI was delayed very considerably, but still occurred. The transport behavior of the system that carries the entering amino acids is compatible with the so-called A system being activated, with entry of the other amino acids taking place, perhaps, by heteroexchange. New protein synthesis is required for this RVI, with addition of the inhibitor cycloheximide sufficient to abolish it.

The research performed by Motais and his colleagues (Guizouarn and Motais, 1999; Guizouarn *et al.,* 2000; Godart *et al.,* 1999) on volume regulation in trout red blood cells is an important addition to such studies. From their work, it is very clear that, in these cells, there are two distinct routes for RVD. One route is called the isosmotic route and occurs when the cells take up water together with solute, such as when urea or an ammonium salt or some hormone is added to the external medium. Such additions drive inward solute, which must be accompanied by water (Guizouarn and Motais, 1999). This route opens up a channel for potassium ions. Motais and his colleagues monitored this ion movement as an influx of the potassium congener rubidium. Entry of K (Rb) is strictly chloride dependent and absolutely requires Cl to enable the cation inflow to occur without a build-up of membrane potential. (Fig. 9A). The other route, called hyposmotic, occurs when the external medium is simply diluted so that water flows in unaccompanied by ions, thus diluting the internal electrolytes (see Fig. 10, where these two routes are depicted, and Fig. 9B for the relevant data on this second route). This route allows inflow of rubidium (i.e., K^+) unaccompanied by Cl^-, and allows the inflow (or outflow) of taurine, sorbitol, and choline. An ingenious experimental design, in which a combination of urea and NH_4NO_3 was used, allowed the volume of the cell to be held constant while the electrolyte content was altered (Fig. 11). The second, chloride-independent, path is partly blocked by DIDS. The volume change has to take place quickly for the system to be turned on (Godart *et al.,* 1999). The taurine constitutes some 50% of the total osmotic material that is released during RVD and it moves as a zwitterion. With this path open, the K efflux that is activated during RVD, and is partly offset by Na influx in this nondiscriminatory pathway (Guizouarn *et al.,* 2000). Physiologically, the trout red cells can be swollen isotonically or hypotonically. When the trout is exposed to deep hypoxia, catecholamines are released that stimulate the erythrocyte's Na/H exchangers leading to an accumulation of electrolytes. This is isotonic volume regulation and allows the fish to survive at low pO_2. But trout can also be exposed to waters of varying salinities,

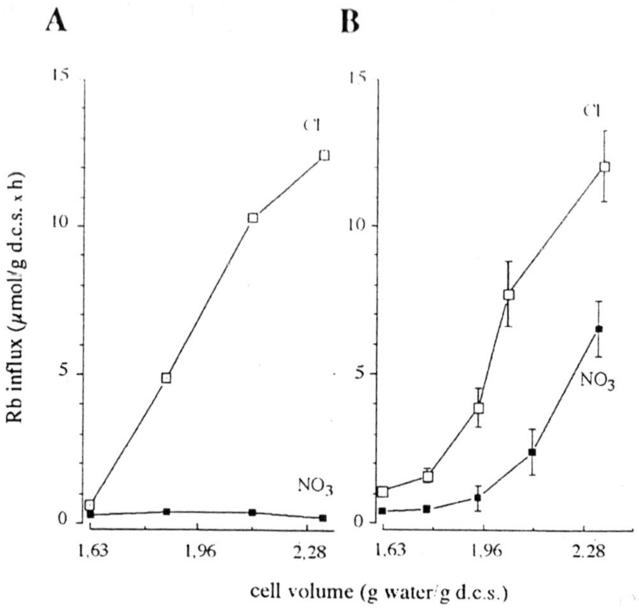

FIG. 9 Potassium fluxes (measured using the congener rubidium) across trout red cell membranes as a function of cell swelling. In (A), the cells are isosmotically swollen, by exposure of the cells to permeant ions. In (B), the cells are hyposmotically swollen, by exposure of the cells to media of reduced tonicities. The fluxes are depicted as a function of the achieved cell water content when the external medium contained either Cl or NO_3. (Taken with permission from Guizouarn and Motais, 1999.)

leading to hyposmotic stress, and the animal must be able to perform RVD when it is at low external osmolarities. Guizouran and Motais (1999) write that

> "when swelling results from an uptake of electrolytes, the best way for a cell to undergo volume regulation is to activate the pathway that selectively mediates a loss of electrolytes [the isotonic route]. Conversely, when swelling results from an entry of water, which dilutes cell electrolytes, the best way is then to activate the taurine pathway [the hypotonic route] by using organic osmolytes to recover volume and by preventing an additional decrease of electrolyte level".

The question of volume regulation plays a particularly critical role in the case of animal liver (Graf and Haussinger, 1996). Here, large changes in the intracellular osmolarity take place during metabolism. It will be remembered that the hormones insulin and glucagon play opposite metabolism-directing roles in the liver. Insulin's function is to trigger an anabolic response which includes stimulation of protein and glycogen synthesis and the uptake of amino acids, while inhibiting glycogenolysis, glycolysis, and proteolysis. In contrast, glucagon acts as a catabolic signal, stimulating glycogenolysis, glycolysis, and proteolysis, while

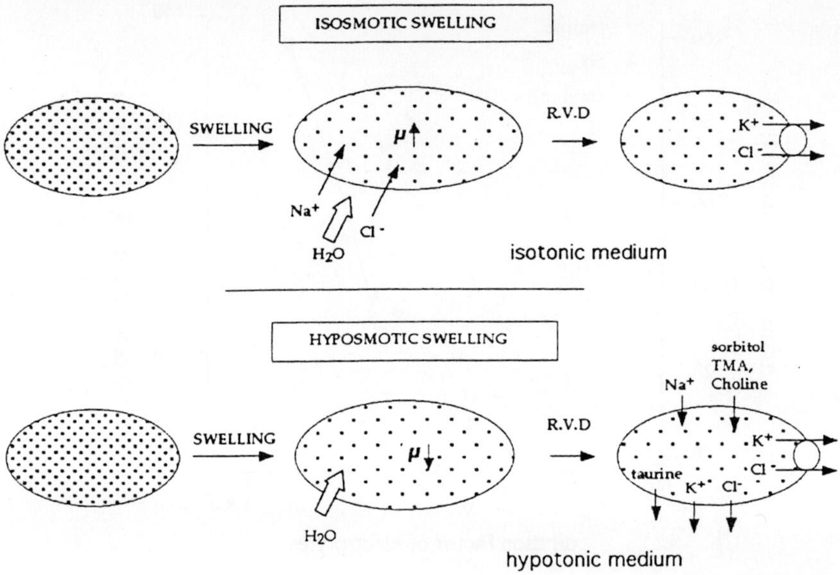

FIG. 10 Schemes for isosmotic (upper figure) and hyposmotic (lower figure) swelling induced by exposure to permeable salts or to hyposomotic media, respectively. In each section, the rightmost figure depicts the pathways that are opened up under the respective condition. (Taken with permission from Guizouarn and Motais, 1999.)

inhibiting protein and glycogen synthesis. Now, these effects of insulin (the conversion of sugars and amino acids, small molecules, to glycogen and proteins, large molecules) bring about a decrease in the osmotic material inside the cell. To compensate for these osmotic consequences of anabolism, insulin also triggers a variety of ionic RVI responses, in addition to its role in stimulating the uptake of amino acids (Graf and Haussinger, 1996). Similarly, glucagon directly stimulates a range of RVD responses including the loss of intracellular potassium ions. The RVD and RVI responses brought about by these hormones are so effective that the insulin-induced "anabolic state" of the liver is characterized by a steady state of high volume, while the cell's response to glucagon is for it to remain in the "catabolic state" of shrinkage (Fig. 12). So closely integrated are these hormonal and volume-related responses that an experimentally imposed swelling of the liver cell, brought about by a reduction in external osmolarity, will itself trigger the anabolic responses that insulin brings about. Again, an external hyperosmolarity, inducing shrinkage and a subsequent RVI, will trigger the catabolic state responses, even in the absence of glucagon. Indeed, Haussinger and his colleagues, in seminal papers on this subject (Graf and Haussinger, 1996; Haussinger and Lang, 1991; Lang *et al.,* 1998) have introduced the concept that the transduction of these

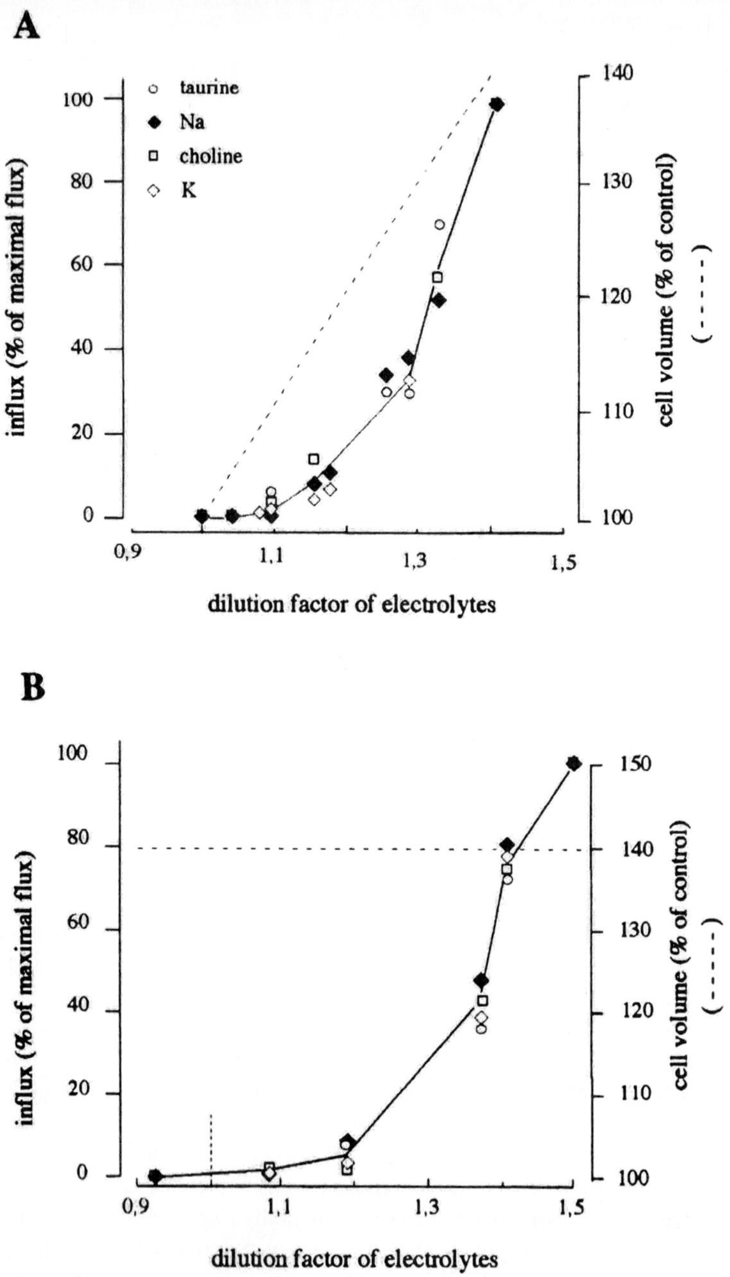

FIG. 11 Fluxes of taurine (○), sodium, (◆), choline (□), and the Cl-independent component of potassium (◇) across trout red cell membranes as a function of the intracellular electrolyte composition. In (A), the swelling took place under hyposmotic conditions, in (B) under isosmotic conditions, as in Fig. 9. The dashed line shows the volume as a percentage of the control value. (Taken with permission from Guizouarn and Motais, 1999.)

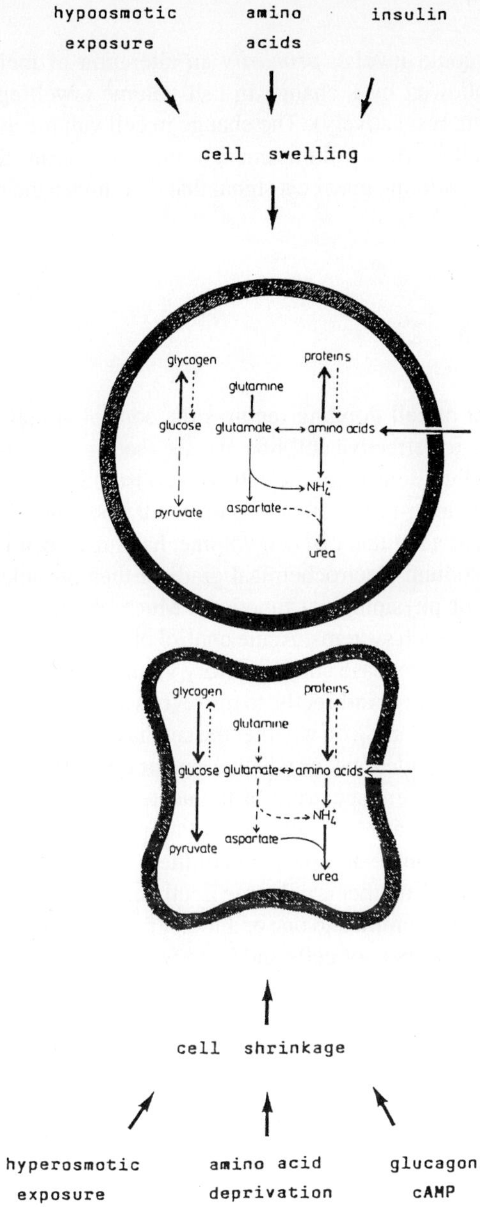

FIG. 12 Schemes for the interaction between hormonal, metabolic, and osmotic controls on cell volume. The upper scheme shows the “anabolic pattern “ brought about by exposure of the liver cells to hyposmotic conditions, or to a supply of amino acids or to the hormone insulin. Glycogen and protein synthesis are stimulated as is amino acid uptake. The lower scheme shows the “catabolic pattern “ brought about by exposure of the liver cells to hyperosmotic conditions, or to a deprivation of amino acids or to the hormone glucagon. Glycogen and protein synthesis are inhibited as is amino acid uptake. (Taken with permission from Haussinger and Lang, 1991.)

hormone signals could involve *primarily* an alteration of membrane transport. This would be followed by a change in cell volume (swelling or shrinkage by insulin or glucagon, respectively). The change in cell volume assists in executing the hormonal stimulus. Thus, volume changes themselves can stimulate metabolic changes, and hence volume may be a signal that determines the metabolic "set" of the cell.

VIII. Conclusion

Parrying the threat of cell flooding requires the control of cell volume, which is thus a prerequisite for effective cell life. We see that, in animals, it is the sodium pump, ejecting sodium ions from the cell with consumption of metabolic energy, which provides the long-term defence against cell flooding. The primary role of the sodium pump, as a regulator of cell volume, has subsequently been adapted so that the imposed sodium electrochemical gradient then provides the basis for an enormous variety of physiological functions. Much of animal physiology arises from the interplay of such systems. Acute control of cell volume, brought about by the regulatory volume decrease and regulatory volume increase systems, uses this sodium gradient, directly or indirectly, to move ions and nonionic metabolites into, and out of, the cell. Depending on whether the cell has a roughly equal permeability for potassium and chloride ions, or whether one or other of these ions is the more permeable, so is a different spectrum of the acute response systems available to the cell. "Cotransporters as Molecular Water Pumps," by Zeuthen and MacAuley, this volume, considers in detail how some of these systems are themselves active pumps for water. In still another article (see Zeuthen, this volume), it is shown how, when such systems are confined to one or another face of the cell in epithelia, water can be moved across a sheet of cells and be moved against an osmotic gradient.

References

Altman, P. L., and Katz, D. D. (1985). *Cell Biol. Vol. I,* FASEB, Bethesda.

Ballanyi, K., Grafe, P., and Ten Bruggencate, G. (1987). Ion activities and potassium uptake mechanisms of glial cells in guinea-pig olfactory cortex slices. *J. Physiol. Lond.* **382,** 159–174.

Biagini, G. A., Kirk, K., Schofield, P. J., and Edwards, M. R. (2000). Role of K^+ and amino acids in osmoregulation by the free-living microaerophilic protozoan *Hexamita inflata. Microbiology* **146,** 427–433.

Carafoli, E. (1991). The calcium pumping ATPase of the plasma membrane. *Annu. Rev. Physiol.* **53,** 531–547.

Coimbra, J., Machado, J., Fernandes, P. L., Ferreira, H. G., and Ferreira, K. G. (1988). Electrophysiology of the mantle of Anodonta Cygnea. *J. Exp. Biol.* **140,** 65–88.

Dall'Asta, V., Bussolati, O., Sala, R., Parolari, A., Alamanni, F., Biglioli, P., and Gazzola, G. C. (1999). Amino acids are compatible osmolytes for volume recovery after hypertonic shrinkage in vascular endothelial cells. *Am. J. Physiol.* **276,** C865–C872.

Dubinsky, W. P., Mayorga-Wark, O., and Schultz, S. G. (1999). Volume regulatory responses of basolateral membrane vesicles from *Necturus* enterocytes: Role of the cytoskeleton. *Proc. Natl. Acad. Sci. U.S.A.* **96,** 9421–9426.

Fagan, M. J., and Saier, M. H. (1994). P-type ATPases of eukaryotes and bacteria: Sequence comparisons and construction of phylogenetic trees. *J. Mol. Evol.* **38,** 57–99.

Garcia-Soto, J. J., and Grinstein, S. (1990). Determinants of the transmembrane distribution of chloride in rat lymphocytes: Role of Cl-HCO3 exchange. *Am. J. Physiol.* **258,** C1108–C1116.

Garner, M. M., and Burg, M. B. (1994). Macromolecular crowding and confinement in cells exposed to hypertonicity. *Am. J. Physiol.* **266,** C877–C892.

Geck, P., and Heinz, E. (1986). The Na-K-2Cl cotransport system. *J. Membr. Biol.* **91,** 91–105.

Gick, G. G., and Ismail-Beigi, F. (1990). Thyroid hormone induction of Na,K-ATPase and its mRNAs in a rat liver cell line. *Am. J. Physiol.* **258,** C544–C551.

Glynn, I. M. (1993). All hands to the sodium pump. *J. Physiol. Lond.* **462,** 1–30.

Godart, H., Ellory, J. C., and Motais, R. (1999). Regulatory volume response of erythrocytes exposed to a gradual and slow decrease in medium osmolality. *Pflügers Arch.-Eur. J. Physiol.* **437,** 776–779.

Graf, J., and Haussinger, D. (1996). Ion transport in hepatocytes: Mechanisms and correlations to cell volume, hormone actions and metabolism. *J. Hepatol.* **24 (Suppl 1),** 53–77.

Guizouarn, H., and Motais, R. (1999). Swelling activation of transport pathways in erythrocytes: Effects of Cl^-, ionic strength, and volume changes. *Am. J. Physiol.* **276,** C210–C220.

Guizouarn, H., Motais, R., Garcia-Romeu, F., and Borgese, F. (2000). Cell volume regulation: The role of taurine loss in maintaining membrane potential and cell pH. *J. Physiol. Lond.* **523,** 147–154.

Haussinger, D., and Lang, F. (1991). Cell volume in the regulation of hepatic function: A mechanism for metabolic control. *Biochim. Biophys. Acta* **1071,** 331–350.

Hernandez, J. A., and Chifflet, S. (2000). Electrogenic properties of the sodium pump in a dynamic model of membrane transport. *J. Membr. Biol.* **176,** 41–52.

Hernandez, J. A., and Cristina, E. (1998). Modeling cell volume regulation in nonexcitable cells: The roles of the Na^+ pump and of cotransport systems. *Am. J. Physiol.* **275,** C1087–C1080.

Herzberg, M. (1974). Interactions between membrane functions and protein synthesis. *Eur. J. Biochem.* **45,** 161–170.

Hille, B. (1992). "Ionic Channels of Excitable Membranes." 2nd Ed. Sinauer Associates, Sunderland, MA.

Hoffmann, E. K., and Ussing, H. H. (1992). Membrane mechanisms in volume regulation in vertebrate cells and epithelia. *In* "Membrane Transport in Biology," Vol. 5 (J. A. Schafer, H. H. Ussing, P. Kristen, and G. H. Giebisch, Eds.), pp. 317–399. Springer-Verlag, Berlin.

Ince, C., Thio, B., van Duijn, B., van Dissel, J. T., Ypey, D. L., and Leijh, P. C. (1987). Intracellular K^+, Na^+ and Cl^- concentrations and membrane potential in human monocytes. *Biochim. Biophys. Acta* **905,** 195–204.

Krogh, A. (1946). The active and passive transport of inorganic ions through the surfaces of living cells and through living membranes generally. *Proc. R. Soc. Lond. Ser. B,* **133,** 140–200.

Lang, F., Busch, G. L., Ritter, M., Volkl, H., Waldegger, S., Gulbins, E., and Haussinger, D. (1998). Functional significance of cell volume regulatory mechanisms. *Physiol. Rev.* **78,** 248–306.

Law, R. O. (1994). Taurine efflux and the regulation of cell volume in incubated slices of rat cerebral cortex. *Biochim. Biophys. Acta* **1221,** 21–28.

Leaf, A. (1956). On the mechanism of fluid exchange of tissues in vitro. *Biochem. J.* **62,** 241–248.

Lew, V. L., and Bookchin, R. M. (1986). Volume, pH, and ion-content regulation in human red cells: Analysis of transient behavior with an integrated model. *J. Membr. Biol.* **92,** 57–74.

Macknight, A. D. C., Gordon, L. G. M., and Purves, R. D. (1994). Problems in the understanding of cell volume regulation. *J. Exp. Zool.* **268,** 80–89.

Maizels, M. (1954). Active cation transport in erythrocytes. *Symp. Soc. Exp. Biol.* **8,** 202–227.

Minton, A. P., Colclasure, G. C., and Parker, J. C. (1992). Model for the role of macromolecular crowding in regulation of cellular volume. *Proc. Natl. Acad. Sci. USA* **89,** 10504–10506.

Parker, J. C. (1993). In defense of cell volume ? *Am. J. Physiol.* **265,** C1191–C1200.

Post, R. L., and Jolly, P. C. (1957). The linkage of sodium, potassium, and ammonium active transport across the human erythrocyte membrane. *Biochim. Biophys. Acta* **25,** 118–128.

Russell, J. M. (2000). Soidum-potassium-chloride cotransport. *Physiol. Rev.* **80,** 212–276.

Saier, M. H. (1994a). Computer-aided analyses of transport protein sequences: Gleaning evidence concerning function, structure, biogenesis, and evolution. *Microbiol. Rev.* **58,** 71–93.

Saier, M. H. (1994b). Convergence and divergence in the evolution of transport proteins. *Bioessays* **16,** 23–29.

Simchowitz, L., Spilberg, I., and De Weer, P. (1982). Sodium and potassium fluxes and membrane potential of human neutrophils: Evidence for an electrogenic sodium pump. *J. Gen. Physiol.* **79,** 453–479.

Sjodin, R. A. (1984). Contributions of electrogenic pumps to resting membrane potentials: The theory of electrogenic potentials. *In* "Electrogenic Transport: Fundamental Principles and Physiological Implications. (M. P. Blaustein and M. Lieberman, Eds.), pp. 105–127. Raven Press, New York.

Stein, W. D. (1990). "Channels, Carriers, and Pumps". Academic Press, San Diego.

Strange, K. (1993). Maintenance of cell volume in the central nervous system. *Pediatr.Nephrol.* **7,** 689–697.

Therien, A. G., and Blostein, R. (2000). Mechanisms of sodium pump regulation. *Am. J. Physiol. Cell Physiol.* **279,** C541–C566.

Toyoshima, C., Nakasako, M., Nomura, H., and Ogawa, H. (2000). Crystal structure of the calcium pump of sarcoplasmic reticulum at 2.6 A resolution. *Nature* **405,** 647–655.

Vanoye, C. G., and Reuss, L. (1999). Stretch-activated single K^+ channels account for whole-cell currents elicited by swelling. *Proc. Natl. Acad. Sci. USA* **96,** 6511–6516.

Vieira, L., Lafuente, E., Gamarro, F., and Cabantchik, Z. I. (1996). An aminoacid channel activated by hypotonically induced swelling of *Leishmania major* promastigotes. *Biochem. J.* **319,** 691–697.

Whalley, D. W., Hool, L. C., Ten Eick, R. E., and Rasmussen, H. H. (1993). Effect of osmotic swelling and shrinkage on Na^+-K^+ pump activity in mammalian cardiac myocytes. *Am. J. Physiol.* **265,** C1201–C1210.

Wilson, J. A. (1979). "Principles of Animal Physiology" 2 Ed. pp. 230–231. Macmillan, New York.

Wilson, T. H. (1954). Ionic permeability and osmotic swelling of cells. *Science* **120,** 104–105.

Zeuthen, T. (1992). From contractile vacuole to leaky epithelia. Coupling between salt and water fluxes in biological membranes. *Biochim. Biophys. Acta* **1113,** 229–258.

Zhang, Z., Lewis, D., Strock, C., Inesi, G., Nakasako, M., Nomura, H., and Toyoshima, C. (2000). Detailed characterization of the cooperative mechanism of Ca(2^+) binding and catalytic activation in the Ca(2^+) transport (SERCA) ATPase. *Biochemistry* **39,** 8758–8767.

Cotransporters as Molecular Water Pumps

Thomas Zeuthen and Nanna MacAulay
Institute of Medical Physiology, The Panum Institute, University of Copenhagen, DK 2200 Copenhagen N, Denmark

Molecular water pumps are membrane proteins of the cotransport type in which a flux of water is coupled to substrate fluxes by a mechanism within the protein. Free energy can be exchanged between the fluxes. Accordingly, the flux of water may be relatively independent of the external water chemical potential and can even proceed uphill. In short, water is being cotransported. The evidence for water cotransport is reviewed with particular emphasis on electrogenic cotransporters expressed in *Xenopus* oocytes under voltage clamped conditions. Phenomena such as uphill water transport, tight coupling between water transport and clamp current, cotransport of small hydrophilic molecules, and shifts in reversal potentials with osmolarity are discussed with examples from the Na^+/glutamate and Na^+/glucose cotransporters. Unstirred layers and electrode artifacts as alternative explanations for such cotransport can be ruled out for both experimental and theoretical reasons. Indeed, substrate fluxes mediated by channels or ionophores generate much smaller water fluxes than those observed with cotransporters. Theoretical models, using reasonable values for the intracellular diffusion coefficient, indicate the presence of only small unstirred layers in the membranes studied.

KEY WORDS: *Xenopus* oocytes, Na^+ coupling, Epithelia, Transport.

I. Introduction

Water crosses cell membranes by several routes: across the lipid bilayer, through specific water channels, and via proteins usually associated with other functions such as uniports and cotransporters. There are also several transport mechanisms for water. In osmotic transport (see Cotransporters as Molecular Water Pumps by Zeuthen and MacAulay, this volume), water is driven by the difference in water chemical potential, and the mechanism can be diffusive or hydraulic. The subject of the present article is, rather, cotransport of water in which the flux of water is coupled to ion or substrate transport within a cotransport protein. Energy for the water transport can be derived from the difference in free energy of the nonaqueous substrates; accordingly, cotransporters may be said to act as molecular water pumps, Fig. 1.

The concept of water cotransport originated from thc simultaneous considerations of two experimental findings (Zeuthen, 1981, 1982, 1983). First, epithelial cells from gall bladder remained hyperosmolar relative to the outside when placed in dilute external solutions. Hyperosmolarities as high as 40 mosm L^{-1} were observed relative to the apical solution, the serosal solution, and the lateral intercellular spaces. Second, osmotically induced influxes or effluxes of water across the cell membrane induced a simultaneous co-flux of K^+ ions. The ratio between K^+ and water fluxes was about 100 mmol L^{-1}. These experiments suggested an answer to the question of how cells could be hyperosmolar relative to the surroundings in a steady state: The coupling between K^+ and water could be a property of a membrane protein. Accordingly, cellular water homeostasis could be viewed as a steady state where the passive osmotic influx of water was matched by an active efflux energized by the ubiquitous K^+ leak.

The suggestion that the Na^+-coupled cotransporters for glucose, alanine, and other organic substrates also act as molecular water pumps (Zeuthen 1994, 1995;

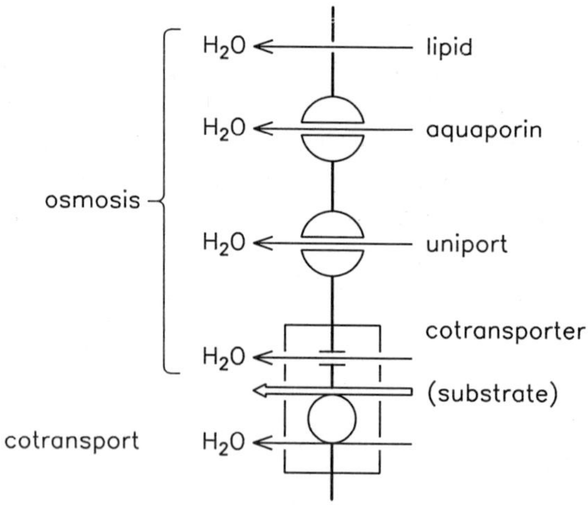

FIG. 1 Pathways and mechanisms of water transport across cell membranes. Water transport takes place across the membrane by several routes and by several mechanisms. The route can be via the lipid bilayer, specific water channels, and proteins which are usually thought of as having other functions: uniports and cotransporters. The transport mechanism for water can be osmosis or, as is the subject of this review, cotransport in a cotransport protein.

Zeuthen and Stein, 1994) has recently been confirmed in studies using the *Xenopus* oocyte expression system. As will be discussed below, the combination of voltage clamping with fast and precise recordings of oocyte volumes has led to accurate definitions of the molecular coupling properties and to the experimental exclusion of possible artifacts such as unstirred layers (Loo *et al.*, 1996; Zeuthen *et al.*, 1997, 2001; Meinild *et al.*, 1998a, 2000; MacAulay *et al.*, 2001). It seems that coupled water transport is a general property of cotransporters of the symport type (for references see Table I).

In the following, we shall review the evidence for the function of cotransporters as molecular water pumps:

1. There is a fixed stoichiometry between water and the nonaqueous substrates under a variety of conditions.
2. Energy can be transferred between the fluxes of the various substrates, as manifested, for example, by the capability for uphill water transport.
3. The rate of cotransport of water is altered abruptly by changes in clamp current.
4. Small hydrophilic molecules are translocated along with the cotransported water.
5. The equilibrium potential of the cotransporter is a function of the water chemical potential.
6. There seem to be no unstirred layer effects.

Finally, we will discuss molecular mechanisms that may explain the cotransport of water.

II. The Coupling Ratio between Water and Substrate Fluxes

Cotransport is characterized by a fixed stoichiometry that is independent of the magnitude and direction of the external driving forces. Thus, the number and types of substrate molecules transported at each turnover are given entirely by the properties of the cotransporter. In a complete description, it must be shown that the driving force originating from each substrate is able to drive the transport, but the properties of the expression system limit the types of experiments that can be done. The situation is more versatile the closer the cotransporter is poised toward its equilibrium state. A wide spectrum of experiments has been possible with the electroneutral K^+/Cl^- (Zeuthen 1991a,b, 1994) and H^+/lactate cotransporters (Zeuthen *et al.*, 1996), both situated in their native tissues (for a review see Zeuthen, 2000). In contrast, the types of energy transfer experiments that can be performed on the electrogenic Na^+-dependent cotransporters expressed in *Xenopus* oocytes are relatively limited. The cotransporters are poised strongly for inward transport due to the inherent Na^+ gradient and the negative intracellular potential. For this reason, most experiments on these transporters have been performed in the kinetic situation far from equilibrium.

A. Substrate Gradients Can Drive Water Transport under Isotonic Conditions

Cotransport of water can be driven by an electrochemical gradient of the non-aqueous substrates under conditions of no transmembrane osmotic gradient. This has been shown for several types of Na^+-dependent electrogenic cotransporters expressed in *Xenopus* oocytes, in particular the glucose transporter SGLT1, the dicarboxylate transporter NaDC1, and the glutamate transporter EAAT1 (for references, see Table I). The *Xenopus* expression system is well suited for the study of water transport in electrogenic cotransporters. Ion transport can be monitored as the clamp current and the water transport followed from the change in oocyte volume. Measurement of the oocyte volume is accurate to 0.03% (equivalent to 0.3 nL), and the effects of osmotic gradients as small as 2.5 mosm L^{-1} are discernible. Furthermore, the external solution can be changed in a few seconds (Zeuthen *et al.*, 1997). An example from EAAT1 is shown in Fig. 2. When the organic substrate was added isotonically to the bathing solution and the intracellular potential was clamped to a negative value, an abrupt increase in inward current (Fig. 2A) and a linear increase in oocyte volume (Fig. 2B) was observed (Fig. 2C represents an experiment in which gramicidin was added; see Section VIII). The correspondence between the substrate-induced current and the water influx was apparent from the fact that the integrated current predicted the change in volume under the assumption that a fixed number of water molecules were coupled to the influx of each unit charge. The number of water molecules transported per turnover of the protein is given in Table I. This coupling ratio was independent of the expression level of the protein, the external driving forces obtained at various clamp voltages (see Fig. 3), Na^+ concentrations, and concentrations of organic substrate.

The electroneutral K^+/Cl^- cotransporter situated in the choroid plexus epithelium has been studied by means of ion-selective microelectrodes. A downhill flux of K^+ can drive fluxes of Cl^- and water, while downhill fluxes of Cl^- can drive K^+ and water fluxes; this applies to both the inward and outward direction. In all types of experiments, the coupling ratio is the same, about 500 water molecules per pair of K^+ and Cl^- ions (Zeuthen, 1991a,b, 1994). Similar results were obtained for the H^+/lactate cotransporter (Zeuthen *et al.*, 1996; Hamann *et al.*, 2000).

B. Osmotic Gradients Can Drive Substrate Transport

Given physiological limits, water chemical potential differences are not very strong energetically compared to, say, electrical potential differences. The concentration of water is proportional to exp(-osm/n_w), where osm is the osmolarity in units

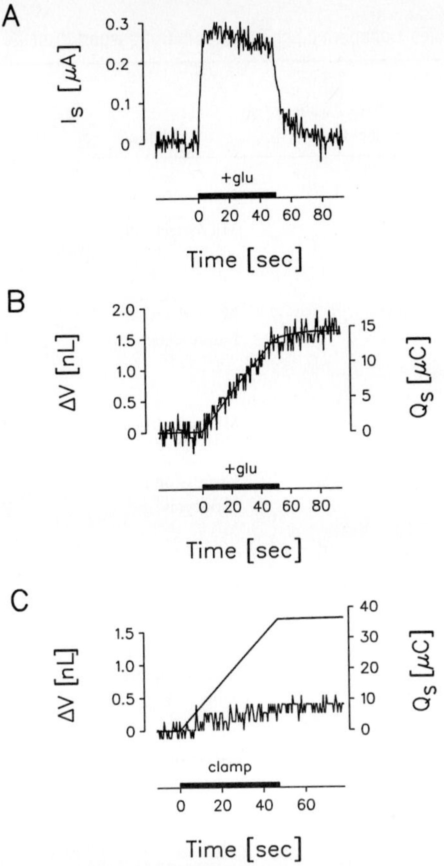

FIG. 2 Substrate induced current and water cotransport under isotonic conditions. (A) In this example, the oocyte expressed the Na^+/glutamate cotransporter EAAT1 and was clamped continuously at −30 mV. The clamp current I_s was induced by adding 200 μmol L^{-1} of L-glutamate abruptly to the bath (black bar). I_s increased rapidly toward a maximum of 280 nA. The experiment was performed at low external Cl^-concentration (6 mmol L^{-1}) to minimize currents via the Cl^- conductance of this cotransporter. (B) I_s was associated with a linear increase in oocyte volume (ΔV, jagged line) indicative of a constant rate of water influx ($J_{H2O} = \Delta V/\Delta t$) of about 30 pL sec^{-1}. The volume increase and I_s were abolished when L-glutamate was removed from the bath. The integrated current (Q_s, smooth line in B) describes the volume changes, if a fixed number of water molecules enter per unitary charge (in the present example 449) is assumed. (C) When the inward Na^+ current was mediated by the ionophore gramicidin, there was no initial change in oocyte volume. The inward current was about 750 nA (not shown) and was obtained by changing the intracellular electrical potential from the resting potential (about −45 mV) to −90 mV by means of the voltage clamp (black bar). The integrated current (smooth line) had no resemblance to the volume changes. The oocyte treated with gramicidin expressed EAAT1 to the same degree as the one used in A and B, but the cotransporter was kept inactive with no substrate present. (Data from MacAulay *et al.,* 2001.)

TABLE I

Number of Water Molecules Transported per Turnover in Cotransport Proteins

Cotransporter Organ Species	Number of water molecules, n.	Method	References
K^+/Cl^-			
Choroid plexus			
Necturus maculosus	500	Microelectrodes	Zeuthen, 1991a,b, 1994
H^+/lactate			
Retina			
Necturus maculosus	500	Microelectrodes	Zeuthen *et al.*, 1996;
Human, cultures	N.d.[a]	Fluorescence	Hamann *et al.*, 2000
Na^+/alanine			
Small intestine			
Necturus maculosus	N.d.[a]	Microelectrodes	Zeuthen, 1995
Na^+/glucose, SGLT1			
Rabbit	265, 310, 424	Expression in *Xenopus* oocytes	Loo *et al.*,1996; Zeuthen *et al.*, 1997, 2001;
Human	210, 264		Meinild *et al.*, 1998a;
Na^+/I^-			
Rat	~200	Expression in *Xenopus* oocytes	Loo *et al.*, 1996
Na^+/Cl^-/GABA			
Human	~200	Expression in *Xenopus* oocytes	Loo *et al.*, 1996
H^+/amino acid			
Plant	~50	Expression in *Xenopus* oocytes	Loo *et al.*, 1996
Na^+/dicarboxylate, NaDC1			
Rabbit	175	Expression in *Xenopus* oocytes	Meinild *et al.*, 2000
Na^+/glutamate, EAAT1			
Human	435	Expression in *Xenopus* oocytes	MacAulay *et al.*, 2001

[a] N.d., not determined.

of osm L^{-1} and n_w is the number of moles of water per liter, about 55. For the Na^+-dependent cotransporters with coupling ratios of 200 to 500 between water and substrates, this means that an osmotic difference of 100 mosm L^{-1} is equivalent to a difference in electrical potential of about 5 mV. Thus osmotic gradients of 25 mosm L^{-1} or higher would be required to obtain measurable changes in substrate fluxes (for calculations see Section VI). Accordingly, the clamp currents of the SGLT1 and the EAAT1 expressed in *Xenopus* oocytes were not affected by extracellular hyper- or hyposmolar osmotic gradients of up to 20 mosm L^{-1} (Meinild *et al.*, 1998a; MacAulay *et al.*, 2001). At abrupt increases in external osmolarities of

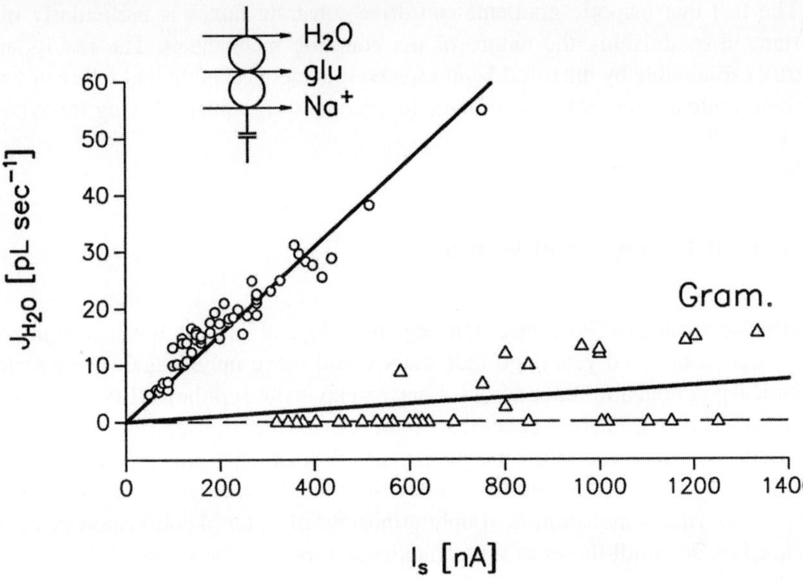

FIG. 3 Cotransport of water J_{H2O} in Na^+/glutamate cotransporters as a function of clamp current I_s. Experiments on EAAT1 expressed in *Xenopus oocytes* (see Fig. 2A and B). I_s was varied by choosing clamp voltages in the range 0 to −110 mV. J_{H2O} (in units 10^{-3} pL sec^{-1}) was a linear function of the clamp current = 77.4*I_s(nA), equivalent to a coupling ratio of 424 water molecules per unit charge (32 measurements in 6 oocytes). The J_{H2O}s obtained by clamp currents in gramicidin-treated oocytes (Gram., see Fig. 2C) were much smaller than those obtained with activated EAAT1.

100 mosm L^{-1}, however, a reduction of about 7% in the magnitude of the inwardly directed clamp current was observed for the EAAT1 (MacAulay *et al.,* 2001). It should be emphasized that this experiment was done under nonequilibrium conditions, and the effect may not be ascribed entirely to changes in water chemical potential but could also be influenced by a shift in apparent substrate affinity.

Osmotic gradients induced significant fluxes of K^+ and Cl^- in the K^+/Cl^- cotransporter located in the choroid plexus epithelium (Zeuthen 1991a,b, 1994). The effect on K^+ fluxes was studied in detail by abrupt increases in the extracellular osmolarity of between 50 and 600 mosm L^{-1}. The K^+ flux was a linearly increasing function of osmotic gradients up to 200 mosm L^{-1}. At higher gradients, the flux saturated. At all gradients tested, the water and ion fluxes were coupled in a ratio of 500 water molecules per K^+ ion. For the H^+/lactate cotransporter situated in the retinal pigment epithelium, the results were similar (Zeuthen *et al.,* 1996). In the presence of lactate, cell shrinkage induced by application of mannitol caused an uphill efflux of H^+, as indicated by a rapid cell alkalinization. In the absence of lactate, a small acidification was observed. By correlating the induced H^+ fluxes and water fluxes, the coupling ratio was found to be 500, analogous to the one found in other types of experiments.

The fact that osmotic gradients can drive substrate fluxes is particularly important in establishing the nature of the coupling mechanism. The results are hardly explainable by unstirred layer effects. It is inconceivable that a flux of water could pile up ions at the membrane to affect any significant driving force (see Section VIII).

III. Uphill Transport of Water

Uphill water transport is of long-standing physiological interest. It was recognized more than a hundred years ago that water could move uphill, against the water chemical potential difference across acinar cell layers and epithelia (Ludwig, 1861; Reid, 1892, 1901). There is general agreement that water transport is secondary to salt transport, but the coupling mechanism and its location is not agreed upon at all (compare the chapters by Zeuthen and by Shachar-Hill and Hill, this volume). In this context, the demonstration of uphill transport of water in cotransport proteins induced by downhill fluxes of the nonaqueous substrates becomes of central importance. It opens the possibility that the coupling between salt and water takes place at the membrane level of the water-transporting cell.

Uphill transport of water in the K^+/Cl^- cotransporter has been demonstrated *in vitro* in the apical membrane of the choroid plexus epithelium. As already mentioned, an inward water flux was generated by means of an inward directed flux of K^+ or Cl^-. A simultaneous change of the transmembrane osmotic gradient adverse to the inward water flux reduced the water influx but did not stop it (Zeuthen, 1994). The strength of the coupled water transport was such that it took a gradient of about 100 mosm L^{-1} to match a water flux generated by a ten-fold increase in the external concentration product of $K^+$and Cl^- ions. Similar results were obtained for the H^+/lactate cotransporter (Zeuthen *et al.*, 1996).

The *Xenopus* oocyte expression system is well suited for the study of uphill water transport via electrogenic cotransporters (Zeuthen *et al.*, 1997). The effect of adverse osmotic gradients on water cotransport can be studied within a range of osmolarities that is of primary physiological interest, up to, say 25 mosm L^{-1}. Uphill transport of water has been demonstrated in the SGLT1 (Zeuthen *et al.*, 1997, 2001; Meinild *et al.*, 1998a), the NaDC1 (Meinild *et al.*, 2000), and the EAAT1 (MacAulay *et al.*, 2001). Consider an experiment where the organic substrate is added simultaneously with mannitol to the external solution (see legend to Fig. 4; here the cotransporter is EAAT1 and the organic substrate glutamate). The water flux is the sum of two contributions: a glutamate-coupled inward flux and an outward osmotic flux. The osmotic component can be obtained independently by adding mannitol under conditions where the cotransporter is inactivated (Fig. 5). If this experimental protocol is repeated at different osmotic challenges, the water flux can be obtained as a function of the osmotic gradient. In this plot, the osmotic water flux is given by the line going through the origin; the cotransport

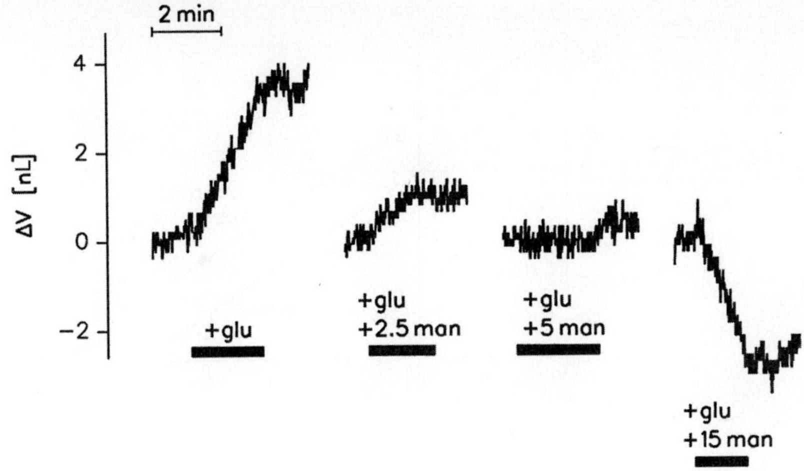

FIG. 4 Uphill transport of water by the Na^+/glutamate cotransporter EAAT1. An EAAT1 expressing oocyte was initially bathed in glutamate-free control solutions under clamped conditions (−30 mV). At the time indicated by the black bar, glutamate (200 μM) was added to the bathing solution (+glu). This induced an immediate swelling of the oocyte (see also Fig. 2A and B). In the following three recordings, the stimulation by glutamate was combined with hyperosmotic challenges of 2.5, 5, and 15 mosm L^{-1} of mannitol (+man). The glutamate-induced current I_s (not shown) was the same, about 300 nA in all four experiments. It appears that the inwardly directed cotransport of water can proceed despite adverse osmotic gradients. In Fig. 5, the rates of water transport are plotted as a function of the osmotic challenge. Data from MacAulay *et al.* (2001).

component appears as the constant vertical displacement between the two curves. It follows that the cotransport component is independent of the osmotic gradient. The fluxes could also be separated into a cotransport and an osmotic component, under hyposmotic conditions. Similar results were obtained for the human SGLT1 (Meinild *et al.*, 1998a).

The insensitivity of the cotransport component to osmotic challenges of the order of ±25 mosm L^{-1} is not surprising. The Gibbs equation (Eq. 4) indicates that it would take adverse osmotic gradients of around 2000 mosm L^{-1} to stop cotransport in the EAAT1 and the SGLT1, given typical clamp voltages and Na^+ concentrations.

IV. Tight Coupling between Water Cotransport and Clamp Current

The high accuracy and time resolution of the oocyte volume recording system (Zeuthen *et al.*, 1997) permit a precise correlation between transport of substrate and water. Instantaneous changes in clamp currents can be induced by changes in

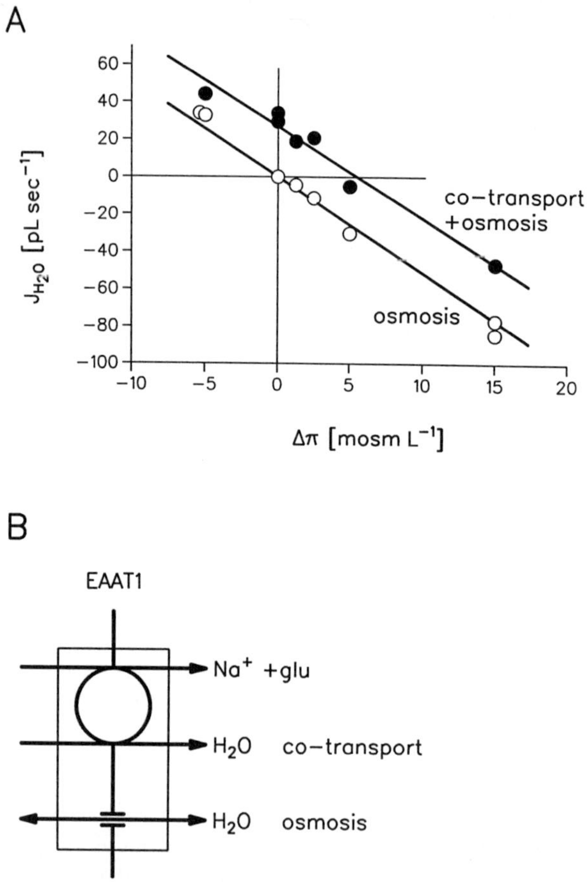

FIG. 5 Two components of water transport in the Na$^+$/glutamate cotransporter, EAAT1. (A) The upper curve (filled circles) represents the water fluxes J_{H2O} obtained by the simultaneous activation of the cotransport and the osmotic water transports by abrupt application of glutamate and mannitol (data from Fig. 5, $J_{H2O} = \Delta V/\Delta t$). J_{H2O} as a function of $\Delta\pi$ is given by the regression line $J_{H2O} = 26.7$ pL sec^{-1} – 4.8 pL sec^{-1} (mosm L^{-1})$^{-1}$. The lower curve (open circles) represents J_{H2O}s obtained from osmotic challenges only; glutamate is present, but the cotransport function is inactivated by the absence of external voltage clamp. J_{H2O} is given by the line −5.5 pL sec^{-1} (mosm L^{-1})$^{-1}$. The two sets of data are from the same oocyte. The slopes of the two lines are similar and represent the L_p of the EAAT1 in the presence of glutamate. The constant vertical displacement represents the cotransport component of the water transport. (B) A model of the protein in which the cotransport and the osmotic water transport work in parallel and independently. Adapted from MacAulay *et al.* (2001).

the clamp voltage, and the resulting changes in oocyte volume can be followed with a time resolution of 2 sec, given the frequency and amplitude of the background noise (Loo *et al.*, 1996; Zeuthen *et al.*, 2001). An example using rabbit SGLT1 is shown in Fig. 6A. To begin the experiment, sugar was added isoosmotically at a given clamp voltage, in this case −80 mV. The water flux and clamp current

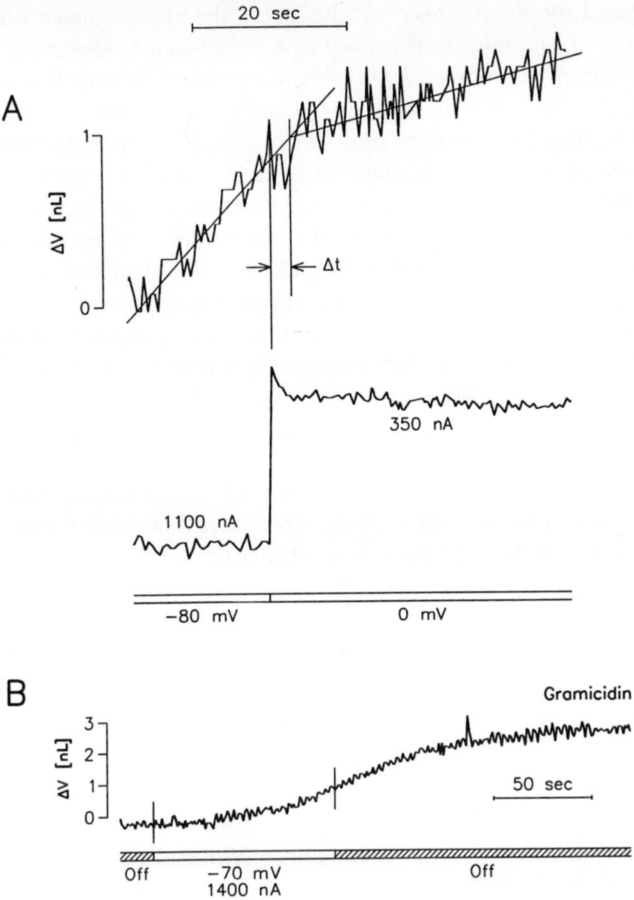

FIG. 6 Tight coupling between J_{H2O} and clamp voltage in an SGLT1-expressing oocyte. (A) Rabbit SGLT1 expressed in *Xenopus* oocyte was clamped to −80 mV in the presence of of 1 mmol L^{-1} sugar in the external solution. This gave a swelling rate of 48 pL sec^{-1} as shown by the regression line; the corresponding I_s was 1100 nA. The rate of swelling and the magnitude of the current are indicative of a coupling ratio of 240 water molecules per unit charge. When the clamp voltage was jumped to 0 mV (within 10 msec), I_s fell to 350 nA and J_{H2O} to 13 pL sec^{-1} as given by the new regression line. The coupling ratio was now 204 water molecules per unit charge, close to the one measured initially. In the present recording, the intercept between the two regression lines lies $\Delta t = 2.5$ sec later than the change in clamp voltage. On average, this intercept took place 1.1 ± 0.7 sec ($n = 18$) after the voltage jump, which is not significantly different from zero. We conclude that the coupling is tight between water and ion transport in the rSGLT1. (B) Volume changes induced by clamp currents mediated by the ionophore gramicidin were slow. The ionophore was inserted into an oocyte expressing rSGLT1 which was kept inactivated by the absence of sugar. When the clamp voltage was switched on (−70 mV, first vertical line), an inward Na^+ current of 1400 nA was induced (current trace not shown). Only after about 30 sec did the oocyte begin to swell; after about 70 sec, J_{H2O} ($= \Delta V/\Delta t$) was about 24 pL sec^{-1}. When the clamp was switched off (second vertical line), the current was abolished, and J_{H2O} gradually decreased toward zero. By comparing A and B, it is seen that volume changes induced by currents are much slower than those induced via cotransport. (Data from Zeuthen *et al.*, 2001.)

were recorded for about 20 sec. At this point, the clamp voltage was changed abruptly to 0 mV (within 10 msec) and new steady-state values of the water flux and clamp current were obtained. On average, the change in water flux was delayed by 1.1 ± 0.7 sec ($n = 18$) compared to the time of the jump in voltage. This delay is not significantly different from zero. The relation between sugar-induced clamp current and clamp voltage was identical to earlier findings (Parent *et al.,* 1992). The ratio between the water flux and the clamp current was the same before and after the change in clamp voltage. At all voltages, the coupling ratio was around 440 water molecules per turnover of the protein (data from Zeuthen *et al.,* 2001).

It is interesting to compare the temporal correlation between water fluxes and currents obtained above with one in which the clamp current is mediated by an ionophore (gramicidin) while the cotransporter is kept inactivated by the absence of sugar; an example is shown in Fig. 6B. By clamping the oocyte to a potential of −70 mV, large inward Na^+ fluxes were generated via the gramicidin, given by inward clamp currents of 1400 nA (Meinild *et al.,* 1998a; Zeuthen *et al.,* 2001). The gramicidin-mediated Na^+ fluxes did not produce any initial influxes of water, and only after an average of about 35 sec were volume changes discernible. When the clamp was switched off, the volume changes receded gradually. It should be noted that gramicidin does not affect the passive water permeability of the SGLT1 expressing oocyte (Zeuthen *et al.,* 1997).

The rapidity of the SGLT1-mediated response compared to the slowness of the gramicidin-induced response demonstrates the tightness of the coupling between the cotransport component of water transport and the turnover of the protein. It also suggests no unstirred layers (see Section VIII).

V. Cotransport of Small Hydrophilic Molecules

It has recently been shown that cotransporters can mediate urea transport by two mechanisms. Cotransporters act as urea channels (see Passive Water Transport in Biological Pores by Zeuthen and MacAulay, this volume), and they have the unique ability of coupling urea transport to substrate cotransport. This was demonstrated for the rabbit SGLT1, the pig low-affinity Na^+–glucose cotransporter (pSGLT3), and the human Na^+–Cl^-–GABA cotransporter (hGAT1) (Leung *et al.,* 2000). For the SGLT1 expressing oocytes in the absence of sugar, there was a four-fold increase in urea uptake relative to that obtained in the native oocytes. The urea uptake was independent of the presence of Na^+, linear with incubation time, and blocked by the specific inhibitor phlorizin. The activation energy was low, about 6 kcal mol^{-1}, compatible with that of passive water transport through aquaporins. When the cotransporter was activated by addition of sugar, the uptake of urea increased proportionally to the rate of Na^+–glucose cotransport. Similar increases in urea transport were observed for the other cotransporters. The amount of urea cotransport was approximately proportional to the molar concentration of urea in the bathing solution. The data suggest that (i) cotransporters contain a selective aqueous pathway that is shared by urea and water and that (ii) water and urea are both cotransported as a result of conformational changes in the protein.

Similar experiments using the Na^+–glutamate transporter EAAT1 have been attempted (see Passive Water Transport in Biological Pores by Zeuthen and MacAulay, this volume). In the EAAT1, the putative water channel is much more conductive to urea and water than are the transporters tested above. Furthermore, the water channel of the EAAT1 opens in the presence of glutamate. As a result, the passive tracer fluxes give a background on which cotransport of urea of the expected magnitude is undetectable.

VI. Thermodynamics of Water Cotransport, Gibbs' Equation, and Reversal Potential E_r

If water is cotransported on equal footing with the other substrates, the Gibbs equation can be used to relate the electrical, chemical, and water chemical potentials across the membrane at equilibrium. Consider a case where a cotransporter transports n_A moles of solutes A of charge z_A, n_B moles of B of charge z_B, and n water molecules (concentration W) from side 1 to side 2 of a membrane; at the same time, it counter-transports n_C moles of C, charge z_C, from side 2 to 1. The electrical potential difference between compartments 1 and 2 are $\Psi_1 - \Psi_2$ or $\Delta\Psi$ (see Fig. 7). At equilibrium

$$\begin{aligned} n_A RT/F \ln(A_1/A_2) + n_B RT/F \ln(B_1/B_2) + nRT/F \ln(W_1/W_2) \\ + (\Psi_1 - \Psi_2)(n_A z_A + n_B z_B) \\ = n_C RT/F \ln(C_1/C_2) + (\Psi_1 - \Psi_2) n_C z_C \end{aligned} \quad (1)$$

where the water concentration W is proportional to $\exp(-osm/n_w)$, with osm equal to the osmolarity in units of osm L^{-1}, and n_w is the number of moles of water per liter, about 55.

For the electroneutral K^+/Cl^- cotransporter situated in a cell membrane, the equation simplifies to

$$[K^+]_i\,[Cl^-]_i\,W_i^n = [K^+]_o\,[Cl^-]_o\,W_o^n \quad (2)$$

With typical intracellular (i) values for K^+ and Cl^- of 80 and 40 mmol L^{-1} and of 2 and 110 mmol L^{-1} in the extracellular (o) compartment, Eq. (2) shows that, at equilibrium, a transmembrane osmotic difference of 320 mosm L^{-1} can exist (Zeuthen, 1994, 2000). This prediction agrees with the experimental finding that it takes osmotic differences of 200 to 300 mosm L^{-1} to stop the water transport mediated by the K^+/Cl^- cotransporter in the epithelial cells of choroid plexus.

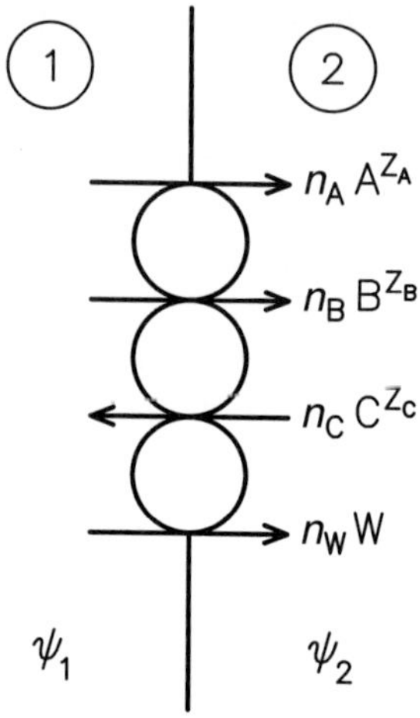

FIG. 7 Thermodynamics of cotransporters are shown. The cotransporter transports n_A moles of solutes A of charge z_A, n_B moles of B of charge z_B, and n water molecules (of concentration W) from side 1 to side 2 of a membrane; at the same time it counter-transports n_C moles of C, charge z_C, from side 2 to 1. The electrical potential difference between compartments 1 and 2 is $\Psi_1 - \Psi_2$. The reversal potential (equilibrium potential) for this cotransporter is analyzed by the Gibbs equation, see text.

Interestingly, this osmotic gradient is close to the upper limit that leaky epithelia can transport water against (for reviews see Zeuthen, 1992, 1995, and article by Zeuthen, this volume).

The Na^+/glucose transporter SGLT1 cotransports two Na^+ ions and 1 glucose molecule. In that case, Eq. (1) becomes

$$2RT\ln(\mathrm{Na}_i/\mathrm{Na}_o) + RT\ln(G_i/G_o) + nRT\ln(W_i/W_o) = 2F(\Psi_o - \Psi_i) \quad (3)$$

Consider the human SGLT1 where the number of water molecules transported per turnover of the protein, n, equals 210 (Meinild *et al.*, 1998a). If the outside Na^+ concentration is 10 times higher than the intracellular concentration, and the glucose concentrations on the two sides (G) are the same, while the intracellular potential is −50 mV relative to the outside, then it can be calculated that the inward water flux would proceed in spite of adverse osmotic differences of up to 2300 mosm L^{-1}. In case the intracellular concentration of glucose were 10 times higher than on the outside, the difference would be 1650 mosm L^{-1}. It follows that the resulting water transport is rather insensitive to the small osmotic differences that may occur physiologically.

In the case of EAAT1, we may assume a stoichiometry with cotransport of 3 Na^+, 1 H^+, 1 glutamate, countertransport of 1 K^+ (Kanner *et al.*, 1982; Nelson *et al.*, 1983; Barbour *et al.*, 1988; Zerangue *et al.*, 1996;), and an *n* of 436 water molecules (MacAulay *et al.*, 2001). In this case, Eq. (1) gives (see also Rudnick, 1998):

$$\begin{aligned}3RT\ln(\mathrm{Na}_i^+/\mathrm{Na}_o^+) + RT\ln(H_i^+/H_o^+) + RT\ln(\mathrm{glu}_i^-/\mathrm{glu}_o^-) \\ - RT\ln(K_i^+/K_o^+) + nRT\ln(W_i/W_o) \\ = 2F(\Psi_o - \Psi_i)\end{aligned} \quad (4)$$

With a membrane potential, $\Psi_\mathrm{i} - \Psi_o$, of -50 *mV*, a ten-fold concentration difference for Na^+, glutamate (inwardly directed), and K^+ (outwardly directed), and no concentration difference for H^+, it can be calculated that the inward water flux would proceed in spite of adverse osmotic gradients of up to 2100 mosm. The Na^+ and the electrical gradients alone would account for 1800 mosm. With the assumption of cotransport of 2 Na^+, the numbers would be 1500 and 900 mosm.

In summary, under normal physiological situations there is sufficient energy in the electrochemical gradients of the nonaqueous substrates to mediate cotransport of water. Cotransport of water alleviates the osmotic problems that would arise from the net transport of osmotic active molecules by the cotransporter. Considering the coupling ratio, it appears that a large fraction of the water required for osmotic equilibrium of the cell is cotransported along with the substrates.

VII. Relation between Reversal Potential E_r and the Transmembrane Osmotic Gradient

In an important study, Parsegian and co-workers (Colombo *et al.*, 1992) described the role of water chemical potential for the oxygen binding in hemoglobin. They used osmotic stress and found that the conformational changes involved in oxygen binding were associated with the uptake of a significant number of water molecules from the external solution, about 65; at oxygen release, the water molecules were returned to solution again. Similar effects have been found for other biological molecules (see chapters by Parsegian, by Rand, and by Kornblatt and Kornblatt, this volume). An electrogenic cotransporter is a convenient system in which to study these effects. It is membrane bound, and its functional state can be estimated from the electrical potential difference between the two compartments defined by the membrane. The Gibbs equation (Eq. 1) defines equilibrium situations where the chemical and osmotic driving forces are exactly balanced by the electrical forces. For electrogenic cotransporters, there are no transport currents under these conditions, and the electrical potential difference is known as the reversal potential E_r.

The dependence of E_r on transmembrane osmotic gradients can be used to determine n, the number of cotransported water molecules. Consider the SGLT1 (or the EAAT1) and assume that the activities of the nonaqueous substrates are kept

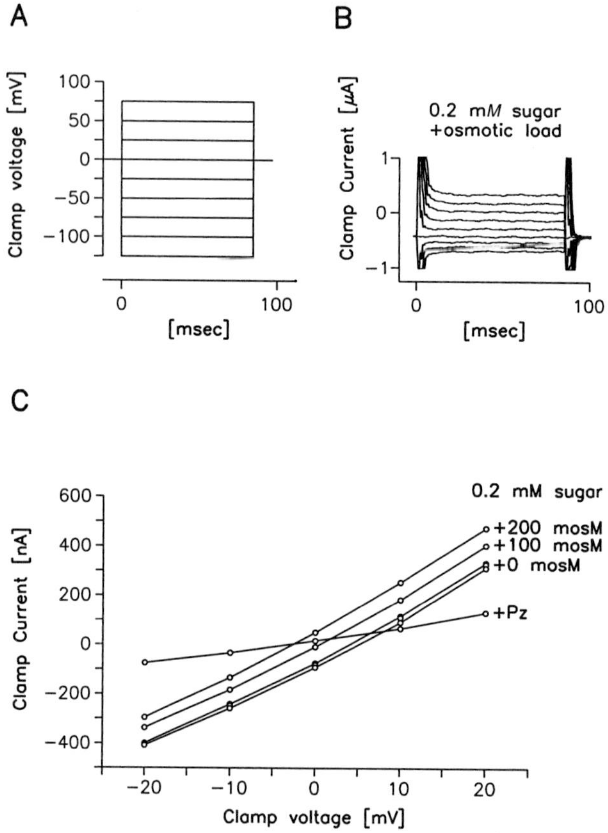

FIG. 8 Reversal potential E_r for rSGLT1 expressed in *Xenopus* oocytes as a function of external osmolarity. Oocytes were initially loaded with sugar (methyl-α-D-glucopyranoside) by clamping for 10 min at a clamp potential of −50 mV at an external sugar concentration of 2 mmol L^{-1}. Control solutions contained (in mmol L^{-1}), 50 NaCl, 1 KCl, 1 $CaCl_2$, 1 $MgCl_2$, 5 Hepes, pH 7.4, 100 urea. (A) During tests, the clamp voltage was jumped from a holding voltage of −50 mV to a series of 9 voltage pulses ranging from +75 to −125 mV (steps of 25 mV lasting 85 msec). (B) The resulting clamp currents were recorded with sugar (0.2 mmol L^{-1}) and various osmotic loads (maintained by addition or removaval of urea) in the external bathing solution. (C) The relation between the steady-state clamp current (I) and clamp voltage (V) for a typical oocyte, only the values near the reversal potential (zero current) are shown. The oocyte was bathed in control solution and tested in solutions with 0.2 mmol L^{-1} of sugar and osmotic loads of 0, 100, or 200 mmol L^{-1} of urea. The I-V curves were obtained after 45 sec in the test solution. The five tests shown were performed in the order +0, +100, +0, +200, +0 mmol L^{-1} urea to confirm reproducibility. Please note that two of the curves obtained with 0 mmol L^{-1} of urea superimpose. The I-V curve marked + Pz was obtained with 200 μmol L^{-1} of the inhibitor phlorizin in the bath. I-V curves obtained with phlorizin were independent of the osmotic load. The experiment shows clearly that a change in external osmolarity (i.e., a transmembrane osmotic gradient) shifts the reversal potential; see Fig. 9.

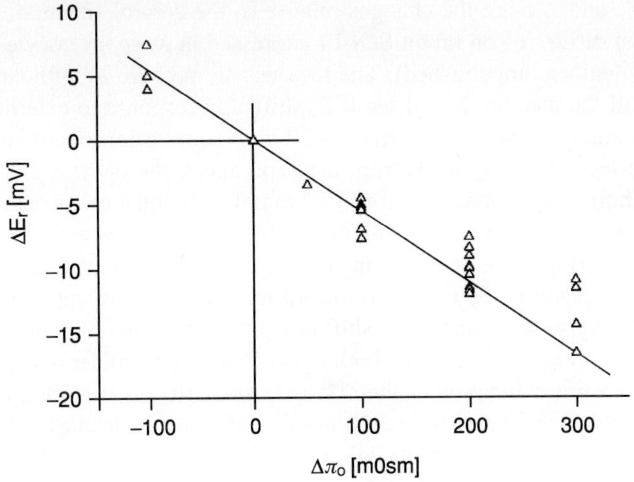

$$\Delta E_r = -n\frac{RTV_W}{2F}\Delta\pi_O$$

FIG. 9 Relation between reversal potential E_r and external osmolarity π_O. The reversal potential E_r for the sugar-dependent component of the rabbit SGLT1-mediated current was determined at different external osmolarities π_O as described in Fig. 8. Values of ΔE_r and $\Delta\pi_O$ are given relative to the values determined under isotonic conditions. In case of water cotransport, Gibbs' equation predicts a linear relation between ΔE_r and $\Delta\pi_O$ given by a proportionality coefficient of $-nRTV_W/2F$; see text. Here, n is the number of water molecules cotransported per turnover of the SGLT1, R the gas constant, T the absolute temperature, V_W the partial molar volume of water, F Faradays constant. Data are from six oocytes. The regression line is based on the data obtained with osmotic challenges of up to $\pm$ 100 mosm L^{-1} and gives n equal to 247 $\pm$ 17. Values for n obtained at the larger gradients tended to be slightly smaller.

constant in both compartments. The Gibbs equation (Eq. 3 and 4) simplifies to

$$nRT\ln(W_i/W_o) = 2F(\Psi_o - \Psi_i) \tag{5}$$

If, for example, the osmolarity of the external solution is increased, an outward transport of water and of positive charge would follow, as both SGLT1 and EAAT1 carry positive charge along with the water. To maintain equilibrium, therefore, the membrane potential would change so as to make the intracellular electrical potential more negative. In Eq. (5), the water concentration W is proportional to $\exp(-osm/n_w)$, where osm is equal to the osmolarity in units of osm L^{-1}, and n_w is the number of moles of water per liter, about 55, or a partial molar volume for water $V_w(= 1/n_w)$ of 18 cm^3/mol. Equation (5) takes the form

$$\Delta E_r = -nRTV_w/2F\Delta\pi \tag{6}$$

where ΔE_r and $\Delta\pi$ are the changes relative to the control situation. An experimental test of Eq. (6) on rabbit SGLT1 expressed in *Xenopus* oocytes is shown in Fig. 8 (Zeuthen, unpublished). The idea was to measure E_r with sugar on the two sides of the membrane and see if E_r shifted in response to external osmotic loads. Accordingly, the oocyte was loaded with sugar prior to experiments (for details see legend to Fig. 8). During the experiment, the oocytes were exposed to subsaturating sugar concentrations (0.2 mmol L^{-1}) and various osmotic loads. Steady-state clamp currents (Fig. 8B) were determined in response to a series of voltage pulses (Fig. 8A). The resulting current–voltage relationships are shown in Fig. 8C. It is seen that E_r is shifted toward more negative values the higher the extracellular hyperosmolarity. The shift in $E_r(\Delta E_r)$ as a function of the external osmolarity is shown in Fig. 9 for both hyper- and hypoosmolar tests. It appears that ΔE_r is a linear function of the external osmolarity, as predicted by Eq. (6). The regression line shows that n, the number of water molecules cotransported per turnover of the protein, equals 247, in good agreement with the coupling ratio determined from kinetic experiments: 260 (Loo *et al.,* 1996); 390 (Zeuthen *et al.,* 1997); 424 (Zeuthen *et al.,* 2001).

VIII. Unstirred Layers and Electrode Artifacts

It might be argued that the mechanism of cotransport of water is in fact entirely osmotic; what appears to be a molecular mechanism is really arising from the presence of unstirred layers. The cotransport-driven flux of substrates must lead to a significant build up of concentration on one side of the membrane and a reduction on the other. The resulting transmembrane osmotic difference might drive the water transport. The effect would be enhanced were the osmotic water permeability of the membrane underestimated and if diffusion were severely restricted inside the cell (Duquette *et al.,* 2001). Several lines of experimental evidence, however, rule out significant unstirred layer effects. For a discussion in regard to the electroneutral cotransporters, see Zeuthen (2000). Experiments with the Na^+-dependent cotransporters and theoretical considerations are reviewed below.

A. Experiments

The relation between transport rates and the build up of unstirred layers can be studied directly in the *Xenopus* oocyte by the insertion of ionophores or channels. If, in the absence of cotransporters, transport is entirely electrodiffusive, water transport can only be driven by unstirred layers. This has been tested with the cation-selective ionophore gramicidin, the anion-selective ionophore nystatin, and the cation channel Connexin 50 (Zeuthen *et al.,* 1997, 2001; Meinild *et al.,* 1998a; MacAulay *et al.,* 2001) and with expressed K^+ channels (Duquette *et al.,* 2001). Clamping the oocyte to a negative potential caused large inward currents of up to four-fold of those obtained in cotransport experiments but did not induce any initial changes in oocyte volume (see Figs. 2C, 3, and 6B). Volume changes were only apparent after about 35 to 60 sec. In comparison, cotransport produced instantaneous changes in water fluxes (Fig. 6A). Now, in terms of unstirred layers and electrode artifacts, the two situations, one where the currents are carried by the ionophore and the one where the current is carried entirely by cotransport, are similar. The only difference is the molecular mechanism responsible for the transport. We must conclude that it is the molecular mechanisms within the cotransporter itself that give rise to the initial phase of the water transport and not unstirred layers or other artifacts such as ion flows from the clamp electrodes.

The above conclusion rests on two assumptions: that the passive water permeability is the same in the ionophore and the cotransport experiments, and that diffusion of Na^+ and the organic substrate is roughly the same in the cytoplasm. First, in most of the studies just cited, ionophores or channels were inserted into oocytes in which the cotransporter was present but inactivated by the absence of organic substrate. This would ensure the same water permeability; the water permeability supplied by the ionophores is insignificant (Zeuthen *et al.,* 1997). Second, there is evidence that the organic substrates (glucose or glutamate) give rise to larger unstirred layers than those estimated for Na^+. This could be concluded from experiments where the water channel aquaporin 1 (AQP1) was coexpressed with the SGLT1 (Zeuthen *et al.,* 2001). The addition of AQP1 increased the water permeability more than 10-fold, thus enabling unstirred layers as small as 0.5 mosm L^{-1} to draw significant water fluxes. Under clamped conditions, application of sugar still elicited an immediate transport of water. This was compared to experiments where water transport was induced by Na^+ currents via gramicidin (while the SGLT1 was kept inactivated by the absence of sugar). The comparison showed that the diffusion of three Na^+ ions via gramicidin, in the absence of sugar transport, gave the same unstirred layer effects as diffusion of two Na^+ ions and one sugar. Accordingly, diffusion of sugar does not give rise to larger unstirred layer effects than the diffusion of Na^+ ions.

The passive water permeability L_p is not influenced significantly by unstirred layers, provided the L_p is derived from the *initial* rate of change in oocyte volume in response to abrupt changes in the external osmolarity. In aquaporin-expressing oocytes (Meinild *et al.,* 1998b), the L_p is independent of the magnitude of the applied osmotic gradient over the range 2.5 to 100 mosm L^{-1}. This independence constitutes strong evidence against unstirred layers (Finkelstein, 1987).

If, however, osmotic challenges are maintained for longer periods, say minutes, the volume changes of oocytes will deviate from linearity. If this results from a build up of an unstirred layer inside the oocyte, it will be determined by those intracellular constituents which have the smallest diffusion coefficient (Dick, 1966; Meinild *et al.*, 1998b). It would be a mistake to attribute this unstirred layer to diffusion of the small ions Na^+, K^+, Cl^- (Lapointe *et al.*, 2001).

When the cotransport of water is studied under equilibrium conditions, unstirred layers arising from substrate diffusion are absent (Section VII). The relation between the equilibrium (or reversal) potential and osmotic gradients in rabbit SGLT1 showed that the number of water molecules cotransported with the two Na^+ and one sugar was around 247 (Fig. 9). This coupling ratio is similar to those estimated in nonequilibrium experiments: 260 (Loo *et al.*, 1996), 390 (Zeuthen *et al.*, 1997), 424 (Zeuthen *et al.*, 2001). This supports the notion that unstirred layer effects are small in the kinetic experiments.

In a recent paper, Duquette *et al.* (2001) present experiments on water transport by the human SGLT1 and explain them by a theoretical model based exclusively on osmosis and unstirred layers. Unfortunately, the equipment employed by Duquette *et al.* did not permit simultaneous changes in external solutions and recording of oocyte volume (see Fig. 2), which makes their study inconclusive. Also, as will be discussed below, the model used by Duquette *et al.* (2001) for interpretation of their data employs unrealistic values for the diffusion coefficient inside the oocyte and for solute fluxes via the clamp electrode. The experiments of Duquette *et al.* (2001) with channel-mediated transport were in agreement with ours as discussed above.

B. Theoretical Considerations

To evaluate the significance of unstirred layers theoretically, it is necessary first to estimate the value of the intracellular diffusion coefficient D. There is no experimental value for D in *Xenopus* oocytes, but it is possible to give some estimates. Baumgartner *et al.* (1999) find the intracellular conductance in *Xenopus* oocytes to be 20% of the extracellular value. This agrees with the data of Cameron *et al.* (1990) who determined a similar fraction of the intracellular water as bulk water. The conductivity depends on the total concentration of mobile ions in the free fraction which are: Na^+, 6–20; K^+, 88–150; Cl^-, 33–62; all in mmol L^{-1}. The total concentration of these ions is only slightly lower than that of the outside solution, i.e., frog's Ringer solution. The picture that emerges is one where 20 to 25% of the volume is available for free diffusion of ions and smaller solutes, while the rest is occupied by organelles such as yolk platelets and mitochondria. A reasonable estimate, then, would be an average intracellular diffusion coefficient D of around 20 to 25% of the free solution value. Given a free solution diffusion coefficient of $1.5 * 10^{-5}$ cm^{-2} sec^{-1} for 0.1 M NaCl (Robinson and Stokes, 1970) a lower estimate of the average intracellular D would be $3.8\ 10^{-6}$ cm^{-2} sec^{-1}. With a D of $4.0\ 10^{-6}$ cm^{-2} sec^{-1} and an unstirred layer thickness as large as the oocyte diameter 1.2 mm, theory (Dainty and House, 1966) would suggest that unstirred layers may account for only 3% of the observed cotransported water fluxes (Loo *et al.*, 1996).

In a theoretical model, Duquette *et al.* (2001) and Lapointe *et al.* (2001) assumed D to be between 1.2 10^{-6} cm^{-2} sec^{-1} and 0.4 10^{-6} cm^{-2} sec^{-1}, which is from 10 to 40 times lower than the free solution diffusion coefficient. This together with the assumption of a major influx of ions from the electrode enabled them to explain some cotransport experiments by simple osmosis. As argued above, this choice of D is unrealistic. In addition, there is no theoretical or experimental basis to support the notion of significant influxes of ions from the current clamp electrode (see below).

C. Electrode Artifacts

Do ion fluxes from the microelectrodes obscure the assessment of the fluxes mediated by the cotransporters? The question would be particularly relevant for electrogenic cotransporters expressed in *Xenopus* oocytes under voltage clamped conditions. To minimize net solute flow, microelectrodes are filled with KCl solutions of a high concentration. During steady-state experiments, this will ensure that the clamp current in the electrode is carried by equal and opposite fluxes of K^+ and Cl^-. The clamp current *per se* does not bring about any net osmotic flux and should not contribute significantly to the intracellular osmolarity.

This argument is supported by experiments. The *total* number of osmolytes entering the oocyte by the SGLT1 and the clamp electrode can be estimated from the steady-state rate of swelling of the oocyte. The number of osmolytes entering the oocyte by the SGLT1 alone can be determined by the clamp current. The difference between these two numbers is an estimate of the flux from the current electrode. Using the data of Duquette *et al.* (2001) on the human SGLT1, it can be calculated that the influx of osmolytes into the oocyte is predominantly determined by the hSGLT1; there is an *efflux* via the electrode that is only about 15% of flux carried by the hSGLT1. A similar calculation using the hSGLT1 data from Zeuthen *et al.* (2001) also showed that the electrode component was insignificant. The excellent correspondence between these two studies strongly suggests that electrode artifacts are small.

IX. Molecular Models and Conclusions

Cotransporters can be viewed as membrane-bound enzymes that catalyze the movement of substrate from one side of a lipid membrane to the other. In continuation of this idea, the behavior of cotransporters as molecular water pumps can be understood in terms of mechanisms well established for enzymes in general (Fig. 10).

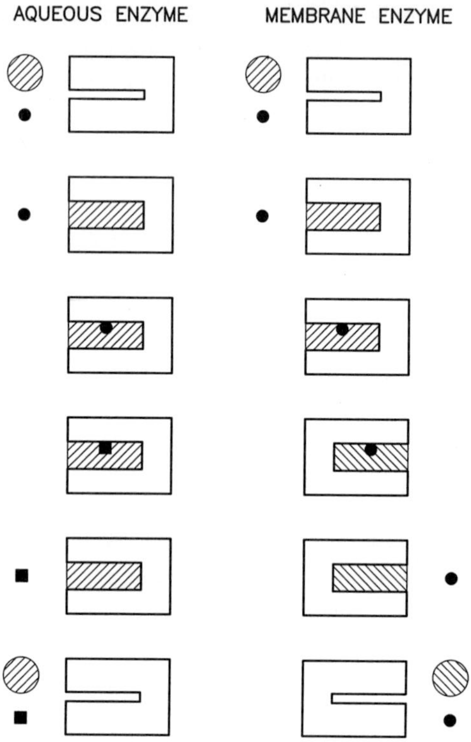

FIG. 10 Molecular model for molecular water pump. An aqueous enzyme (left) opens an aqueous cavity (top) in order for the substrate (●) to gain access to the binding site. The reaction product is subsequently released (bottom, ■). In the process, water (hatched) is taken up from and released to the external medium. For simplicity, we have not included a closure of the enzyme during the enzymatic transformation of the substrate. A membrane-bound enzyme (right) may function as a molecular water pump which mediates cotransport of a nonaqueous substrate (●) and of water (hatched). Hydration of the access volume allows binding of the nonaqueous substrate (top). When this binding is complete, conformational changes shift the permeability barrier from one side of the protein to the other; subsequently the nonaqueous substrate is delivered to the *trans* side. For vectorial transport of water to ensue, it is assumed that the protein then proceeds to a closed state. Consequently, the water contained in the access pathway and on the protein surfaces is delivered to the *trans* bath.

Conformational changes in enzymes involve movements of large protein domains relative to each other. This may lead to changes in the size of the surface of the protein and thereby of the amount of loosely held surface water. Conformational changes may also change the size of water-filled cavities in the proteins (see articles by Parsegian, Rand, and Kornblatt and Kornblatt, this volume). The effects may be studied by monitoring enzymatic activity as a function of the chemical potential for water in the external medium (Parsegian *et al.,* 1993) or by resolution of the three-dimensional crystal structure under various conditions (Qian *et al.,*

1995). Membrane-bound, as well as water-soluble, enzymes have been studied, and a wide range of values for the number of exchangeable water molecules has been determined, from 10 in cytocrome c oxidase (Kornblatt and Hoa, 1990; Kornblatt, 1998; Kornblatt *et al.,* 1998) to more than a thousand in a voltage-gated anion channel (Zimmerberg and Parsegian, 1986). Values of about 100 have been found for the aqueous enzymes hexokinase (Steitz *et al.,* 1981; Rand and Fuller, 1992) and α-amylase (Qian *et al.,* 1995), 60–75 for hemoglobin (Bulone *et al.,* 1991; Colombo *et al.,* 1992), and 40–50 for a K^+ channel (Zimmerberg *et al.,* 1990). Such values are compatible with the water molecules being enclosed in cavities or attached to the protein as surface water. Studies of crystals of the sugar-binding enzyme α-amylase have shown that between 5 and 10% of the protein volume is occupied by water (Qian *et al.,* 1995). This is equivalent to 300 to 500 water molecules, about 90 of which are exchanged with bulk water during conformational changes. From studies of hemoglobin (Colombo *et al.,* 1992), one surface-bound water molecule was found to occupy 10 A^2. In this way, 500 water molecules would occupy a surface of 5000 A^2, equivalent to the surface of a 100-A long, 15-A diameter circular pore. As bulk water, 500 water molecules would occupy the volume of such a pore, namely 15,300 A^3. The K^+/Cl^- cotransporter translocates 500 water molecules per cycle (Zeuthen, 1994). With a molecular weight of the K^+/Cl^- cotransport proteins of 260 kDa (Cherksey and Zeuthen, 1987), this constitutes about 3% of the protein volume.

How could the structural features and conformational changes in cotransport proteins give rise to cotransport of water? One model would be an extension of the mobile barrier model suggested by Mitchell to explain secondary and primary active transport (Mitchell, 1957). Interestingly, Mitchell (1990) did allude to a link between water movements and conformational changes in this model: "If as seems likely, the mobile barrier mechanism involves the opening and closing of a cleft on either side of a substrate-binding domain, one might expect considerable hydrodynamic action as aqueous medium was sucked in and squirted out by the crevices." In this model, vectorial transport of water and the other substrates would take place in the cotransport protein if the phenomena described for the aqueous enzyme took place in a specific order (Fig. 10). First, the nonaqueous substrates gain access to the binding sites by migration into an aqueous cavity that opens to the *cis* side of the protein, while a barrier separates the substrates and the *trans* compartment. The requirement for an aqueous channel has been established for the K^+ channel. The ion migrates in the channel to reach its binding site, and water serves to shield the ionic charges prior to binding (Doyle *et al.,* 1998). Second, the binding of the substrates results in a conformational change. In the mobile barrier model, this causes the barrier to open, exposing the bound substrates to the *trans* compartment. Simultaneously, the *cis* barrier closes again, separating the substrates from the *cis* compartment. Third, the substrates are released into the *trans* compartment. Fourth, in analogy to the aqueous enzymes, the cavity finally closes and water delivered to the *trans* compartment. From this closed configuration, the protein can open again to the *cis* side and a new cycle commence. As a result, the water in the cavity is cotransported from the *trans* to the *cis* side. A similar transport would take place if water were located to the surfaces of the cavity.

In this model, the transport of the nonaqueous substrates cannot take place without the movement of a specific amount of water. In other words, water is a cosubstrate. This means that the osmotic work in transporting the volume of water contributes to the overall energy balance as does the traditional chemical and electrical work. This interrelation is described by the Gibbs equation (Eq. 1). Under conditions of nonequilibrium (transport), the strict coupling inherent in the model ensures that free energy can be transferred from one kind of substrate to another, including the water molecules.

References

Barbour, B., Brew, H., and Attwell, D. (1988). Electrogenic glutamate uptake in glial cells is activated by intracellular potassium. *Nature* **335,** 433–435.

Baumgartner, W., Islas, L., and Sigworth, F. J. (1999). Two-microelectrode voltage clamp of Xenopus oocytes: Voltage errors and compensation for local current flow. *Biophys. J.* **77,** 1980–1991.

Bulone, D., Donato, I. D., Palma-Vittorelli, M. B., and Palma, M. U. (1991). Density, structural lifetime, and entropy of H-bond cages promoted by monohydric alcohols in normal and supercooled water. *J. Chem. Phys.* **94,** 6816–6826.

Cameron, I. L., Merta, P., and Fullerton, G. D. (1990). Osmotic and motional properties of intracellular water as influenced by osmotic swelling and shrinkage of *Xenopus* Oocytes. *J. Cell. Physiol.* **142,** 592–602.

Cherksey, B. D., and Zeuthen, T. (1987). A membrane protein with a K^+ and a Cl^- channel. *J. Physiol.* **387,** 33P.

Colombo, M. F., Rau, D. C., and Parsegian, V. A. (1992). Protein solvation in allosteric regulation: A water effect on hemoglobin. *Science* **256,** 655–659.

Dainty, J., and House, C. R. (1966). Unstirred layers in frog skin. *J. Physiol.* **182,** 66–78.

Dick, D. A. T. (1966). "Cell water." Butterworth, London.

Doyle, D. A., Cabral, J. M., Pfuetzner, R. A., Kou, A., Gulbis, J. M., Cohen, S. L., Chait, B. T., and MacKinnon, R. (1998). The structure of the potassium channel: Molecular basis of K+ conduction and selectivity. *Science* **280,** 69–77.

Duquette, P.-P., Bisonnette, P., and Lapointe, J.-Y. (2001). Local osmotic gradients drive the water flux associated with Na+/glucose cotransport. *Proc. Natl. Acad. Sci. USA* **98,** 3796–3801.

Finkelstein, A. (1987). "Water Movement through Lipid Bilyers, Pores and Plasma Membranes." Wiley-Interscience, New York.

Hamann, S., la Cour, M., Lui, G. M., Bundgaard, M., and Zeuthen, T. (2000). Transport of protons and lactate in cultured human fetal retinal pigment epithelial cells. *Pflügers Arch.* **440,** 84–92.

Kanner, B. I., and Bendahan, A. (1982). Binding order of substrates to the sodium and potassium ion coupled L-glutamic acid transporter from rat brain. *Biochemistry* **21,** 6327–6330.

Kornblatt, J. A. (1998). The water channel of cytochrome c oxidase: Inferences from inhibitor studies. *Biophys. J.* **75,** 3127–3134.

Kornblatt, J. A., and Hoa, G. H. B. (1990). A nontraditional role for water in the cytochrome c oxidase reaction. *Biochemisrty* **29,** 9370–9376.

Kornblatt, J. A., Kornblatt, M. J., Rajotte, I., Hoa, G. H. B., and Kahn, P. C. (1998). Thermodynamic volume cycles for electron transfer in the cytochrome c oxidase and for the binding of cytochrome c to cytochrome c oxidase. *Biophys. J.* **75,** 435–444.

Lapointe, J. Y., Gagnon, M., Porier, S., and Bissonette, P. (2001). The presence of local osmotic gradients accounts for the water flux driven by the Na^+/glucose cotransporter. Proceedings of the IUPS meeting, New Zealand 2001. *J. Physiol.* (in press).

Leung, D. W., Loo, D. F., Hirayama, B. A., Zeuthen, T., and Wright, E. M. (2000). Urea transport by cotransporters. *J. Physiol.* **528.2,** 251–257.

Loo, D. D. F., Zeuthen, T., Chandy, G., and Wright, E. M. (1996). Cotransport of water by the Na^+/glucose cotransporter. *Proc. Natl. Acad. Sci. USA* **93,** 13,367–13,370.

Ludwig, C. (1861). "Lehrbuch der Physiologie des Menschen." C. F. Wintersche Verlagshandlung, Leipzig und Heidelberg.

MacAulay, N., Gether, U., Klaerke, D. A., and Zeuthen, T. (2001). Water transport by the Na^+-coupled glutamate cotransporter. *J. Physiol.* **530,** 367–378.

Meinild, A.-K., Klaerke, D. A., Loo, D. D. F., Wright, E. M., and Zeuthen, T. (1998a). The human Na^+/glucose cotransporter is a molecular water pump. *J. Physiol.* **508.1,** 15–21.

Meinild, A.-K., Klaerke, D. A., and Zeuthen, T. (1998b). Bidirectional water fluxes and specificity for small hydrophilic molecules in aquaporins 0 to 5. *J. Biol. Chem.* **273,** 32,446–32,451.

Meinild, A.-K., Loo, D. D.F, Pajor, A., Zeuthen, T., and Wright, E. M. (2000). Water transport by the renal Na^+/dicarboxylate cotransporter. *Am. J. Physiol.* **278,** F777–F783.

Mitchell, P. (1957). A general theory of membrane transport from studies of bacteria. *Nature* **180,** 134–136.

Mitchell, P. (1990). Osmochemistry of solute translocation. *Res. Microbiol.* **141,** 286–289.

Nelson, P. J., Dean, G. E., Aronson, P. S., and Rudnick, G. (1983). Hydrogen ion cotransport by the renal brush border glutamate transporter. *Biochemistry* **22,** 5459–5463.

Parent, L., Supplison, S., Loo, D. D. F., and Wright, E. M. (1992). Electrogenic properties of the cloned Na^+/glucose cotransporter. *J. Membr. Biol.* **125,** 49–62.

Parsegian, A., Rau, D., and Zimmerberg, J. (1993). Structural transitions induced by osmotic stress. *In*"Membranes, Metabolism, and Dry Organisms" (A. C. Leopold, Ed.). Comstock Publishing Associates, London.

Qian, M., Haser, R., and Payan, F. (1995). Carbohydrate binding sites in a pancreatic α-amylase-substrate complex, derived from X-ray structure analysis at 2.1 Å resolution. *Prot. Sci.* **4,** 747–755.

Rand, R. P., and Fuller, N. L. (1992). Water as an inhibiting ligand in yeast hexokinase. *Biophys. J.* **61,** A345.

Reid, E. W. (1892). Report on experiments upon absorbtion without osmosis. *Br. Med. J.* **1,** 323–326.

Reid, E. W. (1901). Transport of fluid by certain epithelia. *J. Physiol.* **26,** 436–444.

Robinson, R. A., and Stokes, R. H. (1970). "Electrolyte Solutions." Second edition, fifth impression. Butterworths, London.

Rudnick, G. (1998). Ion-coupled neurotransmitter transport: Thermodynamic vs. kinetic determinations of stoichiometry. *Methods Enzymol.* **296,** 233–246.

Steitz, T. A., Shoham, M., and Bennett, W. S., Jr. (1981). Structural dynamics of yeast hexokinase during catalysis. *Phil. Trans. R. Soc. Lond.* **293,** 43–52.

Zerangue, N., and Kavanaugh, M. P. (1996). Flux coupling in a neuronal glutamate transporter. *Nature* **383,** 634–637.

Zeuthen, T. (2000). Molecular water pumps. *Rev. Physiol. Biochem. Pharm.* **141,** 97–151.

Zeuthen, T. (1981). Isotonic transport and intracellular osmolarity in the necturus gallbladder epithelium. *In* "Water Transport across Epithelia." (H. H. Ussing, N. Bindslev, N. A. Lassen, and O. Sten-Knudsen, Eds.). Munksgaard, Copenhagen.

Zeuthen, T. (1982). Relations between intracellular ion activities and extracellular osmolarity in necturus gallbladder epithelium. *J. Membr. Biol.* **66,** 109–121.

Zeuthen, T. (1983). Ion activities in the lateral intercellular spaces of gallbladder epithelium transporting at low external osmolarities. *J. Membr. Biol.* **76,** 113–122.

Zeuthen, T. (1991a). Water permeability of ventricular cell membrane in choroid plexus epithelium from necturus maculosus. *J. Physiol.* **444,** 133–151.

Zeuthen, T. (1991b). Secondary active transport of water across ventricular cell membrane of choroid plexus epithelium of necturus maculosus. *J. Physiol.* **444,** 153–173.

Zeuthen, T. (1992). From contractile vacuole to leaky epithelia. Coupling between salt and water fluxes in biological membranes. *Biochim. Biophys. Acta* **1113,** 229–258.

Zeuthen, T. (1994). Cotransport of K^+, Cl^- and H_2O by membrane proteins from choroid plexus epithelium of Necturus maculosus. *J. Physiol.* **478,** 203–219.

Zeuthen, T. (1995). Molecular mechanisms for passive and active transport of water. *Int. Rev. Cytol.* **160,** 99–161.

Zeuthen, T., Hamann, S., and La Cour, M. (1996). Cotransport of H^+, lactate and H_2O by membrane proteins in retinal pigment epithelium of bullfrog. *J. Physiol.* **497,** 3–17.

Zeuthen, T., Meinild, A.-K., Klaerke, D. A., Loo, D. D. F., Wright, E. M., Belhage, B., and Litman, T. (1997). Water transport by the Na^+/glucose cotransporter under isotonic conditions. *Biol. Cell* **89,** 307–312.

Zeuthen, T., Meinild, A.-K., Loo, D. D. F., Wright, E. M., and Klaerke, D. A. (2001). Isotonic transport by the Na+ -glucose cotransporter SGLT1. *J. Physiol.* **531. 3,** 631–644.

Zeuthen, T., and Stein, W. D. (1994). Co-transport of salt and water in membrane proteins: Membrane proteins as osmotic engines. *J. Membr. Biol.* **137,** 179–195.

Zimmerberg, J., Bezanilla, F., and Parsegian, V. A. (1990). Solute inaccessible aqueous volume changes during opening of the potassium channel of the squid giant axon. *Biophys. J.* **57,** 1049–1064.

Zimmerberg, J., and Parsegian, V. A. (1986). Polymer inaccessible volume changes during opening and closing of a voltage-dependent ionic channel. *Nature* **323,** 36.

General Models for Water Transport across Leaky Epithelia

Thomas Zeuthen
Institute of Medical Physiology, The Panum Institute, University of Copenhagen, DK 2200 Copenhagen N, Denmark

The group of leaky epithelia, such as proximal tubule and small intestine, have several common properties in regard to salt and water transport. The fluid transport is isotonic, the transport rate increases in dilute solutions, and water can be transported uphill. Yet, it is difficult to find common features that could form the basis for a general transport model. The direction of transepithelial water transport does not correlate with the direction of the primary active Na^+ transport, or with the ultrastucture as defined by the location of apical and basolateral membranes, of the junctional complex and the lateral intercellular spaces. The presence of specific water channels, aquaporins, increases the water permeability of the epithelial cell membranes, i.e., the kidney proximal tubule. Yet other leaky epithelia, for example, the retinal pigment epithelium, have no known aquaporins. There is, however, a general correlation between the direction of transepithelial transport and the direction of transport via cotransporters of the symport type. A simple epithelial model based on water permeabilities, a hyperosmolar compartment and restricted salt diffusion, is unable to explain epithelial transport phenomena, in particular the ability for uphill water transport. The inclusion of cotransporters as molecular water pumps in these models alleviates this problem.

KEY WORDS: Choroid plexus, Gallbladder, Proximal tubule, Molecular water pumps, Aquaporins, Cotransporters.

I. Introduction

Salt and water transport across epithelial and endothelial cell layers is central for the maintenance of the internal milieu of the organism. Where large rates of volume transport are required, so-called leaky epithelia are predominant. The majority of these consist of a single layer of cells and gain their name from the high permeability of the paracellular pathway to smaller ions. Leaky epithelia are found throughout the body, in the kidney proximal tubule, the small intestine, choroid plexus in the brain, glandular tissue, and in the eye (see Hamann, this volume). The transport rates are appreciable: In humans, the kidney and intestine epithelia together are responsible for reabsorbing up to 200 liters of fluid each day. Tight epithelia as are found in the distal part of the kidney tubules, the gastric mucosa, and in the colon form another group. These epithelia have relatively low permeabilities to ions via the paracellular pathway and are mainly engaged in facultative (i.e., hormonally controlled) salt and water transport.

This review will deal with the intrinsic ability of leaky epithelia for water transport and how this is linked to metabolism. It is generally held and probably true that the link is via the ubiquitous Na^+/K^+–ATPase, the action of which leads to the build up of ionic gradients across the membrane (see Stein, this volume). The locations and magnitudes of these gradients will, in turn, depend upon the distribution of cotransporters, channels, and the ultrastructure of the cell layer. The crucial question is how the energy invested in the ionic gradients can be transferred to a transepithelial, or vectorial, transport of water.

In terms of fluid transport, leaky epithelia have three fundamental properties in common: isotonic transport, increased transport rates in dilute external solutions, and transport against an osmotic gradient. This consensus suggests the existence of a general model for fluid transport. The group of leaky epithelia also exhibits a high degree of variability. A general model will have to cope with the lack of correlation between, on the one hand, the direction of fluid transport and, on the other, the ultrastructure of the cell layer and the direction of primary active transport of Na^+ by the ATPase. There is, however, a general correlation between the direction of fluid transport and the direction of cotransport in membrane proteins of the symport type. Consequently, a general model of transcellular fluid transport based on the cotransporters will be presented and discussed in relation to the three fundamental properties of fluid transport. Evidently, there is also a general correlation between junctional properties and fluid transport. This has led to the formulation of a paracellular model (see Shachar-Hill and Hill, this volume).

Some reviews on epithelial fluid transport are by House (1974), Diamond (1979), Hill (1980), Spring (1983, 1998), Schafer (1984, 1990), Whittembury (1985), Fischbarg (1989), Tripathi and Boulpaep (1989), Zeuthen (1992, 1995, 1996), Schultz (1998), Snigirevskaya and Komissarchik (2000).

II. Transcellular or Paracellular Transport

The present review focuses on models in which the majority of ion and water transport follows the cellular, as opposed to the paracellular, route (Fig. 1). There are several lines of evidence to supports this; see reviews above. Investigations with

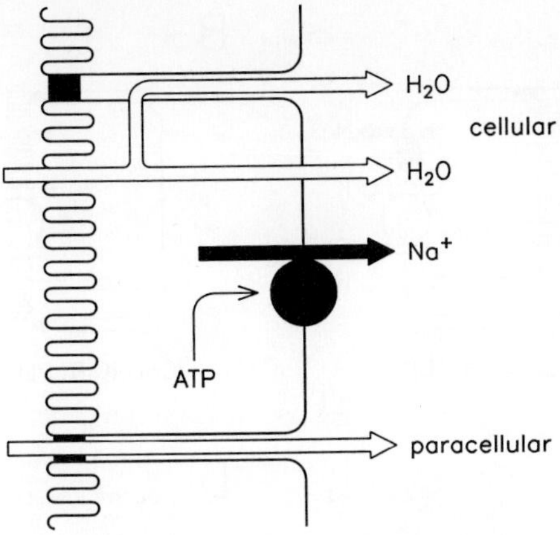

FIG. 1 Pathways for Na^+ and water transport across leaky epithelia. A major unsolved problem in epithelial transport physiology is how the hydrolysis of ATP is linked to transepithelial transport of water. It is generally accepted that ATP provides the energy for a membrane-bound Na^+/K^+-ATPase (filled circle), the activity of which results in the storage of free energy in transmembrane ionic gradients. The coupling between these gradients to water transport by the *cellular* route is the subject of the present article. The possibility for *paracellular* transport is the subject for the chapter by Shachar-Hill and Hill, this volume.

tracers and ion-selective microelectrodes in the gall bladder and the choroid plexus, for example, suggest that a large fraction of the transepithelial Na^+ and Cl^- transport goes through the cells (Graf and Giebish, 1979; Zeuthen and Wright, 1981; Saito and Wright, 1982; Baerentsen *et al.,* 1983; Giraldez, 1984; Reuss, 1984; Hill and Hill, 1987a). Some of the most illustrative evidence was obtained with the ion-selective microelectrodes. When transport across the mucosal membrane was arrested, the rate of decreases in intracellular ion activities mirrored the transepithelial transport. Studies of changes in cell volume can also been taken as evidence for a transcellular route of waterflow (Spring and Hope, 1979; Spring and Ericson, 1982; Ericson and Spring, 1982a,b). Incidentally, this type of experiment precludes significant amounts of recycling of Na^+ (but see Larsen *et al.* 2000). If Na^+ ions were taken up again from the exit bath, the Na^+/K^+ ATPase should transport faster than that estimated from the net transepithelial solute flux. It does not. In case of significant recirculation, the rate of decrease in intracellular Na^+ after the pump is arrested should have been much larger than that given by the transepithelial flux.

III. Lack of Correlation between the Direction of Fluid Transport, Ultrastructure, and Localization of the Na^+/K^+ ATPase

In most cases, the ultrastructure can be defined on the basis of three elements: the apical membrane, the serosal membrane, and the junctional complex. Four types

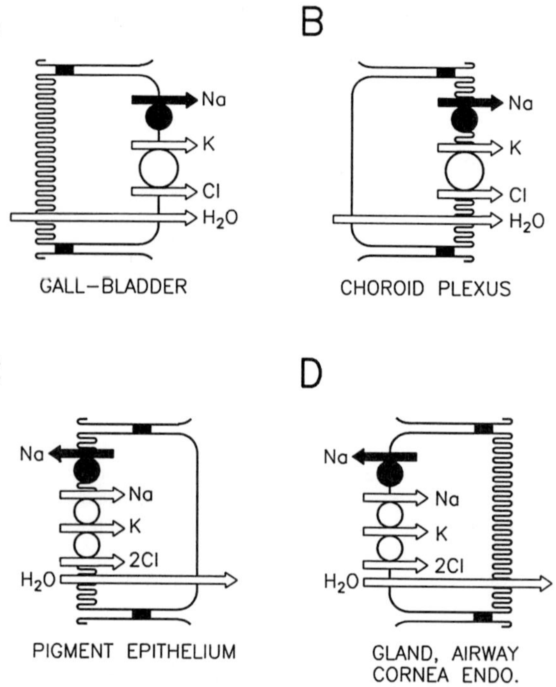

FIG. 2 Correlation between ultrastructure, water transport, and ion transport in different types of epithelia. The ultrastructure is defined by the brush-border (folded), the basolateral membrane, and the junctional complex (black squares). The direction of Na^+ transport by the Na^+/K^+-ATPase is given by the filled arrow. The location and direction of transport is also given for some cotransporters (open arrows). (A) Forward-facing epithelium as found in gallbladder, small intestine, and kidney proximal tubule. (B) Backward-facing epithelia, for example, choroid plexus. (C) Retinal pigment epithelium. (D) Secretory system as found in airway epithelium, acini of glands, or cornea endothelium Three things are noted: First, the ultrastructure is not correlated to the direction of water transport. Second, the direction of the primary active Na^+ transport is not correlated to the direction of the water transport. Third, there is a correlation between the direction of water transport and the direction of cotransport. Adapted from Zeuthen (1996).

of epithelia can be recognized, for examples, see Fig. 2. The epithelia from the gall bladder, the small intestine, and the kidney proximal tubule form one group (Fig. 2A) in which water enters across the apical (or brush-border) membrane and exits across the serosal (or basolateral) membrane. The latter membrane together with the apically located junctional complex defines the lateral intercellular spaces. In these epithelia, the Na^+/K^+ ATPase also pumps Na^+ into the lateral intercellular spaces in the direction of the transepithelial flux of water. In the choroid plexus epithelium (Fig. 2B), the direction of water transport is from basal to apical, opposite to that in Fig. 2A. Water enters across the serosal membrane, while the apical membrane with its microvilli constitutes the exit membrane for water. As in the former group of epithelia, the junctional complex (black squares in Fig. 2) is close to the apical membrane. The Na^+/K^+ATPase is localized to the apical membrane and pumps Na^+ directly into the secreted solution. In the retinal pigment epithelium (Fig. 2C), water transport and ultrastructure are orientated as in the gall bladder group (Fig. 2A). Water enters across the apical (or brush-border) membrane and leaves across the serosal (or basolateral) membrane; the junctions are located toward the apical end. The orientation of the Na^+/K^+ATPase is now, however, opposite to that in Fig. 2A. It is located in the apical membrane and pumps Na^+ in the direction opposite to that of the transepithelial water flux. In the secretory systems such as acini of glands, airway epithelia, and the cornea endothelium (Fig. 2D), the ultrastructure and the direction of water transport are as in the choroid plexus epithelium (Fig. 2B). Water enters across the serosal membrane and leaves across the apical membrane. The orientation of the Na^+/K^+ATPase is opposite to that in Fig. 2B. It is located to the serosal side and pumps Na^+ in the direction opposite to that of the transepithelial water flux.

It appears that all combinations of water fluxes, active Na^+ fluxes, and ultrastructural orientations are found among the group of leaky epithelia. It is unlikely, therefore, that this set of parameters can be the basis of a general model for isotonic fluid transport and uphill water transport, phenomena which are common for the group of leaky epithelia (see Section V). The point can be illustrated by a specific example. Most speculations on the coupling between salt and water are based on models for the gall bladder type of epithelium (for a review, see Spring, 1998). Salt is assumed to accumulate in the lateral intercellular spaces by active transport and to effect water transport by osmosis. The osmolarity in the lateral spaces is a complicated function of the rate of active Na^+ transport, the passive water permeability of the lateral cell membranes. In some cases, the basement membrane and the subepithelial tissues are assumed to restrict salt diffusion. Were this true, it would imply an increased osmolarity in the spaces (see Section VII.A) (Persson and Spring, 1984; and references in Zeuthen, 1992). This model, however, is inapplicable to other types of leaky epithelia. In Fig. 2, there are examples of epithelia where the lateral spaces are orientated opposite to that of the gall bladder, i.e., choroid plexus. Importantly, in the choroid plexus epithelium there is no restricted diffusion, since secretion takes place directly into the exit bath. In epithelia such as cornea endothelium, Na^+ is indeed pumped into the lateral intercellular spaces. Yet the transepithelial water transport is in the opposite direction to that which the accumulation of Na^+ would imply. In other words, the lateral intercellular spaces and the active Na^+ transport into them are orientated exactly opposite to that of the gall bladder.

Dissociation between water transport and Na^+/K^+ATPase activity has also been demonstrated in experiments where the pump is inhibited by ouabain (Hill and Hill, 1987b). Fluid transport exhibited an immediate reduction of one-third at ouabain application, after which it continued for at least an hour. The mechanism for Na^+ exit and the coupling to water transport under these conditions remain to be established.

In conclusion, a simple topological analysis underscores the difficulty of setting up a general model for epithelial fluid transport in these terms. Formidable problems are indeed posed by some epithelia. The stimulated gastric mucosa secretes daily up to 2 liters of HCl solution which is isotonic to blood plasma. The H^+ secretion is mediated by a H^+/K^+–ATPase located to the ventricular surface of the oxyntic cells; transport of K^+ and Cl^- into the lumen may take place via channels (Wolosin and Forte, 1985). The Na^+/K^+–ATPase is located to the serosal membrane and pumps Na^+ into the blood. The juncions are tight to ions and are located toward the ventricular lumen. Isotonic transport in this epithelium is clearly a challenge to any model. Another intriguing example is the hypotonic secretion into the contractive vacuole of the freshwater amoeba as described in the chapter by Allen, this volume.

IV. Correlation between Direction of Water Transport and Localization of Specific Water Channels, Aquaporins

In regard to the distribution of aquaporins (AQPs), leaky epithelia can be divided into three groups. Those which express aquaporins at both membranes, those which only express aquaporins at the membrane across which water is secreted, and those which express no or only insignificant amounts of aquaporins (Fig. 3). As a rule, leaky epithelia that operate in an environment of constant external osmolarity are equipped with aquaporins at both membranes. The kidney proximal tubule has high passive water permeabilities of both the apical and the serosal membrane due to the presence of AQP1 (for reviews, see Borgnia *et al.,* 1999; Verkman *et al.,* 2000). This enables the proximal tubule *in situ* to take advantage of a small transepithelial osmotic difference to increase the rate of fluid absorbtion, in the rat by about 50% (Green and Giebisch, 1984). As in the proximal tubule, the corneal endothelium expresses AQP1 on both membranes (Hamann *et al.,* 1998).

Epithelia that are exposed to high or variable osmolarities apparently do not express significant amounts of aquaporins. The physiological role of the small intestine epithelium and the retinal pigment epithelium is different from that of the kidney proximal tubule. These epithelia face high and variable osmolarities at their apical side (see Fig. 2) and may have to transport water uphill as a part of their normal physiological function. High passive water permeabilities, i.e., a high level of aquaporins, would compromise this barrier function and lead to systemic dehydration, in the case of the intestine, and to detachment of the neuro-retina, in the case of the pigment epithelium (see Hamann, this volume). Accordingly, there seems to be no significant expression of aquaporins in the small intestine. AQP3 has been located to the serosal membrane of cells in the ileum and to a lesser extent in the jejunum, most pronounced near the villus tips (Ramírez-Lorca *et al.,*

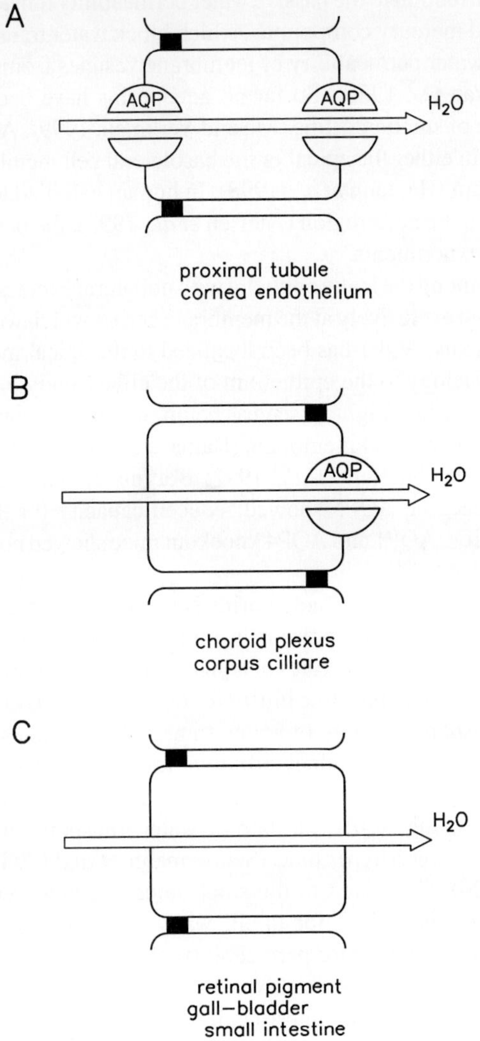

FIG. 3 Distribution of water channels, aquaporins, in leaky epithelia. Three types of distributions are dominating: (A) Aquaporins in both types of membranes; examples are the kidney proximal tubule and the cornea endothelium. (B) Aquaporins in the exit membrane only; examples are the choroid plexus epithelium and the corpus cilliare from eye. Gland acini may be seen as belonging to this group; see text. (C) Epithelia with no known aquaporins: the retinal pigment epithelium. The gall bladder and the small intestine can be considered to belong to this group; see text.

1999). Their contribution to the passive water permeability must be small, however. $HgCl_2$ and related mercury compounds, which block water transport in AQP3, had no effect on the water permeability of membrane vesicles from rat small intestine (Heeswijk and Van Os, 1986). So far no aquaporins have been localized to the apical membrane of small intestine (Ma and Verkman, 1999). Also, no aquaporins have been found in either the apical or the basolateral cell membrane of the retinal pigment epithelium (Hamann *et al.,* 1998). In human gall bladder, AQP1 has been located but only to the neck region (Nielsen *et al.,* 1993), the part usually removed during transport experiments.

Interestingly, among the leaky epithelia with unilateral expression of aquaporins, the protein is found exclusively at the membrane across which water leaves the cell. In the choroid plexus, AQP1 has been localized to the apical membrane (Borgnia *et al.,* 1999), in analogy to the epithelium of the cillary body of the eye (Hamann *et al.,* 1998). Gland acini might be seen as belonging to this group. Salivary glands express AQP5 at their apical membrane (Raina *et al.,* 1995) and AQP1 and AQP4 at their serosal side (Nielsen *et al.,* 1997; Borgnia *et al.,* 1999). But, whereas transgenic mice lacking AQP5 showed reduced capacity for fluid transport and hypertonic secretion, AQP1 and AQP4 knockout mice showed no defects (Ma *et al.,* 1999).

Accordingly, the presence of aquaporins is not a general prerequisite for leaky epithelia; aquaporins may be present in both, in one, or in neither of the two membranes. What, then, are the roles of aquaporins? Comparisons between kidney proximal tubule and small intestine illustrate some of them. (i) One role is to ensure near-isotonic transport. The rate of solute transport determines the maximal rate of fluid transport rate. In the kidney proximal tubule, the presence of aquaporins ensures that water transport can proceed sufficiently fast to ensure isotonicity. In transgenic mice lacking AQP1, the rate of water transport cannot keep up, and the transportate becomes hypertonic (Schnermann *et al.,* 1998; Verkman, 1999; Vallon *et al.,* 2000). Compared to the small intestine, however, the presence of aquaporins reduces the ability for uphill water transport; see Table I. (ii) The high, aquaporin-mediated, water permeability of the kidney brush border allows the water flux into the cells to be driven by an intracellular hyperosmolarity of about 2 mosm L^{-1}. In the small intestine, it takes around 40 mosm L^{-1}, given the relatively low water permeability (for references, see Zeuthen *et al.,* 2001). Thus a much larger energy consumption is required in the small intestine to achieve the same rate of fluid transport as in the kidney. The increase in energy consumption has two components. First, maintenance of a larger ionic gradient (against dissipative leaks). Second, water is moved out of the cell against a higher osmotic gradient. Seen this way, the presence of aquaporins is an energy saving measure.

In conclusion, aquaporins are not mandatory for isotonic transport. In those epithelia where they are present, they reduce the osmotic gradients across the plasma membranes. This cuts the energy cost of isotonic fluid transport significantly but reduces the capacity for uphill water transport.

TABLE I
Water Transport Parameters for Some Leaky Epithelia (See Eq. 1)

Tissue	J_{H2O} [nl cm^{-2}sec^{-1}]	L_p [10^{-2}cm sec^{-1}]	$\Delta\pi_{max}$[mosm L^{-1}]
Kidney proximal convoluted tubule rat	10–65[a]	10–30[b]	20–60[c]
Small intestine			
Rabbit	6.7[d]	1.3[d]	
Rat	25[e]	2.5[e]	50[e]
			150[f]
Dog			200[g]
Gall bladder			
Rabbit	14[h]	0.24[h]	80[h]
Dog			120[i]
Necturus	3.5[j]		
	2.7[k]	1.6[k]	30[k]
	3.3[l]		
		3.5[m]	
Fish	4.1[n]	0.55[n]	40[n]
Choroid plexus			
Rabbit			160[o]

[a]Green and Giebisch (1984, 1989); Weinstein and Windhager (1985); Bomsztyk and Wright (1986). [b]Berry (1983); Green and Giebisch (1984, 1989); Liu *et al.* (1984); Preisig and Berry (1985); [c]Bomsztyk and Wright (1986); Green *et al.* (1991); Frömter *et al.* (1973); Alpern *et al.* (1985). [d]Naftalin and Tripathi (1986). [e]Pappenheimer and Reis (1987). [f]Parsons and Wingate (1961). [g]Grim (1962); Hakim *et al.* (1963). [h]Diamond (1964). [i]Grim and Smith (1957). [j]Zeuthen (1982). [k]Persson and Spring (1982). [l]Reuss (1985). [m]Cotton *et al.* (1989). [n]Diamond (1962). [o]Welch *et al.* (1966).

V. Correlation between the Direction of Water Transport and of Cotransport

The correlation that does exit is that between the direction of water transport and the direction of secondary active ion transport maintained by cotransporters of the symport type. In one group of epithelia (A and B in Fig. 2), there is a K^+/Cl^- cotransporter present in the membrane across which water leaves the cell. In another group (C and D in Fig. 2), a $Na^+/K^+/2Cl^-$ cotransporter is present in the membrane across which water enters the cells.

Given the high intracellular concentrations of K^+ and Cl^-, the K^+/Cl^- cotransporter transports ions in the same direction as the water flow. The cotransporter has been located in gall bladder (Reuss, 1983; Garzia and Armstrong, 1983; Larson and Spring, 1984; Hill and Hill, 1987a), small intestine (Halm *et al.*, 1985), kidney proximal tubule (Baum and Berry, 1984; Gullans *et al.*, 1986; Eveloff and

Warnock, 1987; Sasaki, 1988), and choroid plexus (Johanson *et al.,* 1992, 1993; Zeuthen, 1987, 1994a). The unilateral localization and mode of regulation makes the K^+/Cl^- cotransporter a good candidate for direct participation in the water transport. It is well described that the activity of the cotransporter is increased with cell swelling, that is, the cotransporter is active in regulatory volume decrease (RVD) (see Stein, this volume, and references below). When water enters the cell across the opposite membrane, the ensuing tendency for cell swelling would stimulate the cotransporter and somehow increase the export of water. In this model, fluid transport can be seen as a kind of one-sided or vectorial volume regulation. The K^+ and Cl^- ions can also exit via channels. In the excised frog choroid plexus, it was found that the transport of KCl was shared about equally between channel and cotransport (Zeuthen, 1994a). This fraction may be regulated according to the water balance of the tissue.

In another group of epithelia, there is a $Na^+/K^+/2Cl^-$ cotransporter present in the membrane across which water enters the cell. Given the usual intra- and extracellular concentrations of Na^+, K^+, and Cl^-, the direction of cotransport is inward and coincides with that of the water flow (see Fig. 2). The cotransporter has been located in gland acini (Nauntofte, 1992), airway epithelia (Willumsen *et al.,* 1993), corpus cilliare (see Hamann, this volume), and in the retinal pigment epithelium (LaCour, 1993). The unilateral localization and mode of regulation makes the $Na^+/K^+/2Cl^-$ cotransporter a good candidate for a direct role in transepithelial water transport. The cotransporter is activated by cell shrinkage, that is, the cotransporter is active in regulatory volume increase (RVI) (see Stein, this volume, and references below). When water leaves the cell across the other membrane, cell shrinkage would stimulate the cotransporter, and an increase in water influx would ensue. In this model, fluid transport can again be seen as a one-sided or vectorial volume regulation. The vectorial transport of ions by other cotransporters may also coincide with the direction of water transport. This applies to the H^+/lactate cotransporter in the retinal pigment epithelium. Water and ion transports were found to be directly coupled in this protein (Zeuthen *et al.,* 1996).

In conclusion, the K^+/Cl^- and $Na^+/K^+/2Cl^-$ cotransporters play central roles in the volume regulation of apolar cells (some reviews are by Lauf *et al.,* 1992; Lang *et al.,* 1993; Parker, 1993; O'Neill, 1999). In epithelia, their unilateral localization would suggest a direct role in vectorial water transport. The activity of the cotransporters is under multifactorial control. This is well described for the K^+/Cl^- (Jennings and Schulz, 1991; Kaji and Tsukitani, 1991; Parker *et al.,* 1991; Colclasure and Parker, 1992; Sachs, 1994) and the $Na^+/K^+/2Cl^-$ cotransporter (Lauf, 1985; Haas, 1989; Pewitt *et al.,* 1990a,b; Haas and McBrayer, 1994; see also review, Zeuthen, 1996). Accordingly, the regulation of these cotransporters in epithelia might lead to control of transepithelial transport.

VI. Three General Properties of Leaky Epithelia

A general model for water transport in leaky epithelia should explain three groups of findings: isotonic transport, an increased rate of transport in dilute solutions, and the ability for uphill transport of water. So far, all leaky epithelia tested exhibit these properties (for reviews, see Introduction). The phenomena are most conveniently analyzed given the empirical relation between the transepithelial rate of water transport J_{H2O} and the transepithelial osmotic difference illustrated in Fig. 4A and B. The epithelia from small intestine and kidney proximal tubule are particularly well studied in this respect. In these, transport takes place from the mucosal side (m) into the serosal solution (s), and the expression for J_{H2O} is of the form:

$$J_{\mathrm{H2O}} = J^{V}_{\mathrm{H2O}} - L_p\,(\pi_m - \pi_s) \tag{1}$$

J_{H2O} can be divided into two parts, an active and a passive part. The first term, J^{V}_{H2O}, describes the intrinsic or vital transport capacity of the epithelium. It depends on an intact metabolism and arises as a coupling, within the epithelium, between primary active Na^{+} transport and water fluxes. The second term, $L_p(\pi_m - \pi_s)$, describes the passive osmotic flux across the epithelium and is given by the passive water permeability L_p of the epithelium and the transepithelial osmotic gradient of $\pi_m - \pi_s$ (or $\Delta\pi$). The separation into an active and passive part is supported by the fact that metabolc poissons abolish J^{V}_{H2O} while leaving L_p intact (Diamond, 1962, 1964; Parsons and Wingate, 1961; Green and Giebisch, 1989a,b; Green *et al.*, 1991). Figures 4C and D illustrate the data from the rat kidney proximal tubule and rat small intestine. Data for some mammalian epithelia are given in Table I.

A. Isotonic Transport

Consider an epithelium bathed on the two sides in the same physiological NaCl solution, $\pi_m = \pi_s$ in Eq. (1). Isotonic transport is manifest when the epithelium transports a solution which has the same osmolarity as the bathing solutions (for a review, see House, 1974, Table 10.1). The finding raises at least two (so far unanswered) questions: How does the epithelium generate the J^{V}_{H2O} of Eq. (1) apparently without any transepithelial osmotic gradient? What are the mechanisms that ensure that the solute flux, J_s, and the water flux J_{H2O}, proceed in a ratio that equals the concentration of the bathing solution?

Two properties of isotonic transport should be emphasized. First, when isotonic transport is studied in the unilateral preparation, the epithelium is bathed by the test solution on the mucosal side, while the serosal solution consists of the secreted solution itself. In gall bladder, these experiments reveal small but significant

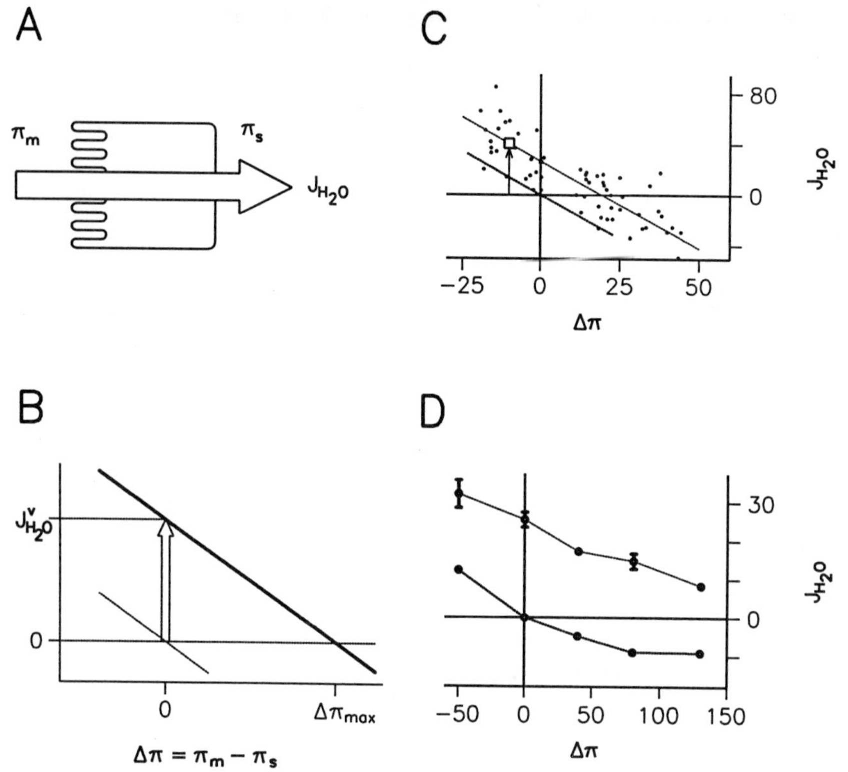

FIG. 4 Relation between transepithelial water flow J_{H2O} and osmotic difference (Eq. 1). (A) In this example, the model epithelium absorbs from the mucosal compartment (osmolarity π_m) into the serosal compartment (osmolarity π_s). (B) The water flux J_{H2O} is a linear function of the osmotic difference across the cell layer, the slope defines the osmotic water permeability, L_p. The osmotic difference required to stop transport is $\Delta\pi_m$. At zero osmotic difference, the water transport rate equals J^V_{H2O}, the inherent or vital transport capacity. Accordingly, J_{H2O} can be divided into a vital and a passive component. If metabolism is stopped, only the passive component remains (thin line going through the origin). (C) Example from rat kidney proximal tubule (redrawn from Green *et al.,* 1991). When metabolism was impaired by cyanide, the L_p was reported to be unchanged (defined by line going through origin). (D) Example from small intestine, plotted from Parsons and Wingate (1961). Here the passive component was obtained by removing glucose. Units in (C) and (D) are mosm L^{-1} for the ordinate and nL cm^{-2} sec^{-1} for the abcissa.

differences between the compositions of the mucosal and the serosal solutions. The Na^+ concentrations are the same, the K^+ and Ca^{2+} concentrations are slightly lower, and the Cl^- concentration slightly higher in the secreted solution (Diamond, 1964; Zeuthen, 1983). The implication is that the secretion is borne into the lateral spaces and is not simply an ultrafiltrate of the mucosal solution. Second, isotonic transport is not obligatory; there are conditions under which leaky epithelia transport anisotonically. Whitlock and Wheeler (1964) exposed rabbit gall bladders to three kinds of conditions: (i) identical hypotonic, isotonic, or hypertonic NaCl solutions on the two sides; (ii) identical solutions containing NaCl plus sucrose on the two sides; and (iii) asymmetrical but isosmotic conditions, NaCl on one side and NaCl plus sucrose on the other. The transportate was only isotonic when the NaCl concentrations were the same on the two sides and when sucrose was absent. All other conditions produced a hypertonic transportate (for a review, see House, 1974).

B. Transport in Dilute Solute Solutions

When epithelia are adapted to transport in dilute solutions, J^V_{H2O} rates go up in proportion to the degree of dilution (Fig. 5A), while the rate of salt transport J_s remains fairly constant (Fig. 5B). This has been studied in gall bladder (Whitlock and Wheeler, 1964; Diamond, 1964; Hill and Hill, 1978; Zeuthen, 1982) and in amphibian kidney proximal tubule (Whittembury and Hill, 1982). It has been suggested that the effect is a result of the cells striving to maintain high intracellular osmolarity, in the face of external dilution. This notion is confirmed by measurements with ion-selective electrodes and microosmometry in *Necturus* gall bladder (Zeuthen, 1981, 1982, 1983). If, for example, the extracellular solution were diluted by a factor of four, the intracellular steady-state osmolarity would only be reduced by a factor of two (Fig. 5C). It turns out that the osmotic driving force across the mucosal membrane is approximately proportional to the dilution of the external solutions. Accordingly, the osmotic influx of water and thereby the steady-state transepithelial flux of water will mirror the dilution. Importantly, the transportate remains isotonic to the diluted external solutions despite the accelerated rates of transport (Fig 5C). What remains to be explained then, is how water is secreted, uphill, from the relatively hyperosmolar cell interior into the hyposmolar serosal solution (Fig. 6).

C. Uphill Water Transport

The concept of transepithelial uphill transport of water is contained in Eq. (1) and in Fig. 4. J_{H2O} decreases with increasing opposing osmotic gradient $(\pi_m - \pi_s)$ and is zero at a certain value, $\pi_m - \pi_s = \Delta\pi_{max}$. With this gradient applied, the inherent ability of the tissue for water transport J^V_{H2O} is matched exactly by an opposing osmotic flow. Neither the rate of active salt transport J_s nor the L_p seems to be affected by the opposing osmotic gradient (Parsons and Wingate, 1961). $\Delta\pi_{max}$ is referred to as the strength of transport (Spring, 1983). Values are in the range 20 to 200 mosm L^{-1}, with the small value obtained in the kidney proximal tubule and the largest in the small intestine (Table I). In situations where

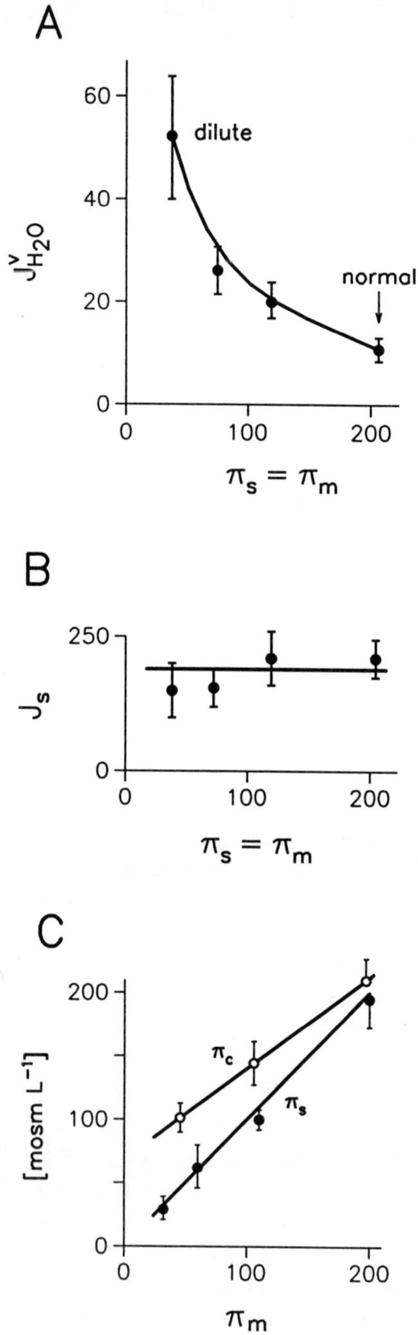
A
60
40
20
0
0
100
200
dilute
normal
$J^v_{H_2O}$
$\pi_s = \pi_m$
B
250
0
0
100
200
J_s
$\pi_s = \pi_m$
C
200
100
0
0
100
200
π_c
π_s
[mosm L^{-1}]
π_m

the osmolarity of the mucosal solution is larger than the serosal ($\pi_m > \pi_s$), while not so large as to stop transport, water is transported across the epithelium from a lower water chemical potential to a higher. This capacity for uphill transport is important physiologically since it enables absorption of water from hyperosmolar solutions. It is not surprising that the capacity is most pronounced in the small intestine, which under physiological conditions may be exposed to high luminal osmolarities.

VII. Models Based on Osmosis

Most models of the coupling between salt and water transport in leaky epithelia are based on the physical process of osmosis combined with hydraulic flow and solute diffusion. The assumption has been that the activity of the Na^+/K^+ ATPase results in an increase in osmolarity of some intraepithelial compartment, typically the cell or the lateral intercellular spaces, or both. Given the passive properties of the various boundaries, water is taken up osmotically and leaves the epithelium driven by a small hydrostatic pressure difference. The scheme fails, however, to provide a general model for the properties discussed above: isotonic transport, increased rate of transport in dilute solutions, and uphill water transport.

A. Isotonic Transport

Assume that the cell is hyperosmolar relative to the compartment from which the absorption takes place, that is, the mucosal compartment in the small intestine or kidney proximal tubule. In steady state, the osmotic entry of water has to be matched by the exit of water across the basolateral membrane. While the entry step is conceptually straightforward, the mechanisms behind the exit of water are not understood at all. If the exit is supposed to take place by osmosis, the serosal compartment must have a higher osmolarity than the cellular compartment. In this

---

FIG. 5 Transport properties at normal and low external osmolarities. (A) Necturus gall bladders were adapted to transport in solutions with normal (200 mosm L^{-1}) and reduced osmolarities (down to 40 mosm L^{-1}) were tested. J_{H2O} (in units of $\mu L\ cm^{-2}\ hr^{-1}$) was larger in dilute solutions and followed an inverse relationship. Bladders were tested as unilateral (or sweating) preparations. Here the serosal solution is identical to the transported solution, and the epithelium is automatically bathed on the two sides in solutions of similar osmolarities $\pi_m = \pi_s$. (B) The rate of NaCl transport J_s (mmol $cm^{-2}\ hr^{-1}$) was not affected by the dilution. (C) The transport was isotonic: the serosal osmolarity π_s was equal to the mucosal osmolarity π_m at all dilutions tested. The intracellular osmolarity π_c was higher than the extracellular osmolarity; this was most pronounced at dilute conditions. Removal of oxygen and poisoning the Na^+/K^+ ATPase abolished the intracellular hyperosmolarity. Data from Zeuthen (1981, 1982).

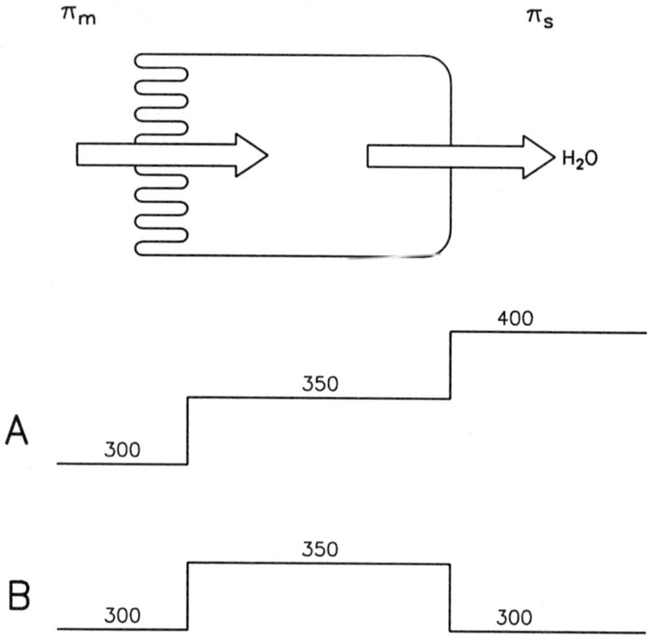

FIG. 6 The central problem for models based on osmosis. (A) If transepithelial water transport is assumed to take place by simple osmosis across both the mucosal and the serosal membrane, the osmotic profile through the cellular pathway must increase in two steps. The magnitudes of the steps depend on the rate of water transport and of the L_ps. The numbers (in mosm L^{-1}) are typical for mammalian small intestine, see Table I. (B) Experimentally, however, the transported solution is found to be isotonic with the mucosal solution, in this case, 300 mosm L^{-1}. The central problem is therefore to explain how water moves, uphill, from the cell into the serosal compartment.

model, transport across the cell is viewed as taking place by two osmotic steps (Fig. 6A). Since the osmolarity of the transportate is found experimentally to be close to that of the apical bath (isotonic transport), there are severe restrictions on the magnitudes of the two osmotic steps. Three modeling strategies have been proposed to cope with this problem: high water permeabilities, subepithelial diffusion restriction, and standing gradients.

In the case of high water permeabilities of the two limiting membranes, only minute osmotic gradients would be needed to maintain a transepithelial transport that was practically isotonic. The scheme fails, however, as a general model. First, water permeabilities are not large for all leaky epithelia. Given the transport parameters for the mammalian proximal tubule (Table I), it would take a transepithelial gradient of 5 mosm L^{-1} to explain transport by osmosis, the gradient being shared about equally between the apical and serosal membranes. Given the accuracy of measurements, this could be considered isotonic transport. In the mammalian small intestine, however, the relatively low water permeabilities would require a total osmotic gradient of more than 100 mosm L^{-1} (Fig. 6A), which is incompatible with isotonic transport. Second, high water permeabilities would impair the ability of the epithelia for uphill transport. In the example of the kidney proximal tubule, an opposing osmotic gradient of 5 mosm L^{-1} would abolish transport. This goes against the finding that it takes between 20 and 60 mosm L^{-1} to stop transport (see Table I).

If diffusion of salt were restricted in the subepithelial tissues, the osmolarity in the exit bath would decrease away from the tissue (Fig. 6B) (Spring, 1999, Fig. 2). The solution abutting the serosal membrane would have a high osmolarity and would generate the osmotic water efflux. The question is whether sufficient restrictions exist. Assume that the combination of basement membranes, connective tissues, and capillaries reduces solute diffusion by one-half compared to free solutions. In view of experimental values, this must be considered a high estimate (Wright and Diamond, 1968; Hénin *et al.,* 1977; Persson and Spring, 1984). In renal proximal tubule, the hydraulic conductivity of the basement membrane was 300 to 800 times greater than that of the cell layer, and was permeable to large molecules such as horseradish peroxidase (Moreno and Diamond, 1972). A simple calculation will illustrate the problem. Consider an osmotic gradient Δosm of 10 mosm L^{-1} that is dissipated over a distance Δx of 100 μm taken as the thickness of the restrictive layer. With a diffusion constant D of 0.8 10^{-5} cm^2 sec^{-1}, Fick's equation ($J = -D\Delta\text{osm}/\Delta x$) implies a solute flux of at least 0.8 10^{-8} osm cm^{-2} sec^{-1}. This is larger than the transepithelial solute fluxes observed (Table I) in the *Necturus* gall bladder, for example, by more than one order of magnitude. The calculation illustrates that diffusion restriction and unstirred layers are unable to maintain hyperosmolarities as large as 10 mosm L^{-1} at the serosal aspect of the cell.

The invaginations of the serosal membranes in epithela such as the gall bladder could constitute a diffusion barrier. This possibility has been formulated qualitatively in the standing gradient hypothesis (Bossert and Diamond, 1967). If a hyperosmolarity can be maintained in the lateral spaces, water will enter osmotically from the cell and be extruded into the serosal solution by a hydrostatic pressure difference (Curran and Macintosh, 1962; Patlak *et al.,* 1963). The hyperosmolarity is supposed to diminish toward the serosal end of the space due to influx of water from the cell. A variety of factors work against the scheme. Lateral spaces are generally too short and have too large a cross section to sustain any hyperosmolarities or standing gradients. The water permeabilities of the basolateral membranes are too low to ensure that the transportate achieves isosmolarity (for a review see Hill, 1980). Accordingly, no hyperosmolarities have been observed in the lateral intercellular spaces of water-transporting epithelia (Zeuthen, 1983; Ikonomov *et al.,* 1985).

B. Transport in Dilute Solute Solutions

Dilute conditions increase the rates of isotonic transport (Fig. 5), augmenting the difficulties outlined above for an explanation based on osmosis and restricted diffusion. The osmotic steps required to move water at sufficient rates would have to be larger, and a hyperosmolarity of the transportate might become detectable. Take the *Necturus* gall bladder as an example. At normal osmolarities, about 200 mosm L^{-1}, the transport could be explained by osmotic steps of a total around 5 mosm L^{-1}, given the values of the passive water permeabilities (Fig. 7A). At a five-fold dilution, the transport rate increases about five-fold (Fig. 5A) (Hill and Hill, 1978; Zeuthen, 1982, 1983). This would require the sum of the osmotic steps to increase to about 25 mosm L^{-1}, yet the transport remains isotonic (Hill and Hill, 1978; Zeuthen, 1981). For isotonicity to ensue, the 25 mosm L^{-1} should be dissipated by diffusion in the subepithelial tissues (Fig. 7B). A calculation similar to the one above shows that this is not possible, given the parameters.

C. Uphill Water Transport

Uphill water transport in the osmotic model has been discussed by Weinstein and Stephenson (1981), Weinstein *et al.* (1981), and Spring (1983) for the gall bladder type of epithelia. When an impermeable osmolyte is added to the mucosal solution, the water transport is arrested when the osmolarity of the mucosal solution is increased by the amount $\Delta\pi_{max}$; see Eq. (1). To null the water flux, the model requires the osmolarity to be the same on the two sides of the cell layer, that is, both the mucosal solution and the solution abutting the serosal surface are hyperosmolar by the amount $\Delta\pi_{max}$ (Fig. 7C). In this situation, the diffusion restriction by the subepithelial structures becomes of pivotal importance; it is this restriction which causes the solute flux J_S to build up the serosal osmolarity. Under conditions of uphill water transport, the requirement for this restriction is even higher than that required in connection with isotonic transport. Consider the *Necturus* gall bladder which is able to transport against an osmotic gradient of 30 mosm L^{-1} (Persson and Spring, 1982). If this gradient is to be dissipated over a distance of 100 μm, and the diffusion coefficient is half that of the free solution, the Fick equation predicts a flux of at least 2.3 10^{-8} mol cm^{-2} sec^{-1}. This is about two orders of magnitude larger than the measured solute flux J_S of 3 10^{-10} mol cm^{-2} sec^{-1} (Table I). In other words, the restriction is not sufficient to retain the salt.

In conclusion, there are fundamental difficulties in explaining isotonic and uphill water transport by a combination of osmosis and restricted diffusion in the subepithelial tissues in epithelia of the gall bladder type. The serosal restrictions are not adequate to build up the required osmotic gradients. It should be emphasized that it would be even more difficult to apply the osmotic model to other types of epithelial cell layers. The choroid plexus epithelium and the corneal endothelium,

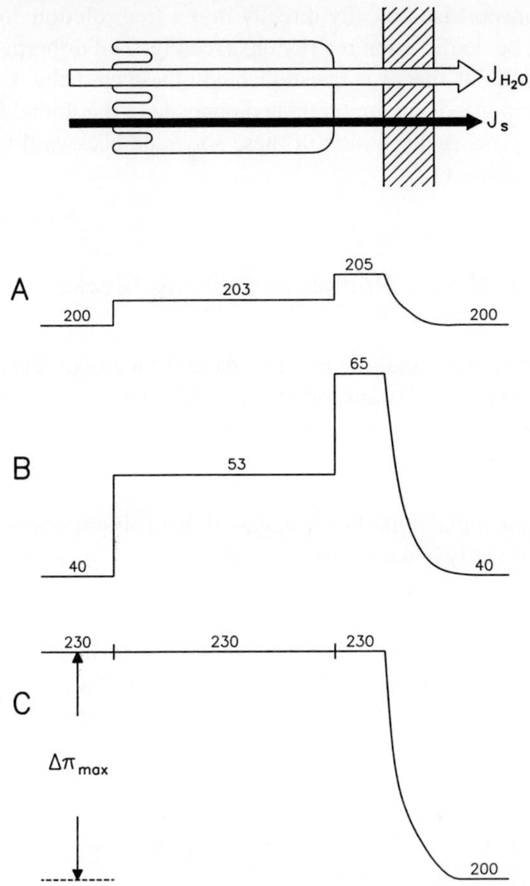

FIG. 7 Model based on osmosis and diffusion constraints. Water transport J_{H2O} is assumed to take place through the cell by two osmotic steps (see Fig. 6A). Solute transport J_s (predominantly NaCl) also takes place through the cell. Diffusion of NaCl is assumed to be restricted in subepithelial structures such as basement membrane and connective tissues (hatched). J_{H2O} is not affected by this layer. The numerical examples are based on data from *Necturus* gall bladder; see Fig. 5 and Table I. (A) Osmolarity profile during transport at normal external osmolarities. Water transport through the cell is maintained by two osmotic steps. The excess osmolarity in the serosal compartment (205 mosm L^{-1}) is dissipated in the by restricted diffusion, and the emerging osmolarity becomes isotonic to the bathing solutions; see also Spring (1999, Fig. 2). The problem is that the NaCl gradient required in the connective tissues is likely to give rise to unrealistically large values of J_s; see text. (B) Under dilute conditions, J_{H2O} goes up. This would require larger osmotic steps, as J_s is unaffected. In the given example, the five-fold dilution gives rise to a five-fold higher J_{H2O} (see Fig. 5A). This would increase the demands on the diffusion restriction five-fold; this seems unrealistic (see text). (C) Transport across *Necturus* gall bladder stops at opposing osmotic gradients of 30 mosm L^{-1}, (Table I). In this situation, the model suggests that the osmotic gradient across the cell layer is abolished, both sides being 230 mosm L^{-1}. The salt transport J_s is the same as under control conditions. Accordingly, 30 mosm L^{-1} has to be dissipated (Spring, 1983). As in (A) and (B), this would place unrealistic demands on the solute restriction, see text.

for example, transport isotonically directly into a free solution compartment, in which there can be no diffusion restrictions (see Fig. 2). Furthermore, the role of the lateral intercellular spaces is questionable in these epithelia, since the spaces are not very pronounced and are orientated opposite to that found for epithelia of the gall bladder type. A discussion of these so-called backward-facing epithelia can be found in House (1974).

VIII. Molecular Water Pumps as Building Blocks

It is clear from the above analysis that the major difficulty of a general epithelial model is to explain the uphill transport of water. While water may be transported by osmosis (downhill) across one membrane, water will have to move uphill, against the water chemical potential gradient, across another aspect of the epithelial barrier (see Fig. 6).

To solve this dilemma, it has been suggested that cotransporters of the symport type participate directly in the coupling of salt and water in epithelia (for reviews, see Zeuthen and Stein, 1994; Zeuthen, 2000). These cotransporters serve as molecular water pumps in which the energy dissipated by the downhill transport of the nonaqueous substrates couples by a mechanism inside the protein to the uphill transport of the water (see Cotransporters as Molecular Water Pumps by Zeuthen and MacAulay, this volume). Accordingly, cotransporters can be incorporated into a general epithelial model to provide capacity for uphill water transport. In the following, two specific examples will be discussed: water influx across brush-border membranes mediated by the Na^+-coupled sugar transporters and water efflux mediated by the K^+/Cl^- cotransporter (Fig. 8).

A. Lessons from Na^+/Glucose Cotransporters Expressed in *Xenopus* Oocytes

The human Na^+/glucose cotransporter hSGLT1 has been found to cotransport about 240 water molecules per turnover (Meinild *et al.*, 1998; Zeuthen *et al.*, 2001). With 2 Na^+ ions and 1 glucose molecule transported, this gives about 80 water molecules per unit osmolyte. For the rabbit cotransporter rSGLT1, estimates are slightly higher, between 260 and 424 water molecules per turnover (Loo *et al.*, 1996; Zeuthen *et al.*, 1997, 2001). Accordingly, the solution transported by the SGLT1 is hypertonic; in mammalian plasma there are about 180 water molecules per unit osmolyte. This points to several roles for the SGLT1. First, it cotransports water. Second, the cotransporter contributes to the build up of the cellular hyperosmolarity that provides the driving force for parallel osmotic import of water. Third, as the SGLT1 has an inherent passive water permeability (see Cotransporters

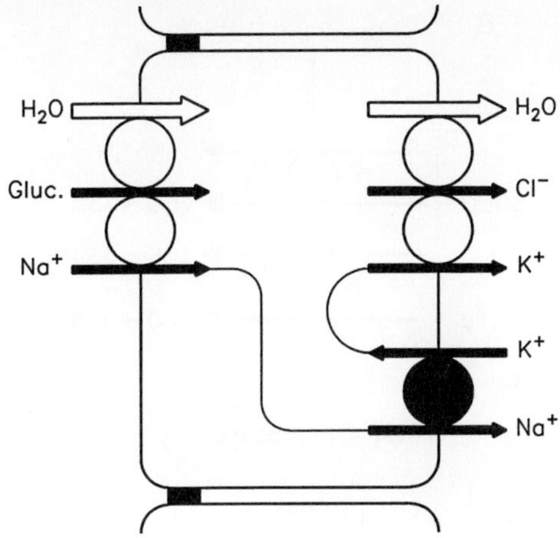

FIG. 8 Molecular water pumps as building blocks in transepithelial water transport. Coupling between water transport in the Na^+/glucose/H_2O and the K^+/Cl^-/H_2O cotransporters and metabolism via the Na^+/K^+ ATPase (filled circle). The pump generates the low intracellular Na^+ concentration and the negative electrical electrical potential which are driving forces for the influx of water via the Na^+/glucose/H_2O cotransporter. The ATPase also generates the high K^+ concentration, which is the driving force for the efflux of water via the K^+/Cl^-/H_2O cotransporter. A complete model should include a variety of other transporters, i.e., aquaporins. (Redrawn from Zeuthen, 2000.)

as Molecular Water Pumps by Zeuthen and MacAulay, this volume), part of the osmotic transport must take place via the SGLT1 itself. This triple role for SGLT1 has relevance for models of transcellular isotonic transport in leaky epithelia. We have recently mimicked isotonic transport across the apical membranes of the absorptive epithelia in small intestine and kidney proximal tubule by expression of SGLT1 and aquaporin 1 (AQP1) in *Xenopus* oocytes (Zeuthen *et al.* 2001).

The apical membrane of the small intestine contains no known aquaporins, and water transport is probably determined by the cotransport proteins and the lipid component. The most complete data are from rat. Here the rate of water transport J_{H2O} under sugar-transporting conditions is about 25 nL cm^{-2} sec^{-1}, and the water permeability L_p about 2.5 10^{-2} cm sec^{-1}, values per epithelial cm^2 (Worman and Field, 1985; Heeswijk and VanOs, 1986; Pappenheimer and Reis, 1987; Dempster *et al.,* 1991; for a review, see Zeuthen, 1996, and Table I). The intestinal apical membrane was modeled by expression of human or rabbit SGLT1 in *Xenopus* oocytes (Zeuthen *et al.,* 2001); the properties of the rat SGLT1 are similar to those of the rabbit and human clones (Wright *et al.,* 1998). At the given expression level, it could be estimated that at 37°C, J_{H2O} would have to be about 20 nL cm^{-2} sec^{-1}, the L_p about 10^{-2} cm sec^{-1}, and the intracellular osmolarity about 15 mosm L^{-1} above the external solution. Thus the J_{H2O} and L_p for the oocyte and the epithelia were quite similar. In the oocyte model, it was determined that the J_{H2O} was divided with about one-third cotransport, one-third osmosis via the SGLT1, and one-third

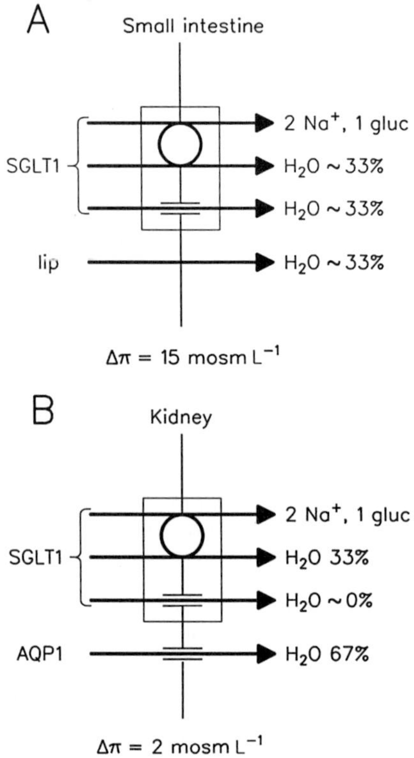

FIG. 9 Components of isotonic fluid transport across a membrane expressing SGLT1 (A) and coexpressing SGLT1 and aquaporin 1 (B). (A) The water transport has three components: one-third is cotransported via a mechanism within the SGLT1, one-third is osmotically driven via the SGLT1, and one-third is osmotically driven through the native lipid membrane (lip). The SGLT1 builds up a relatively large transmembrane osmotic gradient ($\Delta\pi = 15$ mosm L^{-1}) to drive the osmotic flux required for isotonic transport. (B) In analogy to (A), one-third of the water transport takes place by cotransport in the SGLT1. Coexpression of AQP1, however, imparts a large passive water permeability to the membrane. As a consequence, the majority of the osmotic water transport takes place via the AQP1 driven by a small osmotic gradient ($\Delta\pi = 2$ mosm). It should be noted that the rates of Na^+ transport are almost the same in (A) and (B) given by the rate of turnover of the protein. Data from human SGLT1 and human AQP1 adjusted to a temperature of 37°C. Data from Zeuthen *et al.* (2001).

osmosis via the native oocyte membrane; a similar division of the water fluxes could be expected for the brush-border membrane (Fig. 9A). A final evaluation must consider the activation of other cotransporters with properties similar to the SGLT1 and transport in proteins with no capacity for water cotransport (i.e., channels).

The apical membrane of kidney proximal tubule is rich in cotransporters and aquaporins. Apart from the SGLT1, the Na^+-dicarboxylate cotransporter (NaDC1) has also been expressed in the in *Xenopus* oocytes (Meinild *et al.*, 2000) and found to function as molecular water pump. The epithelial cells express AQP1 strongly in the apical (and basolateral) membrane (for a review see Borgnia *et al.*, 1999). Most epithelial data are from rat, where J_{H2O} lies between 10 and 30 nl cm^{-2} sec^{-1}, and the L_p of the apical membrane is about 0.4 cm sec^{-1}, values per epithelial cm^2 (Green and Giebisch, 1984; Gonzáles *et al.*, 1984; Carpi-Medina *et al.*, 1984; Pratz *et al.*, 1986; Heeswijk and VanOs, 1986; Dempster *et al.*, 1991). The apical membrane was modeled by coexpression of human SGLT1 and AQP1 in *Xenopus* oocytes (Zeuthen *et al.*, 2001). The level of expression of SGLT1 in the oocyte was similar to that of the apical membrane, while that of AQP1 was about 2 times lower in the oocyte. In analogy to the small intestine, it was estimated that at 37°C the cotransport component counted for about one-third of the total flux of water. The presence of AQP1 caused the remainder two-thirds of the water transport, the osmotic component, to be mediated via the AQP1 driven by a small osmotic gradient of about 2 mosm L^{-1} (Fig. 9B).

In summary, the mammalian small intestine and the kidney proximal tubule have roughly the same transepithelial rates of Na^+ and water transport per cm^2 of epithelium. They both achieve about one-third of their water uptake by cotransport in the Na^+/glucose transporter. Their major difference is the passive water permeability; the apical membrane of the proximal tubule is 10 times more permeable than the small intestine. This enables the proximal tubule to achieve its osmotic component by a small transmembrane osmotic difference. The intestinal cell requires larger intracellular osmolarities to drive the osmotic water transport. On the other hand, this enables the intestine to transport against larger adverse osmotic gradients, because the osmotic backflux is minimized (Table I).

B. Coupling between Water Efflux and Cotransport of KCl

As discussed in Section VII, the mechanism of water exit from the cell is the central problem for epithelial models. If the cell is hyperosmotic relative to the surroundings, water must leave uphill, against the water chemical potential gradient (Fig. 6B). Cotransport of water in the K^+/Cl^- cotransporter has been suggested to be responsible for this uphill transport (Zeuthen, 1991a,b). This cotransporter is found in the exit membrane of all leaky epithelia engaged in obligatory water transport, for example, proximal tubule and small intestine, (references in Section V) and has been studied in detail in the choroid plexus epithelium from *Necturus maculosus*. Here the cotransporter is situated in the apical membrane facing the cerebrospinal solution and lends itself directly to experiments in which the water transport properties can be determined quantitatively (Zeuthen, 1991a,b, 1994).

The cotransporter is electroneutral and transports one K^+, one Cl^-, and 500 water molecules per turnover. The coupling ratio is independent of external driving forces, and the free energy for transport can be supplied by the electrochemical gradient of each of the ions as well as from the difference in water chemical potential. From the Gibbs equation (see Cotransporters as Molecular Water Pumps, Zeuthen and MacAulay, this volume, Eq. 1), and assuming first-order kinetics, water transport rates can be described by (Zeuthen, 1996, pg. 66):

$$J_{H2O} = B[K_i Cl_i \exp(-nosm_i/n_w) - K_o Cl_o \exp(-nosm_o/n_w)] \quad (2)$$

B was determined from experiments to 1.8 10^{-2} cm sec^{-1} (osm $L^{-1})^{-2}$. Suffix *i* designates the intracellular compartment, *o* the extracellular solution. The molarity for water, n_w, is 55 mol L^{-1}, the coupling ratio for water n is 500.

Given normal cellular conditions, how large is J_{H2O} and how sensitive is J_{H2O} to adverse osmotic gradients? Assume K^+ and Cl^- concentrations in the cell of 80 and 40 mmol L^{-1}, and outside of 2 and 110 mmol L^{-1}, and intra- and extracellular osmolarities of 300 mosm L^{-1} (Fig. 10A). With these values, J_{H2O} calculates to 3.6 nL cm^{-2} sec^{-1}, with the direction out of the cell, and similar to that determined for other amphibian epithelia Table I). It is also close to J_{H2O} observed for bullfrog choroid plexus when surface foldings are taken into account (Wright *et al.*, 1977; Zeuthen and Wright, 1981). To assess the sensitivity of J_{H2O} to adverse osmotic gradients, consider the situation where the intracellular compartment is made hyperosmolar by an impermeable and inert osmolyte. In Fig. 10B, curve 1, J_{H2O} is calculated as a function of the difference in osmolarity across the cell membrane. J_{H2O} is relatively insensitive to osmotic differences within the normal physiological range; an intracellular hyperosmolarity of 30 mosm L^{-1} would change J_{H2O} by only about 10%. It would require hyperosmolarities of around 300 mosm L^{-1} to arrest transport. It is clear that the outward gradients for K^+ and Cl^- provide strong driving forces for the efflux of water. If the change in osmolarity were implemented by addition of a chloride salt, both the Cl^- concentration and the osmolarity would alter. The difference in Cl^- concentration would affect the rate of cotransport and counteract the effects of the osmotic changes. In such experiments, the $K^+/Cl^-/H_2O$ cotransport would appear rather insensitive to osmotic changes (Fig 10B, curve 2). The different responses to osmotic challenges implemented by inert osmolytes and Cl^- salts might explain the finding that the capacity for uphill water in epithelia is higher when the osmotic gradient is achieved by additions of NaCl (Zeuthen, 1992).

C. Properties of a Molecular Model for Water Transport in Leaky Epithelia

The capacity of K^+/Cl^- and Na^+/glucose cotransporters for water transport enables epithelial models to treat uphill water transport as a molecular phenomenon localized to membrane proteins. Consider a simple model based on $K^+/Cl^-/H_2O$ cotransport combined with passive water permeabilities (Fig. 11A). With values from the *Necturus* choroid plexus, the passive water permeability L_p of the membrane

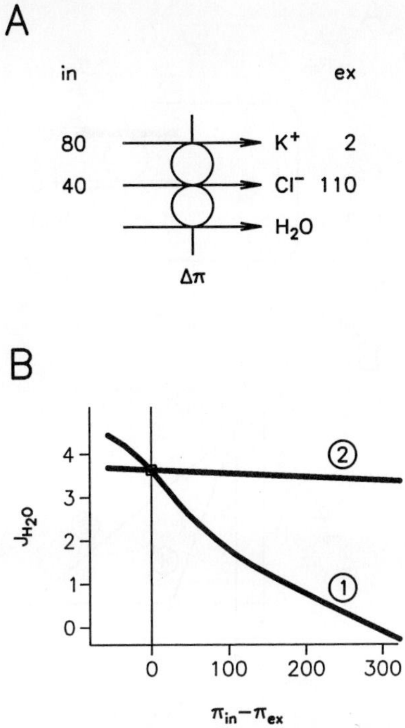

FIG. 10 Properties of $K^+/Cl^-/H_2O$ cotransport. (A) The cotransporter is bathed in (mmol L^{-1}) 80 K^+ and 40 Cl^- on the intracellular side (in) and 2 K^+ and 110 Cl^- on the extracellular side (ex). (B) This gives a rate of water transport J_{H2O} of 3.6 nL cm^{-2} sec^{-1} at zero transmembrane osmotic gradient; see text. If an osmotic gradient is imposed by adding mannitol to the inside, solution J_{H2O} changes, as given by curve 1; about 300 mosm L^{-1} is required to stop the water transport. If $\Delta\pi$ is changed by diluting the outside solution, i.e., removing NaCl, J_{H2O} is less sensitive to the osmotic gradient because the cotransporter is speeded up by the reduction in Cl^- concentration, curve 2. Redrawn from Zeuthen (2000).

across which water enters is 0.6 10^{-2} cm sec^{-1}, while the exit membrane has an L_p of 0.3 10^{-2} cm sec^{-1}, all values per surface area of the flat epithelium (Zeuthen, 1991a,b, 1994). Three epithelial properties can be approached on the basis of this molecular model: The transepithelial rate of water transport J_{H2O}, the ability for uphill transport of water, and isotonic transport under normal and dilute conditions.

The rate of transepithelial water transport is to a first approximation given by the $K^+/Cl^-/H_2O$ cotransporter, which, given normal ion concentrations, provides 3.6 nL cm^{-2} sec^{-1} (Fig. 10B). In a steady state, an intracellular hyperosmolarity of about 20 mosm L^{-1} would bring water into the cell at a rate of 2.4 nL cm^{-2} sec^{-1} across the entry membrane and into the cell at a rate of 1.2 nL cm^{-2} sec^{-1} across the exit membrane (Fig. 11A). This recirculation of water results in a transepithelial transport rate of 2.4 nL cm^{-2} sec^{-1}, slightly smaller than the rate of

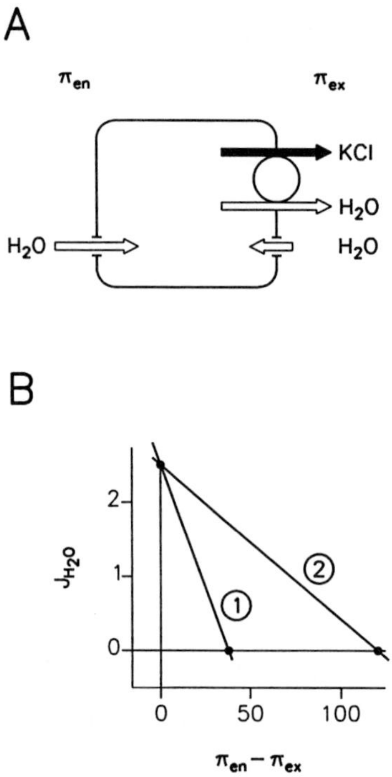

FIG. 11 A simple molecular model for epithelial water transport based upon $K^+/Cl^-/H_2O$ cotransport and osmosis. (A) Water crosses the entry membrane by osmosis and leaves by $K^+/Cl^-/H_2O$ cotransport. There is a passive reuptake of water across the exit membrane. The lengths of the arrows defining the water transports are quantitatively correct for the choroid plexus epithelium of *Necturus maculosus* (Zeuthen 1991a,b, 1994). (B) Prediction of the model for transepithelial water transport. J_{H2O} is calculated as a function of the transepithelial osmotic difference implemented by adding mannitol to the entry side, curve 1, or by the removal of NaCl from the exit side, curve 2.

water cotransport. The calculated net rate compares well to those determined in other amphibian epithelia (Table I). Two points emerge from this analysis. First, the rate of $K^+/Cl^-/ H_2O$ cotransport is the upper limit for J_{H2O}. Second, if entry of water had been by cotransport, for example, by Na^+/glucose/H_2O transport, the intracellular osmolarity required for the influx of water would have been smaller, and so would the degree of recirculation of water across the exit membrane.

How well does the simple molecular model predict uphill water transport? Consider the case where the ability for uphill transport is tested by the addition of an inert solute, say mannitol, to the entry side of the epithelium. When the osmolarity has been increased by $\Delta\pi_{max}$, J_{H2O} is zero (Section VI.C, Table I). In this situation, the intracellular osmolarity equals the one of the entry bath, and the osmotic reuptake of water across the exit membrane matches exactly the efflux maintained by $K^+/Cl^-/H_2O$ cotransport. A rough measure of $\Delta\pi_{max}$ can therefore be obtained from the water permeability of the exit membrane (L_{pe}) and the assumption that the intracellular osmolarity has increased by the same amount as the external one. Accordingly, J_{H2O} ceases when $\Delta\pi_{max}L_{pe}$ equals the rate of $K^+/Cl^-/H_2O$ cotransport. With the parameters used above, $\Delta\pi_{max}$ is estimated to 40 mosm L^{-1}, which compares well to the lower range of values in Table I. This estimate of $\Delta\pi_{max}$ is a minimum one. The increase in intracellular osmolarity will result from increases in intracellular K^+ and Cl^- activities, which in turn increases the rate of $K^+/Cl^-/H_2O$ transport. This effect would be most pronounced if the uphill water transport were tested by additions of NaCl instead of mannitol to the entry bath or, alternatively, by dilutions of the exit bath. This is possibly why the larger estimates for $\Delta\pi_{max}$ in Table I (around 150 mosm L^{-1}) are obtained from changes in NaCl concentrations. Testing the simple molecular model (Fig. 11A) by removal of NaCl from the exit bath gives an estimate for $\Delta\pi_{max}$ of about 110 mosm L^{-1} (Fig. 11B, curve 2). This corroborates the finding that rabbit choroid plexus under the conditions of NaCl dilution can transport against osmotic gradients of 160 mosm L^{-1} (Welch, 1966). Similar considerations apply to the other leaky epithelia listed in Table I, except the *Necturus* gall bladder. In a simple model, the water permeabilities of this epithelium are too high to sustain uphill water transport.

The simple molecular model does not address the question of isotonic transport at either normal or dilute solution. In the model, the tonicity of the transportate is determined by the $K^+/Cl^-/H_2O$ transporter and the Na^+/K^+ATPase . Since the cotransporter delivers a KCl solution which is roughly isotonic to plasma, and since the ATPase exchanges 2 K^+ ions with 3 Na^+ ions, the resulting transportate becomes hyperosmolar. One possibility for attaining isotonic transport would be reuptake of Na^+ across the exit membrane. This has been discussed elsewhere (Zeuthen, 1996).

IX. Conclusions

Leaky epithelia share a wide range of transport properties: isotonic fluid transport, uphill transport of water, and increased rates of water transport when adapted to low osmolarities. It is tempting therefore to look for some common features that correlate to the direction of water transport. These might constitute the basis for a general transport model. The ultrastructure, defined by the brush-border, basolateral membrane, and junctional complex cannot be part of a general model. Among the epithelia, there is no correlation between the distribution of these structural elements and the direction of water transport. Neither does the distribution of the Na^+/K^+ATPase or of the aquaporins conform to the direction of water transport. The only general finding is that the direction of transport in cotransport protein of the symport type coincides with the direction of transepithelial water transport. This, together with the finding that these symporters may function as molecular water pumps, suggests a role for these proteins in a general model for salt–water coupling in leaky epithelia.

Traditional epithelial models are based on the existence of a hyperosmolar compartment within the cell layer. It follows that the limitations of this compartment must impose a certain diffusion restriction on the ions. Two situations can be envisaged: If the hyperosmolar compartment is localized to the cell interior, the diffusion restriction is sufficient; but water has to leave the cell uphill, against the direction of the water chemical potential. If, on the other hand, the compartment is located outside the cell, the diffusion restriction is not large enough to keep the hyperosmolarity sufficiently high. The problem is particularly pronounced in situations where the epithelium transports against a transepithelial osmotic gradient.

Cotransporters working as molecular water pumps could be important building blocks in epithelial models; dissipative (downhill) fluxes of ions in these proteins energize significant uphill fluxes of water. The presence of cotransporters in the membrane across which water enters (i.e., Na^+/glucose) and in the exit membrane (i.e., K^+/Cl^-) would alleviate the problems inherent in the traditional models based on osmosis alone. A simple molecular model combining cotransport of water in the K^+/Cl^- cotransporter with osmotic transport would mimic uphill water transport for a variety of leaky epithelia.

Acknowledgment

Useful discussions with Professor W. D. Stein and Dr. D. Klaerke are gratefully acknowledged. Svend Christoffersen is thanked for artwork.

References

Alpern, R. J., Howlin, K. J., and Preisig, P. A. (1985). Active and passive components of chloride transport in the rat proximal convoluted tubule. *J. Clin. Invest.* **76,** 1360–1366.

Baerentsen, H., Giraldez, F., and Zeuthen, T. (1983). Influx mechanisms for Na^+ and Cl^- across the brush border membrane of leaky epithelia: A model and microelectrode study. *J. Membr. Biol.* **75,** 205–218.

Baum, M., and Berry, C. A. (1984). Evidence for neutral transcellular NaCl transport and neutral basolateral chloride exit in the rabbit proximal convoluted tubule. *J. Clin. Invest.* **74,** 205–211.

Berry, C. A. (1983). Water permeability and pathways in the proximal tubule. *Am. J. Physiol.* **245,** F279–F294.

Bomsztyk, K., and Wright, F. S. (1986). Dependence of ion fluxes on fluid transport by rat proximal tubule. *Am. J. Physiol.* **250,** F680–F689.

Borgnia, M., Nielsen, S., Engel, A., and Agre, P. (1999). Cellular and molecular biology of the aquaporin water channels. *Annu. Rev. Biochem.* **68,** 425–458.

Bossert, W. H., and Diamond, J. M. (1967). Standing-gradient osmotic flow. A mechanism for coupling of water and solute transport in epithelia. *J. Gen. Physiol.* **50,** 2061–2083.

Carpi-Medina, P., Lindemann, B., González, E., and Whittembury, G. (1984). The continuous measurement of tubular volume changes in response to step changes in contraluminal osmolarity. *Pflügers Arch.* **400,** 343–348.

Colclasure, G. C., and Parker, J. C. (1992). Cytosolic protein concentration is the primary volume signal for swelling-induced [K-Cl] cotransport in dog red cells. *J. Gen. Physiol.* **100,** 1–10.

Cotton, C. U., Weinstein, A. M., and Reuss, L. (1989). Osmotic water permeability of necturus gallbladder epithelium. *J. Gen. Physiol.* **93,** 649–679.

Curran, P. F., and Macintosh, J. R. (1962). A model system for biological water transport. *Nature* **193,** 347–348.

Dempster, J. A., Hoek van, A. N., deJong, M. D., and VanOs, C. H. (1991). Glucose transporters do not serve as water channels in renal and intestinal epithelia. *Pflügers Arch.* **419,** 249–255.

Diamond, J. M. (1962). The mechanism of solute tranport by the gall-bladder. *J. Physiol.* **161,** 474–502.

Diamond, J. M. (1964). Transport of salt and water in rabbit and guinea pig gall bladder. *J. Gen. Physiol.* **48,** 1–14.

Diamond, J. M. (1979). Osmotic water flow in leaky epithelia. *J. Membr. Biol.* **51,** 195–216.

Ericson, A.-C., and Spring, K. R. (1982a). Coupled NaCl entry into Necturus gallbladder epithelial cells. *Am. J. Physiol.* **243,** C140–C145.

Ericson, A.-C., and Spring, K. R. (1982b). Volume regulation by necturus gallbladder: Apical Na^+-H^+ and Cl^--HCO_3^- exchange. *Am. J. Physiol.* **243,** C146–C150.

Eveloff, J., and Warnock, D. G. (1987). K-Cl transport systems in rabbit renal basolateral membrane vesicles. *Am. J. Physiol.* **252,** F883–F889.

Fischbarg, J. (1989). On the theory of solute-solvent coupling in epithelia. *In* "Water Transport in Biological Membranes" (G. Benga, Ed.), pp. 153–167. CRC Press.

Frömter, E., Rumrich, G., and Ullrich, K. J. (1973). Phenomenologic description of Na^+, Cl^- and HCO_3^- absorption from proximal tubules of the rat kidney. *Pflügers Arch.* **343,** 189–220.

Garzia, A., and Armstrong McD, W. (1983). KCl cotransport: A mechanism for basolateral chloride exit in necturus gallbladder. *J. Membr. Biol.* **76,** 173–182.

Giraldez, F. (1984). Active sodium transport and fluid secretion in gallbladder epithelium of necturus. *J. Physiol.* **348,** 431–455.

Gonzáles, E., Carpi-Medina, P., Linares, H., and Whittembury, G. (1984). Osmotic water permeability of the apical membrane of proximal straight tubular (PST) cells. *Pflügers Arch.* **402,** 337–339.

Graf, J., and Giebish, G. (1979). Intracellular sodium activity and sodium transport in necturus gallbladder epithelium. *J. Membr. Biol.* **47,** 327–355.

Green, R., and Giebisch, G. (1984). Luminal hypotonicity: A driving force for fluid absorption from the proximal tubule. *Am. J. Physiol.* **246,** F167–F174.

Green, R., and Giebisch, G. (1989a). Osmotic forces driving water reabsorption in the proximal tubule of the rat kidney. *Am. J. Physiol.* **257,** F669–F675.

Green, R., and Giebisch, G. (1989b). Reflection coefficients and water permeability in rat proximal tubule. *Am. J. Physiol.* **257,** F658–F668.

Green, R., Giebisch, G., Unwin, R., and Weinstein, A. M. (1991). Coupled water transport by rat proximal tubule. *Am. J. Physiol.* **261,** F1046–F1054.

Grim, E. (1962). Water and electrolyte flux rates in the duodenim, jejunum, ileum and colon and effects of osmolarity. *Am. J. Digest. Dis.* **7,** 17–27.

Grim, E., and Smith, G. A. (1957). Water flux rates across dog gallbladder wall. *Am. J. Physiol.* **191,** 555–560.

Gullans, S. R., Avison, M. J., Ogino, T., Shulman, R. G., and Giebish, G. (1986). Furosemide-sensitive K^+ efflux induced by glucose in the rabbit proximal tubule. *Kidney Int.* **29,** 396–396.

Haas, M. (1989). Properties and diversity of (Na-K-Cl) cotransporters. *Annu. Rev. Physiol.* **51,** 443–457.

Haas, M., and McBrayer, D. G. (1994). Na-K-Cl cotransport in nystatin-treated tracheal cells: Regulation by isoproterenol, apical UTP, and $[Cl]_i$. *Am. J. Physiol.* **266,** C1440–C1452.

Hakim, A., Lester, R. G., and Lifson, N. (1963). Absorption by an in vitro preparation of dog intestinal mucosa. *J. Appl. Physiol.* **18,** 409–413.

Halm, D. R., Krasny, E. J., and Frizzell, R. A. (1985). Electrophysiology of flounder intestinal mucosa. I. Conductance properties of the cellular and paracellular pathways. *J. Gen. Physiol.* **85,** 843–864.

Hamann, S., Zeuthen, T., La Cour, M., Nagelhus, E. A., Ottersen, O. P., Agre, P., and Nielsen, S. (1998). Aquaporins in complex tissues: Distribution of aquaporins 1–5 in human and rat eye. *Am. J. Physiol.* **274,** C1332–C1345.

Heeswijk, M. P. E., and Van Os, C. H. (1986). Osmotic water permeabilities of brush border and basolateral membrane vesicles from rat renal cortex and small intestine. *J. Membr. Biol.* **92,** 183–193.

Hénin, S., Cremaschi, D., Schettino, T., Meyer, G., Donin, C. L. L., and Cotelli, F. (1977). Electrical parameters in gallbladders of different species. Their contribution to the origin of the transmural potential difference. *J. Membr. Biol.* **34,** 73–91.

Hill, A. (1980). Salt-water coupling in leaky epithelia. *J. Membr. Biol.* **56,** 177–182.

Hill, A. E., and Hill, B. S. (1987a). Steady-state analysis of ion fluxes in necturus gall-bladder epithelial cells. *J. Physiol.* **382,** 15–34.

Hill, A. E., and Hill, B. S. (1987b). Transcellular sodium fluxes and pump activity in necturus gall-bladder epithelial cells. *J. Physiol.* **382,** 35–49.

Hill, B. S., and Hill, A. E. (1978). Fluid transfer by necturus gall bladder epithelium as a function of osmolarity. *Proc. R. Soc. Lond.* **200,** 151–162.

House, C. R. (1974). "Water Transport in Cells and Tissues." Edward Arnold, London.

Ikonomov, O., Simon, M. X., and Frömter, E. (1985). Electrophysiological studies on lateral intercellular spaces of necturus gallbladder epithelium. *Pflügers Arch.* **403,** 301–307.

Jennings, M. L., and Schulz, R. K. (1991). Okadaic acid inhibition of KCl cotransport. *J. Gen. Physiol.* **97,** 799–818.

Johanson, C. E., Murphy, V. A., and Dyas, M. (1992). Ethacrynic acid and furosemide alter Cl, K and Na distribution between blood, choroid plexus, CSF and brain. *Neurochem. Res.* **17,** 1079–1085.

Johanson, C. E., Palm, D. E., Dyas, M. L., and Knuckey, N. W. (1993). Microdialysis analysis of effects of loop diuretics and acetazolamide on chloride transport from blood to CSF. *Brain Res.* **641,** 121–126.

Kaji, D. M., and Tsukitani, Y. (1991). Role of protein phosphatase in activation of KCl cotransport in human erythrocytes. *Am. J. Physiol.* **260,** C178–C182.

LaCour, M. D. (1993). Ion transport in the retinal pigment epithelium: A study with double barrelled ion-selective microelectrodes. *Acta Ophthalmol.* **71,** 1–32.

Lang, F., Ritter, M., Völkl, H., and Häussinger, D. (1993). The biological significance of cell volume. *Renal Physiol. Biochem.* **16,** 48–65.

Larsen, E. H., Nedergaard, S., and Ussing, H. H. (2000). Role of lateral intercellular space and sodium recirculation for isotonic transport in leaky epithelia. *Rev. Physiol. Biochem. Pharm.* **141,** 153–212.

Larson, M., and Spring, K. R. (1984). Volume regulation by necturus gallbladder: Basolateral KCl exit. *J. Membr. Biol.* **81,** 219–232.

Lauf, P. K. (1985). $K^+ : Cl^-$ cotransport: Sulfhydryls, divalent cations, and the mechanism of volume activation in a red cell. *J. Membr. Biol.* **88,** 1–13.

Lauf, P. K., Bauer, J., Adragna, N. C., Fujise, H., Martin, A., Zade-Oppen, M., Ryu, K. H., and Delpire, E. (1992). Erythrocyte K-Cl cotransport: Properties and regulation. *Am. J. Physiol.* **263,** C917–C932.

Liu, F., Cogan, M. G., and Rector, Jr., F. C. (1984). Axial heterogeneity in the rat proximal convoluted tutuble. II. Osmolarity and osmotic water permeability. *Am. J. Physiol.* **247,** F822–F826.

Loo, D. D. F., Zeuthen, T., Chandy, G., and Wright, E. M. (1996). Cotransport of water by the Na^+/glucose cotransporter. *Proc. Natl. Acad. Sci. USA* **93,** 13,367–13,370.

Ma, T., Song, Y., Gillespie, A., Carlson, E. J., Epstein, C. J., and Verkman, A. S. (1999). Defective secretion of saliva in transgenic mice lacking aquaporin-5 water channels. *J. Biol. Chem.* **274,** 20,071–20,074.

Ma, T., and Verkman, A. S. (1999). Aquaporin water channels in gastrointestinal physiology. *J. Physiol.* 317–326.

Meinild, A.-K., Klaerke, D. A., Loo, D. D. F., Wright, E. M., and Zeuthen, T. (1998). The human Na^+/Glucose cotransporter is a molecular water pump. *J. Physiol.* **508(1),** 15–21.

Meinild, A.-K., Loo, D. F. F., Pajor, A., Zeuthen, T., and Wright, E. M. (2000). Water transport by the renal Na^+/dicarboxylate cotransporter. *Am. J. Physiol.* **278,** F777–F783.

Moreno, J. H., and Diamond, J. M. (1972). Cation permeation mechanisms and cation selectivity in "tight junctions" of gallbladder epithelium. *In* "Membranes" (G. Eisenman, Ed.), pp. 383–515. Marcel Dekker, Inc. New York.

Naftalin, R. J., and Tripathi, S. (1986). The roles of paracellular and transcellular pathways and submucosal space in isotonic water absorption by rabbit ileum. *J. Physiol.* **370,** 409–432.

Nauntofte, B. (1992). Regulation of electrolyte and fluid secretion in salivary acinar cells. *Am. J. Physiol.* **263,** G823–G837.

Nielsen, S., Smith, B. L., Christensen, E. I., and Agre, P. (1993). Distribution of the aquaporin CHIP in secretory and resorptive epithelia and capillary endothelia. *Proc. Natl. Acad. Sci. USA* **90,** 7275–7279.

Nielsen, S., King, L. S., Christensen, B. M., and Agre, P. (1997). Aquaporins in complex tissues. II. Subcellular distribution in respiratory and glandular tissues of rat. *Am. J. Physiol.* **273,** C1549–C1561.

O'Neill, W. C. (1999). Physiological significance of volume-regulatory transporters. *Am. J. Physiol.* **276,** C995–C1011.

Pappenheimer, J. R., and Reis, K. Z. (1987). Contribution of solvent drag through intercellular junctions to absorption of nutrients by the small intestine of the rat. *J. Membr. Biol.* **100,** 123–136.

Parker, J. C. (1993). In defense of cell volume? *Am. J. Physiol.* **265,** C1191–C1200.

Parker, J. C., Colclasure, G. C., and McManus, T. J. (1991). Coordinated regulation of shrinkage-induced Na/H exchange and swelling-induced [K-Cl] cotransport in dog red cells. *J. Gen. Physiol.* **98,** 869–880.

Parsons, D. S., and Wingate, D. L. (1961). The effect of osmotic gradients on fluid transfer across rat intestine in vitro. *Biochim. Biophys. Acta* **46,** 170–183.

Patlak, C. S., Goldstein, D. A., and Hoffman, J. F. (1963). The flow of solute and solvent across a two-membrane system. *J. Theoret. Biol.* **3,** 420–442.

Persson, B.-E., and Spring, K. R. (1982). Gallbladder epithelial cell hydraulic water permeability and volume regulation. *J. Gen. Physiol.* **79,** 481–505.

Persson, B.-E., and Spring, K. R. (1984). Permeability properties of the subepithelial tissues of Necturus gallbladder. *Biochim. Biophys. Acta* **772,** 135–139.

Pewitt, E. B., Hedge, R. S., and Palfrey, H. C. (1990a). [3H]bumetanide binding to avian erythrocyte membranes. Correlation with activation and deactivation of Na/K/2Cl cotransport. *J. Biol. Chem.* **265,** 14,364–14,370.

Pewitt, E. B., Hedge, R. S., Haas, M., and Palfrey, H. C. (1990b). The regulation of Na/K/2Cl cotransport and bumetanide binding in avian erythrocytes by protein phosphorylation and dephosphorylation. Effects of kinase inhibitors and okadaic acid. *J. Biol. Chem.* **265,** 20,747–20,456.

Pratz, J., Ripoche, P., and Corman, B. (1986). Osmotic water permeability and solute reflection coefficients of rat kidney brush-border membrane vesicles. *Biochim. Biophys. Acta* **861,** 395–397.

Preisig, P. A., and Berry, C. A. (1985). Evidence for transcellular osmotic water flow in rat proximal tubules. *Am. J. Physiol.* **249,** F124–F131.

Raina, S., Preston, G. M., Guggino, W. B., and Agre, P. (1995). Molecular cloning and characterization of an aquaporin cDNA from sallivary lacrimal, and respiratory tissues. *J. Biol. Chem.* **270,** 1908–1912.

Ramírez-Lorca, R., Vizuete, M. L., Venero, J. L., Revuelta, M., Cano, J., Ilundáin, A. A., and Echevarría M. (1999). Localization of aquaporin-3 mRNA and protein along the gastrointestinal tract of Wistar rats. *Pflügers Arch.* **438,** 94–100.

Reuss, L. (1983). Basolateral KCl co-transport in a NaCl-absorbing epithelium. *Nature* **305,** 723–726.

Reuss, L. (1984). Independence of apical membrane Na^+ and Cl^- entry in necturus gallbladder epithelium. *J. Gen. Physiol.* **84,** 423–445.

Reuss, L. (1985). Changes in cell volume measured with an electrophysiologic technique. *Proc. Natl. Acad. Sci. USA* **82,** 6014–6018.

Sachs, J. R. (1994). Soluble polycations and cationic amphiphiles inhibit volume-sensitive K-Cl co-transport in human red cell ghosts. *Am. J. Physiol.* **266,** C997–C1005.

Saito, Y., and Wright, E. M. (1982). Kinetics of the sodium pump in the frog choroid plexus. *J. Physiol.* **328,** 229–243.

Sasaki, S. (1988). KCl co-transport across the basolateral membrane of rabbit renal proximal straight tubules. *J. Clin. Invest.* **81,** 194–199.

Schafer, J. A. (1984). Mechanisms coupling the absorption of solutes and water in the proximal nephron. *Kidney Int.* **25,** 708–716.

Schafer, J. A. (1990). Transepithelial osmolality differences, hydraulic conductivities, and volume absorption in the proximal tubule. *Annu. Rev. Physiol.* **52,** 709–726.

Schnermann, J., Chou, C.-L., Ma, T., Traynor, T., Knepper, M. A., and Verkman, A. S. (1998). Defective proximal tubular fluid reabsorbtion in transgenic aquaporin-1 null mice. *Proc. Natl. Acad. Sci. USA* **95,** 9660–9664.

Schultz, S. G. (1998). A century of (epithelial) transport physiology: From vitalism to molecular cloning. *Am. J. Physiol.* **274,** C13–C23.

Snigirevskaya, E. S., and Komissarchik, Y. Y. (2000). Structural correlates of the transepithelial water transport. *Int. Rev. Cytol.* **198,** 203–275.

Spring, K. R. (1983). Fluid transport by gallbladder epithelium. *J. Exp. Biol.* **106,** 181–194.

Spring, K. R. (1998). Routes and mechanism of fluid transport by epithelia. *Annu. Rev. Physiol.* **60,** 105–119.

Spring, K. R. (1999). Epithelial fluid transport—A century of investigation. *News Physiol. Sci.* **14,** 92–98.

Spring, K. R., and Ericson, A.-C. (1982). Epithelial cell volume modulation and regulation. *J. Membr. Biol.* **69,** 167–176.

Spring, K. R., and Hope, A. (1979). Fluid transport and the dimensions of cells and interspaces of living necturus gallbladder. *J. Gen. Physiol.* **73,** 287–305.

Tripathi, S., and Boulpaep, E. L. (1989). Mechanisms of water transport by epithelial cells. *Q. J. Exp. Physiol.* **74,** 385–417.

Vallon, V., Verkman, A. S., and Schnermann, J. (2000). Luminal hypotonicity in proximal tubules of aquaporin-1-knockout mice. *Am. J. Physiol.* **278,** F1030–F1033.

Verkman, A. S. (1999). Lessons on renal physiology from transgenic mice lacking aquaporin water channels. *J. Am. Soc. Nephrol.* **10,** 1126–1135.

Verkman, A. S., Yang, B., Song, Y., Manley, G. T., and Ma, T. (2000). Role of water channels in fluid transport studied by phenotype analysis of aquaporin knockout mice. *Exp. Physiol.* **85S,** 233S–241S.

Weinstein, A. M., and Stephenson, J. L. (1981). Models of coupled salt and water transport across leaky epithelia. *J. Membr. Biol.* **60,** 1–20.

Weinstein, A. M., Stephenson, J. L., and Spring, K. R. (1981). The coupled transport of water. *In* "Membrane Transport, New Comprehensive Biochemistry" (K. L. Bonting, and J. dePont, Eds.), pp. 311–351. Elsevier, Amsterdam.

Weinstein, A. M., and Windhager, E. E. (1985). Sodium transport along the proximal tubule. *In* "The Kidney: Physiology and Pathophysiology" (D. W. Seldin, and G. Giebisch, Eds.), pp. 1033–1062. Raven Press, New York.

Welch, K. (1966). Volume flow across choroidal ependyma of the rabbit. *Am. J. Physiol.* **210,** 232–236.

Whitlock, R. T., and Wheeler, H. O. (1964). Coupled transport of solute and water across rabbit gallbladder epithelium. *J. Clin. Invest.* **43,** 2249–2265.

Whittembury, G. (1985). Mechanisms of epithelial solute-solvent coupling. *In* "The Kidney: Physiology and Pathophysiology" (D. W. Seldin, and G. Giebish, Eds.), pp. 199–214. Raven Press, New York.

Whittembury, G., and Hill, B. S. (1982). Fluid reabsorption by Necturus proximal tubule perfused with solutions of normal and reduced osmolarity. *Proc. R. Soc. Lond.* **215,** 411–431.

Willumsen, N. J., Davis, C. W., and Boucher, R. C. (1993). Asymmetric water permeabilities of human airway epithelial cell membranes. *In* "Isotonic Transport in Leaky Epithelia. Alfre Benzon Symposium 34" (H. H. Ussing, J. Fischbarg, O. Sten-Knudsen, E. H. Larsen, and N. J. Willumsen, Eds.), pp. 298–307. Munksgaard, Copenhagen.

Wolosin, J. M., and Forte, J. G. (1985). K^+ and Cl^- conductances in the apical membrane from secreting oxyntic cells are concurrently inhibited by divalent cations. *J. Membr. Biol.* **83,** 261–272.

Worman, H. J., and Field, M. (1985). Osmotic water permeability of small intestinal brush-border membranes. *J. Membr. Biol.* **87,** 233–239.

Wright, E. M., and Diamond, J. M. (1968). Effects of pH and polyvalent cations on the selective permeability of gall-bladder epithelium to monovalent ions. *Biochim. Biophys. Acta* **163,** 57–74.

Wright, E. M., Loo, D. D. F., Panayotova-Heiermann, M., Hirayama, B. A., Turk, E., Eskandari, S., and Lam, J. (1998). Structure and function of the Na^+/glucose cotransporter. *Acta Physiol. Scand.* **16,** 1217–1224.

Wright, E. M., Wiedner, G., and Rumrich, G. (1977). Fluid secretion by the frog choroid plexus. *Exp. Eye Res. Suppl.* **25,** 149–155.

Zeuthen, T. (1981). Isotonic transport and intracellular osmolarity in the necturus gall-bladder epithelium. *In* "Water Transport across Epithelia. Alfred Benzon Symposium 15" (H. H. Ussing, N. Bindslev, N. A. Lassen, and O. Sten-Knudsen, Eds.), pp. 313–331. Copenhagen, Munksgaard.

Zeuthen, T. (1982). Relations between intracellular ion activities and extracellular osmolarity in necturus gallbladder epithelium. *J. Membr. Biol.* **66,** 109–121.

Zeuthen, T. (1983). Ion activities in the lateral intercellular spaces gallbladder epithelium transporting at low external osmolarities. *J. Membr. Biol.* **76,** 113–122.

Zeuthen, T. (1987). The effects of chloride ions on electrodiffusion in the membrane of a leaky epithelium. *Pflügers Arch.* **408,** 267–274.

Zeuthen, T. (1991a). Water permeability of ventricular cell membrane in choroid plexus epithelium from necturus maculosus. *J. Physiol.* **444,** 133–151.

Zeuthen, T. (1991b). Secondary active transport of water across ventricular cell membrane of choroid plexus epithelium of necturus maculosus. *J. Physiol.* **444,** 153–173.

Zeuthen, T. (1992). From contractile vacuole to leaky epithelia. Coupling between salt and water fluxes in biological membranes. *Biochim. Biophys. Acta* **1113,** 229–258.

Zeuthen, T. (1994a). Cotransport of K^+, Cl^- and H_2O by membrane proteins from choroid plexus epithelium of Necturus maculosus. *J. Physiol.* **478,** 203–219.

Zeuthen, T. (1994b). "Coupling of Salt and Water Fluxes in Epithelial Cell Membranes." D.Sc Thesis, University of Copenhagen, pp. 1–62.

Zeuthen, T. (1995). Molecular mechanisms for passive and active transport of water. *Int. Rev. Cytol.* **160,** 99–161.

Zeuthen, T. (1996). "Molecular Mechanisms of Water Transport," pp. 1–170. Springer, Berlin R. G. Landes Company, Texas.

Zeuthen, T. (2000). Molecular water pumps. *Reviews of Physiology, Biochemistry and Pharmacology.* **141,** 97–151.

Zeuthen, T., Hamann, S., and La Cour, M. (1996). Cotransport of H^+, lactate and H_2O by membrane proteins in retinal pigment epithelium of bullfrog. *J. Physiol.* **497,** 3–17.

Zeuthen, T., Meinild, A.-K., Klaerke, D. A., Loo, D. D. F., Wright, E. M., Belhage, B., and Litman, T. (1997). Water transport by the Na^+/glucose cotransporter under isotonic conditions. *Biol. Cell* **89,** 307–312.

Zeuthen, T., Meinild, A.-K., Loo, D. D. F., Wright, E. M., and Klaerke, D. A. (2001). Isotonic transport by the Na^+-glucose cotransporter SGLT1. *J. Physiol.* **531.3,** 631–644.

Zeuthen, T., and Stein, W. D. (1994). Co-transport of salt and water in membrane proteins: Membrane proteins as osmotic engines. *J. Membr. Biol.* **137,** 179–195.

Zeuthen, T., and Wright, E. M. (1981). Epithelial potassium transport: Tracer and electrophysiological studies in choroid plexus. *J. Membr. Biol.* **60,** 105–128.

Paracellular Fluid Transport by Epithelia

Bruria Shachar-Hill and Adrian E. Hill*
Physiological Laboratory, Cambridge University, Cambridge CB2 3EG, United Kingdom

The evidence that a major fraction of water crosses the paracellular route during isotonic fluid transfer is reviewed together with a description of the theory and experimental results derived from extracellular probe studies. Four transporting epithelia which have been studied using the method are gallbladder, intestine, Malpighian tubule, and salivary gland. It is concluded that paracellular probe flows are not due to simple convection generated by osmotic flow through the junctions but are generated by active fluid transport within the junction: a mechano-osmotic process. The geometry of the pathway involved would indicate that some salt accompanies the paracellular fluid, representing a hypo-osmotic flow. Transport of salt by the cell route, which may be accompanied by some water, represents a hypertonic flow. The problem then becomes one of balancing the two to produce an isotonic transportate. We suggest, using recent data from knockout mice, that some aquaporins are functioning in different epithelial tissues as osmo-comparators within a feedback loop that regulates the paracellular fluid flow rate. This results in an overall quasi-isotonic transport by the epithelium. The model is applied to forward-facing systems such as proximal tubule and backward-facing systems such as exocrine glands.

KEY WORDS: Epithelia, Fluid transport, Isotonic flow, Junctions, Paracellular pathway, Dextrans, Salivary gland, Intestine, Malpighian tubule, Gallbladder.

*Dr. Hill may be reached via e-mail at aeh1@cam.ac.uk.

I. Introduction

The problem of isotonic fluid transport across epithelia is, once again, an actively debated topic. This is because the consensus on local osmosis as the fundamental underlying mechanism has broken down, leaving a confused state of affairs. For a process that is vital for homeostasis and osmoregulation in animals, this is unsatisfactory.

Part of the confusion stems from the fact that a crucial problem needs to be confronted: Is there a basic mechanism which operates to allow isotonic transfer in all fluid transporting epithelia as there is, for example, a basic mechanism in nerve conduction and in muscle contraction? The general assumption is that there is, and therefore any mechanism proposed has to account for both secretion and absorption in a wide variety of epithelia. This also implies the integration of the diverse structural features of fluid transporting epithelia within a single functional model. These features include

(i) The properties of junctional complexes, comprising the tight junction or *zonula occludens* and adjacent adhering junction or *zonula adherens,* here referred to together as the "junction"
(ii) The lateral intercellular space system or "interspace," lined by the basolateral membrane (and which may be of constant or increasing cross section from the junction to the cell base)
(iii) The polarized cells which actively transport in one direction due to the disposition of the Na:K pump, and
(iv) The direction of the fluid transport which may be from apical (mucosal) to basolateral (serosal) bath or the reverse, that is, "forward-facing" or "backward-facing" as described in Section III.A. Surprisingly, polarity of (iii) and (iv) may differ (Section III.A). (These features are shown in the diagrams of Fig. 6.a,b.)

Many recent reviews (Spring, 1998, 1999; Larsen *et al.,* 2000) do not present a unified scheme that incorporates the available experimental and structural evidence. A common pattern is to select one absorbing epithelium as the model and then to refer to other systems only in general terms.

An integrated approach was attempted by Diamond and Bossert (1967), who presented an attractive local osmotic scheme termed "the standing gradient model." This depended on some basic assumptions which have since proved implausible to many: for instance, an interspace geometry whose dimensions are suitable for effective osmotic coupling, a possible confinement of the Na/K pump to the closed end of the interspace, and the ionic tightness of the junctional complex. As a result, this hypothesis as originally presented is not now considered valid. In its essentials, however, it still encapsulates the basic idea of local osmosis: a flow of water across the cell into the interspace driven by an osmotic gradient of salt together with the requisitely high osmotic permeability of the interspace membranes to allow effective equilibration within the interspace leading to an overall isotonic flow. Most mechanisms currently proposed to explain isotonic transport include variants of local osmotic equilibration. Many include an additional or subsidiary role for the paracellular pathway and some of these involve extensive ion recirculation within this route. In one scheme, for example, ions are pumped through the cell, while there is a backflux to the source bath through the junctions (without water) ensuring that salt and water are transferred across the tissue in isotonic proportions (Ussing and Eskesen, 1989; Larsen *et al.,* 2000). Although the contribution of this paracellular pathway to the overall fluid flow is generally not specified (Spring, 1998), the implication is that any flow through it would be generated only by osmosis; however, all isotonically transporting epithelia have junctions of low electrical resistance and are ionically leaky, which mitigates against osmotically driven flows (this point is considered in Section II.C below).

The high osmotic permeability of epithelial cells has been attributed to the presence of special water channels, the aquaporins. Recently, however, the extensive use of gene-knockout mice has facilitated the realization that although aquaporins may provide a high membrane osmotic permeability, the epithelial fluid transport does not in many cases depend on their presence and can proceed in many cases at high rates without such high water permeabilities (Verkman *et al.,* 2000b). The role of aquaporins will be assessed in Section III.A and subsequent sections.

As well as simple membrane osmotic equilibration, membrane molecular pumps including a wide range of cotransporters have also been proposed as a major pathway for fluid transport (Zeuthen and Stein, 1994; Zeuthen, 2000). One scheme is that water enters a space in the cotransporter protein and is then released on the transmembrane face during a conformation change driven by the Na^+ or glucose free energy gradient (or that of other ions). Such pumping has been demonstrated by transporters expresed in oocytes, and this mechanism is effectively a secondary active transport of water (Loo *et al.,* 1996). How this mechanism may function in an overall scheme involving both membranes is at present unclear, though it may operate in conjunction with other mechanisms (see also Zeuthen article, this volume). In this context, it has been suggested that cotransporters work in parallel with aquaporins in the membrane to assist the isotonic transport of water in epithelia (Zeuthen, 2000).

In this chapter, we argue that by treating the junctional complex as an active volume-transporting organelle that is dynamically regulated by the cell, the paracellular system becomes a major route for fluid transport. This scheme can account for both forward and backward fluid transport and allows switching between them in the same tissue. (Such switching is essential for explaining pathological states like cholera and other gastrointestinal diseases where fluid transport is reversed.) The ionic and solute transporting systems are divided between the cellular and paracellular paths in a tissue-specific way which again is regulated by the cells. The osmotic water permeability of either the cellular or the paracellular route becomes much less important as the route for water flow than in cellular osmotic models, and the role of aquaporins in cellular osmotic balance can be newly assessed. Initially, therefore, it is important to discuss the experimental evidence for the paracellular transport of water.

II. Water Flow Routes Revealed by Probes

The use of paracellular probes to measure fluid velocities is based upon the assumption that a probe molecule traverses the paracellular pathway through permeable junctions and not via the cellular route. Virtually all the isotonic fluid transporting epithelia have tight junctions that are leaky to a greater or lesser extent. If the molecular dimensions and shape of the probes are known reasonably accurately, then it is possible to extract a great deal of information about the channels which traverse the junctions and to determine elementary aspects of the transport process that comprise (i) the fraction of the overall fluid flow required to accompany the probe (of any specific radius) and (ii) the nature of the convection, i.e., whether the probe is dragged through the channels by the water flow or whether both are propelled by a different mechanism. Those epithelia that have been studied in this way show that most of the water flow is via the junctional route in normal isotonic volume transport and that it is not compatible with simple convection. In other words, the results do not show that there is an osmotic water flow component across the junction and do not show that the probes get swept along by it through a channel.

A. The Basis of Probe Studies

Probe studies are based on knowing several important properties. First, the molecular shapes of the probes are essential information. If the molecules are extended thread-like polymers, they become entangled and are not cleanly convected, and the size of the molecule does not bear a simple relation to the size of the channel through which it passes. Most probes are either small isodiametric molecules or dextrans, which have been shown (over the range used) to be quasi-spherical (Shachar-Hill and Hill, 1993). Ideally, probe molecules should be ellipsoids with a small axial ratio.

Second, it is immensely helpful to any precise analysis if the probes form a homologous series. Dextrans are such a series. They can be made radioactive by covalent ^{3}H-labeling, and such a probe sample contains molecules of very many different sizes. When applied to one side of an epithelium during normal isotonic fluid secretion (the source bath), the fluid can be sampled or collected (in the case of a gland) on the other side (the sink bath) and then fractionated by chromatography, enabling a whole spectrum of probes to be analyzed in a single flux experiment.

If the probes are chosen so that they are relatively chemically inert, then the difference between probes depends upon their size rather than differences in their chemical reactivities. The probe fluxes lie on curves when plotted against probe radius; curves can be extrapolated either to zero flux or to zero radius—especially if they are linear over the initial part of their range (which they seem to be, experimentally). When an extrapolation to zero flux is made, this indicates the dimensions of the smallest channel through which probes will pass. When an extrapolation to zero radius is made, this indicates the real velocity of the fluid; no hinderance (selectivity or drag effects) can be offered to a convecting or diffusing molecule of zero size, or size equivalent, to that of water.

Third, unstirred-layers (USLs) must be taken into account. A traditional objection to probe studies is simply that USLs confuse the issue. If there is fluid flow across the epithelium, then this motion is also present in the baths adjacent to the tissue; the probe is swept up to one side of the epithelium (membranes and junction) and swept away at the downstream side, creating a concentration gradient. This has been dubbed "pseudo-solvent drag"or the "sweeping away effect." In reality, this is not a problem at the apical face because the cell surface of planar epithelia can be very well stirred—indeed, so fast that the flow can be rendered turbulent, which destroys any USL there. The subepithelium cannot be stirred internally but, knowing the properties of the layer, calculations of the pseudo-solvent drag effect show it to be utterly negligible at the slow fluid flow rates encountered with transporting epithelia (Pedley and Fischbarg, 1980; Aldis, 1983; Shachar-Hill and Hill, 1993).

Fourth, the experimental setup has to allow for removal of the diffusive contribution to the probe fluxes. Probes move in either direction as the result of both convective and diffusive forces. Here convective is used in a general way to indicate movement of probes by interaction with the movement of other molecules through the paracellular system; it appears that this cannot be simply interpreted as being swept along by water flow in an open channel. For a planar epithelium, or a cylindrical one (see below), the probe fluxes can be measured in both directions, i.e., with the water flow (forward flux) or against it (back flux). By performing two such measurements on the tissue and subtracting one from the other, the net flux is obtained. The diffusive contributions from each unidirectional flux will be equal (per unit concentration difference, the probes being uncharged in the first instance) and cancel out, leaving the convective component only. For exocrine glands, the back fluxes cannot be measured, for obvious reasons, but the diffusive flux can be subtracted away. The diffusion coefficient D of any one of the probes, if it were in free solution, would be expected to show a dependence upon molecular radius r_s known as the Stokes–Einstein relationship:

$$D = RT/6\pi\eta r_s \tag{1}$$

where η is a the viscosity. The flux J_s through any channel or pathway, per unit concentration difference, is proportional to D and the area of cross section A, or

$$J_s = k/r_s \tag{2}$$

where k is the overall coefficient of proportionality ($ART/6\pi\eta$). It is an experimental fact that the unidirectional flux across all epithelia tested so far with a range of probes shows a tail going off to high molecular weights; this is evident from about 6 Å (Fig. 4A). This must be a large channel that lets high MW probes across in both directions. It has been speculated that this is most probably due to holes in the epithelium made by cell damage (or replacement) and that 1 cell in 6000 with a channel 0.4 μm would suffice (Van Os and Wright, 1979). The tail in exocrine glands, where only a forward flux from perfusate to secretion can be investigated, fits Eq. (2) very closely—a hyperbolic relationship. This indicates that there are indeed channels in which diffusion occurs as if in free solution; Eq. (2) fits this part of the curve very well, indicating that the probes have virtually free diffusion through a very large channel. When it is fitted and subtracted from the unidirectional flux in the rat submandibular gland, the linear relationship between permeation and radius shown in Fig. 4B is obtained. One cannot do more with a glandular system, but the procedure is uncomplicated and theoretically straightforward, and yields a linear curve in common with other epithelia (see Section II.B). Once one has such a flux curve for the probes as a function of radius, this must represent a paracellular convective curve free from any diffusional components. Here, convective is used as a general term for any nondiffusional movement, which may include simple hydraulic convection or more elaborate schemes. It involves a coupled flux of water and the solutes through the junction and interspace due to the fact that probes (dextrans) are extracellular and cannot be transported across cell membranes. A full treatment of the process, including drag effects, is quite complex (Hill and Shachar-Hill, 1993; Shachar-Hill and Hill, 1993), but there is a quite simple form of the argument that makes all the salient points.

Across the epithelium, there is a flux of any particular probe species from the source bath (apical in the case of a forward-facing or absorbing epithelium) into the sink bath. The source bath is a probe solution of concentration C (perfusate). The isotonic fluid flow across the whole epithelium (J_v) may be considered to include a fluid flow-through junction and interspace equal to j_v. If the convective transport of probe across the epithelium from the source bath is j_s, as measured experimentally by, say, radioactive labeling, then

$$j_s = j_v C\theta \tag{3}$$

where θ is a drag or hindrance coefficient defining in very general terms the selectivity. If the probe radius approaches that of water, $\theta \rightarrow 1.0$, i.e., the probe is passing with no hindrance, compared to the solvent. If the probe is larger in radius than that of the channel, then $\theta \rightarrow$ zero, and the paracellular flux is blocked.

Dividing Eq. (3) by $J_v C$, we have

$$j_s / J_v C = \theta j_v / J_v \tag{4}$$

where j_v/J_v is the fraction of the total fluid transport that is passing the junction and interspace. The three parameters j_s, J_v, and C are all experimental values. If we have a curve of $j_s/J_v C$ versus radius from a homologous series of probes, we can extrapolate the curve to zero (or water) radius, in which case, $\theta \rightarrow 1.0$, and $j_s/J_v C$ is then equal to the fraction of the epithelial fluid transport passing the junctions. For an exocrine gland, the source bath is represented by a perfusate C, and the ratio of transported probe to total volume j_s/J_v is the secreted probe concentration C_e. Thus for the gland

$$C_e / C = \theta j_v / J_v \tag{5}$$

where C_e/C is equal to the secretion:perfusate probe ratio for any probe radius. Again, if we have a quasi-linear relationship between C_e/C and radius, we can extrapolate ($r_s \rightarrow 0$; $\theta \rightarrow 1.0$) to find the fraction of volume transport passing through the paracellular pathway.

To date, there are four epithelial systems for which these data have been collected, using probes as described above (Figs. 1–4). They all show the linear relationship referred to at low radius. This very linearity is also interesting because it sets strict limits on the sorts of processes that can be involved in the water:probe transfer (see Section II.D).

B. Probe Studies on Various Systems

1. Gallbladder

In Fig. 1, the results are shown for experiments using labeled dextrans as probes across *Necturus* gallbladder epithelium prepared as a planar sheet (Shachar-Hill and Hill, 1993). This epithelium has been the subject of much study, and the overall fluid transport rate is known and is isotonic within very small limits. This tissue is rather extraordinary in that, unlike rabbit (and other) gallbladders, it is very permeable indeed to a range of large solutes. In addition, the junctions, when examined by transmission electron microscopy, reveal a wide spacing between membranes in the junction (Hill and Hill, 1978). The ratio $j_s/J_v C$, equivalent to the fraction of volume crossing the junction and interspace, is shown as a function of probe radius. It can be seen that the intercept $j_s/J_v C$ at small radius is close to 1.0 indicating, in terms of the analysis given above, that in this epithelium most of the fluid is passing through the junctions. As mentioned in the preceding section, when an extrapolation of the curve is made to zero ordinate (equivalent to $j_v/J_v = 0$) the value of the radius axis is a cut-off radius through which probes cannot pass; this is close to 38 Å here and agrees with the general solute leakiness

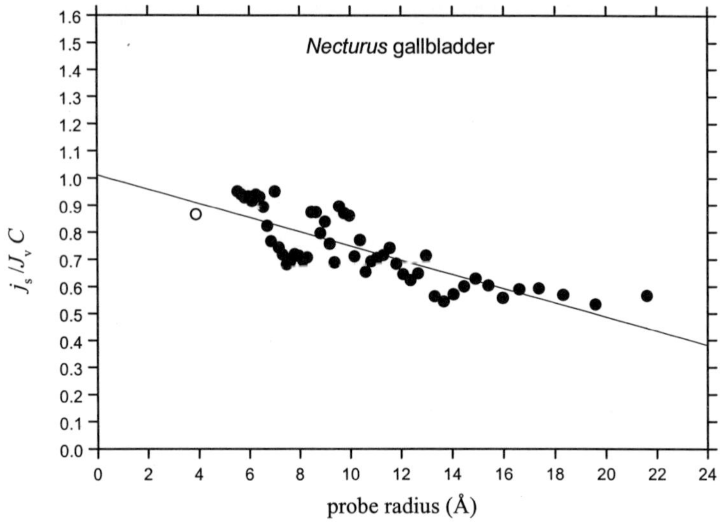

FIG. 1 The fraction of volume flow $j_s/J_v C$ accompanying probe dextrans in *Necturus* gallbladder epithelium as a function of radius. Mannitol was used as an additional probe (open circle). Redrawn from Shachar-Hill and Hill (1993).

and the measured anatomical width of the junction. A striking aspect of this curve is its linearity.

2. Intestine

Necturus gallbladder epithelium is constitutive; in other words, the rate of fluid transport cannot apparently be modulated by metabolites during its normal functioning *in vivo*. Fluid transport in the *Necturus* intestine, however, has a basal level which can be increased by up to 300% by the addition of nutrient metabolites such as sugars and amino acids. When labeled dextrans are used as probes during isotonic transport, the pattern is more complicated and depends upon whether the metabolite is present or not. In Fig. 2, $j_s/J_v C$ is shown as a function of probe radius for *Necturus* intestine stimulated by 20 mM alanine (Hill and Shachar-Hill, 1997). The intercept at low radius is close to unity both with or without alanine, however, again indicating that the net volume absorption is paracellular. The interesting and somewhat surprising results at higher radii are explicable in terms of bidirectional fluid convection of the probes and are discussed below (Section IV.C).

3. Malpighian Tubule

Figure 3 shows the results of a series of experiments on the invertebrate *Rhodnius* where the excretory organ, the Malpighian tubule, is formed of a tubular epithelium which secretes isotonically at a high rate when stimulated with

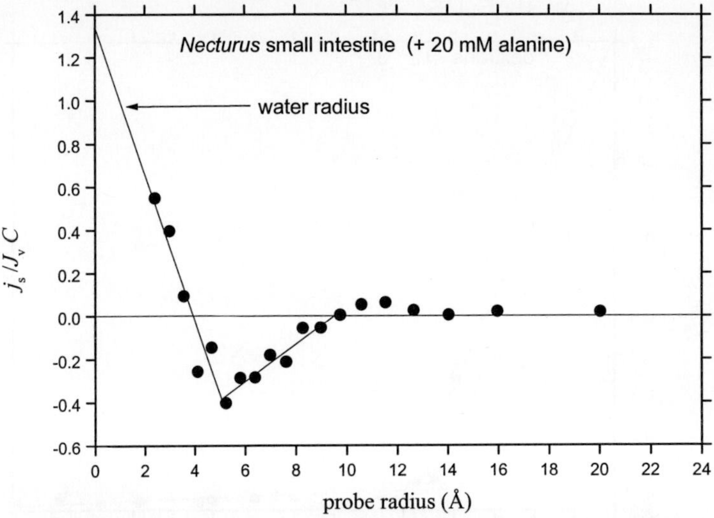

FIG. 2 The fraction of volume flow $j_S/J_v C$ accompanying probe dextrans in *Necturus* small intestine epithelium as a function of radius. The linear fit to the data between 0 and 5 Å indicates an intercept near 1.0 at the water radius. Subtraction of this section from the data reveals a back-flux between 5 and 10 Å. Beyond 10 Å, there are no net fluxes. Redrawn from Hill and Shachar-Hill (1997).

5-hydroxytryptamine. The tubule can be detached and perfused inside and out, similar to a renal tubule (O'Donnell *et al.,* 1982). This enables bidirectional, and hence net, fluxes to be measured with solutions of labeled probes (Hernandez *et al.,* 1995). Most of the smaller probes in this study differ chemically from each other, that is, they are not a homologous series, although the larger ones are dextrans. The intercept is again close to unity (1.1–1.2), indicating that the major fraction of volume flow in the stimulated state is a paracellular water flux. There are small but marked net fluxes of the higher MW dextrans which show a systematic decline with rising radius. In this preparation, it is possible that the stirring (created by the flow of probe solutions from the intertubular perfusion pipettes) was not enough to destroy effectively the USLs, and consequently there was a slight intraluminal depletion of probes in operation leading to slight asymmetry. Otherwise, it is difficult to explain net fluxes of such large and inert molecules.

4. Salivary Gland

Figure 4 presents similar experiments with the whole rat submandibular gland, perfused *in vitro*. The fluid secretion is started by adding acetylcholine to the perfusate which is circulating through the arterial cannulation. The glandular lumen cannot of course be perfused because it represents a series of blind-ended spaces; the fluxes are therefore only unidirectional ones in the direction of fluid transport.

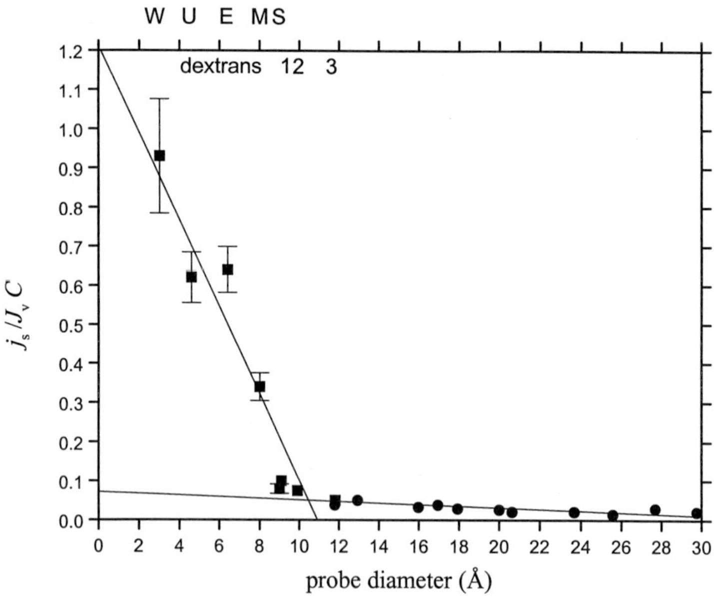

FIG. 3 The fraction of volume flow $j_s/J_v C$ accompanying probe dextrans in *Rhodnius* Malpighian tubular epithelium as a function of radius. The first eight probes are shown above: water, W; urea, U; erythritol, E; mannitol, M; sucrose, S; dextran fractions, 1, 2, & 3. The straight line for these probes is a linear regression $j_s/J_v C = 1.211–0.111 \times$ radius, which is not significantly different from 1 at the radius of water (1.3Å). The larger probes (dextrans) have a slope not significantly different from zero. Redrawn from Hernandez *et al.* (1995).

In Fig. 4A, $j_s/J_v C$ (equal to the secretion:perfusate ratio C_e/C for a gland) is plotted against radius. The pattern of these unidirectional fluxes is very similar in all epithelia that have been studied—the curve is L-shaped with a high MW tail running up to quite large radii. This indicates that there are basically two channels involved: a small selective one and a very large indiscriminate one, the latter probably representing relatively large but infrequent spaces in the epithelial structure created by cell replacement. The high MW tail must represent the diffusion path through a very large channel because, although the size of the fluxes through this channel are not very large, they are fitted very well by a Stokes–Einstein curve as described above (Eq. 1) indicating virtually unhindered diffusion. $j_s/J_v C$ as a function of radius is shown for this system in Fig. 4B in which the convective flux remaining after the computed diffusion curve has been subtracted from the original. It reveals a curve rising sharply to near 1.0 at the water radius, again indicating both a quasi-linear relationship to radius and a very high flow of water through the junction and interspace route (Murakami *et al.*, 2001).

The submandibular salivary gland also shows a different pattern which is independent of the radius of the probe but which can be demonstrated with any

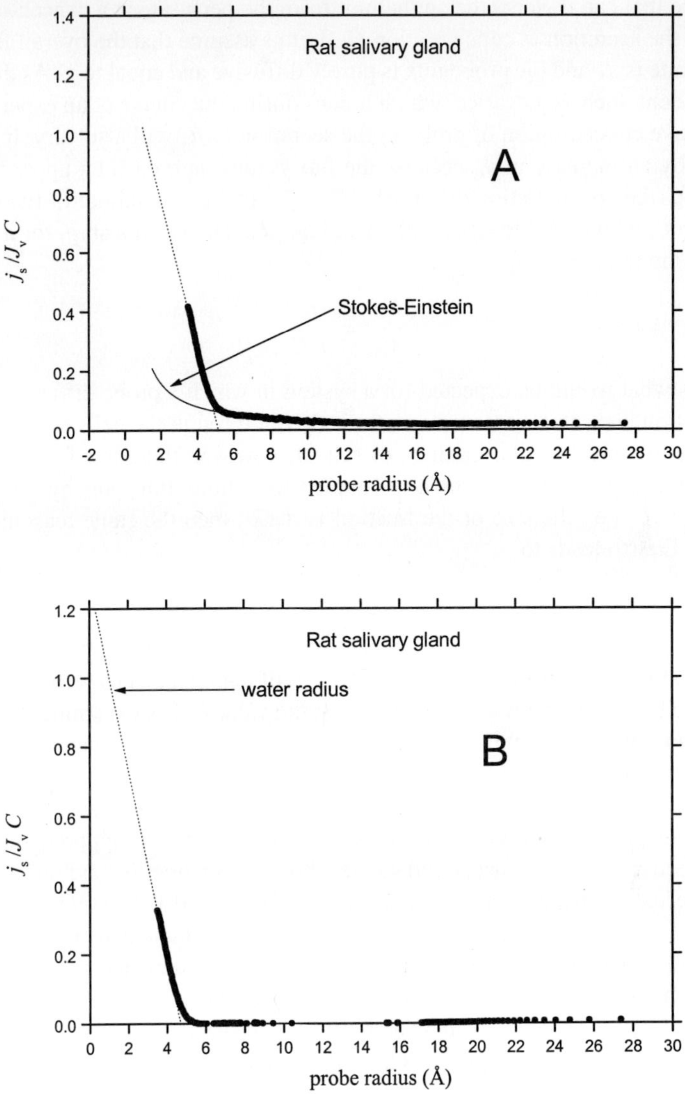

FIG. 4 (A) The fraction of volume flow j_s/J_vC accompanying probe dextrans in rat submandibular salivary gland epithelium as a function of radius. The dextran salivary flow is a unidirectional flux and therefore contains a diffusive component which can be fitted to a Stokes–Einstein relation. (B) Subtraction of the Stokes–Einstein curve from (A) leaves a dataset with a value of j_s/J_vC near 1.0 at the water radius (Murakami *et al.,* 2001).

molecule that can traverse the epithelium from the perfusate at a concentration C to enter the secretion at concentration C_e. Let us assume that the overall fluid secretion rate is J_v and the probe flux is purely diffusive and equal to j_s. As these are independent, then as J_v varies (which it does during the course of an experiment), the relative concentration of probe in the secretion C_e/C will also vary. It will be diluted by an increase in J_v because the flux j_s must approach an upper limit as $J_v \to \infty$ (the concentration difference $(C - C_e)$ attains a maximum). Eventually, the ratio C_e/C will fall to small values at high J_v. The relationship, for a simple lumped model, is given by

$$\frac{C_e}{C} = \frac{j_s}{J_v C} \tag{6}$$

which is what would be expected for a system in which a probe diffuses through the junction and interspace while water is driven through the cells by osmosis in response to a transcellular gradient of salt. If, however, there is a fraction of the paracellular flow of probe which is coupled to volume transport by a coupling coefficient θ, i.e., the size of the fraction is $J_v C\theta$, then the same reasoning that leads to Eq. (6) leads to

$$\frac{C_e}{C} = \frac{j_s}{J_v C} + \theta \tag{7}$$

Plots of C_e/C against the fluid flow rate J_v will not lead to a hyperbolic relation as before but to a curve which flattens out at the value θ. Curves similar to Eq. (7) have been obtained for almost all the solutes with which the submandibular gland system has been tested (Case *et al.,* 1985), including those for which the paracellular system is the only possible route across the epithelium, such as dextrans (Murakami *et al.,* 2001). Such results give strong independent support to the model of convective transfer of water and solutes through the junction and interspace.

The results from these four epithelial systems (Section II.B.1–4) are summarized in Table I along with three somewhat earlier studies in which there are quantitative data on the fraction of water crossing the paracellular system, determined by using assorted molecules as probes and applying either convection theory (Hill and Hill, 1978; Steward, 1982) or Kedem–Katchalsky equations (Case *et al.,* 1985); in this study, the ratio $j_s/J_v C$ is interpreted as a filtration coefficient $(1 - \sigma)$.

C. Osmotic Flow across Junctions Must Be Very Small

If probe studies reveal that water is mainly traversing the junctions during isotonic flow in leaky epithelia, from apical to basolateral bath or vice versa, it is natural to ask whether the flow is driven by osmosis. This question has been raised many times in the past, but the usual assumption is that osmotic flow can make little contribution. As this is such an central point, it is important to expand on it.

TABLE I
Paracellular Probe Studies on Fluid Transporting Epithelia

System	Probes	Fluxes	Paracellular water fraction
Necturus gallbladder (Hill and Hill, 1978)	Sucrose	Unidirectional	0.98
Rabbit gallbladder (Steward, 1982)	Various	Unidirectional	High
Rabbit mandibular salivary gland (Case *et al.*, 1985)	Various	Unidirectional	High
Necturus gallbladder (Shachar-Hill and Hill, 1993)	Dextrans	Bidirectional	1.0
Rhodnius Malpighian tubule (Hernandez *et al.*, 1995)	Various plus dextrans	Bidirectional	1.1
Necturus intestine (Hill and Shachar-Hill, 1997)	Dextrans	Bidirectional	0.98–1.1
Rat submandibular gland (Murakami *et al.*, 2001)	Dextrans	Unidirectional	0.96–1.1

Compared with the transcellular flow, the *fraction* of total epithelial water traversing the junction and interspace must be very small. It is on the basis of the probe fluxes through channels in the junction that the water flows have been assessed. If this water flow is osmotic, then it must be driven by an osmotic gradient of salt through these channels which also let through the larger probes and which therefore must be leaky. Osmotic flow is the product of three factors: the area A, the driving force $\Delta\pi$, and the osmotic permeability P_{os} (where the latter is the product of the hydraulic permeability P_f and the reflexion coefficient σ), i.e.

$$J = A\Delta\pi P_{os} \tag{8}$$

It is convenient to consider these three separately. The cross-sectional area of the junctions is very small indeed compared to that of the cells: the linear extent of junction per cm^2 of epithelium is about 10^3 cm; the junction width is less than 10^{-7} cm; the total junctional area is therefore of the order of 10^{-4} or a ten-thousandth of the epithelial surface per cm^2. The driving forces are quite small: from measurement of the ionic concentration differences between baths and interspaces with microelectrodes, the transjunctional osmotic difference cannot exceed 1–2 mOsm (Zeuthen, 1983; Ikonomov *et al.*, 1985). If the channels in the junctions carrying the dextrans are about 6 Å radius as found in experiments (Figs. 2–4; in *Necturus* gallbladder; in Fig. 1, they are larger), then taking the overall junction length ($\sim$0.5 μm)) into account, P_{os} can be calculated and the osmotic flow would be of the order of 0.001 $\mu L/cm^2$.hr. This is minute compared with the actual fluid flow rates and reflects the fact that leakiness to salt combined with a small area is not a good recipe for an osmotic flow route.

In a study that visualized the gradients of an intercellular marker trapped in the interspace system of a strain of MDCK epithelium (Kovbasnjuk *et al.,* 1998a), the fluid velocity was calculated from the marker gradient and local concentration by inserting them into a steady-state convection–diffusion equation

$$J = vC - dC/dx = 0 \tag{9}$$

It was apparent that fluid entered the lateral spaces but that the velocity at the blind end adjacent to the junction was near-zero in the absence of applied osmotic gradients, indicating that little fluid crossed the junctions. This result is in good accord with what one would expect from such a system. The MDCK epithelium is not an isotonically transporting one, being probably derived from distal tubule or collecting duct (Herzlinger *et al.,* 1982); the flow rate is very small, about 1 μL/cm^2.hr (Timbs and Spring, 1996), and the transportate Na concentration can be calculated to be about 1.6×10^{-10} mol/cm^2.sec from the cell dimensions and transport rates given (Kovbasnjuk *et al.,* 1995; Timbs and Spring, 1996). The transportate must be about 350% of the Ringer Na molarity (0.142 M). This result is what one might expect for an epithelium driven by osmotic gradients alone and functioning by standing-gradient osmosis in the interspaces: the transport is very hypertonic, and the junctional system is not actively transporting fluid. It is only the existence of an active junctional fluid transport mechanism in other epithelial systems which presumably assures overall isotonic transport. Osmotic gradients were also applied across the epithelium, but the interspace fluid velocity adjacent to the junction still remained near-zero, indicating that junctional osmotic flow is indeed too small to measure. However, the junctions in this MDCK epithelium are not impermeable seals through which osmotic water flow cannot occur (i.e., truly tight junctions) because this preparation is a low resistance epithelium (60 Ω.cm^2) which is moderately permeable to ions and presumably to water (Kovbasnjuk *et al.,* 1998b).

D. Paracellular Fluid Flow Requires a Nonosmotic Mechanism

If fluid is crossing the epithelium by the junction and interspace route but osmosis is ruled out as a probable mechanism, it is essential to determine what the nature of the flow process could be. It is here that a more detailed consideration of the characteristics of the probe transport is required. It reveals that the relation between the fluxes of the probes and water must depend upon movements within the junctions.

It turns out that it is very difficult to envisage the flow of probes (or any equivalent species) as being a simple convective flow in which there is a movement of water set up and the probes or other solutes are swept or dragged along as molecules in pipe flow. For this type of quite precisely defined flow mechanism, the net flux of probe solutes J_{net} is given in convection theory (Shachar–Hill and Hill, 1993) by

$$J_{net} = vAGS \tag{10}$$

in which v is the fluid velocity through the convective channel whose area is A. vA is thus the fluid flow rate (as a volume) and G and S are factors described below. From the available area through which the flow can occur, the velocity required in *Necturus* gallbladder to convect the probes has to be quite large—orders of magnitude larger than the velocity obtained by dividing the observed fluid flow by the total junctional area (Hill and Shachar-Hill, 1993).

G represents a "drag" factor. In Eq. (3), a similar factor θ was used to represent a general hindrance effect (also radius-dependent), but here G has a precise form for the special case where particles are dragged through a channel by hydraulic flow, and the force on them depends on their size in relation to that of the channel. The relation is complex but has been calculated for spheres in different geometries of channel. Clearly, when the solute size reaches that of the channel, then $G = 0$; and when it can enter with the same size as water, $G = 1$, i.e., it has no effect. In theory, plots of G against radius are all nonlinear no matter what the channel geometry—whether they are cylindrical pores or parallel surfaces (slits), for example. From the systems investigated above, we have seen that the relationship between the probe fluxes J_s and radius is linear to a very good approximation, which would seem to rule out simple channel flow with concomitant drag.

S is the other factor which has to be taken into account and which describes how molecules partition from a bath into a small space such as a channel. This has nothing to do with the flow of fluid, being a static effect. The center of molecules will be unable to explore all of a restricted space open to a bath; the centers are not able to approach the wall of the space closer than a distance equal to the molecular radius, and this effect is greater the larger its radius. For any sized molecule, there is an excluded volume around the walls, an effect which is not felt in the free solution space of the bath. The effect is that the concentration of molecules, measured as their number per total volume of the space, falls with increasing radius. The linearity, as before with the drag factor G, is dependent upon the channel geometry—it is linear with radius only for slit channels. The linearity of S is bounded by two extremes: when the channel half-width is equal to the probe molecular radius, it is excluded completely, and $S = 0$ (essentially a cut-off radius); when the probe radius is very small, $S \rightarrow 1.0$, and the concentration in the channel is equal to that in the bath. It appears, therefore, that the observed linearity of the net flux curves of probes with radius imposes the following restrictions: the flow is one with no apparent drag on molecules, but they have to enter a compartment.

How can the absence of significant osmotic flow across the junction and the problems of a simple convective model be reconciled with the experimental fact that, in these epithelia, inert probes are moved in a net flow together with water? We require a plausible mechanism that (i) moves water and small molecules in a coupled manner, (ii) evades drag effects, i.e., is not simple convective flow-through

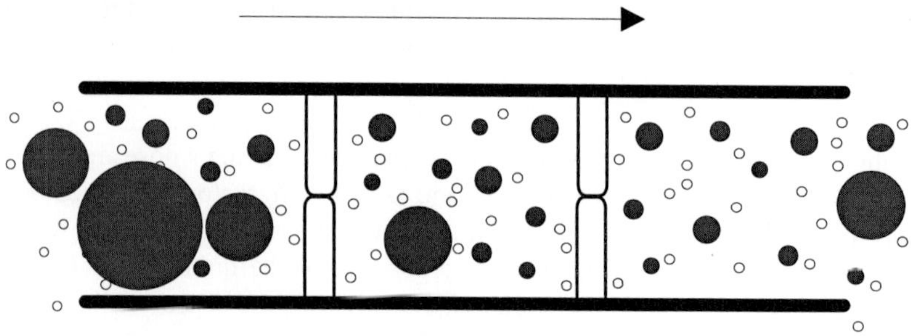

FIG. 5 The junctional transport of molecules (filled circles) and water (open circles) by compartments within the junction. The precise mechanism of admission and release cannot be deduced at this stage, but closure and effective linear transport are required by the data (see text, Section II.D). For a discussion of possible schemes, see Hill and Shachar-Hill (1993). The ionic leak pathway is not shown.

a channel, and (iii) obeys an S factor which is linear with radius. In Fig. 5, such a mechanism is depicted diagrammatically in cross section, but it already embodies several features which are compatible with what we know about junctions.

The mechanism consists of more than one compartment in series. The first of these (there may be a large number in parallel around the cell circumference) opens to admit water and probes, here a family of molecules of different radii. The filling of the compartment is according to the S factor. There is then closure followed by translocation of the compartment along the system in the direction of fluid flow; an analogy with a railway system is not at all out of place. This translocation moves all the water and other molecules in the compartment along with the same velocity but cannot discriminate between them on the basis of their sizes; thus the system behaves as if there were no drag effects in operation and $G = 1$ for all species. By analogy, all passengers are transported at the same rate when the compartment door is closed. To release probe molecules, compartments have to finally open to the other bath (which may be a lateral intercellular space in communication with the basolateral bath in a forward-facing epithelium). This transport has some similarities with pinocytosis and with possible molecular water pumping by cotransporters in the membrane; but the former requires enormous membrane turnover and would not discriminate between molecules at all (because the compartments are so large), while the latter operates between bath and cytoplasm and could not accommodate probes at all (because the compartments are so small).

Elements of the system have to effect transfer from bath to bath. Again, we can derive properties of the system from the experimental data; taking perfectly plausible values for the dimensions of the junctions and the fluid transfer rates, it is clear that no membrane component can move with the required velocity. If the linear extent of junction in the plane of the membrane is about 10^3 cm and the junctional width is of the order of 100 Å, to accommodate a fluid transport of

40 μL/cm^2.hr the fluid velocity through the junctional system would have to be in excess of 100 μm/sec. While it is perfectly possible for fluids to move at this speed through small channels, it is inconceivable that any membrane component could move laterally at this speed through a fluid membrane. It is possible, however, for lateral peristaltic movements to propel fluid at this speed and to involve no movements in the plane of the membrane at all. We are therefore led to postulate some form of microperistalsis—but one in which the opposite sides come together to effect a virtual closure. There are several peristaltic models that can be devised to do this.

If we relate this sytem to the known structures between and adjacent to epithelial cell membranes, there are obvious similarities.

(i) The junctions are structures formed within parallel-sided membranes; membranes of adjacent cells are not generally parallel, but in the region of the junction they are *locally* parallel.
(ii) They contain serially repeating structures, the intramembrane strands, which seem to divide the junction length into domains which often show internal volume in many electron micrographs (and under certain conditions can show swelling or blistering).
(iii) They connect with the cell interior via a complex architecture linked to both actin and myosin (Madara, 1998; Mitic and Anderson, 1998), which would provide a possible coupling of motion to an energy source.

Obviously, such a mechanism requires energy—the junctions of isotonically transporting epithelia in this scheme are energized organelles. We do not know what this energy source might be, other than that is most probably draws ultimately from the ATP pool; but there may be intermediate transducers.

Finally, there is the problem of the ionic conductance of junctions—the universal observation that the tight junctions of isotonically transporting epithelia, belying their name, are low-resistance structures that have lead to their description as leaky tight junctions. There is no convincing theory as to why they should be leaky; in many tight epithelia, the salt transport rates are as great as in most leaky ones, but the junctions are high-resistance structures. Together with leakiness goes selectivity; many junctions are more permeable to one ion (usually a cation), although in many systems this is only relative. The backflux curves of probes (i.e., contrary to fluid flow) in most epithelia that have been examined show that there are pathways which account for the ion permeability but which mediate unidirectional fluxes far greater than the convective *net* fluxes. This indicates that these pathways are not part of the convective model system discussed above. It is not clear whether a charge in the junction controls the selectivity to ions (it is probable and would have a marked effect); but the high-conductance path through the junctions, in intimate association and in parallel with the fluid convection channels, may serve to equilibrate the cation and anion fluxes across the active junction and interspace fluid transport system, independent of the cellular route.

III. Solutes Probably Take Two Routes

In a paracellular model of fluid transfer, there is no restriction a priori on the route of solutes. The only desired result is that the fluid secreted, or absorbed, should be nearly isotonic, or that the osmolarity of a bath (such as the CSF or the fluid film on airway epithelium) be maintained. If there is a junctional pump in operation, it must transfer some ions with the water (to consider first only the saline environment, without the transfer of metabolites such as glucose or amino acids). A central problem is then that of controlling the combination of this flow with that of the cellular route, which has generally been assumed to be the only significant one.

A. The Cellular Route and Aquaporins

There is a generally accepted model for forward-facing or backward-facing epithelia. This nomenclature has arisen to describe fluid transport from apical to basolateral bath (gallbladder, proximal tubule) or vice versa (exocrine glands) with a Na:K pump on the basolateral membrane in these two cases.[1]

In the forward-facing model (Koefed-Johnsen and Ussing, 1958), Na^+ is pumped directly into the basolateral bath, K^+ recirculates across the basolateral membrane, and Cl^- mainly traverses the junctions or the cells via a transporter to achieve electroneutrality. In the backward-facing model (Silva *et al.,* 1977; Hannafin *et al.,* 1983), the pump, the primary ATP-linked ion transporter, is pumping ions toward the basal bath. Na^+ recirculates at the basolateral membrane, reentering the cell with K^+ and Cl^- on a NaK2Cl cotransporter. Cl^- is driven through the cell, and leaves the apical membrane cell via a channel or accompanies a fraction of the K^+ by a KCl apical cotransport. In this case, the cell is mainly a Cl^- ion transporter, while the co-ion deficiency is compensated by a flow of Na^+ ion through the junctions from the basolateral to the apical side. In the absence of any appreciable fluid transport by the junctional system, there would be water flowing transcellularly to accompany this salt, driven by osmosis. It is the combined ability of the apical and basolateral membrane systems to achieve near-isotonic equilibration that is the subject of debate.

When aquaporins came onto the scene, it seemed to many in the field of epithelial fluid transport that they were obviously designed to facilitate water flow (by osmosis) across the membranes, and approaches based upon ideas of paracellular fluid transport, however reasonable their experimental basis, seemed less plausible. Water channels in epithelia, which were originally postulated as the

[1]There are other isotonically transporting vertebrate epithelia, not considered here, which have expressed the Na:K pump on the apical membrane but which transport fluid in either direction, i.e., in the apical-to-basolateral direction and are therefore forward-facing (retinal pigmented epithelium) or backward-facing (choroid plexus). All these four systems have to be incorporated into a final scheme of fluid transport, whatever the underlying mechanism (see Zeuthen article, this volume).

basis for hormone-regulated permeability increases in amphibian bladder, were later identified as AQP2, also present in mammalian collecting duct (Deen *et al.*, 1994). However, it was discovered that humans lacking aquaporin-1 (AQP1, the Colton antigen), the aquaporin most widely distributed in tissues and cells, manifest no obvious pathology (Preston *et al.*, 1994). These findings raised the question of the precise role of AQPs in epithelial fluid transport.

In the last few years, this problem has been addressed by two approaches: a very extensive search for new aquaporins with their localization (Borgnia *et al.*, 1999) and the production of transgenic knockout animals (AQP +/– and –/– mice) lacking specific AQPs. The first line of research has identified nine AQPs to date in animals (excluding AQP0 from the lens), and demonstrated their wide distribution in most organs and tissues. The second approach has enabled the role of individual aquaporins in fluid transport to be investigated in the whole animal and in specific tissues and organs (Verkman *et al.*, 2000b). All the AQP-knockout mice produced to date are viable and many are grossly normal. If they play a central role in epithelial isotonic fluid transport, this is a perplexing finding, and their role is proving to be a more complex issue.

Lack of AQP1 does not abolish fluid absorption in the proximal tubule of the kidney, although it reduces the osmotic water permeability of the tubular wall by 80% (Schnermann *et al.*, 1998). In accord with this, it was previously shown that applying $HgCl_2$, a potent inhibitor of AQP1, to the luminal surface of perfused proximal tubules did not affect rates of fluid transport (Wareing *et al.*, 1996a,b). Mice lacking AQP1 and AQP3 showed a decreased ability to concentrate urine (Ma *et al.*, 1998, 2000b). This indicates that aquaporins have a role in kidney function, which may be related to osmotic balance but not specifically to isotonic transport. The case of AQP5-knockout mice is of particular interest because two different epithelia which both transport isotonically are affected very differently. Salivary secretion in these mice is hyperosmotic and of lower volume than the near-isotonic secretion from wild-type mice. However, the isotonic transport performed by lung alveolar epithelium of these knockout mice is unaffected (Ma *et al.*, 2000a; Verkman *et al.*, 2000a).

There is no simple way to explain these results within the paradigm of fluid transport as a simple osmotic process. But if the AQPs are not simply involved in a unitary task of facilitating cellular osmotic equilibration across cell membranes, they may affect different tissues in other ways. The examples given above indicate that aquaporins are involved in fluid balance and movement but apparently do not simply provide the passive channels required for osmotic equilibration.

B. Aquaporins as Osmo-sensors.

A possible role for aquaporins is to act as osmo-sensors. An osmo-sensor could sense the osmotic pressure because it contains a region within its core to which only water has access; when the osmotic pressure of the bath is raised, there is an extra free energy cost in moving water into this core from the bath. Another way of regarding this is that when in contact with a bath containing osmolytes at osmotic pressure π, the water in the core is under a tension $-\pi$. Now, if the molecular structure can respond to this internal tension or free energy difference by undergoing a conformational change—from state A to state B, for example—then the A:B ratio is a measure of the osmotic pressure of the bath. There are many examples of molecules which undergo two-state changes in conformation, such as enzymes, redox-couples, and ligand-binding molecules, which show these changes in ratio between their two states (Zeuthen, 1995; Rand *et al.,* 2000; see Cotransporters as Molecular Water Pumps by Zeuthen and MacAulay, this volume and Rand, this volume). There are also an increasing number of ion channels which signal osmotic stress by changing their conductance state (e.g., the vanilloid receptor-related channel (Liedtke *et al.,* 2000) and many others).

Aquaporins have not been shown to possess gated states; they do not appear to show changes in their water channel conductance. However, they do possess quaternary structure in the membrane, and this may have significance for the control of other cell functions in response to changes in tonicity adjacent to the membrane. This is an important possibility that would assign them a place in epithelial fluid transport which is distinct from acting as simple water conduits. In the discussions below of selected epithelial systems, we have indicated how this concept can be used to integrate paracellular flow with cellular ion transport to achieve near-isotonic flows.

C. The Paracellular Route and Aquaporins

The junctional transport revealed by the probe studies will produce a hypotonic fluid. This is because all solutes, including ions, are larger than water; the compartments of Fig. 5 fill, as described above, according to the partitioning factor S. This, the parallel-sided compartment of the junction has the linear form

$$S = 1 - \frac{r_s}{r_{ch}} \tag{11}$$

so that for a channel of half-width r_{ch} equal to 5 Å (similar to many of the epithelial systems studied with probes, considered previously), the factor S for an ion of hydrated radius 2 Å (for example) is 0.6. For water of radius 1.5 Å in this channel, the factor S is 0.7. The ratio of ion to water compared to that in the source bath would therefore be 0.6/0.7 or 86% hypotonic. In short, the junction must select for water over salt, or any other solutes larger than water. If the junctional compartments are charged, which they may be, the effect will be even greater. The two ions, mainly Na^+ and Cl^-, will not be transported at the same rate, and the ionic leakiness of the junctional pathway (separate from and parallel to the compartments) may here function to reduce the potential difference which might otherwise retard the process.

There must therefore be two inputs to the composition of the transportate: a hypertonic flow through the cells and a hypotonic flow through the junction and interspace system. The role of volume transport thus falls on the junction, but the strength and osmolarity of the transportate are determined in this model by the relative contributions of the two pathways. (The potential mixing of these two flows must be very intimate, if we consider the geometry of the interspace and basolateral membrane system. The interspace system usually contains a very large fraction of the total membrane of the cell, and the fluid volume within it is in series with the junction. Furthermore, the interspace often leads diagonally into the basal infoldings; in transmission electron micrographs, these two systems often appear to be separate but in reality are rarely so.) The ionic composition of the transportate may include a contribution from the transport processes of the cellular route.

What is certainly critical for any isotonically transporting epithelium of this nature is some mechanism by which the osmolarity is controlled within quite narrow limits. It is just possible that the separate elements have flows whose magnitude and selectivity are set intrinsically, and isotonicity of the transportate is achieved automatically according to a fixed ratio. This is unlikely, however, because salt and water transport rates can be varied over quite large ranges, while the tonicity of the transportate remains close to that of the source bath. The *in vivo* source bath may be intrinsically variable (such as intestine); the overall salt and water transports may be modulated by hormones or neurotransmitters (such as salivary glands); or the source bath may be manipulated *in vitro* (such as gallbladder). In all of these systems, it is difficult if not impossible to devise a scheme which allows for this flexibility with fixed-ratio elements. What is required is a third element: a system that monitors the osmolarity and adjusts it by altering the ratio of paracellular to cellular fluid. In other words, an osmo-comparator with an output signal that controls the paracellular:cellular ratio, as shown in Fig. 6A. A simplified analogy to such a system would be a thermostat on a mixed flow system like a shower; here, temperature stands in for tonicity. If we regard the cell route, which is hypertonic, as a "hot" flow, and the junctional route, which is hypotonic, as the "cold" flow, then opening a hot flow will cause the thermostat to adjust the cold flow until the mixture matches the required temperature. The volume flow rate is dependent upon both the initial hot flow setting and the temperature of the cold water.

The cellular route and the junctional route have to be working together, and the proportion of total solute passing the cell may vary from one epithelial system to another. If there is an osmo-sensor in the membrane, comparing either the source bath with the cytoplasm, or the transportate with the cyptoplasm—this membrane could be apical or basolateral, depending on the epithelium. In a backward-facing epithelium such as a gland, the apical membrane would be the site of an osmo-comparator, as shown in Fig. 6B. In either case, the output would use a cell signaling pathway to regulate the rate of junctional pumping. The cell would have, in the steady state, a tonicity intermediate between source and transportate, irrespective

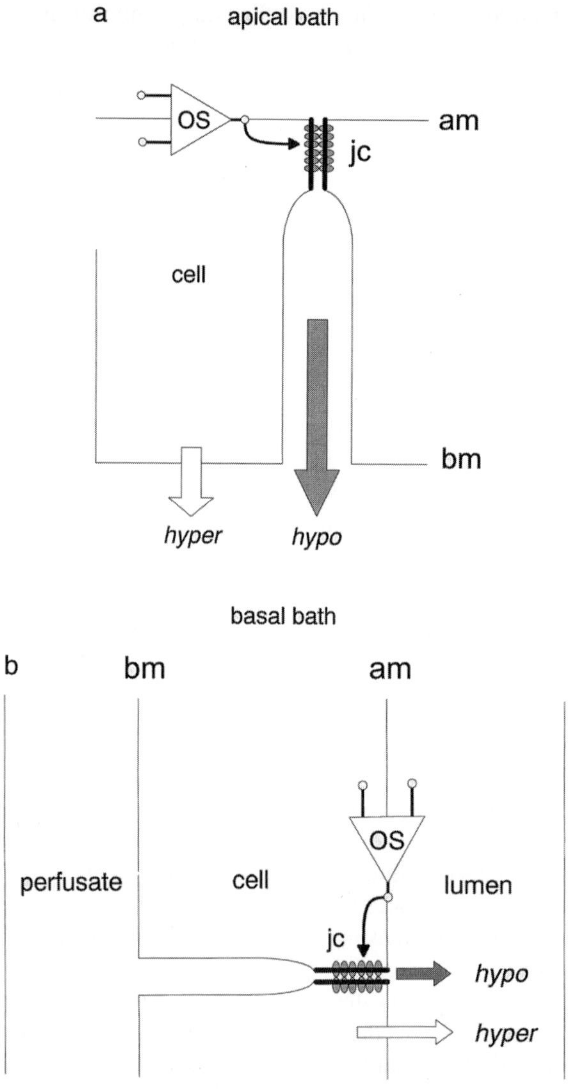

FIG. 6 (A) A scheme for coupling paracellular and cellular flows to achieve near-isotonic fluid transport in a forward-facing epithelium. An osmotic sensor molecule (OS) in the apical membrane (am) acts as an osmo-comparator between apical (source) bath and cell. When the tonicity of the transportate bathing the basolateral membrane (bm) is above that of the apical bath, the tonicity of the cell is intermediate, due to partial equilibration across the two membranes. The OS output then signals an up-regulation in fluid transport by the junctional complex (jc) until the combined effect of the increased paracellular flow (hypotonic) and transcellular flow (hypertonic) brings the cell tonicity down close to that of the apical bath, i.e., the transportate becomes quasi-isotonic (see Sections III.C and IV.A). (B) The control scheme of (A) applied to a backward-facing epithelium, such as the salivary gland. Here, the osmo-comparator (OS) is also assigned tentatively to the apical membrane. The

of its geometry and membrane area. When quasi-isotonicity between source and transportate is achieved by mixing the flows in the right ratio, the tonicity difference between cell and source or sink would be small, and the system would be in a stable steady state.

An interesting feature of such a system is that, when quasi-isotonicity is reached, there is little if any tonicity difference across the cell to drive water osmotically through the apical and basolateral membranes. Although the osmotic permeability of the cell membranes may not be high enough for the cellular route to achieve isotonic transport by itself, when the steady state is reached this becomes irrelevant; there is no substantial flow other than that through the junctional pathway, in spite of the presence of aquaporins contributing to the P_{os} of the cell membranes. This explains why, in those studies which have revealed almost complete paracellular flow of water as described above, the cellular osmotic route through its water channels is not making any sizable contribution for lack of a transepithelial gradient.

IV. Application to Some Epithelial Systems

In view of the discussion above, an obvious candidate for the osmo-comparator would be an aquaporin, specific members of this family being usually limited to one of the membranes, depending upon the epithelium. The observation that knockout mice strains (−/−) for a particular aquaporin retain the ability to transport fluids at high rates while often manifesting changes in fluid osmolarity would support this idea. In what follows, we discuss two epithelia for which data of this kind are available in terms of this model: proximal tubule (A) and salivary gland (B). For a third system, the small intestine (C), we discuss possible control of the junctional flow systems by a cotransporter.

A. Renal Proximal Tubule

The proximal tubule deals with the largest isotonic volume transfer in the body, and as such provides a crucial test for any model aiming to explain the mechanism underlying this process. The epithelium is forward facing, the Na/K pump is facing in the same direction as the fluid flow, and tubular fluid transport has been

---

paracellular flow (hypotonic) and transcellular flow (hypertonic) combine to form the luminal exudate which bathes the apical membrane. The cell tonicity is biased close to that of the perfusate (plasma) by partial osmotic equilibration over the extensive basolateral membrane. The osmocomparator signal is generated between exudate and cell (see Sections III.C and IV.B).

investigated widely. The proximal tubule has been perhaps the most debated fluid transport epithelium so far; there have been earlier claims that a large fraction of salt crosses the junctional system (Fromter and Ullrich, 1973; Weinstein, 1987) as well as water (Andreoli and Watkins, 1979). It is not clear what is currently the status of these views, as they are not widely held. It is also clear that no significant osmotic gradient has been demonstrated in the intercellular spaces of isolated functional tubules to date. In addition, there are studies in particular systems which indicate that transcellular osmosis is not applicable; studies on fluid absorption in perfused tubules from *Necturus* kidney show that near-isotonic absorption occurring at both normal and low osmolarities cannot be accounted for by osmotic equilibration across the tubule wall (Whittembury and Hill, 1982). Probe studies have not been extended to proximal tubules yet, but the available data on tubular absorption fit well within a junctional transport mechanism.

In order to achieve the observed rate of fluid absorption by osmosis, the cell membranes must have a high osmotic permeability. It has been shown that this high permeability is conferred by the presence of AQP1 in the apical tubule cell membrane. Studies of the tubule apical membrane have shown that AQP1 is responsible for at least 80% of P_{os}, the osmotic permeability (Zhang *et al.,* 1993; Schnermann *et al.,* 1998). However, the addition of $HgCl_2$, a potent blocker of AQP1, to the apical membranes of transporting rat proximal tubule via the perfused lumen (under conditions where neither the luminal nor plasma osmolarity could alter significantly) did not inhibit the fluid transport rate (Wareing *et al.,* 1996a,b). This argues against AQP1 participating in the steady-state fluid transport process as water channels.

In studies of AQP1-knockout mice (AQP−/−), it was found that a lack of AQP1 caused a reduction of 78% in the apical P_{os} and also caused a decrease in fluid transport rate of 50% (Schnermann *et al.,* 1998). Taken by themselves, these figures show that volume transport—though affected by AQP1—is not inhibited as expected by its absence. These studies do not support the assumption that a high osmotic permeability is essential for isotonic fluid absorption in the kidney. In a separate study with AQP1-knockout mice, the osmolarity of fluid collected from segments of late proximal tubule was compared with that of the plasma (which was different in the two animals) in both wild-type and knockout mice (Vallon *et al.,* 2000). In wild-type tubules (AQP1 +/+), the difference was 9–12 mOsm but in knockout ones (AQP1 −/−) it had increased to 39 mOsm. This is taken as evidence for an increased hypertonicity of absorption by the tubule.

If the tonicity of the transportate is not greatly altered, although the fluid transport rate is halved, then the net salt transfer by the epithelium must be greatly reduced. There are two possible explanations for this. (i) The phenotypic absence of AQP1 has severely down-regulated the Na:K pump in the basolateral membrane, which is unlikely. (ii) A major fraction of the salt flow is created by convection in the junctional pumping system rather than being transferred entirely by the Na:K pump. In this situation, the overall salt:water transport ratio is determined by osmo-control of the junctional transport system (Fig. 6A). A decrease in the junctional pump rate, as would be expected from the removal of AQP control, would lead to a fall in both fluid transport and salt transport with a modest rise in hypertonicity of the transportate—as seems to be observed.

It is apparent from these studies that AQP1 has a role in regulating the rate of absorption in the proximal tubules, but the precise mechanism is unclear. From the above, it is reasonable to conclude that osmosis is unlikely to provide the basic mechanism for isotonic absorption in the proximal tubule. A junctional fluid transport mechanism does not depend on any gradients being maintained or on a high osmotic permeability of the cell membrane. It would be regulated by the cells, and volume flow (and possibly composition) can be adjusted via signaling systems which are known to be associated with the junctions (Madara, 1998; Mitic and Anderson, 1998).

B. Exocrine Gland

In the salivary gland (as in the lacrimal gland) of the mouse, the fluid transport is across a backward-facing epithelium. A problem with such systems is that there is no obvious coupling space for transcellular osmotic equilibration. If ions are actively transported by the cell over the apical membrane, the only forward coupling space is the lumen with a very minute membrane area. Osmotic coupling cannot occur by unstirred layers in a unilateral system (i.e., one in which the sink bath is created by the epithelial secretion itself, as in a salivary gland) without violating the convection–diffusion equation. The osmolarity gradient that drives secretion must represent the difference between the secretion and the cytoplasm.

AQP5 has been localized in this apical membrane (Ma *et al.,* 1999), and knockout mice lacking this aquaporin have been generated. AQP1 and AQP4 are also present in the gland, but their absence does not affect either the volume or the composition of saliva. When saliva production is stimulated by pilocarpine, the fluid flow rate in knockout animals (AQP5 −/−) decreases to about 40% of the wild type (AQP5 +/+) and the osmolarity rises to 420 mOsm, i.e., near 125% hypertonic. Thus, as with the proximal tubule, it seems that fluid secretion can proceed without this aquaporin, but its tonicity is affected.

We should note also that the salt transport is affected very substantially; the product of the volume flow and concentration in stimulated AQP5 −/− strains is $0.4 \times 125\%$, i.e., 50% reduced compared to the wild type. Amylase and protein secretion in these serous glands were normal, showing that the pilocarpine receptors and the stimulation of the gland cells were not affected. It is clear that, in a manner similar to that for the proximal tubule described above, the results cannot be fitted into a simple scheme of transcellular osmosis. They can be explained, however, by a model of paracellular fluid transport by the junctional complex, as in Fig. 6B. If the role of AQP5 is not to act primarily as a passive water channel but to control the tonicity of the secretion, then the effect of removal of this control by knockout is understandable.

Here AQP5 may be acting as a comparator of tonicity between cell and secretion, in which case its location on the apical membrane is necessary. As in all these systems, its role would be to up-regulate the paracellular junction pump (via a cell signaling system) in response to an increase in disparity between cytoplasm and acinar (luminal) transportate. The addition of hypotonic junctional secretion to the transcellular salt flow would bring the overall secretion back to near-isotonicity. Removal of AQP5 would remove primary control of isotonicity in this feedback loop so that fluid can still be produced but at an uncontrolled tonicity. In this model, the cell tonicity closely tracks that of the plasma by the considerable relative area of basolateral membrane—a process which would be aided by other AQPs in this membrane.

In the lacrimal gland, there is also AQP5 in the apical membrane, but its removal by knockout in mice has little effect upon the volume secretion rate or tonicity of the tear fluid. A similar removal of AQPs 1, 3, and 4 also has no effect. As a result of this, it has been assumed that there can be no involvement of this AQP5 (and the others) in the fluid secretion process in this gland. These findings add a level of complexity and arbitrariness to any simple osmotic model of its function, but the aquaporin may have a function in controlling the tonicity of fluid bathing the eye surface as part of a wider feedback loop. It very much depends (a point little discussed here without recourse to a quantitative model) on the set-point to which a junctional transport system is operating without AQP control: it may be near-isotonic in some systems. What it does demonstrate is that modulation of P_{os} of the membranes has little effect and that fluid secretion cannot be a simple osmotic flow without invoking the paracellular route.

C. Small Intestine

Fluid absorption and recirculation in the intestine is widely accepted as a phenomenon with secretion probably occurring in the crypt and absorption in the villous epithelia of mammals (for a review of this evidence, which is piecemeal and difficult to obtain, see Donowitz and Welsh, 1987). Some amphibian GI tracts, such as that of *Necturus,* do not possess crypts. When nutrients are being absorbed, it would seem that a considerable measure of recirculation must be taking place. The absorption of 1 mol glucose or alanine, generated from hydrolysis of starch or protein, for example, proceeding via a Na^+-linked transporter in the apical membrane would require at least 1 mol Na and 6 liters of water under isotonic conditions in mammals and 10 liters in amphibia. Ingestion of these quantities of Na and water are not obligatory, and they could only be made available through recirculatory flows of both. In Fig. 2, where *Necturus* small intestine is transporting dextran probes in the presence of alanine, the curves have the interesting properties in that

TABLE II
Properties of *Necturus* Intestine Fluid Circuits

		Water fluxes from dextran analysis			
		Unidirectional flow μL/cm^2.hr	Cut-off radius (Å)	Net flow μL/cm^2.hr	Volume flow measurements μL/cm^2.hr
MS	Ringer	51.95	6.09	16.74 ± 1.4	15.15 ± 1.2
SM		35.21	10.82		
MS	Ringer+	58.82	5.08	33.84 ± 5.9	34.52 ± 5.0
SM	20 mM L-alanine	24.98	9.53		

MS (mucosal to serosal); SM (serosal to mucosal)

they show negative portions. When a certain radius of probe is being convected (between about 3 and 10 Å), there is a *negative* flux, i.e., a secretion from basolateral to apical bath. This has been broken down into two linear curves, one representing absorption and the other secretion, which superimpose to give the pattern as seen in Fig. 2 (Hill and Shachar-Hill, 1997). Thus the probes reveal that two pathways, each with its own selectivity, are operating simultaneously. The absorptive fluid path has a cut-off between 5 and 6 Å, while the secretion path value is 9–10 Å, and these selectivities are not altered to any extent by alanine stimulation, as shown in Table II.

It is apparent that the stimulation of uptake caused by alanine is due to an increase of the MS fluid circuit and a concomitant decrease in the SM circuit. These changes are not large fractions of the preexisting rates, but together they add up to a rise in the net absorptive rate of about 200%. Luminal metabolites and ingested salts would be removed quickly into the portal circulation and replaced by plasma saline, while the luminal volume would decrease (Fig. 7). The changes in the two limbs of the circuit are in opposite directions, and it is not ruled out by the data that these changes may be about equal in magnitude. If this is so, it would be consistent with a scheme in which fluid flow in the junctions of some cells is reversed, so that flow in the MS limb gains in magnitude at the expense of the SM limb; but this remains only a possibility. The existence of two opposed pathways implies that the tissue is heterogeneous in junctional function. It is quite possible that various compounds (toxins) would alter the MS:SM ratio in favor of drastic secretion, such as occurs in disease states. The scale of this heterogeneity is not yet clear, and opposing limbs of the fluid circuit may be confined to the junctions of adjacent cells, or to adjacent cell groups.

We have suggested (Hill and Shachar-Hill, 1997) that a function of the Na-alanine transporter may be that of regulating the paracellular flow rate of water, alanine, and ions through the junctional system (via a cell-signaling system) in addition to that of stimulating Na and alanine transport through the cells; the same would apply to glucose via SGLT1 and GLUT1 transporters, which may also be accompanied by some water (Loo *et al.,* 1996, 1999). In the context of the debate

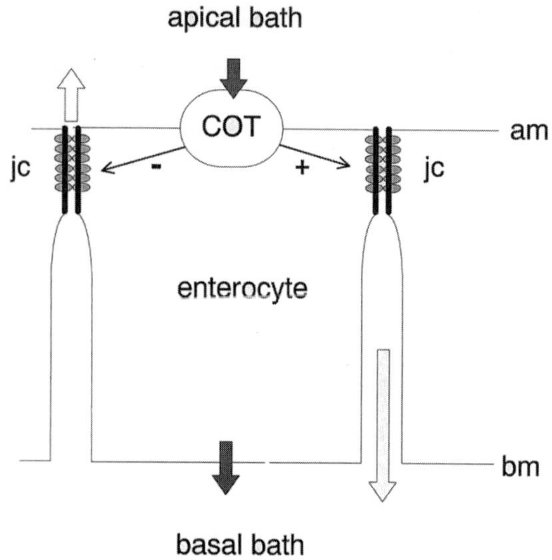

FIG. 7 Fluid transfer in small intestine in which absorption is stimulated by metabolites such as alanine. Experiments show that fluid recirculation occurs, measured by bidirectional movement of probes through the different sets of junctions (jc). It is suggested that alanine exerts its effect through binding to the Na-alanine cotransporter, altering the proportion of bound: unbound states and signaling to up-regulate the absorbing junctions and down-regulate the secretory ones. Alanine is removed from the basal bath by the circulation. At higher absorptive rates, an increasing fraction of alanine is taken with water and salt through the junctions (see Section IV.C).

about the paracellular transport of glucose in small intestine (Pappenheimer and Reiss, 1987; Pappenheimer, 1993) and the fraction transported by this route as a function of concentration, it is possible that not only is the glucose being transported by the junction pump but this process is up-regulated by increasing luminal glucose; the net result is that the paracellular fraction increases with rising glucose concentration. Any model that endeavors to explain the paracellular glucose transport as a convection with osmotically driven water across the junction system is open to the general restrictions on junctional osmosis examined above; the osmotic flow would have to be driven by ions through channels large enough to accommodate glucose molecules which would have a low reflexion coefficient (and therefore a low P_{os}).

The transporter, it should be remembered, is also a receptor existing in a ratio of bound:unbound states capable of initiating cell signaling. This regulation would be similar to that proposed for osmo-sensors (possibly AQPs) in other tissues considered above, with the difference that it is solute dependent. However, this scheme, as shown in Fig. 7, may differ from most other epithelial systems in another major aspect—the fluid transport is not required to be isotonic to plasma; in fact it cannot always be so. If intestinal preparations are set up between isotonic baths, then isotonic absorption is observed. But when hypertonic saline is ingested, or polymers are enzymatically broken down to osmotically active monomers, all the water and solutes are absorbed, but these are (*in toto*) hypertonic to plasma. In contrast, pure water loading of the intestine results in an overall hypotonic absorption with respect to plasma.

The regulation of the isotonicity of transportate with respect to plasma is not mandatory as in many other systems, because of recirculation. In colonic absorption, there is also no requirement for isotonic absorption or for recirculation. This may be the explanation for the general absence of a key aquaporin such as AQP1 from the enterocytes of intestinal epithelia and the sporadic distribution of others (Koyama *et al.,* 1999; Ma and Verkman, 1999). AQP4-knockout mice (−/−), in which this aquaporin has been denied expression from the basolateral membrane of fluid transporting cells, show unimpaired fluid transport in the stomach (Wang *et al.,* 2000a) and in the colon (Wang *et al.,* 2000b). It should be stressed here that although the presence of these aquaporins confers enhanced water permeability, this is not required for absorption, indicating a junctional transport of fluid. It is also possible in any system that, although AQPs are acting as regulators of the junction transport system (under certain conditions) under normal conditions, the set-point of the system without AQP intervention is close to that required to ensure isotonicity.

V. Summary: Mechano-osmosis as the Basis of Fluid Transport

We have considered the accumulating experimental and theoretical bases for a model of fluid-transporting epithelia in which water and some solutes pass through the junction and interspace, probably as a hypotonic transportate. This combines with a hypertonic transfer of ions (with some water) crossing the cell to achieve an overall fluid transport which is near-isotonic.

This paracellular flux cannot be osmotic, due to the minute area and ionic leakiness of the tight junctions. An analysis of the flow of probes through the junction and interspace system reveals that the fluid must move by a mechano-osmotic mechanism involving elements of the junctional complex; both the energy for this process and its control have their origin in the cell.

The rate at which the junctional fluid transport operates may be controlled by receptors or osmo-sensors at one or both epithelial membranes. These would act to coordinate the two pathways. When near-isotonicity is achieved, there is no osmotic gradient, and little if any osmotic water flows across the cell. Paracellular water flow is then dominant, but its rate is due to a matching process and therefore a function of the transcellular flux of ions. Thus, although the task may be shared in a complex way, the junctional pump is a key player.

Acknowledgments

We should like to acknowledge the collaboration of Dr. Masataka Murakami, Okasaki, and support from the Monbusho International Scientific Project "Coupling Mechanisms of Fluid Secretion and Exocytosis" in the work on salivary gland fluid secretion. Our thanks are extended to Dr. Yair Shachar-Hill for a critical reading of the manuscript.

References

Aldis, G. K. (1983). "Concentration Fields Near Transporting Epithelia." PhD. Thesis, Cambridge University.

Andreoli, T. E. S. J. A. T. S. L., and Watkins, M. L. (1979). Solvent drag component of Cl- flux in superficial proximal straight tubules: Evidence for a paracellular component of isotopic fluid absorption. *Am. J. Physiol.* **237,** F455–F462.

Borgnia, M., Nielsen, S., Engel, A., and Agre, P. (1999). Cellular and molecular biology of the aquaporin water channels. *Annu. Rev. Biochem.* **68,** 425–458.

Case, R. M., Cook, D. I., Hunter, M., Steward, M. C., and Young, J. A. (1985). Trans-epithelial transport of nonelectrolytes in the rabbit mandibular salivary-gland. *J. Membr. Biol.* **84,** 239–248.

Deen, P. M. T., Verdijk, M. A. J., Knoers, N., Wieringa, B., Monnens, L. A. H., Vanos, C. H., and Vanoost, B. A. (1994). Requirement of human renal water channel aquaporin-2 for vasopressin-dependent concentration of urine. *Science* **264,** 92–95.

Diamond, J. M., and Bossert, W. H. (1967). Standing-gradient osmotic flow: A mechanism for coupling of water and solute transport in epithelia. *J. Gen. Physiol.* **50,** 2061–2083.

Donowitz, M., and Welsh, M. J. (1987). Regulation of mammalian small intestinal electrolyte secretion. *In* "Physiology of the Gastointestinal Tract. Vol. 2" (L. R. Johnson, Ed.), pp. 1351–1388. Raven Press, NY.

Fromter, E. R. G., and Ullrich, K. J. (1973). Phenomenologic description of Na+, Cl−, and HCO3− absorption from proximal tubules of the rat kidney. *Pflügers Arch.* **343,** 189–220.

Hannafin, J., Kinne-Safran, E., Friedman, D., and Kinne, R. (1983). Presence of a sodium–potassium chloride cotransport system in the rectal gland of *Squalus acanthias*. *J. Membr. Biol.* **75,** 73–83.

Hernandez, C. S., Gonzalez, E., and Whittembury, G. (1995). The paracellular channel for water secretion in the upper segment of the Malpighian tubule of *Rhodnius prolixus*. *J. Membr. Biol.* **148,** 233–242.

Herzlinger, D. A., Easton, T. G., and Ojakian, G. K. (1982). The MDCK epithelial-cell line expresses a cell-surface antigen of the kidney distal tubule. *J. Cell Biol.* **93,** 269–277.

Hill, A. E., and Hill, B. S. (1978). Sucrose fluxes and junctional water flow across Necturus gall bladder epithelium. *Proc. R. Soc. Lond. B.* **200,** 163–174.

Hill, A. E., and Shachar-Hill, B. (1993). A mechanism for isotonic fluid-flow through the tight junctions of Necturus gallbladder epithelium. *J. Membr. Biol.* **136,** 253–262.

Hill, A. E., and Shachar-Hill, B. (1997). Fluid recirculation in Necturus intestine and the effect of alanine. *J. Membr. Biol.* **158,** 119–126.

Ikonomov, O., Simon, M., and Fromter, E. (1985). Electrophysiological studies on lateral intercellular spaces of *Necturus* gallbladder epithelium. *Pflügers Arch.* **403,** 301–307.

Koefed-Johnsen, V., and Ussing, H. H. (1958). The nature of the frog skin potential. *Acta Physiol. Scand.* **42,** 298–308.

Kovbasnjuk, O., Chatton, J. Y., Friauf, W. S., and Spring, K. R. (1995). Determination of the Na permeability of the tight junctions of MDCK cells by fluorescence microscopy. *J. Membr. Biol.* **148,** 223–232.

Kovbasnjuk, O., Leader, J. P., Weinstein, A. M., and Spring, K. R. (1998a). Water does not flow across the tight junctions of MDCK cell epithelium. *Proc. Natl. Acad. Sci. USA* **95,** 6526–6530.

Kovbasnjuk, O. N., Szmulowicz, U., and Spring, K. R. (1998b). Regulation of the MDCK cell tight junction. *J. Membr. Biol.* **161,** 93–104.

Koyama, Y., Yamamoto, T., Tani, T., Nihei, K., Kondo, D., Funaki, H., Yaoita, E., Kawasaki, K., Sato, N., Hatakeyama, K., and Kihara, I. (1999). Expression and localization of aquaporins in rat gastrointestinal tract. *Am. J. Physiol. Cell Physiol.* **45,** C621–C627.

Larsen, E. H., Sorensen, J. B., and Sorensen, J. N. (2000). A mathematical model of solute coupled water transport in toad intestine incorporating recirculation of the actively transported solute. *J. Gen. Physiol.* **116,** 101–124.

Liedtke, W., Choe, Y., Marti-Renom, M. A., Bell, A. M., Denis, C. S., Sali, A., Hudspeth, A. J., Friedman, J. M., and Heller, S. (2000). Vanilloid receptor-related osmotically activated channel (VR- OAC), a candidate vertebrate osmoreceptor. *Cell* **103,** 525–535.

Loo, D. D. F., Hirayama, B. A., Meinild, A. K., Chandy, G., Zeuthen, T., and Wright, E. M. (1999). Passive water and ion transport by cotransporters. *J. Physiol. Lond.* **518,** 195–202.

Loo, D. D. F., Zeuthen, T., Chandy, G., and Wright, E. M. (1996). Cotransport of water by the Na+/Glucose cotransporter. *Proc. Natl. Acad. Sci. USA* **93,** 13,367–13,370.

Ma, T., Yang, B., Gillespie, A., Carlson, E. J., Epstein, C. J., and Verkman, A. S. (1998). Severely impaired urinary concentrating ability in transgenic mice lacking aquaporin-1 water channels. *FASEB J.* **12,** 1924.

Ma, T. H., Fukuda, N., Song, Y. L., Matthay, M. A., and Verkman, A. S. (2000a). Lung fluid transport in aquaporin-5 knockout mice. *J. Clin. Invest.* **105,** 93–100.

Ma, T. H., Song, Y. L., Gillespie, A., Carlson, E. J., Epstein, C. J., and Verkman, A. S. (1999). Defective secretion of saliva in transgenic mice lacking aquaporin-5 water channels. *J. Biol. Chem.* **274,** 20,071–20,074.

Ma, T. H., Song, Y. L., Yang, B. X., Gillespie, A., Carlson, E. J., Epstein, C. J., and Verkman, A. S. (2000b). Nephrogenic diabetes insipidus in mice lacking aquaporin-3 water channels. *Proc. Natl. Acad. Sci. USA* **97,** 4386–4391.

Ma, T. H., and Verkman, A. S. (1999). Aquaporin water channels in gastrointestinal physiology. *J. Physiol. Lond.* **517,** 317–326.

Madara, J. L. (1998). Regulation of the movement of solutes across tight junctions. *Annu. Rev. Physiol.* **60,** 143–159.

Mitic, L. L., and Anderson, J. M. (1998). Molecular architecture of tight junctions. *Annu. Rev. Physiol.* **60,** 121–142.

Murakami, M., Shachar-Hill, B., Hill, A. E., and Steward, M. C. (2001). The paracellular component of water flow in the rat submandibular gland. *J. Physiol. Lond.* **537,** 899–906.

O'Donnell, M. J., Aldis, S. K., and Maddrell, S. H. P. (1982). Measurements of osmotic permeability in the Malpighian tubules of an insect, *Rhodnius prolixus*. *Proc. R. Soc. Lond. B.* **216,** 267–277.

Pappenheimer, J. R. (1993). On the coupling of membrane digestion with intestinal absorption of sugars and amino-acids. *Am. J. Physiol.* **265,** G409–G417.

Pappenheimer, J. R., and Reiss, K. Z. (1987). Contribution of solvent drag through intercellular junctions to absorption of nutrients by the small intestine of the rat. *J. Membr. Biol.* **100,** 123–136.

Pedley, T. J., and Fischbarg, J. (1980). Unstirred layer effects in osmotic flow across gallbladder epithelium. *J. Membr. Biol.* **54,** 89–102.

Preston, G. M., Smith, B. L., Zeidel, M. L., Moulds, J. J., and Agre, P. (1994). Mutations in Aquaporin-1 in phenotypically normal humans without functional CHIP water channels. *Science* **265,** 1585–1587.

Rand, R. P., Parsegian, V. A., and Rau, D. C. (2000). Intracellular osmotic action. *Cell. Mol. Life Sci.* **57,** 1018–1032.

Schnermann, J., Chou, C. L., Ma, T. H., Traynor, T., Knepper, M. A., and Verkman, A. S. (1998). Defective proximal tubular fluid reabsorption in transgenic aquaporin-1 null mice. *Proc. Natl. Acad. Sci. USA* **95,** 9660–9664.

Shachar-Hill, B., and Hill, A. E. (1993). Convective fluid-flow through the paracellular system of Necturus gallbladder epithelium as revealed by dextran probes. *J. Physiol. Lond.* **468,** 463–486.

Silva, P., Stoff, J., Field, M., Fine, L., Forrest, J. N., and Epstein, F. H. (1977). Mechanism of active chloride secretion by shark rectal gland: Role of Na-K-ATPase in chloride transport. *Am. J. Physiol.* **233,** F298–F306.

Spring, K. R. (1998). Routes and mechanism of fluid transport by epithelia. *Annu. Rev. Physiol.* **60,** 105–119.

Spring, K. R. (1999). Epithelial fluid transport—A century of investigation. *News Physiol. Sci.* **14,** 92–98.

Steward, M. C. (1982). Paracellular non-electrolyte permeation during fluid transport across rabbit gallbladder epithelium. *J. Physiol. (Lond.)* **322,** 419–439.

Timbs, M. M., and Spring, K. R. (1996). Hydraulic properties of MDCK cell epithelium. *J. Membr. Biol.* **153,** 1–11.

Ussing, H. H., and Eskesen, K. (1989). Mechanism of isotonic water transport in glands. *Acta Physiol. Scand.* **136,** 443–454.

Vallon, V., Verkman, A. S., and Schnermann, J. (2000). Luminal hypotonicity in proximal tubules of aquaporin-1-knockout mice. *Am. J. Physiol. Renal Physiol.* **278,** F1030–F1033.

Van Os, C. H. W. G., and Wright, E. M. (1979). Volume flows across gallbladder epithelium induced by small hydrostatic and osmotic gradients. *J. Membr. Biol.* **49,** 1–20.

Verkman, A. S., Matthay, M. A., and Song, Y. L. (2000a). Aquaporin water channels and lung physiology. *Am. J. Physiol. Lung Cell. Mol. Physiol.* **278,** L867–L879.

Verkman, A. S., Yang, B. X., Song, Y. L., Manley, G. T., and Ma, T. H. (2000b). Role of water channels in fluid transport studied by phenotype analysis of aquaporin knockout mice. *Exp. Physiol.* **85,** 233S–241S.

Wang, K. S., Komar, A. R., Ma, T. H., Filiz, F., McLeroy, J., Hoda, H., Verkman, A. S., and Bastidas, J. A. (2000a). Gastric acid secretion in aquaporin-4 knockout mice. *Am. J. Physiol. Gastrointest. Liver Physiol.* **279,** G448–G453.

Wang, K. S., Ma, T. H., Filiz, F., Verkman, A. S., and Bastidas, J. A. (2000b). Colon water transport in transgenic mice lacking aquaporin-4 water channels. *Am. J. Physiol. Gastrointest. Liver Physiol.* **279,** G463–G470.

Wareing, M., Cockburn, M. F., and Green, R. (1996a). Lack of effect of luminal HgCl2 on fluid reabsorptive rate (J(v)) in the PCT of the anesthetized rat. *Kidney Int.* **50,** 1788.

Wareing, M., Cockburn, M. F., and Green, R. (1996b). Lack of effect of luminal mercury(II)chloride (HgCl2) on fluid reabsorptive rate in the proximal convoluted tubule (PCT) of the anaesthetized rat. *J. Physiol. Lond.* **493P,** P67–P68.

Weinstein, A. M. (1987). Convective paracellular solute flux: A source of ion–ion interaction in the epithelial transport equations. *J. Gen. Physiol.* **89,** 501–518.

Whittembury, G., and Hill, B. S. (1982). Fluid reabsorption by Necturus proximal tubule perfused with solutions of normal and reduced osmolality. *Proc. R. Soc. Lond. B.* **215,** 411–431.

Zeuthen, T. (1983). Ion activities in the lateral intercellular spaces of gallbladder epithelium transporting at low external osmolarities. *J. Membr. Biol.* **76,** 113–122.

Zeuthen, T. (1995). Molecular mechanisms for passive and active-transport of water. *Int. Rev. Cytol. Survey Cell Biol.* **160,** 99–161.

Zeuthen, T. (2000). Molecular water pumps. *Rev. Physiol. Biochem. Pharmacol.* **141,** 97–151.

Zeuthen, T., and Stein, W. D. (1994). Cotransport of salt and water in membrane proteins: Membrane proteins as osmotic engines. *J. Membr. Biol.* **137,** 179–195.

Zhang, R. B., Skach, W., Hasegawa, H., Vanhoek, A. N., and Verkman, A. S. (1993). Cloning, functional-analysis and cell localization of a kidney proximal tubule water transporter homologous to Chip28. *J. Cell Biol.* **120,** 359–369.

Osmoregulation and Contractile Vacuoles of Protozoa

Richard D. Allen and Yutaka Naitoh
Department of Microbiology and Pacific Biomedical Research Center,
University of Hawaii, Honolulu, Hawaii 96822

Protozoa living in fresh water are subjected to a hypotonic environment. Water flows across their plasma membrane since their cytosol is always hypertonic to the environment. Many wall-less protozoa have an organelle, the contractile vacuole complex (CVC), that collects and expels excess water. Recent progress shows that most, if not all, CVCs are composed of a two-compartment system encircled by two differentiated membranes. One membrane, which is often divided into numerous vesicles and tubules, contains many proton-translocating V-ATPase enzymes that provide an electrochemical gradient of protons and which fuses only with the membrane of the second compartment. The membrane of the second compartment lacks V-ATPase holoenzymes, expands into a reservoir for fluid storage, and is capable of fusing with the plasma membrane. It is this second compartment that periodically undergoes rounding ("contraction"), setting the stage for fluid expulsion. Rounding is accompanied by increased membrane tension. We review the current state of knowledge on osmolarity, ion concentrations, membrane permeability, and electrophysiological parameters of cells and their contractile vacuoles, where these criteria are helpful to our understanding of the function of the CVC. Effects of environmental stresses on the CVC function are also summarized. Finally, other functions suggested for CVCs based on molecular and physiological studies are reviewed.

KEY WORDS: Contractile vacuole complex, Cytosolic osmolarity, Electrophysiological parameters, Environmental stress, Ion activity, Membrane water permeability, Membrane tension, Osmoregulation, Protozoa, V-ATPase

0074-7696/02 $35.00

I. Water Influx as a Problem for Fresh-Water Protists

Fresh-water protozoa and algae live in an environment in which their cytosolic osmolarity is maintained higher than that of their surroundings. This is also true of many, if not all, multicellular eukaryotes living in fresh water such as the fresh-water sponge, coelenterates, and zoospores of some fungi. This means that water has the potential to diffuse into these cells or tissues causing the concentrations of their cytosolic proteins and other organic and inorganic constituents to be diluted. To maintain osmotic and ionic homeostasis, these cells and tissues must have a mechanism for regulating the diluting effect of this incoming water.

In many algae and fungi, the cells have an extensive cell wall which has a very limited capacity to expand. In these cells, water enters until the increased hydrostatic pressure balances osmotic pressure. This hydrostatic pressure is referred to as turgor pressure. In other cells, which lack a rigid nonexpandable wall, the more fragile plasma membrane would presumably stretch until it ruptured if water continuously flowed into the cell without any compensating backward flow. Many of these cells have a special organelle, usually referred to as the contractile vacuole (CV), the CV complex (CVC), or the water expulsion vesicle, which accumulates and expels this excess water (Fig. 1). (In this review, the whole organelle will be referred to as the CVC and its water-storing portion as the CV.) The membrane surrounding this organelle periodically fuses with the cell membrane to create an opening to the cell's exterior through which the fluid is expelled.

Though these organelles have been objects of microscopic and physiological interest to scientists for many years, it has only been in the last decade that significant headway has been made in our understanding of how these organelles might function. This includes how such organelles are able to sequester excess water from

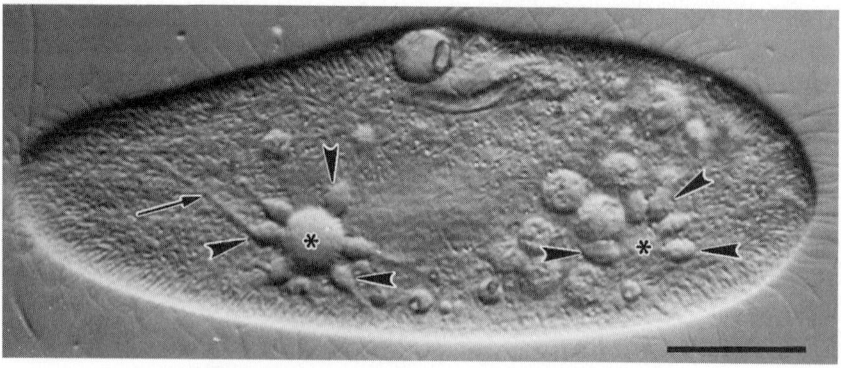

FIG. 1 The CVCs (*) of *Paramecium* viewed under differential-interference contrast microscopy. The CVC to the left is in the filling stage, and the one to the right has expelled its fluid. Arrow, collecting canal; arrowheads, ampullae. Scale bar = 25 μm. From Hausmann and Allen (1977).

the cytosol, how this organelle can fuse with the cell membrane, how it expels its content to the cell's exterior, how its membrane reseals, and how it is able to constantly collect more fluid. In this process, some initial insights into membrane-associated enzymes, *in situ* electrophysiology of the organelle complex, and the mechanics and biophysics of the *in vivo* and *in vitro* CV membrane have been acquired. A summary of recent results on the functioning of the CVC will be emphasized in this chapter as represented in some of the more commonly studied protozoa. For reviews on this topic, see Patterson (1980) and Zeuthen (1992).

A. Osmolarity and Ion Contents of the Cytosol

The osmolarity of the cytosol has been determined or estimated in a few species of protozoa (see list in Prusch, 1977). In *Amoeba proteus,* the average osmolality of the cytosol compared to the external medium (pond water) was determined to be 101 and 8 mOsm, respectively (Schmidt-Nielsen and Schrauger, 1963). A freezing point depression technique was used for these osmolality determinations. Other protozoa have a cytosolic osmolality estimated to be from 45 to 117 mOsm, while the osmolality of fresh water is typically below 7 mOsm. Thus the osmolarity of the cytosol is usually much higher than that of the fresh-water environment, which means that these cells are subjected to a constant osmotic pressure that allows water to flow inward, if we assume that their surface membranes are permeable to water. In protozoa, the intracellular K^+ concentration remains around 30 mmol kg^{-1} even under varied environmental conditions (Dunham and Kropp, 1973; Prusch, 1977), while the Na^+ concentration is less constant and is 20 mmol kg^{-1} or lower. The intracellular Cl^- concentration for *Amoeba proteus* is 9.7 mmol kg^{-1} where the external Cl^- is 0.16 mmol kg^{-1} (Prusch and Dunham, 1972), and for *Chaos carolinensis* it is 16.5 mmol kg^{-1} with an extracellular Cl^- concentration of 1 mmol L^{-1} (Bruce and Marshall, 1965).

B. Membrane Permeability to Water

The plasma membrane of protozoa, like other cells, is known to be readily permeable to water. Protozoa such as *Amoeba proteus* (Mast and Fowler, 1935) and *Chaos carolinensis* (Belda, 1943) were observed to swell and shrink osmotically when the external medium is changed, for example, when nonelectrolytes are added to the surrounding medium. For comparative purposes, water permeability constants, a measure of the ability of water to pass through a membrane, have been calculated for the plasma membrane. For *Amoeba proteus* and *Chaos carolinensis,* constants of 0.026 and 0.02 μm/atm/min were estimated (Mast and Fowler, 1935; Belda, 1943), and for *Acanthamoeba castellanii,* a constant of 0.04 μm/atm/min was estimated (Pal, 1972). For the ciliate *Vorticella marina,* which was transferred to diluted sea water at a temperature of 14 to 17°C, a permeability constant of about 0.1 μm/atm/min was found (Kitching, 1948).

In *Paramecium multimicronucleatum,* the rate that water actually enters the cell depends, as expected, on the difference between the internal and external osmolarities. However, the water permeability constant of the plasma membrane changes as cells adapt to increases in external osmolarities. For external osmolalities of 4 and 64 mOsm, the permeability varies from 0.172 to 0.304 μm/atm/min (to obtain cm sec^{-1}, this unit should be multiplied by 22.6×10^{-4}); between 84 and 144 mOsm, it drops to between 0.049 to 0.075; and between 164 and 204 mOsm, it is only 0.009 to 0.016 (Stock *et al.,* 2001). Thus the properties of the plasma membrane change as changes in the external osmolarity give rise to increases in cytosolic osmolarity. The water permeability of the cell membrane decreases as the cytosolic osmolarity increases. This change may protect cells that have become adapted to hyperosmotic environments from very rapid influxes of water, should they encounter a sudden change to a strongly hypoosmotic environment.

C. The Protozoan Keeps Its Cytosol Hyperosmotic to Its Environment

Fresh-water protozoa are capable of adapting to and living in a wide range of osmotic environments all the way from hypoosmotic environments of near 0 mOsm to a hyperosmotic environment of 200 mOsm. The fresh-water ciliate *Paramecium multimicronucleatum* can be adapted over a 24-hr period to environments ranging from 4 to 200 mOsm (Stock *et al.,* 2001). Longer periods may permit an even wider range of adaptations. To remain active, *Paramecium* adjusts its cytosolic osmolarity to keep its cytosol hypertonic to the surrounding medium. Even marine ciliates appear to maintain their cytosol hypertonic to their environment (Kaneshiro *et al.,* 1969). This ensures that osmotic pressure will cause water to flow into the cell thereby maintaining a positive internal hydrostatic pressure. An internal hydrostatic pressure is apparently a prerequisite for life in these fresh-water protozoa.

II. The Osmoregulatory Organelle of Fresh-Water Cells: Its Structure and Activity Cycle

A. Amoeboid Cells

1. *Amoeba Proteus*

The amoeba, *Amoeba proteus,* has one contractile vacuole which, when filled, is located in the uroid (tail) region of the cell. There is no identifiable permanent pore or surface indentation through which the CV will always expel its contents.

Rather, the CV when filled seems to be propelled against the plasma membrane by some sort of ectoplasmic pressure (Akbarieh and Couillard, 1988). These two membranes then fuse, and the CV contents are ejected through the open pore by the pressure of the surrounding protoplasm (Wigg *et al.,* 1967). During expulsion, the membrane of the CV, viewed by light microscopy, collapses into a "systolic mass," which, viewed in electron micrographs, consists of numerous pentilaminar vesicles (Akbarieh and Couillard, 1988). Pentilaminar vesicles are presumably composed of a unit membrane collapsed against itself. Cytosolic organelles such as mitochondria and even the endoplasmic reticulum (ER) are kept to the periphery of this mass while strands of microfilaments and ribosomes, along with the vesicles, are the only recognizable cytosolic components to be found within the mass itself. The first sign of a new CV is the appearance of a small contributory vacuole that grows in size as it fuses with other vesicles within the mass. During this coalescence phase, the vacuole moves forward in the cell where it comes to lodge behind the nucleus. The vacuole continues to enlarge during its continuous growth phase. The CV will then move back toward the uroid region prior to expulsion. A typical cycle at 20°C lasts 300 sec, and the average presystolic diameter achieved is 26.5 μm, so the rate of fluid accumulation measures on average 35 $\mu m^3\ sec^{-1}$ (Couillard *et al.,* 1985).

In *Amoeba proteus,* the excreted CV fluid has been determined to be hypotonic (32 mOsm) to that of the cytosol (101 mOsm) (Schmidt-Nielsen and Schrauger, 1963), but the concentrations of ionic species within the CV have not been determined.

The membranes of the CV were found to be coated on their cytosolic face (McKanna, 1973a) with peg-shaped elements (6.5 $\times$ 15 nm). This coat, though its composition has not yet been confirmed for *Amoeba,* is probably composed of a layer of V_1 subcomplexes of V-type proton pumps (V-ATPases), as similar coats are known to be present on the CV membranes of *Dictyostelium* (Heuser *et al,* 1993), *Paramecium* (Fok *et al.,* 1995), and the zoospores of oomycetes (Mitchell and Hardham, 1999). Thus these coats must be involved in pumping protons into the vesicles and the contributory vacuoles and in setting up an electric potential gradient across the membranes of the CVC (see Section III.B.1).

The growth of the amoeboid CV, according to Akbarieh and Couillard (1988), follows a process of fluid flow into the pentilaminar vesicles which transforms these vesicles into concentric vesicles and finally into the larger single-wall vesicles that ultimately fuse with the contributory vacuoles of the CV itself. However, without additional inflow of fluid this process would produce a very flaccid vacuole. Thus it has been suggested that either active fluid uptake continues across late phases of the CV membrane or the membrane is removed in the form of small vesicles budding from the CV. Another possibility, not yet considered for amoeba, is that the membrane is effectively segregated from the final planar CV membrane by some other means, such as membrane tubulation, to allow the formation of the mature spherical CV usually observed in the light microscope (see Section III.B.2).

2. *Chaos Carolinensis*

Another amoeba studied in some detail is the giant amoeba, *Chaos carolinensis* (formerly called *Chaos chaos*). Its cytosolic and vacuolar osmolality as well as the concentrations of its principal ions have been measured (Riddick, 1968). This cell differs from *A. proteus* in that it has multiple CVs rather than one CV. The average osmolality of the cytosol of *Chaos* was found to be 117 mOsm and that of its vacuolar fluid 51 mOsm. In these experiments, the osmolality of the medium was less than 2 mOsm. The concentrations of Na^+ and K^+ in the CV averaged 19.9 and 4.6 mM, respectively, while in the cytosol Na^+ and K^+ concentrations averaged 0.57 and 31 mM, respectively. Thus Na^+ was reduced in the cytosol and enriched in the CV, while the reverse was true for K^+. These two ions were thought to account for most, if not all, osmotically active cations of the vacuole. To explain the hypotonicity of the CV fluid, it was proposed (Riddick, 1968) that coalescence of a large number of originally isotonic vesicles, which had become hypotonic as Na^+ was pumped into the vesicles while K^+ was pumped out of the vesicles at a high rate, could have formed the hypo-osmotic mature CV.

3. *Acanthamoeba Castellanii*

The small soil amoeba *Acanthamoeba castellanii* has been used as a model eukaryotic cell especially for physiological and biochemical studies of the cytoskeleton. This cell contains a CV (Fig. 2) that shows a positive reaction for alkaline phosphatase along the luminal surface of its membrane (Bowers and Korn, 1973). The only other CV that is known to have alkaline phosphatase activity is found in another small soil amoeba, *Dictyostelium* (see Section II.A.4).

The presence of reaction product for alkaline phosphatase emphasizes the continuity between the CV membrane and the many small tubules and vesicles that surround the CV (Bowers and Korn, 1968, 1973). Such labeled membranes are

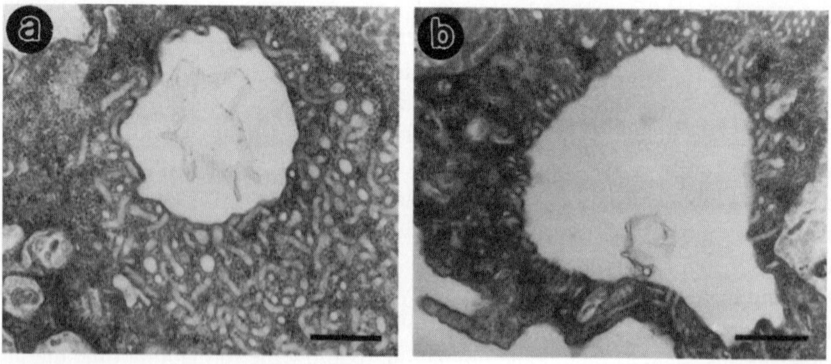

FIG. 2 The CVC of intact *Acanthamoeba* cells. (a) and (b) Numerous tubules are next to and continuous with the CV. Prepared by chemical fixation. Scale bar = 1 μm. From Nolta and Steck (1994).

found for the most part in the immediate vicinity of the CV, indicating that these tubules and vesicular profiles were part of a continuous system of interconnected membranes rather than part of a pool of disconnected and diffusable vesicles. Thus the appearance of the CVC in *Acanthamoeba,* as in other protozoa, supports a view that CV vesiculation into hundreds of vesicles may not occur following fluid expulsion as previously supposed (Patterson, 1980). The membrane system of this amoeba may actually collapse into a system of interconnected tubules that reexpand into the CV again upon accumulation of fluid within their lumens.

4. *Dictyostelium* Discoideum

Amoeba of the slime mold lack a cell wall; consequently they too require CVs to maintain their water balance (Fig. 3). A cell usually contains one or two CVs that, like those of *Acanthamoeba,* become cytochemically labeled on their luminal membrane surfaces for alkaline phosphatase in the presence of appropriate substrate (Quiviger *et al.,* 1978). These 1- to 3-μm diameter CVs are now known to be surrounded by a network of anastomosing tubules linking together smaller saccules (Heuser *et al.,* 1993) that do not label for alkaline phosphatase. The full arrangement of this CVC may best be visualized in cells that have been disrupted, chemically fixed, quick-frozen, and finally, freeze-dried (Heuser *et al,* 1993). Some of the larger tubular and saccular components have also been visualized

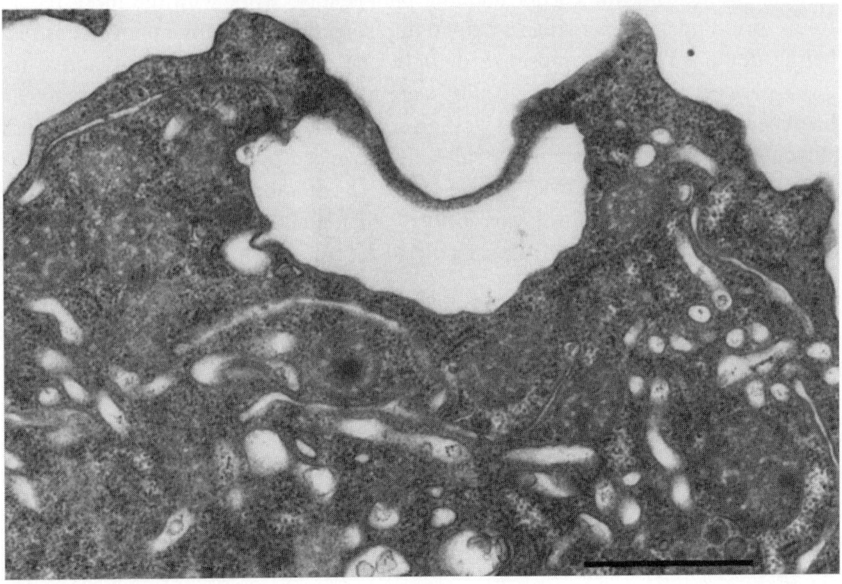

FIG. 3 The CVC of an intact *Dictyostelium* cell. Prepared by conventional chemical fixation for electron microscopy. Scale bar = 1 μm. From Nolta and Steck (1994).

at the light microscope level by interference-reflection microscopy (Gingell *et al.*, 1982; Heuser *et al.*, 1993) and by immunofluorescence microscopy (Heuser *et al.*, 1993; Fok *et al.*, 1993). Some antibodies that have been used to investigate the CVC of *Dictyostelium* are those raised against calmodulin and the C isoform of myosin-I (Doberstein *et al.*, 1993; Zhu and Clarke, 1992), α-actinin (Furukawa and Fechheimer, 1994), and subunits of the V-ATPase (Heuser *et al.*, 1993; Fok *et al.*, 1993) that are particularly abundant in this organelle.

Formation of a CV appears to occur by the swelling of a bladder and by the incorporation of some of the surrounding tubules into this bladder. The discharge of the enlarged vacuole seems to lead to the reverse of this process, with excess membrane being returned to narrow tubules and small saccules. Though no measurements were given, the tubules appear to be $\sim$ 40 nm in diameter in the published figures of Heuser *et al.* (1993). No actin was detected adhering to these structures in freeze-dried images even though cytoskeletal elements are readily seen located against the plasma membrane. Though actin has yet to be demonstrated, myosin-IC, an unconventional myosin, is present associated with the CV membrane of *Dictyostelium* (Zhu and Clarke, 1992). Heuser *et al.* (1993) postulated that the ability of these cisternae to "contract" may be a function of the persistent lipid asymmetries in the two monolayers of the encompassing CV membrane rather than from a contraction of a typical actomyosin system. Until morphological evidence for an apposed cytoskeletal system can be obtained, we tend to favor the speculation first proposed by Heuser *et al.* (1993, in a figure caption) that contraction may arise directly from the composition of the membrane itself (see Section III.B.2).

When comparing the earlier alkaline phosphatase work with the more recent freeze-dried images, one area remained unresolved. The alkaline phosphatase labeling demonstrated the presence of only one or two phosphatase-reactive regions per cell, whereas many saccules were found in each cell. If any saccule can become a contractile vacuole by swelling and by the presumptive incorporation of surrounding tubules and small saccules, as suggested by Heuser *et al.* (1993), the question remaining is: In what part of the CVC's membrane system does alkaline phosphatase exist when the CV is in its expulsion and early filling stages? Does alkaline phosphatase move from saccule to saccule or, in fact, exist in only two bladders that enlarge into the CVs?

This problem was resolved by Nolta and Steck (1994) who showed by their membrane isolation studies that the CVC of *Dictyostelium* is composed of not one type of membrane, as Heuser *et al.* (1993) had concluded, but two distinctly different membrane pools. The heavier pool of vesicles is decorated by the V-ATPase particles. Called "acidosomes" (Nolta *et al.*, 1993; Nolta and Steck, 1994), a term first coined by Allen and Fok (1983) for vesicles that when fused with phagosomes of *Paramecium* bring about the acidification of phagosome lumens (Ishida *et al.*, 1997), these CVC-derived membranes are more properly called the decorated spongiome. The principal function of the electrogenic V-ATPases of these membranes is to establish a positive membrane potential rather than to acidify the CVC

(Section III.B.1). A second pool of membranes contains the alkaline phosphatase activity, certain noncatalytic polypeptides that make up part of the V-ATPase, and other polypeptide bands not found in the decorated spongiome (Nolta and Steck, 1994). These two membranes also have different lipid compositions. The decorated spongiome has only 0.3 μg of lipid/mg of protein while the vacuoles have 1 μg of lipid/mg of protein as well as more sterols and sphingolipids than the smooth spongiome.

B. Flagellated Cells

1. Algal and Nonphotosynthetic Flagellates

Most types of algal flagellates that live in fresh water and that lack a cell wall have a CVC (Hausmann and Patterson, 1984). In the algal flagellate, *Vacuolaria virescens,* a CVC cycle takes about 60 sec at 23° C when the cell is growing in defined medium (Heywood, 1973). Nonphotosynthetic flagellates, for example, the nonpathogenic trypanosomatid-like *Leptomonas collosoma,* also have a CV apparatus (Linder and Staehelin, 1979). The CVC of flagellates normally consists of a central vacuole surrounded by, or continuous with, 40- to 120-nm cylindrical (Fig. 4) or "flattened tubules" (Aaronson and Behrens, 1974; Denning and Fulton, 1989; Hellebust *et al.,*

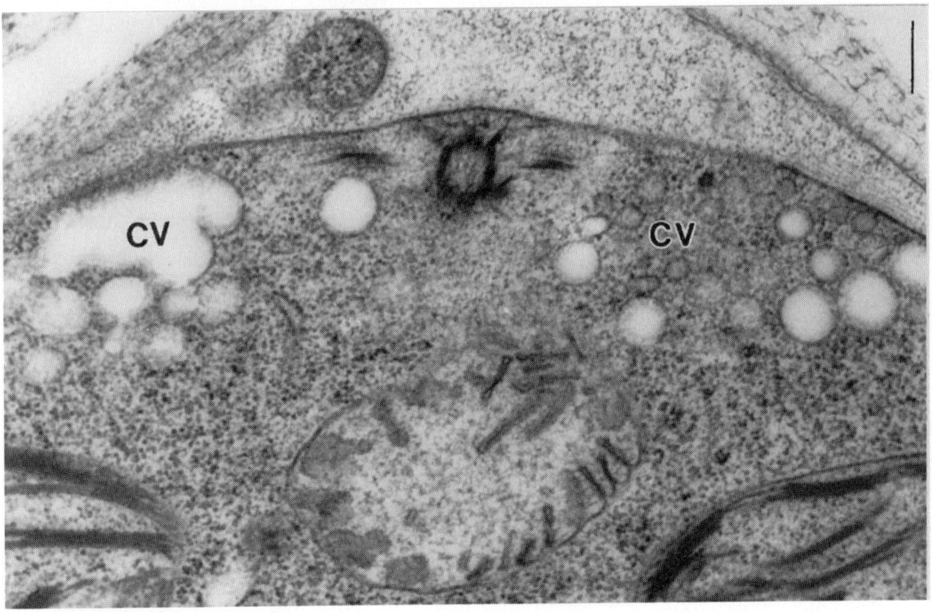

FIG. 4 The CVC (CV) of *Chlamydomonas reinhardtii* under hyperosmotic conditions. Prepared by cryofixation and cryosubstitution. Scale bar = 0.25 μm. From Luykx *et al.* (1997a).

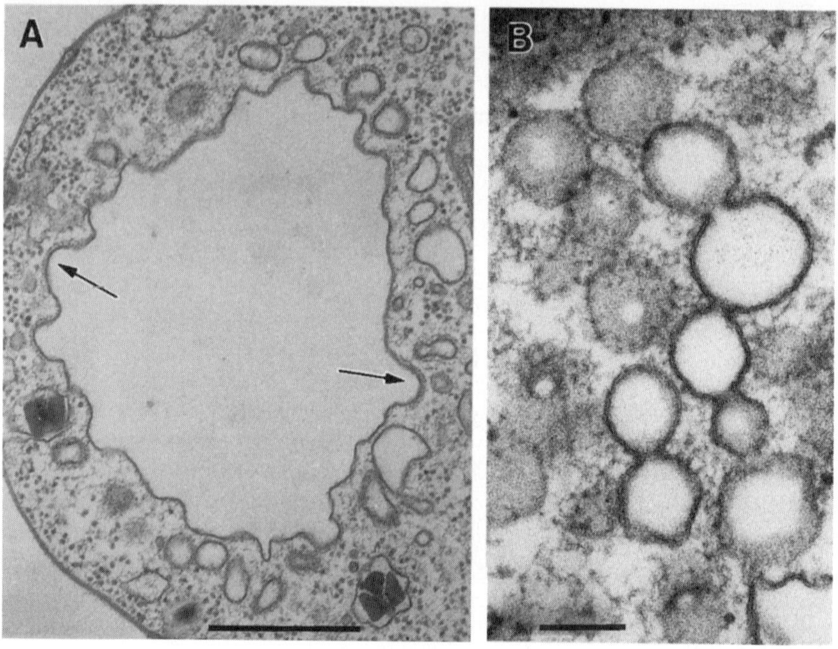

FIG. 5 The CVC of *Cyanthomonas,* a cryptophycean flagellate. (A) A CV and surrounding spongiome that consists of coated and uncoated vesicles. Coated membrane is attached to the CV (arrows). Bar, 0.5 μm. (B) Uncoated vesicles form chains of interconnected membrane. Scale bar = 0.1 μm. From Hausmann and Patterson (1981).

1989; Linder and Staehelin, 1979; Luykx *et al.,* 1997a; Patterson and Hausmann, 1981) coated to various extents on their cytosolic surface with globular protein complexes (Fig. 5A) (Hausmann and Patterson, 1981; Luykx *et al.,* 1997a; Weiss, 1983) or polygonal coats (Heywood, 1978) that, when present, are sometimes likened to clathrin coats. Some of these membrane projections, particularly the flattened tubules, also contain a lining on the luminal side of their membranes (Wessel and Robinson, 1979).

The CV is said to form and fill by the fusion of the fluid-containing decorated vesicles of the spongiome (Hausmann and Patterson, 1984). How the fluid enters the vesicles has not been determined. It has been speculated that some of the coats seen on the spongiome might be clathrin (Patterson and Hausmann, 1982), but the presence of clathrin on the CV membrane needs further confirmation by rigorous biochemical or immunological techniques. In *Dictyostelium,* the presence of clathrin heavy chain was necessary to keep the cell from swelling when the cell was exposed to a hypotonic solution (O'Halloran and Anderson, 1992). However, although mutant cells without a gene for clathrin heavy chain did not produce CVs, the cells recovered their normal appearance after 5 hr in the hypotonic solution. It is not clear if clathrin has a primary or secondary affect on CV activity, if it affects the CVC directly, or if a general cellular process is changed that also affects osmoregulation.

After CV filling is completed, a timing mechanism is said to induce contraction of putative cytoskeletal elements that are thought to surround the vacuole and cause the vacuole to round up, thereby inducing fusion with the plasma membrane (Hausmann and Patterson, 1984). The hydrostatic pressure of the cytosol is said to cause the vacuolar contents to be expelled from the cell. The CV membrane then collapses and is usually thought to fragment into numerous vesicles to complete the cycle. Like the vesiculation of its membrane, the extent to which the CV membrane actually fuses with the plasma membrane is open to different interpretations depending on the cell under study. In some cases, the fluid seems to be discharged through hydrophilic transmembrane channels that cross both membranes (Weiss *et al,* 1977). Another scheme suggests that, in cells such as *Leptomonas,* microchannels of around 20 nm in diameter form across the apposed but unfused membranes (Linder and Staehelin, 1979). Displacements in a plaque of electron-opaque material that fills the 25-nm gap that lies between the CV membrane and the plasma membrane may produce these microchannels. Such plaques and small 15-nm microchannels have also been reported in *Chlamydomonas reinhardtii* (Luykx *et al.,* 1997a) although Luykx *et al.* (1997b) also show one image, in a mutant cell, reported to be of an expanded CV pore cryofixed during fluid discharge and subsequently processed by the freeze-substitution technique.

Some of the membranes of the CVC of *Leptomonas,* which were freeze-fractured following cryofixation of chemically unfixed cells, contain intramembranous particles of distinct patterns and concentrations. This spongiome membrane supports a double helical row of 11-nm diameter particles on its E fracture-face, while its P fracture-face has randomly arranged 6- to 13-nm diameter intramembrane particles. In light of more recent work on several other cells, the 11-nm particles are probably the transmembrane V_0 sectors of V-ATPases, so that such membrane comprises the decorated spongiome of the CVC of this cell. *Leptomonas* also has a three-dimensional lattice of smooth spongiome (Linder and Staehelin, 1980) from which the CV itself may be formed.

In our view, at least two areas of the forgoing schemes of CV activity need to be studied in more detail in the amoeba and in flagellates before the CV cycle of these cells can be completely understood. More information must be obtained on (1) what causes the rounding process and expulsion of the CV. Is this caused by contraction of a cytoskeleton, or is it due to the inherent properties of the CV membrane itself? (2) To what extent does the CV membrane vesiculate after the collapse of the CV? Does the collapsed CV membrane in fact vesiculate (see Fig. 5B) or does it form chains of connected vesicle-like compartments? Thin sections make it appear that the CV membrane, such as that of *Dictyostelium,* vesiculates; but freeze-dried images show that the CV membrane forms a network of tubules and cisternae or bladders (Heuser *et al.,* 1993). It is often very difficult to tell if vesicles have actually separated from a collapsed CV after the CV has been thin-sectioned, as the connections may be out of the section plane. Techniques such as (1) quick-freezing and freeze-fracturing, as used by Linder and Staehelin (1979), (2) freeze-drying of disrupted cells, as used by Heuser *et al.* (1993), or (3) labeling the lumen of the CV with an electron-opaque content marker as occurs in the alkaline phosphatase incubations of soil amoeba or when horseradish peroxidase enters the CV during CV expulsion (de Chastellier *et al.,* 1978) should be used to see to what extent the CV is, in fact, discontinuous with the surrounding vesicles. Freeze-substitution is known to be an excellent method to preserve ultrastructure in its near native state, but even with this method membranes, and thus intermembrane connections, are not always easily resolved.

2. Zoospores of Oomycetes

Oomycetes are a class of organisms traditionally classified as fungi which have close affinities to algae and which include many important plant pathogens such as late blight of the potato. These fungi produce motile, wall-less zoospores that possess a CV (Fig. 6) that is used, presumably, for osmoregulation (Hoch and Mitchell, 1972) or to eliminate excess fluid to compensate for the osmotic influx of water.

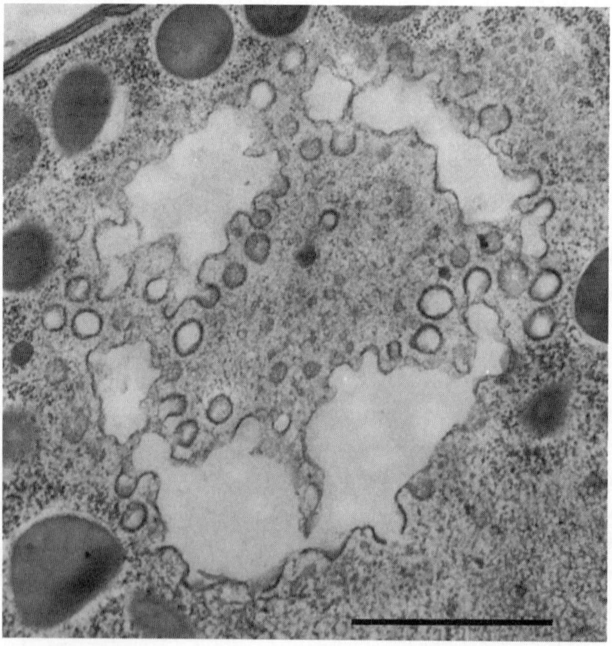

FIG. 6 The CVC of *Phytophthora nicotianae* zoospore prepared by rapid freezing and freeze substitution. Many evaginations and vesicular profiles have a coated appearance. Scale bar = 1 μm. From Mitchell and Hardham (1999).

A recent study of the CV in zoospores of *Phytophthora nicotianae* (Mitchell and Hardham, 1999) describes the development, the activity, and the disappearance of the activity of this organelle as zoospores encyst. The CV lies at the anterior end of the zoospore and has a mean cycle time of about 6 sec. The CV consists of a bladder and some surrounding homogeneous cytoplasm that in electron micrographs consists of short coated tubules and coated vesicles. Expulsion occurs as a rapid phase ($\sim$ 0.08 sec) followed by a slower ($\sim$ 1.0 sec) phase. The cytosol around the CV remains differentiated and of nearly constant size. After 2.4 sec, a small vacuole reappears as an angular vacuole within this mass and grows in size over the next 2 sec. Once the CV becomes circular, contraction will occur in about 0.6 sec. During zoospore development, pulsation begins within the sporangium, continues in the released zoospore, and ceases after the zoospore loses its flagella and encysts.

The fate of the CV membrane following expulsion has been open to question since Cho and Fuller (1989) suggested that the CV membrane of these zoospores might become part of the plasma membrane following expulsion. In their recent study, Mitchell and Hardham (1999) showed that, while both the plasma membrane and the CVC membrane have P-type proton pumps, only the CVC and, in this organelle, only the decorated spongiome membrane labels with V-ATPase antibodies. The CV membrane itself does not label with V-ATPase antibodies. This argues for a biochemical and physiological separation of the CVC's spongiome membrane from that of the plasma membrane and also for a CVC composed of two types of membranes, one bearing V-ATPase and one without V-ATPase from which the CV is formed (Fig. 7).

C. Ciliated Protozoa

1. *Paramecium Multimicronucleatum* and *P. Tetraurelia*

One of the most extensively studied CV systems to date is that in the ciliate *Paramecium*. This ciliate also contains one of the largest as well as one of the most complex CV systems known (Allen, 2000, feature article). A *Paramecium* cell has two CVs, one in the anterior and one in the posterior half of the cell (Fig. 1). These CVs seem to be able to operate independent of each other.

A CVC consists of a central CV (Fig. 8A) which is held in place against a permanent surface indentation of the plasma membrane, called the CV pore, by a system of microtubular ribbons that, in *P. multimicronucleatum,* consists of 20 to 25 microtubules per ribbon that originate adjacent to a pore microtubular collar. The pore collar is composed of helically arranged microtubules that wrap tightly around the cytosolic surface of the plasma membrane indentation that forms the pore (McKanna, 1973b). The cytoskeletal ribbons radiate from the pore, and each ribbon forms the frame along which is attached an elaborate membrane system

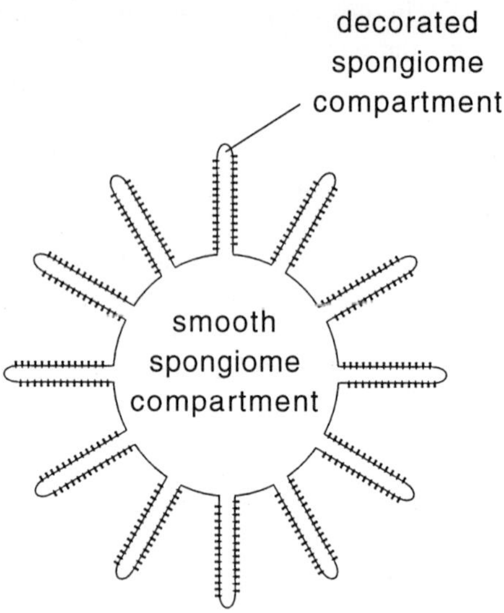

FIG. 7 Drawing of a generalized CVC showing the relationship of the decorated membrane to the smooth membrane from which the CV forms.

of smooth spongiome, that is divisible into a collecting canal, an ampulla, and a three-dimensional array of tubular membranes. This smooth spongiome is linked to the membrane of the decorated spongiome (Schneider, 1960; Hausmann and Allen, 1977). There are usually 5 to 10 microtubular ribbons and thus 5 to 10 radial arms found in each CVC in the cell. In cross section (Fig. 8B), a radial arm consists of a central membrane tube, the collecting canal, which communicates regularly through longitudinally arranged rows of 40 nm diameter openings, spaced at 75-nm intervals along a row (Fig. 8C) (Hausmann and Allen, 1977), with an elaborate three-dimensional system of highly curved membrane. The bilayer of this membrane branches so regularly that at times it appears to crystallize (McKanna, 1976; Tominaga *et al.,* 1999) into a cubic-like bilayer (Landh, 1995). Lying along each collecting canal and covered by the smooth spongiome network are segments of the now forked microtubular ribbon (Hausmann and Allen, 1977; Allen and Fok, 1988).

Peripheral to the encircling smooth spongiome is a second system of membrane tubules that are frequently organized into bundles of straight 50-nm tubes (Fig. 9). Lumens of these bundles open to the smooth spongiome through short ducts (Allen and Fok, 1988). At their opposite ends, the 50-nm tubes end blindly. None of these membrane systems opens directly to the ER, although the ER invariably lies near the CVC.

The decorated spongiome labels with antibody to V-ATPases (Fig. 10A) (Fok *et al.,* 1995). The CV, the collecting canals, and the smooth spongiome do not label

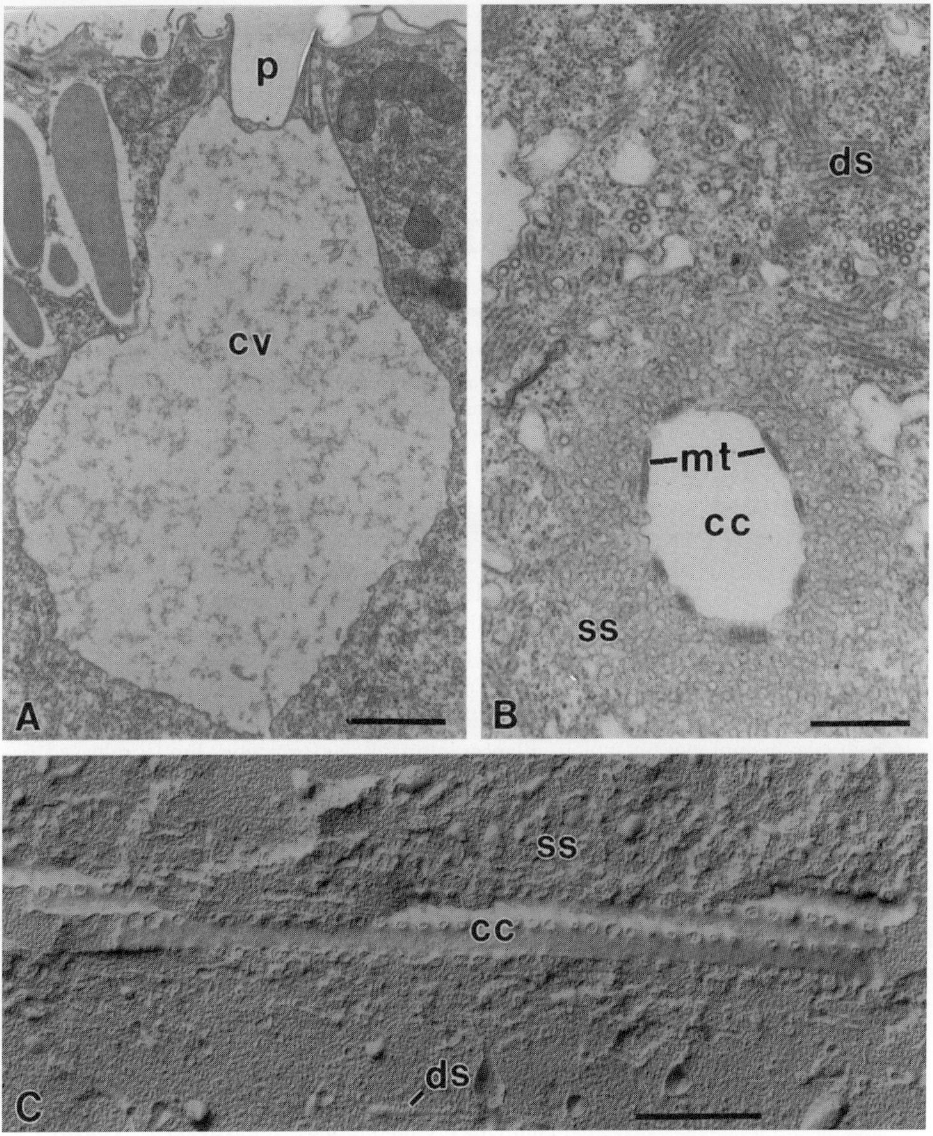

FIG. 8 CVC of *Paramecium*. (A). The filled CV (cv) is affixed to the CV pore (p) by microtubular ribbons. Bar, 2 μm. From Allen and Fok (1988). (B) Radial arm in cross section. ds, Decorated spongiome; ss, smooth spongiome; mt, microtubular cytoskeleton; cc, collecting canal. Bar, 1 μm. From Allen *et al.* (1990). (C) Freeze-fracture replica of collecting canal and spongiomes. Connections between the smooth spongiome (ss) and the collecting canal (cc) are shown here as rows of fractured necks. The decorated spongiome (ds) lies peripheral to the smooth spongiome. Scale bar = 0.5 μm. From Hausmann and Allen (1977).

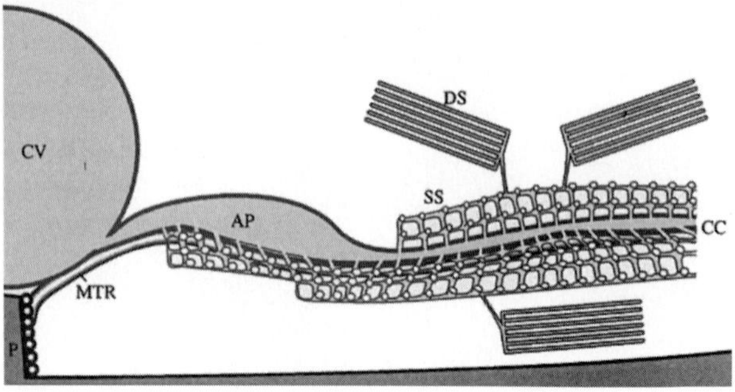

FIG. 9 Cutaway drawing of a portion of the CVC of *Paramecium* showing its pore (P), an ampulla (AP), and a collecting canal (CC) supported by a microtubular ribbon (MTR). The smooth spongiome (SS) is arrayed into a three-dimensional lattice and is continuous through ducts with the decorated spongiome (DS) that contains the vacuolar proton pump. All membranes in red are part of the smooth spongiome and those in green are part of the decorated spongiome. Modified from Tominaga *et al.* (1999). (See also color insert.)

with these V-ATPase antibodies. These latter membranes share an uncharacterized antigen (Fig. 10B) with the plasma membrane (Fok *et al.,* 1995; Ishida *et al.,* 1996) which does not appear to be present in the decorated spongiome. Thus the CVC of *Paramecium,* like the CVC in other species, consists of at least two membrane systems (Fig. 9): one that contains V-ATPases and one that shares antigens with the pellicular membranes but which collapses into a tubular form when fluid is released from its lumen (Allen and Fok, 1988; Naitoh *et al.,* 1997b). This occurs whenever the CV expels its content to the outside or when the ampullae, the expandable reservoirs of the collecting canals located at the CV end of these canals, fuse with the CV membrane and empty their contents into the CV.

Thus in *Paramecium,* a CV cycle takes the fluid that is continuously flowing into the radial arms and transfers this fluid to the CV for release from the cell. The fluid, at first, is accumulated in the ampullae at the time that the CV membrane is disconnected from the radial arms. The arms are disconnected from the CV (Tominaga *et al.,* 1998a) just before the CV fuses with the plasma membrane at a time when the CV is seen to round up into a sphere (Patterson, 1977). Following expulsion, the CV membrane dissociates from the plasma membrane by membrane fission, and the ampullae then fuse again with the CV membrane (Tominaga *et al.,* 1998a). The CV membrane and the radial arms, during a normal cycle, remain connected to the microtubular ribbons (Hausmann and Allen, 1977). This association is probably crucial for ensuring efficient re-fusion of the CV with the membrane along the radial arms. Disruption in the membrane–microtubular associations normally occurs only during events leading to cell division and then only in the more distal regions of the radial arms (Fig. 11A) (Allen *et al.,* 1990).

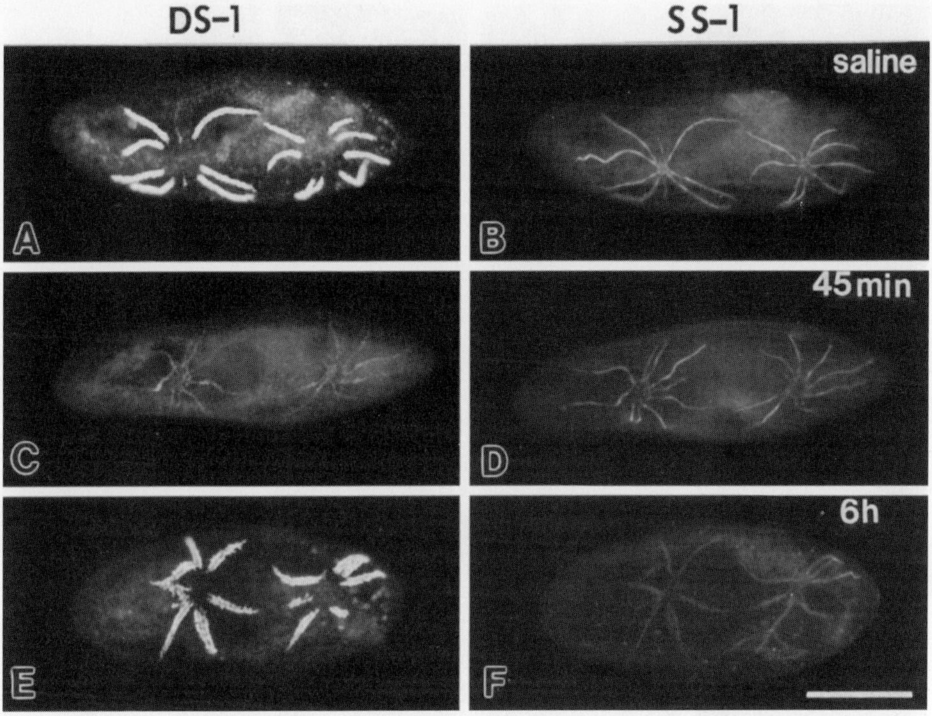

FIG. 10 Immunofluorescence of the CVCs of *Paramecium* subjected to hyperosmotic stress. (A), (C), and (E) are exposed to IgG-FITC-complexed DS-1 mAb that is now known to bind to V-ATPase-rich membranes. (B), (D), and (F) are exposed to IgM-Texas Red-complexed SS-1 mAb that binds to the membranes of smooth spongiome and pellicles. (A) and (B) CVC in cell adapted to standard saline. From Fok *et al.* (1995). (C) and (D) The CV in a cell acclimated to 150 mM sorbitol added to axenic medium (230 mOsm final osmolality) for 45 min. (E) and (F) After 6 hr of continuous incubation in the 150 mM sorbitol, the cell's decorated spongiome has recovered along with much of the CV's fluid segregation activity. (C) to (F) from Ishida *et al.* (1996). Scale bar for (A)–(F) = 50 μm.

The CV membrane of *Paramecium* itself may sometimes separate into segments following expulsion, potentially as many segments as there are microtubular ribbons. Each ampulla then re-fuses with the CV membrane that remains attached to its ribbon. Successive fusions continue until all ampullae are reunited with the CV membrane. If the CV membrane has split into several segments, then stepwise re-fusion will produce one CV again (Tominaga *et al.,* 1998b). This may account for the stepwise CV filling process that is sometimes observed in the cell. In cultures of *P. multimicronucleatum* adapted to salt solution (Soldo *et al.,* 1966), the complete cyclic process requires ~10 sec on average (Ishida *et al.,* 1993) to produce CVs of 13 μm in diameter, thus the fluid output is ~7.7 pL/CV/min. However, the time required for this cycle is directly influenced

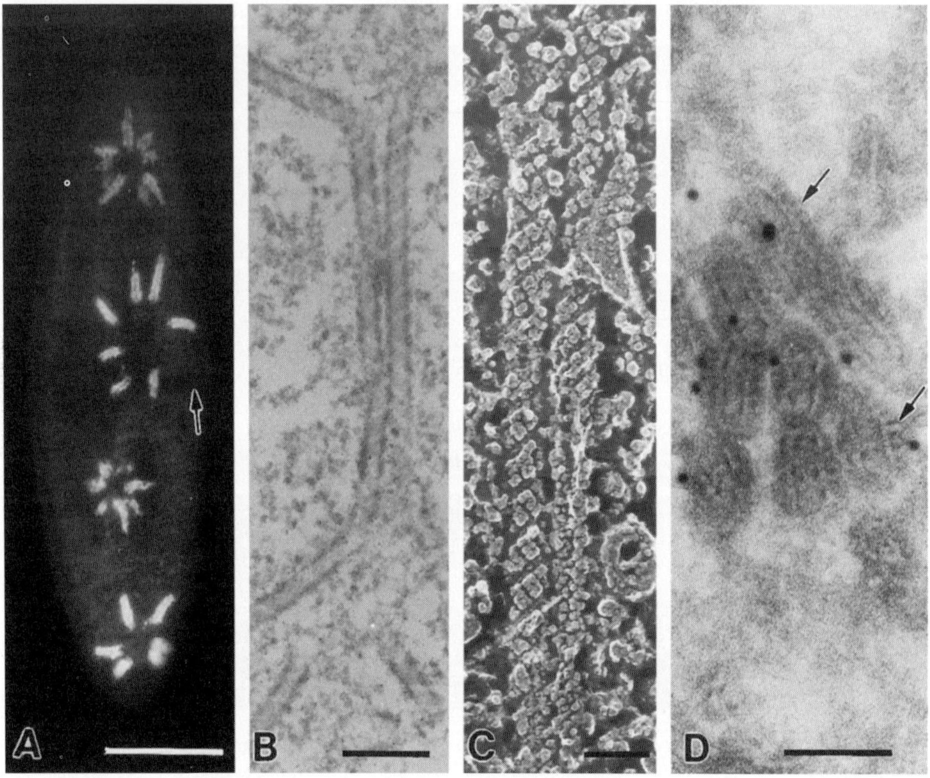

FIG. 11 CVC membrane enriched with V-ATPases. (A) New CVCs in *Paramecium* develop anterior to old CVCs. Old CVCs undergo retrenchment of their decorated spongiomes as the cell's division furrow develops (arrow). From Allen *et al.* (1990). Scale bar = 50 μm. (B) Transmission electron micrograph of decorated tubules in *Paramecium*. Scale bar = 0.2 μm. (C) Quick-freeze deep-etched decorated tubules showing V-ATPase V_1 sectors on the cytosolic surface of the 50-nm tubules. (D) Glutaraldehyde fixed and cryosectioned 50-nm tubules are covered with V_1 V-ATPase sectors (arrows) that are immunogold labeled with V-ATPase specific DS-1 monoclonal antibodies. Scale bar for (C)–(D) = 0.1 μm. (B) to (D) from Fok *et al.* (1995).

by the osmolarity of the surrounding medium (Ishida *et al.,* 1996; Stock *et al.,* 2001; see Section IV).

3. Other Ciliates—Tetrahymena, Vorticella, and Euplotes

CVCs of other ciliates are similar to those of *Paramecium,* although frequently somewhat different morphologically. *Tetrahymena* has a pair of pores near its posterior pole, usually attached by microtubular ribbons to a single CV (Elliott and Bak, 1964; Cameron and Burton, 1969; Elliott and Kennedy, 1973). The major difference in the construction of its CVC appears to be attributable to the reduced extent of the ribbons of microtubules that arise from the pore. Thus, these CVCs lack the long radial arms, extending deep into the cytoplasm, along which the spongiomes are organized. Tubules of smooth spongiome are attached directly to the CV throughout the CV cycle. The smooth spongiome is attached to a 50-nm peg-bearing decorated spongiome (R. D. Allen, unpublished). That the CV dissociates from the decorated spongiomes during fluid expulsion will need to be confirmed, preferably by capacitance measurements (Section III.C). Previous publications have reported that the vacuole and tubules of the CVC are continuous with the ER, but this observation is problematic and needs to be studied in greater detail; such a morphology was shown not to be the case in the ciliate *Pseudomicrothorax* (Hausmann, 1983).

Another modification of a CVC in ciliates includes a CVC that lacks a permanent pore. This is apparently true of the CVC of *Euplotes raikovi* (Dallai *et al.*, 1985). CVs such as those in *Vorticella,* on the other hand, empty into the cell's cytopharyngeal region through an elongated tube (R. D. Allen, unpublished).

Much of the early work on the CVC of *Tetrahymena* involved the CVC's possible role in osmoregulation. For a review of this work, refer to Dunham and Kropp (1973). The CVC of this cell, besides eliminating excess water, is thought to be used to eliminate excess Na^+.

A few puzzling observations made earlier on the CVs of *Tetrahymena* and *Amoeba* can now be placed into the context of what is known about the CVC of *Paramecium*. Dunham and Stoner (1969) observed that the pellicle of *Tetrahymena* around the CV pore becomes indented just before fluid expulsion. These indentations are not always accompanied by fluid expulsion. In addition, Cameron and Burton (1969) observed a widening of the pore in *Tetrahymena* as the CV filled. In another study, isolated CVs from *A. proteus* were observed to contract when exposed to 15 mM ATP (Prusch and Dunham, 1970). Indentation of the pellicle, setting the size of the pore, and "contraction" of an isolated CV under the stimulation of added ATP all match events that occur during the brief rounding phase of the CV cycle in *Paramecium*. These events may all be under the control of a unique increase in membrane tension that results in rounding of the CV, which occurs spontaneously at regular intervals in isolated CVs. This presumably also occurs in the whole cell just before CV expulsion (discussed in Section III.B.2).

III. Function of the Contractile Vacuole

A. CV Fluid: Ions and Osmolarity

The fluid content of the contractile vacuole is presumed to be mainly water. However, its precise contents have not been completely determined in any cell. As stated in Section II.A.2, the osmolality of the CV of *Chaos carolinensis* was found to be 51 mOsm, and it was found that Na^+ was present in the CV at a concentration of 19.9 mM, while K^+ was present in a concentration of only 4.6 mM (Riddick, 1968). The osmolality of the CV fluid in *Amoeba proteus* was determined to be 32 mOsm as opposed to the cytosol that was 101 mOsm (Schmidt-Nielsen and Schrauger, 1963). The concentrations of individual ions have not been determined in the CV of this cell.

In *Dictyostelium,* attempts have been made to investigate the contents of the CV by studying the ion channels found in the CV membrane. Yoshida *et al.* (1997) isolated from cells a membrane fraction enriched in CV markers. This membrane was then incorporated into a planar lipid bilayer where its channels could be studied. Using a patch clamp technique, single K^+ channel currents were observed that could have been derived from channels originally in a CV membrane. These workers suggested that K^+ ions may first enter the CV and then later be retrieved through the observed channels, leaving the fluid more hypotonic to the cytosol than it was at the beginning.

In *Paramecium,* our preliminary results on the contents of the *in situ* CV indicate that the CV has a K^+ concentration of 50 to 60 mM, twice that of the cytosol (Stock, Allen, and Naitoh, unpublished observations). This is much higher than that reported for the CV of *Chaos carolinensis* and may reflect either species differences or differences in techniques used for ion determination. In *Paramecium,* we determined the concentration using a K^+ ion-selective microelectrode inserted directly into the living CV. These inconsistencies do point to a critical need to reconsider earlier work, and to extend such studies to a much wider range of organisms, to determine precisely the ionic composition of the CVs of different organisms. The pH of the CV contents of *Paramecium* was only ~7.0 when measured by an ion-selective microelectrode containing an ionophore cocktail selective for protons (Tominaga, Allen, and Naitoh, unpublished observations). The Na^+ concentration was not determined, as the standard saline solution bathing the cells did not contain Na^+. Only traces of NH_4^+ and HCO_3^- ions were found in the CV, but Cl^- was present in a concentration of 7 to 15 mM (Tominaga, Allen, and Naitoh, unpublished observations).

Based on the earlier results for *Chaos* and *Amoeba,* it appears that the CV is hypotonic to the cytosol. We have not yet been able to determine the osmolarity of the CV of *Paramecium.* If CV content is generally hypotonic to the cytosol, then the CV membrane would need to be relatively impermeable to water passage for it to be able to store water long enough to expel much of its accumulated content.

How water moves into the CV still remains unsolved. If the CV is indeed hypotonic, it will be necessary to abandon the idea that water enters the CV purely by passive diffusion, since the requirement that the CV fluid be hyperosmotic to the cytosol would not be fulfilled. In such a case, we would need to look for a mechanism that actively pumps water into the CV or that could move protons and hydroxyl ions and/or bicarbonate ions into the CV where these ions could be converted into water. Water pumps do exist in nature. Zeuthen has reported two such pumps: one in the *Necturus* choroid plexus epithelium that cotransports one ion each of K^+ and Cl^- together with 500 molecules of water (Zeuthen, 1994; Zeuthen and Stein, 1994), and a second, the human Na^+-glucose cotransporter, that pumps 210 water molecules for every two Na^+ ions and one sugar molecule (Meinild *et al.,* 1998; see Cotransporters as Molecular Water Pumps by Zeuthen and MacAulay, this volume). Another candidate for moving water across membranes would be the presence of an aquaporin-like protein. Such a protein has been identified in *Dictyostelium discoideum,* but its location and role in this cell are unknown (Flick *et al.,* 1997).

B. Membranes

1. Membranes with Proton Pumps

The CVC is now known to be composed of two distinctly different populations of membranes, a membrane rich in V-ATPases (Fig. 11B, C, D) and a second population which lacks intact V-ATPase holoenzyme complexes but may contain some of its individual polypeptides (Nolta and Steck, 1994). This second membrane population also contains many other proteins not found in the V-ATPase-containing membranes; some of these are shared with the plasma membrane.

The V-ATPase-rich membrane was first recognized to be different from the other membranes seen in electron micrographs, since this membrane was seen to be decorated on its cytosolic surface by 15-nm diameter particles or "pegs" (McKanna, 1974, 1976). This was reported to be true of all contractile vacuoles, whether obtained from amoeba, flagellates, ciliates, or fresh-water sponges. Thus, even before it was known that the pegs were in fact the V_1 sectors of V-ATPases (Heuser *et al.,* 1993; Fok *et al.,* 1995), many organisms having contractile vacuole complexes were known to have these peg-containing membranes which came to be called the decorated spongiome (Patterson, 1980). In 1993 it was demonstrated for *Dictyostelium discoideum* that such pegs on the saccules and tubules of the CVC became aggregated under the influence of antibodies raised to V-ATPase polypeptides of chromaffin cells (Heuser *et al.,* 1993). Monoclonal antibodies raised to a 68-kDa polypeptide of the V_1 sector and a 100-kDa integral membrane protein of the V-ATPase of *Dictyostelium* (Fok *et al.,* 1993) immunolabeled most membranes of the CVC reticulum of this amoeba, demonstrating that these membranes were indeed covered by polypeptides of the V-ATPases.

The V-ATPase enzyme of lower eukaryotes that have a CV has been most thoroughly characterized in *Dictyostelium* (Padh *et al.,* 1989). In this cell, one-third of its total pH 7.0 ATPase activity is inhibited by 7-chloro-4-nitrobenzo-2-oxa-1,3-diazole (NBD-Cl), an inhibitor of some V-ATPases. The bulk of the NBD-Cl sensitive ATPase was found on an isopycnic sucrose gradient to be located in a buoyant fraction of cell homogenates. This ATPase had most of the properties of the V-ATPase of higher organisms. For example, interactions with several ATPase inhibitors confirm that this ATPase belongs to the vacuolar (V) class of proton pumps. This is especially clear from this V-ATPase's reaction to macrocyclic lactones, bafilomycin A and concanamycins A and B, which are particularly potent inhibitors of V-ATPases (Bowman *et al.,* 1988; Woo *et al.,* 1992). Like other V-ATPases, the optimal activity for this V-ATPase occurred at pH 8.0; it was MgATP-dependent, and its nucleotide preference was similar to V-ATPases of other organisms (Padh *et al.,* 1989). Treatment of *Dictyostelium* with concanamycin A resulted in this cell's inability to withstand hypotonic shock. Large vesicles appeared in the cell, but these were probably swollen lysosomes or endosomes, as they contained fluid phase markers (Temesvari *et al.,* 1996). In *Paramecium,* treatment with concanamycin B produced no visible change in the V-ATPase-rich membranes even though the CV activity was dramatically reduced (Fig. 12A) (Fok *et al.,* 1995). The V-ATPase holoenzymes do not disassemble as they do following treatment with DS-1 antibody (Fig. 12B), hypertonic stress (Fig. 10C, D), and cold stress (Section IV).

The V-ATPases are always electrogenic (Wieczorek *et al.,* 1999), and their H^+-ATPase activity works against a membrane electrochemical gradient for protons. The isolated enzyme from *Dictyostelium* is activated by KCl in a coupling buffer. In the presence of nigericin (a cation exchanger) and Cl^-, its activity increased

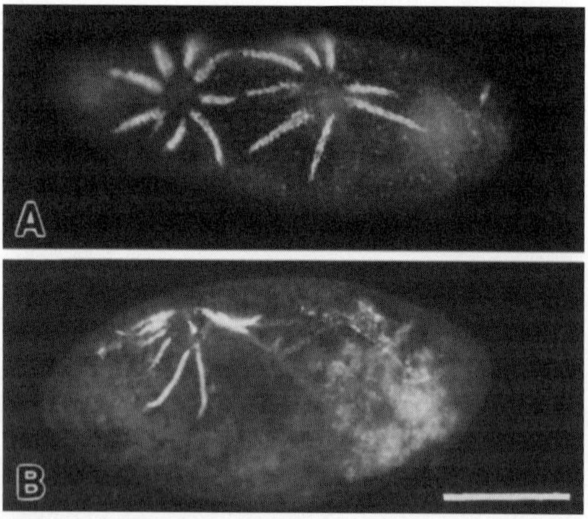

FIG. 12 Concanamycin B and DS-1 mAb injection. (A) Cell exposed to the V-ATPase specific drug concanamycin B (30 nM) for 30 min had greatly inhibited CVC activity without any visible effect on the labeling of V-ATPases of the decorated spongiome. From Fok *et al.* (1995). (B) Cell injected with DS-1 mAb (22 pg/cell) and fixed after 45 min. The CV nearest the injection site shows a breakdown of its decorated spongiome. (Courtesy of M. Ishida.) Scale bar = 50 μm.

20-fold (Padh *et al.,* 1989). A parallel Cl^- conductance pathway may be used for the Cl^- that stimulates the activity of this enzyme, if Cl^- is actually used to stimulate *in vivo* the V-ATPase of the CV membrane in *Dictyostelium.*

The V-ATPase is found both (1) in the endosome/lysosome membrane system, where it leads to acidification of these organelles, and (2) in CVs where its principal function seems to be to establish an electrochemical gradient across the CV membrane, CVs being not known to become measurably acidic. However, the concentration of V-ATPases in the decorated spongiome membrane is much higher than in the endosome/lysosome system. In *Dictyostelium,* 90% of the V-ATPase activity is thought to be part of the CVC, as only 10% is associated with the endosome/lysosome system (Rodriquez-Paris *et al.,* 1993). In *Paramecium,* we estimated that the decorated spongiome, which is composed of 50-nm diameter tubules, has 465 V-ATPases per micrometer of tubule length, which means that there are 6.6×10^5 V-ATPases for every μm^3 of luminal volume in the decorated tubules or about one pump per luminal volume that is equivalent to the size of one V_1 sector (Tominaga *et al.,* 1998a). Such a large number of V-ATPases for a relatively small luminal volume, that do not produce an acid lumen, has to mean that the V-ATPases are used principally to establish and sustain a constant electrochemical gradient. This gradient may then supply the energy to drive the movement of ions across the CVC membrane. In *Paramecium multimicronucleatum,* the membrane potential is kept at a constant +60 to +80 mV inside the CVC compared to the cytosol (Tominaga *et al.,* 1998a) (see Section III.C).

The gene for the 100-kDa protein of the V-ATPase in *Dictyostelium* has now been cloned, and its putative protein product has 815 amino acids, giving a molecular weight of 93 kDa (Liu and Clarke, 1996). The larger M_r on gels may be due to the glycosylation of one asparagine residue. These authors concluded that there was only one isoform of this transmembrane protein, so that this polypeptide would be the same in membranes of both the CVC and the endosome/lysosome system. A proteolipid subunit associated with the V-ATPase of *Dictyostelium discoideum* has also been cloned (Xie *et al.,* 1996). Whether this is the proteolipid that forms the proton pore or an accessory proteolipid is not known, but its sequence is similar to proteolipids of V-ATPases of other organisms. The deduced protein product has 196 amino acids and a M_r of 20,148.

A second proton pump has been identified by immunogold labeling on the CV membranes of *Chlamydomonas reinhardtii* (Robinson *et al.,* 1998). This pump, like the V-ATPase, is also found on the lytic vesicles of this algal cell, but there is a much higher antigen labeling on the CV. In quick-frozen, deep-etched cells, the membrane of the filled CV, as well as small vesicles in the CV region, had a random distribution of 9- to 11.5-nm diameter particles. The published figure suggests that these small vesicles may be interconnected by membrane. Though Robinson *et al.* (1998) concluded the particles might be V_1 sectors of proton pumps, this interpretation is unlikely, since the membrane of the CV was obviously fractured along its interior, making it more likely that the V_1 sectors were removed during fracturing. Thus these particles may be the V_0 sectors. Nearby intermediate size vesicles were covered with 6- to 7.5-nm particles, which is the approximate size of profiles of H^+ -pyrophosphatases (Robinson *et al.,* 1998).

Besides containing all of the V-ATPase-rich membrane, the dense fraction from an immunogold density-shift gradient-centrifugation experiment of *Dictyostelium* also contained prominent electrophoretic bands of ~44- and ~33-kDa proteins. Although the 44-kDa species had an electrophoretic mobility similar to that of actin (Nolta and Steck, 1994), no experiments were performed that actually show it to be actin.

2. Membranes That Undergo Cyclic Changes in Tension and Fusion

V-ATPase-containing membranes of the CVC do not fuse with the plasma membrane and apparently are not used to store fluid. Thus a second population of membranes called the smooth spongiome (Patterson, 1980) serves as a reservoir for the fluid that enters the CVC. This membrane not only fuses with the V-ATPase-containing membrane but is also capable of periodically fusing with the plasma membrane. There is no compelling evidence that it ever fuses with the ER or with the endomembranes of the endosomal/lysosomal system. This membrane may have more species variability in its composition than does the V-ATPase-rich membrane. For example, this membrane is known to be associated with alkaline phosphatase (Bowers and Korn, 1973; Quiviger *et al.,* 1978) and alkaline phosphodiesterase enzymes only in *Dictyostelium* and *Acanthamoeba* (Padh *et al.,* 1989; Nolta *et al.,* 1991). However, testing for the presence of such enzymes may not have been done yet in other organisms. In *Dictyostelium,* it has been suggested that such enzymes may be anchored to the CV membrane by a glycosylphosphatidylinositol (GPI) molecule, (Nolta and Steck, 1994).

In *Dictyostelium,* an immunogold density-shift gradient-centrifugation technique was used to separate the V-ATPase-rich membranes from the smooth membranes (Nolta and Steck, 1994). The upper gradient contained the alkaline phosphatase activity and also contained many other proteins including a band of 280 kDa and a band of 57 kDa, a triplet complex centered at 48 kDa, and a doublet at 37 and 35 kDa. In immunoblots, noncatalytic bands 2 and 3 of the V-ATPase ladder of eight polypeptides were found in the bouyant fraction and the CV-containing subfractions. Some fractions had band 1 of this ladder in great excess. Thus the CV membrane appears to lie in the buoyant fractions, while the satellite- and tubular V-ATPase-containing membranes are in the dense fraction. Besides this protein content, the lighter CVC membrane has a much higher phospholipid content (1 μg versus 0.3 μg of lipid/mg of protein), and the sterol and sphingolipid content is higher in the CV compared to the saccule and tubular membranes (Nolta and Steck, 1994).

Another conclusion from the above work is that the CV membrane does not normally vesiculate into numerous small vesicles following fluid expulsion, as alkaline phosphatase was never found associated with such small vesicles (Nolta and Steck, 1994). Thus, like the CV of *Paramecium,* the CVs of *Dictyostelium* and *Acanthamoeba* can be expected to revert into a tubular form during CV expulsion, and would probably remain in continuity with each other, and so form one or a few continuous masses of interconnected tubules, rather than a large pool of isolated vesicles.

Even the small CVC of *Phytophthora* has a two-membrane system, a smooth membrane that shares a P-type ATPase with the plasma membrane and a V-ATPase-rich membrane (Mitchell and Hardham, 1999) that lacks P-type ATPases. The V-ATPase-containing membrane does not form the CV.

Like the amoeba and the fungal zoospores, *Paramecium* also has a two-chambered CVC system. In this cell, there is a large excess of V-ATPase holoenzyme-free membrane that under unusual situations can be utilized to produce a huge CV. Normally, a CV in an interphase cell of *Paramecium* measures around 13 μm in diameter. However, if for some reason fusion between the CV and the plasma membrane at the pore does not occur, fluid will continue to flow into the CV and it will double its volume before the next rounding phase leads to another attempt at fusion (Tominaga *et al.,* 1998b; Stock *et al.,* 2001). In fact, vacuoles of 40 to 50 μm in diameter can be formed from the excess smooth spongiome that makes up the ampullae, collecting canals, and the crystalline-like smooth spongiome (Tominaga *et al.,* 1999).

A unique feature of this type of membrane is that when its luminal hydrostatic pressure is reduced, this membrane will revert first into a mat of 40-nm tubules (Fig. 13A, B) and then, given enough time, the membrane will collapse into a three-dimensional crystalline-like array composed of a highly curved bilayer (Fig. 13C). The mechanisms that might produce these highly dramatic changes in bilayer shape are as yet in an early stage of theoretical investigation and modeling (Hui and Sen, 1989; Döbereiner *et al.,* 1999; Lipowsky, 1999).

Recent studies on isolated CVs have shown that this type of membrane is distinct in another way from any other membrane reported in nature. This membrane is capable of undergoing cycles of rounding and relaxing, apparently without the aid of a surrounding contractile meshwork. This phenomenon helps to justify the use of the name "contractile vacuole" for this organelle, although the CV is not contractile in the sense that it contracts in order to actively expel its content. When the CV is extruded from a disrupted *Paramecium* cell (Tominaga *et al.,* 1998b; Tani *et al.,* 2000) into a small amount of minimally diluted cytosolic solution, the CV as well as fragments of the smooth spongiome will periodically round and relax (Fig. 14). This process presumably continues until energy is depleted.

If the *in vitro* CV is bisected with a microneedle, both parts of the CV will continue to round and relax (Fig. 15), but they will not remain in phase with each other (Tani *et al.,* 2000). This implies that there is no universal timing mechanism (master pacemaker) in the cell's cytosol that determines the sequence of rounding and relaxing of all membranes upon a given signal. Rather, each fragment of

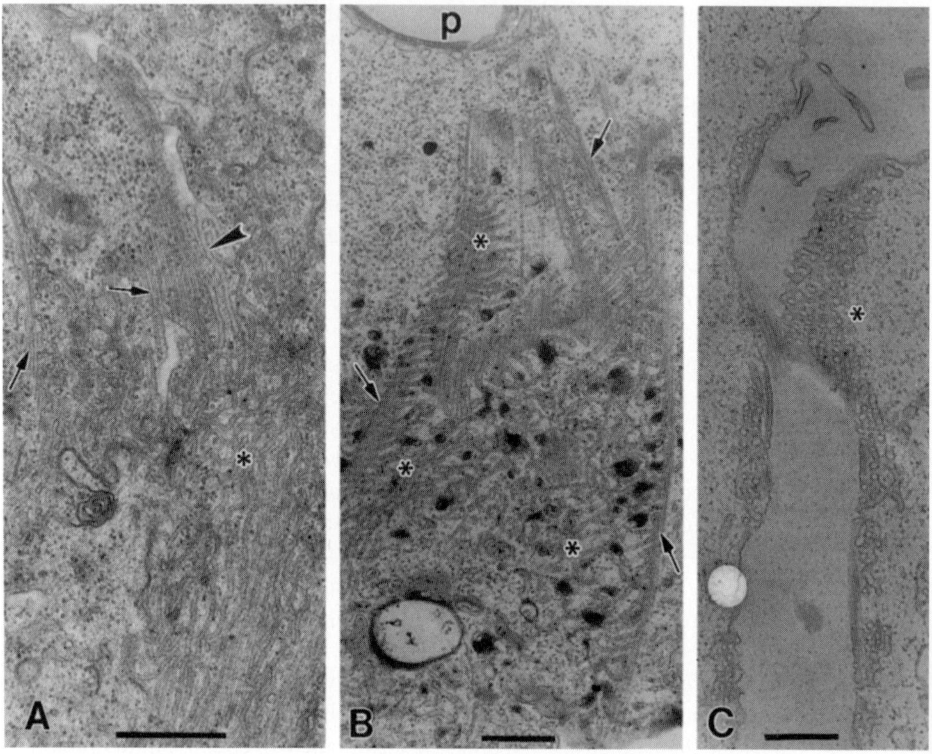

FIG. 13 Nonplanar CVC membranes of the smooth spongiome. (A) Tubules (*) form from the *in vivo* CV membrane during fluid expulsion. Arrowhead indicates a 40-nm tubule forming from a cisterna next to a microtubular ribbon. Arrows indicate microtubules. From Naitoh *et al.* (1997b). (B) Tubules (*) also form in disrupted cells following CV fluid expulsion. p, CV pore; arrows indicate microtubules. (C) Crystalline-like membrane (*) along the collecting canals. All scale bars = 0.5 μm. (B) and (C) From Tominaga *et al.* (1999).

membrane determines its own rounding and relaxing cycle (Tani *et al.,* 2000). What initiates this cycle appears to be an increase in membrane tension. If suction is applied through a micropipette to any part of the membrane that surrounds the smooth spongiome-derived vesicle, the rounding/relaxing cycle will immediately begin a new (extra) cycle by causing the vesicle to round (Tani *et al.,* 2000). Suction, that is, membrane tension, therefore has reset the cycle.

To understand the forces involved in *in vivo* rounding and relaxing of the CV membrane in *Paramecium,* the association between the microtubular ribbons and the smooth spongiome during a cycle was examined to see how the CV membrane is altered at this junction. After using changes in the membrane potential to trigger fixation at a predetermined stage of the cycle, abundant 40-nm tubules were found next to the CV cytoskeleton of microtubular ribbons during rounding (Tominaga

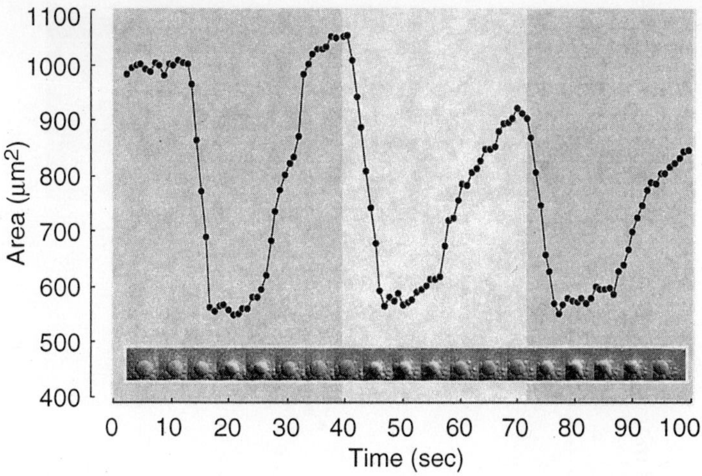

FIG. 14 Rounding–relaxing cycles of an *in vitro* CV. The apparent change in surface area of the CV is graphed above, and video images are shown below. The center of each video frame corresponds to the time when the frame was acquired in seconds. From Tani *et al.* (2000).

et al., 1999). At stages where the CV was more relaxed, the diameter of the tubules and the shape of the CV membrane were more variable. We concluded that an association of the CV membrane with microtubular ribbons promotes or enhances the tendency of the planar membrane to tubulate.

Finally, we have now been able to show that, during rounding, the tension of the CV membrane increases 35-fold over the tension of a relaxed CV membrane (Tani *et al.*, 2001). This was determined by developing a microcantilever to directly

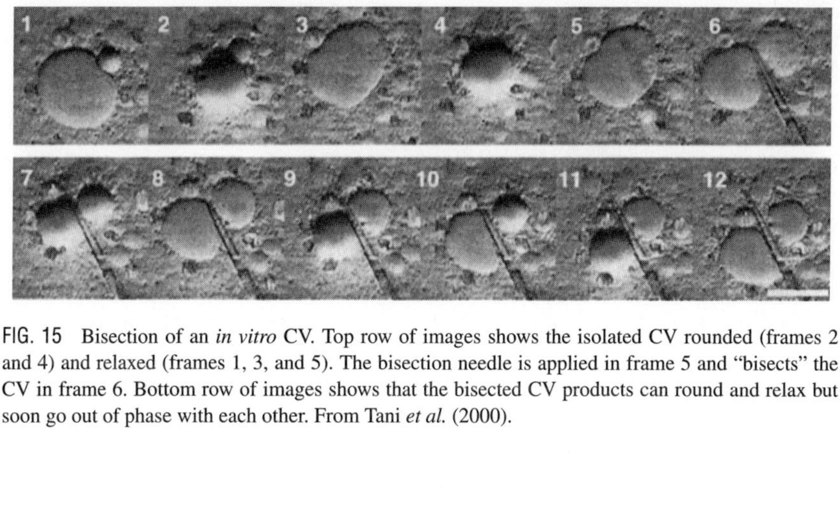

FIG. 15 Bisection of an *in vitro* CV. Top row of images shows the isolated CV rounded (frames 2 and 4) and relaxed (frames 1, 3, and 5). The bisection needle is applied in frame 5 and "bisects" the CV in frame 6. Bottom row of images shows that the bisected CV products can round and relax but soon go out of phase with each other. From Tani *et al.* (2000).

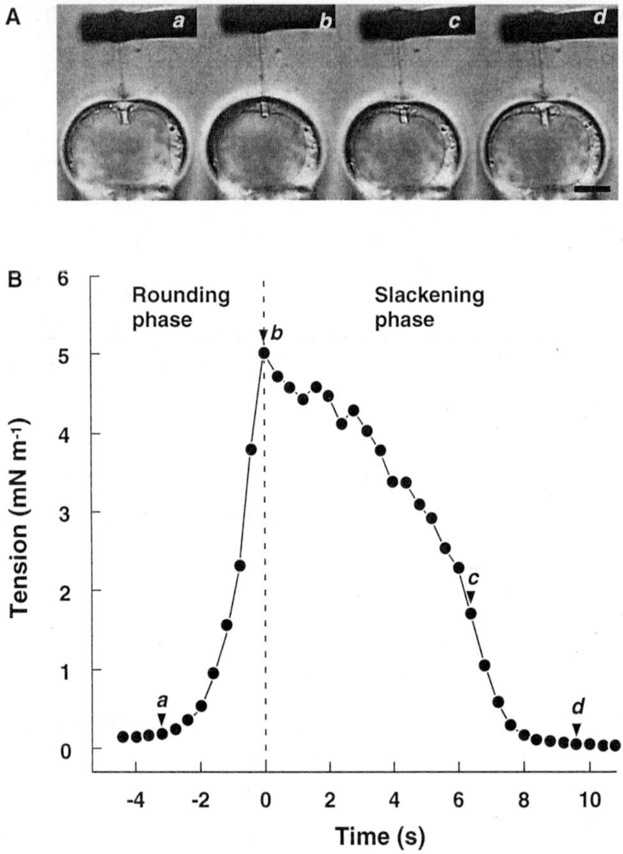

FIG. 16 Membrane tension of the *in vitro* CV membrane. (A) Video images of the degree of depression formed by a thin glass rod mounted on a flexible arm of a microcantilever used to measure changes in tension of the CV membrane during a rounding–relaxing cycle. The letters in (A) correspond to times in (B) when tension was calculated. (B) The graph of the tensions calculated from the force exerted against the cantilever. From Tani *et al.* (2001).

measure the force produced by a rounding CV (Fig. 16) and then converting this force into tension. Tension increased from 0.13 to 5 mN m^{-1}. This suggests that the increased tubulation that reduces the effective surface area of the CV around the fluid is somehow linked to a rise in membrane tension.

Thus, the smooth spongiome membrane has the unique ability to undergo a 35-fold increase in tension *in vitro* which, *in vivo,* may be initiated by an interaction of the CV membrane with the microtubular cytoskeleton. This increase in tension is important for at least three events that set the stage for fluid expulsion: (1) the fusion of the CV with the plasma membrane (Tominaga *et al.,* 1998a), (2) the

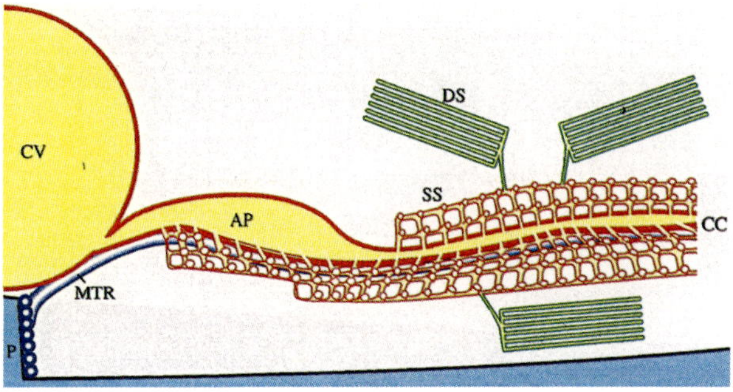

FIG 13.9. Cutaway drawing of a portion of the CVC of *Paramecium* showing its pore (P), an ampulla (AP), and a collecting canal (CC) supported by a microtubular ribbon (MTR). The smooth spongiome (SS) is arrayed into a three-dimensional lattice and is continuous through ducts with the decorated spongiome (DS) that contains the vacuolar proton pumps. All membranes in red are part of the smooth spongiome and those in green are part of the decorated spongiome. Modified from Tominaga *et al.* (1999).

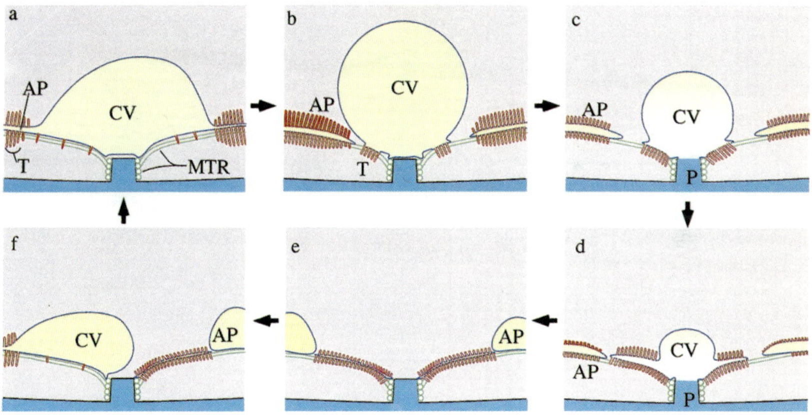

FIG 13.20. A cartoon illustrating a CVC cycle in *Paramecium.* This cartoon emphasizes the tubulation of the smooth spongiome that occurs in association with microtubular ribbons. (a) Filling of the CV; (b) rounding of the CV resulting in dissociation of arms; (c) opening of the pore and fluid expulsion; (d) tubulation of CV membrane; (e) pore closure and ampullar swelling; (f) CV membrane refuses with ampullae, stepwise. AP, ampulla; T, tubule; MTR, microtubular ribbon; P, CV pore. From Tominaga *et al.* (1999).

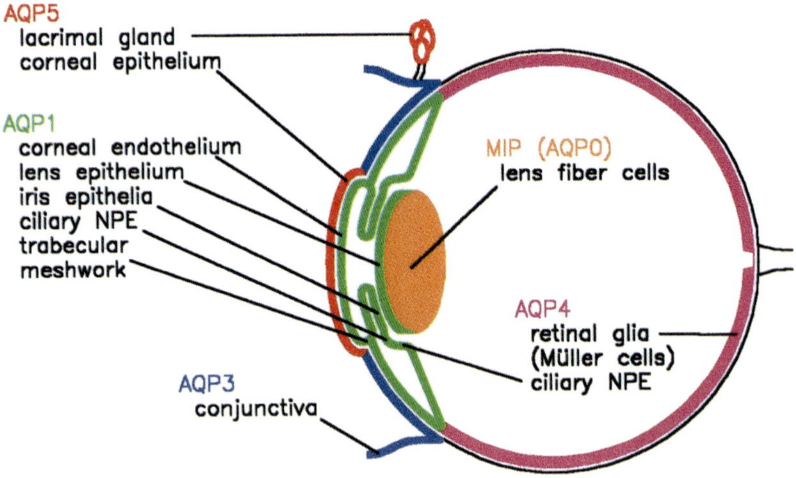

FIG 14.8. Expression of aquaporins in eye. MIP, major intrinsic protein of lens or AQP0. From Hamann *et al.* (1998), with modifications.

setting of the size of the pore opening and thereby the effective rate of expulsion (Naitoh *et al.,* 1997b), and (3) the separating of the CV from the bulk of the membrane along the radial arms (Tominaga *et al.,* 1999).

C. *In Situ* Electrophysiology of the CV

Electrical events associated with CV activity were first recorded from *Paramecium caudatum* (Yamaguchi, 1960) and *Amoeba proteus* (Josefsson, 1966; Prusch and Dunham, 1967). In both cases, hyperpolarization of the cell's membrane was recorded when the vacuole expelled its content. An explanation of this hyperpolarization of the plasma membrane is that the CV in *P. caudatum,* as well as in *Amoeba proteus,* probably remains attached to the electrogenic pumps of the CVC even after the CV has fused with the plasma membrane. Recent work on *P. multimicronucleatum* does not show such a hyperpolarization potential (Tominaga *et al.,* 1998a); consequently, the earlier work may have somehow disturbed the connections of the radial arms with the CV so the electrogenic enzymes do not become disconnected from the CV in these cells. Prusch and Dunham (1967) reported that when the microelectrode was positioned directly in the CV of *A. proteus,* a large depolarization was observed during fluid expulsion. They recorded a resting potential for the plasma membrane of –65 mV and a potential of –45 to –55 mV in the CV with respect to the outside of the cell. Thus the CV was more positive than the cytosol by 10 to 20 mV. When the CV expelled its contents, the plasma membrane became hyperpolarized by as much as 20 mV, indicating that electrogenic complexes remained in association with the plasma membrane during this time.

Small transient potential changes across the plasma membrane were also observed in conjunction with the expulsion of the contractile vacuole in the ciliate *Vorticella convallaria* (Shiono *et al.,* 1980). These changes could be recorded even when the microelectrode was as much as 30 μm away from the CV. Changes of 0 to10 mV and lasting 0.25 to 0.6 sec were measured and showed that the CVC of *Vorticella* must also contain electrogenic generators.

As mentioned before, the CV utilizes an electrochemical gradient to facilitate its major function, the accumulation of excess water (Ishida *et al.,* 1993). Only recently has it been possible to measure the *in vivo* membrane potential of the CV *in situ* (Fig. 17). Tominaga *et al.* (1998a) inserted a fine-tipped microelectrode through the plasma membrane and then into the CV of living *Paramecium multimicronucleatum.* It was determined that the membrane potential remained constant with reference to the cytosolic potential of a CV, during the filling phase, over as many as eight filling–expulsion cycles. The absolute level of this potential varied somewhat from CV to CV, but the average was around 60 mV positive (Fig. 17iii). However, just before expulsion, the membrane potential dropped dramatically to reach 10 mV when the pore opened. After the pore had closed, the

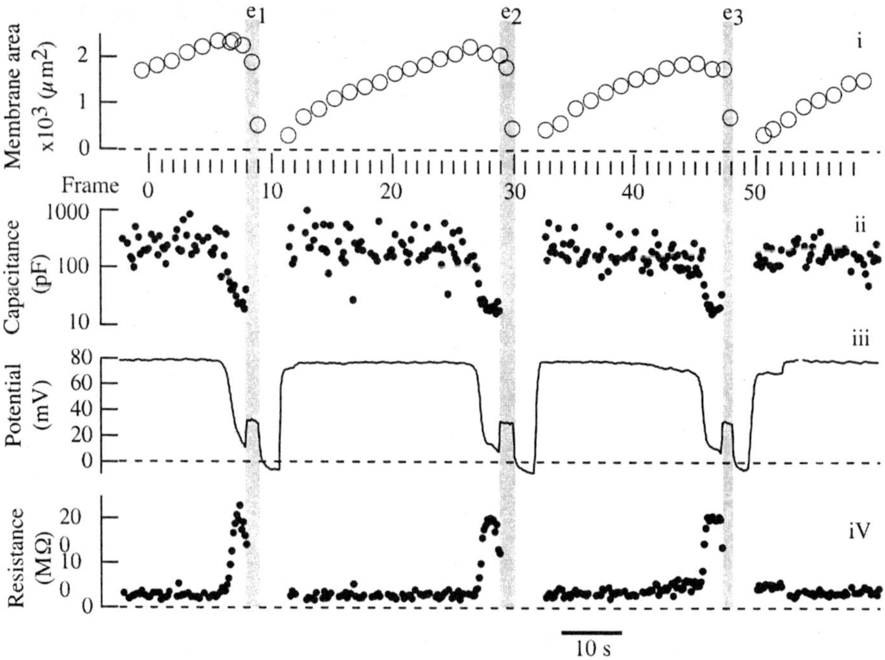

FIG. 17 Electrophysiology of the CVC of *Paramecium*. (i) The CV membrane area. (ii) The organelle's input capacitance. (iii) The organelle's membrane potential with reference to the cytosol. (iv) The organelle's input resistance. Three fluid expulsion phases are labeled e_1, e_2, and e_3. From Tominaga *et al.* (1998a).

membrane potential measured 0 mV and then became slightly negative for a second or two before it quickly rose again to its previous plateau of about +60 mV. Thus, it is possible to conclude that the electrogenic generators, the V-ATPases, became disconnected from the CV before the pore opened.

What causes these extreme fluctuations in membrane potential was clarified when the capacitance and resistance of the CV membrane were determined along with the electric potential. Input capacitance is a measure of the surface area of the membrane involved in producing the passive potential across the membrane. By superimposing a series of square wave pulses into the CV, while recording the membrane potential, it was possible to determine the input capacitance and input resistance from the shapes of the potential shifts that were recorded (Tominaga *et al.*, 1998a). Input capacitance measurements of 210 pF during filling decreased to 21 pF during the rounding phase (Fig. 17ii), and the input resistance measurement of the rounded CV (295 MΩ) was significantly higher than the resistance of the CVC as a whole (31 MΩ) (Fig. 17iv). This showed that the surface area of the CVC remained constant during most of the filling phase, even while the CV grew in size (Fig. 17i), but dropped precipitously following the drop in potential just before fluid expulsion. This ~10-fold drop in capacitance and the ~10-fold increase in resistance occurred even though the maximum diameters of the CVs did not change significantly (Fig. 17i). The simplest explanation for this is that the CV separated from that part of the smooth spongiome that is bound to the microtubules along the radial arms. The concomitant drop in membrane potential indicates that the electrogenic generators (i.e., the V-ATPases) are no longer connected to the CV, so all, or at least most, active electrogenic sites must be located within the membranes of the radial arms.

By dividing specific membrane capacitance, 1.0 μF cm^{-1}, into the input capacitance during fluid-filling, the total surface area of the CVC can be estimated to be 18×10^3 μm^2. This value was close to the figure of 14×10^3 μm^2 estimated from electron microscopic measurements (Tominaga *et al.,* 1998a). Thus the total membrane area of the two CVCs of a *Paramecium* cell, including both the decorated and smooth spongiome systems, is about half that of the plasma membrane (Stock *et al.,* 2000). This membrane system seems to be sufficient in area to easily handle the task of collecting and expelling the water entering the cell through the plasma membrane.

D. Studies of Ion Movement into Isolated CV Vesicles

Giglione and Gross (1995) studied a light membrane fraction of vesicles enriched in CV membranes, isolated from *Dictyostelium*. However, as in all such studies, one must be concerned about what this fraction really contains. Does it contain principally CV membrane of the CVCs or of CV membrane plus V-ATPase-bearing membranes of the CVC, and to what extent are the V-ATPase-bearing acidosomes of the lysosomal/endosomal system also present? These vesicles formed a proton gradient (their pH was decreased) when the bathing solution contained Cl^- ions and the membrane potential of the vesicles was decreased. The presence of bicarbonate ions also reduced the vesicle membrane potential but, unlike experiments where Cl^- was present, a proton gradient was not formed, suggesting that these ions cross the membrane and combine with the protons that are pumped into the vesicles by V-ATPases. Giglione and Gross (1995) concluded that bicarbonate may be conducted into the CV of *Dictyostelium* where it is attracted by the positive electric potential developed across the CV membrane that is generated by the V-ATPases. Bicarbonate would then reduce the inside-positive membrane potential by combining with the protons, and this would account for the fact that a large chemical gradient of protons did not appear.

E. Proteins with Other Functions on the CVC Membranes

Other molecules associated with CV membranes of *Dictyostelium* include calmodulin (Zhu and Clarke, 1992; Zhu *et al.,* 1993), α-actinin (Furukawa and

Fechheimer, 1994) and RabD (Bush *et al.,* 1994). Calmodulin plays many roles in cells. Its role in the CV may be to regulate ion channels such as Ca^{2+}-gated K^+ and Na^+ channels (reviewed in Preston *et al.,* 1991), if such exist in the CV membrane (Zhu *et al.,* 1993), or to regulate Ca^{2+} (Moniakis *et al.,* 1999).

Doberstein *et al.* (1993) developed an antibody to a synthetic peptide, identical to the phosphorylation site of myosin-IC, that was then found to label both the CV and the plasma membrane of *Acanthamoeba.* Zhu and Clarke (1992) showed that this antibody also labels the CV of *Dictyostelium.* When Doberstein *et al.* (1993) introduced this antibody into *Acanthamoeba,* the cell's CV activity was substantially inhibited. These authors concluded that myosin-IC was part of an actomyosin-based contraction system involved in CV fluid expulsion. Another plausible interpretation is that myosin-IC is needed to move the CV to the plasma membrane and/or for events leading to the recognition and fusion of the CV with the plasma membrane. As the interpretation of this report is open to question, we believe it is premature to discard the results of Wigg *et al.* (1967) who concluded that expulsion is driven mostly by cytosolic pressure. We concluded that cytosolic pressure is, indeed, the driving force for fluid expulsion in the CV of *Paramecium* (Naitoh *et al.,* 1997a) and suspect this will be true of most CVs.

A Rab4-like protein, called RabD, has been identified in *Dictyostelium* (Bush *et al.,* 1994). This protein is a member of the large family of small GTPases involved in vesicle trafficking. Its gene was cloned and its protein sequence determined. RabD has about 57% amino acid identity with human Rab4 and except for one amino acid, its effector domain is identical to the effector domain of human Rab4. RabD was localized by immunofluorescence microscopy to the tubular units of the CVC but not to the CV "bladders." It is assumed that RabD is involved in the trafficking of V-ATPase-rich membranes, but this has yet to be studied in any cell. In *Paramecium,* we know that hypertonic and cold stress result in the disruption of the V-ATPase-rich membranes (Section IV) and in their disappearance from around the smooth spongiome (Fig. 10C, D). This ability to break away from the rest of the CV, when subjected to stress, may account for the need for membrane trafficking proteins such as RabD on these tubular membranes. Presumably, a complete complement of recognition proteins, SNARES, and fusion proteins may also be associated with these membranes.

CVs of *Dictyostelium* have also been implicated in a nonclassical protein transport mechanism for moving the 24-kDa Ca^{2+}-dependent cell–cell adhesion molecule, DdCAD-1, from the cytosol to the cell surface during development (Sesaki *et al.,* 1997). Isolated CVs import soluble DdCAD-1 into their lumens in a ATP-stimulated manner, and this protein is then deposited onto the cell surface during CV expulsion.

Another possible function of the CVC in *Dictyostelium* is in Ca^{2+} regulation (Milne and Coukell, 1988, 1989; Rooney and Gross, 1992; Rooney *et al.,* 1994). A gene for a P-type Ca^{2+}-ATPase (PAT1) was cloned, and its protein product was colocalized with bound calmodulin to membranes of the CV (Moniakis *et al.,*

1995). This enzyme may have an important role in regulating Ca^{2+} homeostasis in this soil amoeba, particularly when the cell is under stress (Moniakis *et al.*, 1999).

IV. Adaptations to a Changing External Environment

In short-term experiments, protozoa show an inverse relation between the osmotic concentration of the medium to which the protozoan is subjected and the rate of vacuolar output. *Acanthamoeba castellanii,* subjected to increasing concentrations of nonpenetrating solutes such as sucrose, showed little CV activity when the sucrose added to the external medium reached 70 mM. The internal osmotic pressure was estimated to be 70 mM over that of the medium. At this point, the cell's internal osmolarity was assumed to be nearly isotonic with the external medium, as CVC activity essentially stopped (Pal, 1972).

In a much more comprehensive study on the effects of both short-term and longer-term adaptations to environments of different osmolarities, it was observed in *Paramecium* that, as expected, the rate of fluid expulsion increased when the cell was placed in very hypotonic solutions, and this rate of expulsion was less when the cell was exposed to a less hypotonic solution. Expulsions of the CV then fell to zero in isotonic or hypertonic solutions. For short-term changes in external osmolarity, the time required for the cell to adjust to a strongly hypotonic solution was shorter than the time required to adjust to a less hypotonic solution, and the adjustment time became very long in isotonic or hypertonic solutions (Stock *et al.*, 2001).

The *Paramecium* cell always adjusts its cytosolic osmolarity to be hypertonic to the environment. The rate of fluid expulsion is regulated to keep the cytosol hypertonic both in the presence of a hypotonic environment as well as in more hypertonic environments. With the cell placed in a hypertonic environment, the CV activity initially stopped, but after 12 hr, the CV activity resumed even though the cell was kept continuously in the hypertonic solution. The cytosolic osmolarity had increased and remained hypertonic to the outside. Thus the cell maintained its ability to acquire water even under high hypertonic conditions (Stock *et al.*, 2001). These increases in cytosolic osmolarity of *Paramecium,* however, did not proceed linearly but occurred stepwise (Fig. 18). After adapting to an increased hypertonic environment, CV activity would always begin again.

Thus, at least in *Paramecium,* it appears that CV activity must be maintained even though the cell is placed in an environment which is much higher in osmolarity than that in which it normally lives and for which it would seem, *a priori,* that water need not be eliminated from the cell. Thus the CVC may be needed not only to regulate water but also to eliminate some waste substance (Stock *et al.*, 2001) or for transport of some product to the cell surface. This additional function would presumably require that the CV remain active even at abnormally high external osmolarities.

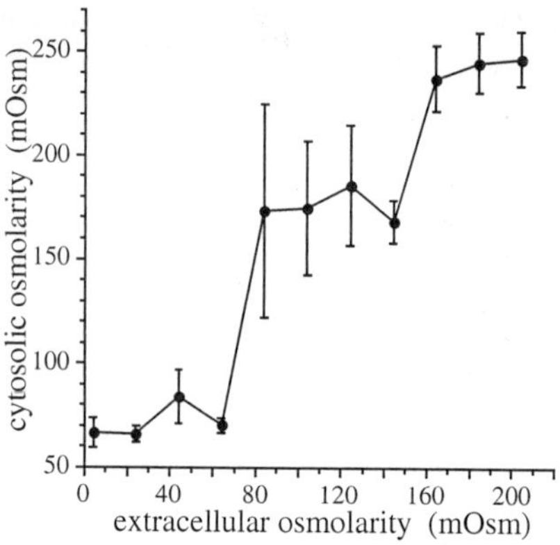

FIG. 18 The cytosolic osmolarity of *Paramecium* changes as a function of the external medium in a stepwise manner. The cell keeps its CVC functional even at high external osmolarities. From Stock *et al.* (2001).

Osmotic stress caused by the addition of 150 mMsorbitol to the axenic medium, producing a final osmolarity of 230 mOsm, caused the CVs of *Paramecium* to cease activity (Ishida *et al.,* 1996). By labeling with fluorochrome-complexed monoclonal antibodies, it was possible to show that the V-ATPase holoenzymes of the decorated spongiomes had been disassembled and were presumably no longer functional (Fig. 10C, D). In fact, intact 50-nm tubules were no longer visible in electron micrographs. Many small vesicles which did not exhibit immunogold labeling with the V-ATPase antibodies were present in the region normally occupied by the 50-nm tubules. The smooth spongiome, consisting of the collecting canals, CV, ampullae, and crytalline-like membranes of the radial arms, was still present in an unaltered morphology. Thus the cell responded to hypertonic stress by disassembling the V-ATPase-rich membrane, that is, not only by releasing the V_1 sectors from the holoenzymes but by vesiculating the V-ATPase-rich membrane. Hypertonic stress causes vesiculation of CV membranes also in *Dictyostelium* (Zhu and Clarke, 1992) and *Chlamydomonas* (Luykx *et al.,* 1997a).

However, when *Paramecium* was returned to its normal axenic medium, the CV activity began to return within 2 hr as did the decorated tubules. Even if the cell remained in the hypertonic medium, CV activity started to return at 3 hr (Fig. 10E, F) and stabilized by 10 hr at a somewhat lower level than when in axenic medium alone (Ishida *et al.,* 1996). A model for the formation of the 50-nm tubules from planar membrane containing V_0 sectors would suggest that the first step in the return of CV activity would be the attachment of V_1 sectors to transmembrane V_0 sectors. Once attached to the V_0 sectors to form holoenzymes, the V_1 sectors may then link to other V_1 sectors in such a way as to form a double row of enzymes that curve into a helical path that pulls the lipid bilayer out into a straight 50-nm tubule (Fig. 19) (Allen, 1995).

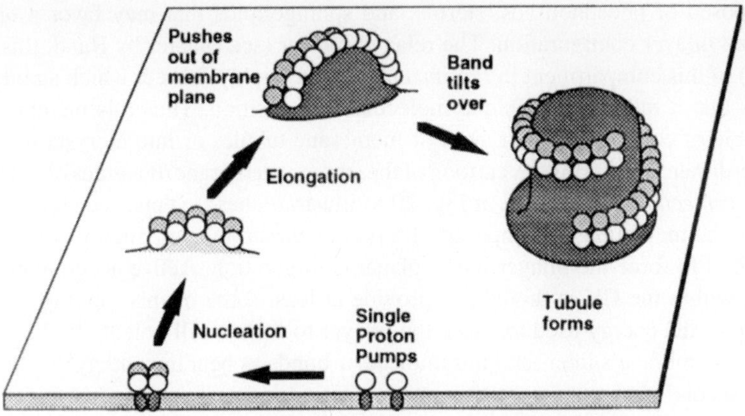

FIG. 19 A cartoon illustrating a possible model for 50-nm tubule formation of the decorated spongiome in *Paramecium*. The major assumption is that V_1 sectors or V-ATPase holoenzymes associate laterally within the plane of the membrane, resulting in rows of enzymes that favor a helical pattern. From Allen (1995).

Thus, at least part of the recovery time of the CV from hyperosmotic stress (Ishida *et al.,* 1996) and cold stress (Fok *et al.,* 1995) consists of the time it takes for the V_1 sectors to reattach to V_0 sectors and to link together laterally to re-form the functioning 50-nm tubules, the CVC's electrogenic engines.

V. Conclusions

Although CVCs vary greatly in size and complexity, they all seem to fit a general plan of structure and activity. CVCs are all constructed on a two-membrane or a two-compartment plan. The first compartment is surrounded by a membrane studded with V-ATPases and has a rather constant, reduced, and essentially non-expandable volume. The second compartment lacks the complete and functional V-ATPases, and its lumen expands with the accumulation of fluid. The V-ATPase-rich compartment in the smallest CVCs may be reduced to simple evaginations of the CV membrane, while in ciliates such as *Paramecium* this part of the CVC is constructed of an elaborate array of bundles of 50-nm tubules. The V-ATPases on the membrane of this compartment are electrogenic. The protons accumulate in the ionic form only to the extent that they form an internal positive charge. In *Paramecium* this charge is +60 mV with respect to the cytosol. The membrane of this compartment does not fuse with the plasma membrane, nor does this compartment serve as a reservoir to store fluid.

The second compartment serves as the fluid reservoir and exhibits unique dynamics not found in other membranes. In *Dictyostelium,* this membrane is composed of phospholipids, sterols, and sphingolipids that may favor a highly curved bilayer configuration. The relaxed bilayer (see chapter by Rand, this volume) of this compartment in *Paramecium* is also highly curved, which again suggests that it may have a unique molecular composition. Thus this membrane is capable of collapsing into a mass of membrane tubules or into a crystalline-like three-dimensional array. (A cartoon of the smooth membrane dynamics of the CVC of *Paramecium* is presented in Fig. 20 to illustrate these events; see color plate.) Since the membrane is composed of a highly curved bilayer, energy is probably required to force the bilayer into a planar configuration. Active accumulation of fluid within the CV is thought to provide at least some of this input of energy. Some of the energy used to cause the bilayer to flatten will potentially be stored in the membrane's inter- and intramolecular bonds as bending energy.

A second phenomenon requires an additional input of energy. Just before the CV fuses with the plasma membrane, the CV rounds up into a sphere. This phenomenon is best observed and studied in isolated CVs of *Paramecium* where rounding/relaxing cycles continue for many minutes if essentially undiluted cytosol and a supply of ATP remain around the *in vitro* CV. The mechanism producing this rounding process appears to be located within the bilayer itself, or possibly lies within a layer of closely associated peripheral membrane proteins.

The possibility that dynamin, a protein thought to be involved in fission of clathrin-coated pits, might be somehow involved in CV membrane tubulation was considered. However, in a study of *Dictyostelium,* in which the gene for dynamin A was lacking, the CV of this cell remained fully functional (Wienke *et al.,* 1999).

During rounding, the tension of the CV membrane in *Paramecium* increases 35-fold. This increase in tension is presumably important for promoting fusion of the CV membrane with the plasma membrane, in setting the rate of expulsion by determining the diameter of the pore, and for separating the CV from the microtubular-bound membranes along the radial arms.

Once fusion with the plasma membrane has occurred, the CV collapses. Fluid excretion is propelled largely by cytosolic pressure (Naitoh *et al.,* 1997a). During this process, the CV disappears from view in the light microscope. Disappearance is due to the return of the planar membrane to a tubular topography that forms an unresolvable mass of interconnected membranes. This change in membrane topography is potentially driven, in part, by the release of bending energy stored in the planar membrane.

Following rounding, the tension of the *in vivo* CV membrane presumably decreases, and fission of the CV membrane from the plasma membrane becomes possible. Relaxation of tension may allow the CV membrane to approach and to fuse again with the smooth membrane of the collecting canal, still attached to the microtubules along the radial canal. The period of increased tension in *in vitro* CVs is about the same length of time as that occupied by rounding, CV expulsion, and the reappearance of a constant membrane potential in a CV that is penetrated with a microelectrode. One effect of inserting a microelectrode into the cell is to

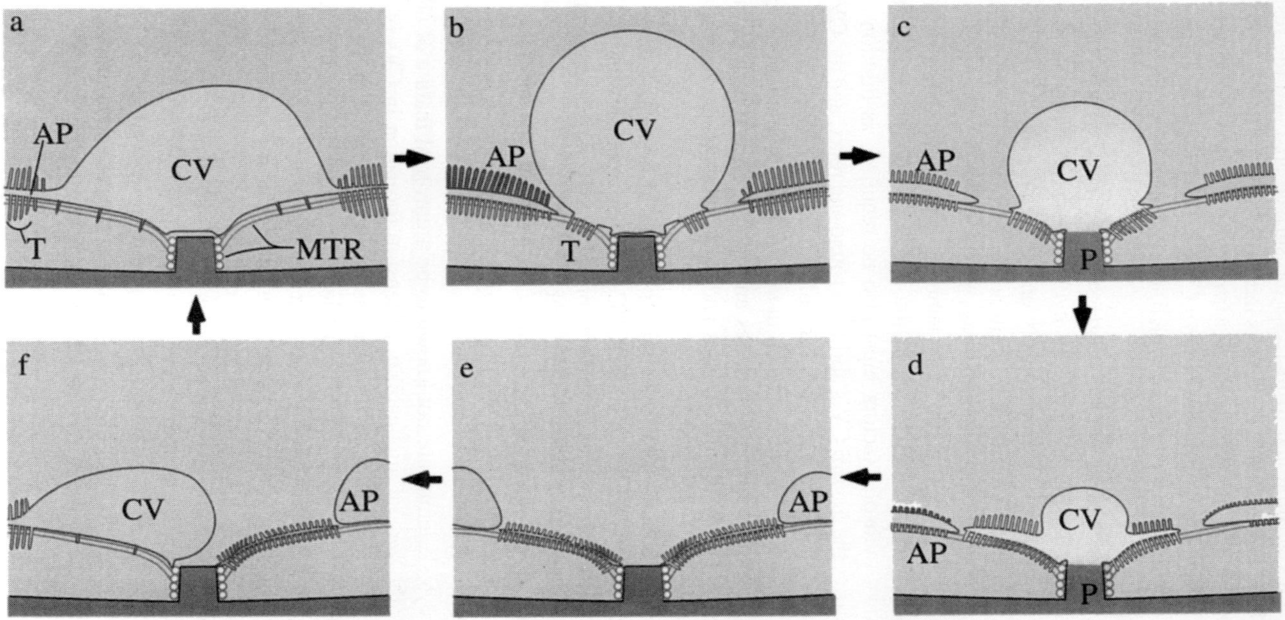

FIG. 20 A cartoon illustrating a CVC cycle in *Paramecium*. This cartoon emphasizes the tubulation of the smooth spongiome that occurs in association with microtubular ribbons. (a) filling of the CV; (b) rounding of the CV resulting in dissociation of arms; (c) opening of the pore and fluid expulsion; ; (d) tubulation of CV membrane; (e) pore closure and ampullar swelling; (f) CV membrane re-fuses with ampullae, stepwise. AP, ampulla; T, tubule; MTR, microtubular ribbon; P, CV pore. From Tominaga *et al.* (1999). (See also color insert.)

quadruple the CV cycle time and to increase the surface area and the resulting volume of the CV. Thus the total cycle time is increased from ~10 to ~45 sec. The cycle time of tension development of the *in vitro* CV is also around 45 sec. Thus cycling of tension development in the *in vitro* CV may represent the unregulated default process of tension development which, *in vivo,* usually occurs at a more rapid pace.

Thus, one critical mechanism yet to be understood is what causes this profound increase in membrane tension and how this cyclic tension increase and decrease are regulated. This mechanism does not rely on the electrogenic V-ATPases nor does it rely on changes in luminal fluid volume. *In vitro* experiments show that neither V-ATPases nor fluid increase are necessary for rounding/relaxing cycles to continue. An association between the CV membrane and microtubular ribbons is probably involved in *in vivo* tension regulation in *Paramecium.* ATP is required, and other unknown cytosolic factors must also be required, as dilution of the cytosol soon destroys the *in vitro* CV's ability to round and relax, and the CV ends up in a state of rigor, that is, in a rounded state of high tension. Thus, energy input is important for relaxing membrane tension in *in vitro* CVs. One might conclude, then, that tubulation of membranes requires an input of energy both to initiate tubulation in association with microtubules and to bring about the relaxation of membrane tension.

Though the above scheme of CV activity is based in large part on what is known about the CVC of *Paramecium,* the CVCs of other organisms may follow a similar scheme. Indeed, the essential observations of a two-membrane system has been confirmed in *Dictyostelium* (Nolta and Steck, 1994) and zoospores (Mitchell and Hardham, 1999), and the apparent absence of an actomyosin-like contractile system influenced Heuser *et al.* (1993) to suggest another mechanism for CV contraction in *Dictyostelium.*

Microtubular cytoskeletons are not known to be a universal feature of CVs in most cells. Thus the general applicability of tension formation in CVs of other cells and the role of microtubular interactions during the initiation of tension development in these other cells should be studied where possible. For the present, we are assuming, until it is proven otherwise, that all CVs share similar fundamental mechanisms of activity and function. These common mechanisms can be identified and understood only if comparative studies are carried out.

Although considerable progress in understanding the structure of the CVC and especially the dynamics of its membrane has been made, we still have a way to go to understand the CVC's role in osmoregulation. The small size of most CVCs makes it difficult to sample the CV's fluid and ion content and to determine its osmolarity. Earlier work relied on micropuncture to collect the CV fluid whose osmolarity could then be determined by freezing-point depression techniques. Riddick (1968), for instance, used a helium-glow photometer to determine the concentrations of Na^+ and K^+ in the collected fluid. Apparently, no one had successfully attempted to repeat these observations on other protozoa until our attempts to do so in *Paramecium*. We are using the somewhat demanding technique of ion-selective microelectrodes to determine the individual ion species in the CV. While we have preliminary results, these remain incomplete and may change as more data are obtained. Clearly, much work remains to be done to determine the composition of the CV. Knowing what ions the CV contains, and the relative concentrations of these ions in the CVC as compared to that in the cytosol, will help greatly in the search for specific ion channels, cotransporters, and antiporters in the CVC's membranes. Only then will it be possible to understand the actual mechanisms used by protozoa in carrying out osmoregulation.

Acknowledgments

Studies on the CVC from our laboratory over the past 12 years were carried out by Drs. M. Ishida, T. Tominaga, T. Tani, C. Stock, and A. K. Fok. M. S. Aihara has provided valued technical support. NSF has provided financial support, most recently through Grant MCB 98-09929.

References

Aaronson, S., and Behrens, U. (1974). Ultrastructure of an unusual contractile vacuole in several chrysomonad phytoflagellates. *J. Cell Sci.* **14,** 1–9.

Akbarieh, M., and Couillard, P. (1988). Ultrastructure of the contractile vacuole and its periphery in *Amoeba proteus:* Evolution of vesicles during the cycle. *J. Protozool.* **35,** 99–108.

Allen, R. D. (1995). Membrane tubulation and proton pumps. *Protoplasma* **189,** 1–8.

Allen, R. D. (2000). The contractile vacuole and its membrane dynamics. *BioEssays* **22,** 1035–1042.

Allen, R. D., and Fok, A. K. (1983). Nonlysosomal vesicles (acidosomes) are involved in phagosome acidification in *Paramecium. J. Cell Biol.* **97,** 566–570.

Allen, R. D., and Fok, A. K. (1988). Membrane dynamics of the contractile vacuole complex of *Paramecium. J. Protozool.* **35,** 63–71.

Allen, R. D., Ueno, M. S., Pollard, L. W., and Fok, A. K. (1990). Monoclonal antibody study of the decorated spongiome of contractile vacuole complexes of *Paramecium. J. Cell Sci.* **96,** 469–475.

Belda, W. H. (1943). Permeability to water in *Pelomyxa carolinensis*. III. The permeability constant for water in *Pelomyxa carolinensis*. *Salesianum* **38,** 17–24.

Bowers, B., and Korn, E. D. (1968). The fine structure of *Acanthamoeba castellanii* I. The trophozoite. *J. Cell Biol.* **39,** 95–111.

Bowers, B., and Korn, E. D. (1973). Cytochemical identification of phosphatase activity in the contractile vacuole of *Acanthamoeba castellanii. J. Cell Biol.* **59,** 784–791.

Bowman, E. J., Siebers, A., and Altendorf, K. (1988). Bafilomycins: A class of inhibitors of membrane ATPases from microorganisms, animal cells and plant cells. *Proc. Natl. Acad. Sci. USA* **85,** 7972–7976.

Bruce, D. L., and Marshall, J. M. Jr. (1965). Some ionic and bioelectric properties of the ameba *Chaos chaos*. *J. Gen. Physiol.* **49,** 151–178.

Bush, J., Nolta, K., Rodriguez-Paris, J., Kaufmann, N., O'Halloran, T., Ruscetti, T., Temesvari, L., Steck, T., and Cardelli, J. (1994). A Rab4-like GTPase in *Dictyostelium discoideum* colocalizes with V-H^+-ATPases in reticular membranes of the contractile vacuole complex and in lysosomes. *J. Cell Sci.* **107,** 2801–2812.

Cameron, I. L., and Burton, A. L. (1969). On the cycle of the water expulsion vesicle in the ciliate *Tetrahymena pyriformis*. *Trans. Am. Microsc. Soc.* **88,** 386–393.

Cho, C. W., and Fuller, M. S. (1989). Observations of the water expulsion vacuole of *Phytophthora palmivora*. *Protoplasma* **149,** 47–55.

Couillard, P., Forget, J., and Pothier, F. (1985). The contractile vacuole of *Amoeba proteus*. III. Vacuolar response to phagocytosis. *J. Protozool.* **32,** 333–338.

Dallai, R., Luporini, P., and Salvatici, P. (1985). Plasma membrane specialization at the discharge site of the excretory pore-less vacuole of the ciliate *Euplotes raikovi*. *Tissue Cell* **17,** 309–320.

de Chastellier, C., Quiviger, B., and Ryter, A. (1978). Observations of the functioning of the contractile vacuole of *Dictyostelium discoideum* with the electron microscope. *J. Ultrastruct. Res.* **62,** 220–227.

Denning, G. M., and Fulton, A. B. (1989). Electron microscopy of a contractile-vacuole mutant of *Chlamydomonas moewusii* (Chlorophyta) defective in the late stages of diastole. *J. Phycol.* **25,** 667–672.

Döbereiner, H.-G., Selchow, O., and Lipowsky, O. (1999). Spontaneous curvature of fluid vesicles induced by trans-bilayer sugar asymmetry. *Eur. Biophys. J.* **28,** 174–178.

Doberstein, S. K., Baines, I. C., Wiegand, G., Korn, E. D., and Pollard, T. D. (1993). Inhibition of contractile vacuole function *in vivo* by antibodies against myosin-I. *Nature* **365,** 841–843.

Dunham, P. B., and Kropp, D. L. (1973). Regulation of solutes and water *Tetrahymena*. *In* "Biology of *Tetrahymena*" (A. M. Elliott, Ed.), pp. 165–198. Dowden, Hutchinson and Ross, Stroudsburg, Pennsylvania.

Dunham, P. B., and Stoner, L. C. (1969). Indentation of the pellicle of *Tetrahymena* at the contractile vacuole pore before systole. *J. Cell Biol.* **43,** 184–188.

Elliott, A. M., and Bak, I. J. (1964). The contractile vacuole and related structures in *Tetrahymena pyriformis*. *J. Protozool.* **11,** 250–261.

Elliott, A. M., and Kennedy, J. R. (1973). Morphology of *Tetrahymena*. *In* "Biology of *Tetrahymena*" (A. M. Elliott Ed.), pp. 57–87. Dowden, Hutchinson and Ross, Stroudsburg, Pennsylvania.

Flick, K. M., Shaulsky, G., and Loomis, W. F. (1997). The *wacA* gene of *Dictyostelium discoideum* is a developmentally regulated member of the MIP famity. *Gene* **195,** 127–130.

Fok, A. K., Clarke, M., Ma, L., and Allen, R. D. (1993). Vacuolar H^+-ATPase of *Dictyostelium discoideum*. A monoclonal antibody study. *J. Cell Sci.* **106,** 1103–1113.

Fok, A. K., Aihara, M. S., Ishida, M., Nolta, K. V., Steck, T. L., and Allen, R. D. (1995). The pegs on the decorated tubules of the contractile vacuole complex of *Paramecium* are proton pumps. *J. Cell Sci.* **108,** 3163–3170.

Furukawa, R., and Fechheimer, M. (1994). Differential localization of α-actinin and the 30 kD actin-bundling protein in the cleavage furrow, phagocytic cup, and contractile vacuole of *Dictyostelium discoideum*. *Cell Motil. Cytoskel.* **29,** 46–56.

Giglione, C., and Gross, J. D. (1995). Anion effects on vesicle acidification in *Dictyostelium*. *Biochem. Mol. Biol. Int.* **36,** 1057–1065.

Gingell, D., Todd, I., and Owens, N. (1982). Interaction between intracellular vacuoles and the cell surface analysed by finite aperture theory interference reflection microscopy. *J. Cell Sci.* **54,** 287–298.

Hausmann, K. (1983). Zur Organisation der kontraktilen Vakuole von *Pseudomicrothorax dubius* (Ciliophora, Kinetofragminophorea, Hypostomatia). *Arch. Protistenk.* **127,** 319–325.

Hausmann, K., and Allen, R. D. (1977). Membranes and microtubules of the excretory apparatus of *Paramecium caudatum*. *Cytobiologie* **15,** 303–320.

Hausmann, K., and Patterson, D. J. (1981). Involvement of smooth and coated vesicles in the function of the contractile vacuole complex of some cryptophycean flagellates. *Exp. Cell Res.* **135,** 449–453.

Hausmann, K., and Patterson, D. J. (1984). Contractile vacuole complexes in algae. *In* "Compartments in Algal Cells and Their Interaction" (W. Wiessner, D. Robinson, and R. C. Starr, Eds.), pp. 139–146. Springer-Verlag, Heidelberg.

Hellebust, J. A., Mérida, T., and Ahmad, I. (1989). Operation of contractile vacuoles in the euryhaline green flagellate *Chlamydomonas pulsatilla* (Chlorophyceae) as a function of salinity. *Mar. Biol.* **100,** 373–379.

Heuser, J., Zhu, Q., and Clarke, M. (1993). Proton pumps populate the contractile vacuoles of *Dictyostelium* amoebae. *J. Cell Biol.* **121,** 1311–1327.

Heywood, P. (1973). Nutritional studies on the Chloromonadophyceae: *Vacuolaria virescens* and *Gonyostomum semen*. *J. Phycol.* **9,** 156–159.

Heywood, P. (1978). Osmoregulation in the alga *Vacuolaria virescens*. Structure of the contractile vacuole and the nature of its association with the Golgi apparatus. *J. Cell Sci.* **31,** 213–224.

Hoch, H. C., and Mitchell, J. E. (1972). The ultrastructure of zoospores of *Aphanomyces euteiches* and of their encystment and subsequent germination. *Protoplasma* **75,** 113–138.

Hui, S.-W., and Sen, A. (1989). Effects of lipid packing on polymorphic phase behaviour and membrane properties. *Proc. Natl. Acad. Sci. USA* **86,** 5825–5829.

Ishida, M., Aihara, M. S., Allen, R. D., and Fok, A. K. (1993). Osmoregulation in *Paramecium:* The locus of fluid segregation in the contractile vacuole complex. *J. Cell Sci.* **106,** 693–702.

Ishida, M., Fok, A. K., Aihara, M. S., and Allen, R. D. (1996). Hyperosmotic stress leads to reversible dissociation of the proton pump-bearing tubules form the contractile vacuole complex in *Paramecium*. *J. Cell Sci.* **109,** 229–237.

Ishida, M., Aihara, M. S., Allen, R. D., and Fok, A. K. (1997). Acidification of the young phagosomes of *Paramecium* is mediated by proton pumps derived from the acidosomes. *Protoplasma* **196,** 12–20.

Josefsson, J.-O. (1966). Some bioelectrical properties of *Amoeba proteus*. *Acta Physiol. Scand.* **66,** 395–405.

Kaneshiro, E. S., Dunham, P. B., and Holz, G. G. (1969). Osmoregulation in a marine ciliate, *Miamiensis avidus*. I. Regulation of inorganic ions and water. *Biol. Bull.* **136,** 63–75.

Kitching, J. A. (1948). The physiology of contractile vacuoles. V. The effects of short-term variations of temperature on a freshwater peritrich ciliate. *J. Exp. Biol.* **25,** 406–420.

Landh, T. (1995). From entangled membranes to eclectic morphologies: Cubic membranes as subcellular space organizers. *FEBS Lett.* **369,** 13–17.

Linder, J. C., and Staehelin, L. A. (1979). A novel model for fluid secretion by the trypanosomatid contractile vacuole apparatus. *J. Cell Biol.* **83,** 371–382.

Linder, J. C., and Staehelin, L. A. (1980). The membrane lattice: a novel organelle of the trypanosomatid flagellate *Leptomonas collosoma*. *J. Ultrastruct. Res.* **72,** 200–205.

Lipowsky, R. (1999). From membranes to membrane machines. *In* "Statistical Mechanics of Biocomplexity" (D. Reguera, J. M. G. Vilar, J. M. Rubi, Eds.), Lecture Notes in Physics, Vol. **527,** pp 1–22. Springer, Berlin.

Liu, T., and Clarke, M. (1996). The vacuolar proton pump of *Dictyostelium discoideum:* Molecular cloning and analysis of the 100 kDa subunit. *J. Cell Sci.* **109,** 1041–1051.

Luykx, P., Hoppenrath, M., and Robinson, D. G. (1997a). Structure and behavior of contractile vacuoles in *Chlamydomonas reinhardtii*. *Protoplasma* **198,** 73–84.

Luykx, P., Hoppenrath, M., and Robinson, D. G. (1997b). Osmoregulatory mutants that affect the function of the contractile vacuole in *Chlamydomonas reinhardtii*. *Protoplasma* **200,** 99–111.

Mast, S. O., and Fowler, C. (1935). Permeability of *Amoeba proteus* to water. *J. Cell. Comp. Physiol.* **6,** 151–167.

McKanna, J. A. (1973a). Membrane recycling: Vesiculation of the amoeba contractile vacuole at systole. *Science* **179,** 88–90.

McKanna, J. A. (1973b). Fine structure of the contractile vacuole pore in *Paramecium*. *J. Protozool.* **20,** 631–638.

McKanna, J. A. (1974). Permeability modulating membrane coats. I. Fine structure of fluid segregation organelles of peritrich contractile vacuoles. *J. Cell Biol.* **63,** 317–322.

McKanna, J. A. (1976). Fine structure of fluid segregation organelles of *Paramecium* contractile vacuoles. *J. Ultrastruct. Res.* **54,** 1–10.

Meinild, A.-K., Klaerke, D. A., Loo, D. D. F., Wright, E. M., and Zeuthen, T. (1998). The human Na^+–glucose cotransporter is a molecular water pump. *J. Physiol.* **508,** 15–21.

Milne, J. L., and Coukell, M. B. (1988). Isolation and characterization of a plasma membrane calcium pump from *Dictyostelium discoideum*. *Biochem. J.* **249,** 223–230.

Milne, J. L., and Coukell, M. B. (1989). Identification of a high-affinity Ca^{2+} pump associated with endocytotic vesicles in *Dictyostelium discoideum*. *Exp. Cell Res.* **185,** 21–32.

Mitchell, H. J., and Hardham, A. R. (1999). Characterisation of the water expulsion vacuole in *Phytophthora nicotianae* zoospores. *Protoplasma* **206,** 118–130.

Moniakis, J., Coukell, M. B., and Forer, A. (1995). Molecular cloning of an intracellular P-type ATPase from *Dictyostelium* that is up-regulated in calcium-adapted cells. *J. Biol. Chem.* **270,** 28,276–28,281.

Moniakis, J., Coukell, M. B., and Janiec, A. (1999). Involvement of the Ca^{2+}-ATPase PAT1 and the contractile vacuole in calcium regulation in *Dictyostelium discoideum*. *J. Cell Sci.* **112,** 405–414.

Naitoh, Y., Tominaga, T., Ishida, M., Fok, A. K., Aihara, M. S., and Allen, R. D. (1997a). How does the contractile vacuole of *Paramecium multimicronucleatum* expel fluid? Modelling the expulsion mechanism. *J. Exp. Biol.* **200,** 713–721.

Naitoh, Y., Tominaga, T., and Allen, R. D. (1997b). The contractile vacuole fluid discharge rate is determined by the vacuole size immediately before the start of discharge in *Paramecium multimicronucleatum*. *J. Exp. Biol.* **200,** 1737–1744.

Nolta, K. V., and Steck, T. L. (1994). Isolation and initial characterization of the bipartite contractile vacuole complex from *Dictyostelium discoideum*. *J. Biol. Chem.* **269,** 2225–2233.

Nolta, K. V., Padh, H., and Steck, T. L. (1991). Acidosomes from *Dictyostelium*. Initial biochemical characterization. *J. Biol. Chem.* **266,** 18,318–18,323.

Nolta, K. V., Padh, H., and Steck, T. L. (1993). An immunocytochemical analysis of the vacuolar proton pump in *Dictyostelium discoideum*. *J. Cell Sci.* **105,** 849–859.

O'Halloran, T. J., and Anderson, R. G. W. (1992). Clathrin heavy chain is required for pinocytosis, the presence of large vacuoles, and development in *Dictyostelium*. *J. Cell Biol.* **118,** 1371–1377.

Padh, H., Lavasa, M., and Steck, T. L. (1989). Characterization of a vacuolar proton ATPase in *Dictyostelium discoideum*. *Biochim. Biophys. Acta* **982,** 271–278.

Pal, R. A. (1972). The osmoregulatory system of the amoeba, *Acanthamoeba castellanii*. *J. Exp. Biol.* **57,** 55–76.

Patterson, D. J. (1977). On the behaviour of the contractile vacuoles and associated structures of *Paramecium caudatum* (Ehrbg.). *Protistologica* **13,** 205–212.

Patterson, D. J. (1980). Contractile vacuoles and associated structures: their organization and function. *Biol. Rev.* **55,** 1–46.

Patterson, D. J., and Hausmann, K. (1981). The behaviour of contractile vacuole complexes of cryptophycean flagellates. *Brit. Phycol. J.* **16,** 429–439.

Patterson, D. J., and Hausmann, K. (1982). Coated vesicle hererogeneity in some cryptophycean flagellates. *Cell Biol. Int. Rep.* **6,** 513–521.

Preston, R. R., Kink, J. A., Hinrichsen, R. D., Saimi, Y., and Kung, C. (1991). Calmodulin mutants and Ca^{++}-dependent channels in *Paramecium*. *Annu. Rev. Physiol.* **53,** 309–319.

Prusch, R. D. (1977). Protozoan osmotic and ionic regulation. *In* "Transport of Ions and Water in Animals" (B. L. Gupta, R. B. Moreton, J. L. Oschman, and B. J. Wall, Eds.), pp. 363–377. Academic Press, London.

Prusch, R. D., and Dunham, P. B. (1967). Electrical and contractile properties of the isolated contractile vacuole of *Amoeba proteus*. *J. Gen. Physiol.* **50,** 1083.

Prusch, R. D., and Dunham, P. B. (1970). Contraction of isolated contractile vacuoles from *Amoeba proteus*. *J. Cell Biol.* **46,** 431–434.

Prusch, R. D., and Dunham, P. B. (1972). Ionic distribution in *Amoeba proteus*. *J. Exp. Biol.* **56,** 551–563.

Quiviger, B., de Chastellier, C., and Ryter, A. (1978). Cytochemical demonstration of alkaline phosphatase in the contractile vacuole of *Dictyostelium discoideum*. *J. Ultrastruct. Res.* **62,** 228–236.

Riddick, D. H. (1968). Contractile vacuole in the amoeba, *Pelomyxa carolinensis*. *Am. J. Physiol.* **215,** 736–740.

Robinson, D. G., Hoppenrath, M., Oberbeck, K., Luykx, P., and Ratajczak, R. (1998). Localization of pyrophosphatase and V-ATPase in *Chlamydomonas reinhardtii*. *Bot. Acta* **111,** 108–122.

Rodriguez-Paris, J. M., Nolta, K. V., and Steck, T. L. (1993). Characterization of lysosomes isolated form *Dictyostelium discoideum* by magnetic fraction. *J. Biol. Chem.* **268,** 9110–9116.

Rooney, E. K., and Gross, J. D. (1992). ATP-driven Ca^{2+}/H^+ antiport in acid vesicles from *Dictyostelium*. *Proc. Natl. Acad. Sci. USA* **89,** 8025–8029.

Rooney, E. K., Gross, J. D., and Satre, M. (1994). Characterization of an intracellular Ca^{2+} pump in *Dictyostelium*. *Cell Calcium* **16,** 509–522.

Schmidt-Nielsen, B., and Schrauger, C. R. (1963). *Amoeba proteus:* Studying the contractile vacuole by micropuncture. *Science* **139,** 606–607.

Schneider, L. (1960). Elektronenmikroskopische Untersuchungen über das Nephridialsystem von *Paramaecium*. *J. Protozool.* **7,** 75–90.

Sesaki, H., Wong, E. F. S., and Siu, C.-H. (1997). The cell adhesion molecule DdCAD-1 in *Dictyostelium* is targeted to the cell surface by a nonclassical transport pathway involving contractile vacuoles. *J. Cell Biol.* **138,** 939–951.

Shiono, H., Hara, R., and Asai, H. (1980). Spontaneous membrane potential changes associated with the zooid and vacuolar contractions in *Vorticella convallaria*. *J. Protozool.* **27,** 83–87.

Soldo, A. T., Godoy, G. A., and van Wagtendonk, W. J. (1966). Growth of particle-bearing and particle-free *Paramecium aurelia* in axenic culture. *J. Protozool.* **13,** 492–497.

Stock, C., Allen, R. D., and Naitoh, Y. (2001). How external osmolarity affects the contractile vacuole complex activity, the cytosolic osmolarity and the water permeability of the plasma membrane in *Paramecium multimicronucleatum*. *J. Exp. Biol.* **204,** 291–304.

Tani, T., Allen, R. D., and Naitoh, Y. (2000). Periodic tension development in the membrane of the *in vitro* contractile vacuole of *Paramecium multimicronucleatum:* Modification by bisection, fusion and suction. *J. Exp. Biol.* **203,** 239–251.

Tani, T., Allen, R. D., and Naitoh, Y. (2001). Cellular membranes that undergo cyclic changes in tension: Direct measurement of force generation by an in vitro contractile vacuole of *Paramecium multimicronucleatum*. *J. Cell Sci.* **114,** 785–795.

Temesvari, L. A., Rodriguez-Paris, J. M., Bush, J. M., Zhang, L., and Cardelli, J. A. (1996). Involvement of the vacuolar proton-translocating ATPase in multiple steps of the endo-lysosomal system and in the contractile vacuole system of *Dictyostelium discoideum*. *J. Cell Sci.* **109,** 1479–1495.

Tominaga, T., Allen, R. D., and Naitoh, Y. (1998a). Electrophysiology of the *in situ* contractile vacuole complex of *Paramecium* reveals its membrane dynamics and electrogenic site during osmoregulatory activity. *J. Exp. Biol.* **201,** 451–460.

Tominaga, T., Allen, R. D., and Naitoh, Y. (1998b). Cyclic changes in the tension of the contractile vacuole complex membrane control its exocytotic cycle. *J. Exp. Biol.* **201,** 2647–2658.

Tominaga, T., Naitoh, Y., and Allen, R. D. (1999). A key function of non-planar membranes and their associated microtubular ribbons in contractile vacuole membrane dynamics is revealed by electrophysiologically controlled fixation of *Paramecium*. *J. Cell Sci.* **112,** 3733–3745.

Weiss, R. L. (1983). Coated vesicles in the contractile vacuole/mating structure region of *Chlamydomonas*. *J. Ultrastruct. Res.* **85,** 33–44.

Weiss, R. L., Goodenough, D. A., and Goodenough, U. W. (1977). Membrane particle arrays associated with the basal body and with contractile vacuole secretion in *Chlamydomonas*. *J. Cell Biol.* **72,** 133–143.

Wessel, D., and Robinson, D. G. (1979). Studies on the contractile vacuole of *Poterioochromonas malhamensis* Peterfi. I. The structure of the alveolate vesicles. *Eur. J. Cell Biol.* **19,** 60–66.

Wieczorek, H., Brown, D., Grinstein, S., Ehrenfeld, J., and Harvey, W. R. (1999). Animal plasma membrane energization by proton-motive V-ATPases. *BioEssays* **21,** 637–648.

Wienke, D. C., Knetsch, M. L. W., Neuhaus, E. M., Reedy, M. C., and Manstein, D. J. (1999). Disruption of a dynamin homologue affects endocytosis, organelle morphology, and cytokinesis in *Dictyostelium discoideum*. *Mol. Biol. Cell* **10,** 225–243.

Wigg, D., Bovee, E. C., and Jahn, T. L. (1967). The evacuation mechanism of the water expulsion vesicle ("contractile vacuole") of *Amoeba proteus*. *J. Protozool.* **14,** 104–108.

Woo, J., Shinohara, C., Sakai, K., Hasumi, K., and Endo, A. (1992). Inhibition of the acidification of endosomes and lysosomes by the antibiotic concanamycin B in macrophage J774. *Eur. J. Biochem.* **207,** 383–389.

Xie, Y., Coukell, M. B., and Gombos, Z. (1996). Antisense RNA inhibition of the putative vacuolar H^+-ATPase proteolipid of *Dictyostelium* reduces intracellular Ca^{2+} transport and cell viability. *J. Cell Sci.* **109,** 489–497.

Yamaguchi, T. (1960). Studies on the modes of ionic behavior across the ectoplasmic membrane of *Paramecium*. I. Electric potential differences measured by intracellular microelectrode. *J. Fac. Sci. Univ. Tokyo, Sect. IV* **8,** 575–591.

Yoshida, K., Ide, T., Inouye, K., Mizuno, K., Taguchi, T., and Kasai, M. (1997). A voltage- and K^+-dependent K^+ channel from a membrane fraction enriched in contractile vacuole of *Dictyostelium discoideum*. *Biochim. Biophys. Acta* **1325,** 178–188.

Zeuthen, T. (1992). From contractile vacuole to leaky epithelia. Coupling between salt and water fluxes in biological membranes. *Biochim. Biophys. Acta* **1113,** 229–258.

Zeuthen, T. (1994). Cotransport of K^+, Cl^- and H_20 by membrane proteins from choroid plexus epithelium of *Necturus maculosus*. *J. Physiol.* **478,** 203–219.

Zeuthen, T., and Stein, W. D. (1994). Cotransport of salt and water in membrane proteins: Membrane proteins as osmotic engines. *J. Membr. Biol.* **137,** 179–195.

Zhu, Q., and Clarke, M. (1992). Association of calmodulin and an unconventional myosin with the contractile vacuole complex of *Dictyostelium discoideum*. *J. Cell Biol.* **118,** 347–358.

Zhu, Q., Liu, T., and Clarke, M. (1993). Calmodulin and the contractile vacuole complex in mitotic cells of *Dictyostelium discoideum*. *J. Cell Sci.* **104,** 1119–1127.

Molecular Mechanisms of Water Transport in the Eye

Steffen Hamann
Department of Medical Physiology, The Panum Institute, University of Copenhagen, DK-2200 Copenhagen N, Denmark

The four major sites for ocular water transport, the corneal epithelium and endothelium, the ciliary epithelium, and the retinal pigment epithelium, are reviewed. The cornea has an inherent tendency to swell, which is counteracted by its two surface cell layers, the corneal epithelium and endothelium. The bilayered ciliary epithelium secretes the aqueous humor into the posterior chamber, and the retinal pigment epithelium transports water from the retinal to the choroidal site. For each epithelium, ion transport mechanisms are associated with fluid transport, but the exact molecular coupling sites between ion and water transport remain undefined. In the retinal pigment epithelium, a H^+-lactate cotransporter transports water. This protein could be the site of coupling between salt and water in this epithelium. The distribution of aquaporins does not suggest a role for these proteins in a general model for water transport in ocular epithelia. Some water-transporting membranes contain aquaporins, others do not. The ultrastructure is also variable among the cell layers and cannot be fitted into a general model. On the other hand, the direction of cotransport in symporters complies with the direction of fluid transport in both the corneal epi- and endothelium, as well as the ciliary epithelium and retinal pigment epithelium.

KEY WORDS: Water, Eye, Cotransport, H_2O homeostasis, Epithelia, Aquaporins.

I. Introduction

The eye is unique in having a large volume of extravascular tissue fluid low in proteins. Large colloid molecules absorb or scatter light, so the optical advantage of having relatively protein-free intraocular fluids is obvious. Fluid in the eye is

0074-7696/02 $35.00

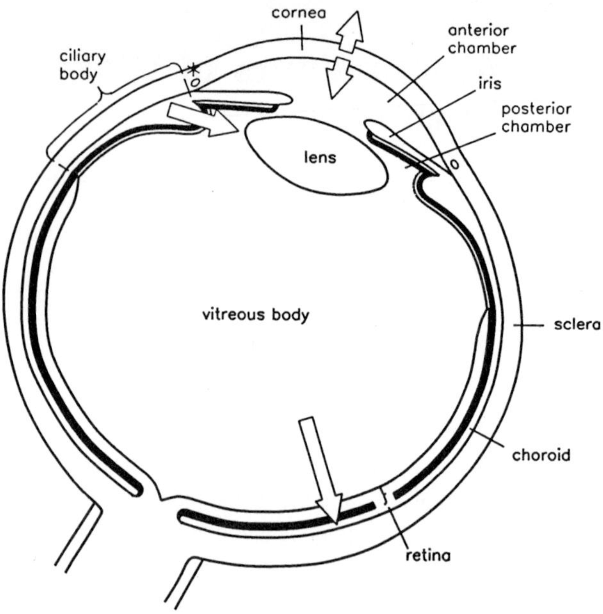

FIG. 1 Horizontal section of the human eye. The globe of the eye consists of three main layers surrounding the transparent refractive media. The outermost, protective tunic is made up of the sclera and the translucent cornea. The intermediate coat is mainly vascular, and consists of the choroid, ciliary body, and iris. The innermost layer is the retina, which contains the nervous elements responsible for vision. The spaces within the eye are filled by the aqueous humor, contained in the anterior and posterior chambers, and the vitreous body in the large space behind the lens and ciliary body. The major H_2O transports are depicted by open arrows. *, Schlemm's canal.

not stagnant. To keep up with the many metabolically active intraocular tissues, there is a continuous movement of fluid between compartments within the eye, and between the eye and the systemic circulation. A diagram of the eye with the major pathways for water transport is shown in Fig. 1.

There are four major pathways for ocular water transport: two in the cornea, one in the ciliary body, and one in the retina. Each pathway has its own physiological characteristics. In the cornea, two cell layers are responsible for water transport. The endothelium and epithelium maintain transparency and neutralize the swelling tendency of the stroma caused by passive water uptake through the two cell layers. The endothelium expels water into the anterior chamber. The epithelium prevents invasion of pathogens and uptake of excess fluid by the stroma, and plays a role in regulation of corneal hydration. The secretion of aqueous humor (see legend to Fig. 1) takes place across the ciliary epithelium. The constant secretion and drainage of aqueous help to inflate the globe of the eye and to maintain an adequate intraocular pressure. Aqueous humor also serves important metabolic functions for avascular tissues in the anterior segment of the eye. Retinal water transport is determined mainly by the retinal pigment epithelium (RPE), with its unique juxtaposition between the neural retina and the choroid. A variety of transport mechanisms in the membranes of this epithelium carry out the vectorial transport of nutrients, ions, and fluid between the outer retina and the choroid. Furthermore, this epithelium supports the outer retina by regulating the chemical composition of the extracellular space of the photoreceptors.

The driving forces and pathways for epithelial water transport in the eye are not completely understood. Water transport across epithelial membranes may occur by three mechanisms: Passive diffusion across the phospholipid bilayer, passive flow through water channels (aquaporins), and water transport linked to secondary active cotransport of ions. While there is general agreement about the two first mechanisms, the nature of the latter mechanism is debated. For several cotransporters, however, there is now good evidence for such a coupling (for a review, see Zeuthen, 2000, and Cotransporters as Molecular Water Pumps by Zeuthen and MacAulay, this volume). This review will discuss the physiological significance of ocular water transport by the three mechanisms. Since ion transport mechanisms provide the sites of primary investment of energy, it will be important to review these first for each epithelium. Secondly, water transport will be reviewed with a special emphasis on possible coupling mechanisms with the ion transport.

II. Cornea

The structure of the cornea is illustrated in Fig. 2. Up to 90% of its thickness is composed of the stroma (CS) (Kaye, 1969), which is bounded externally

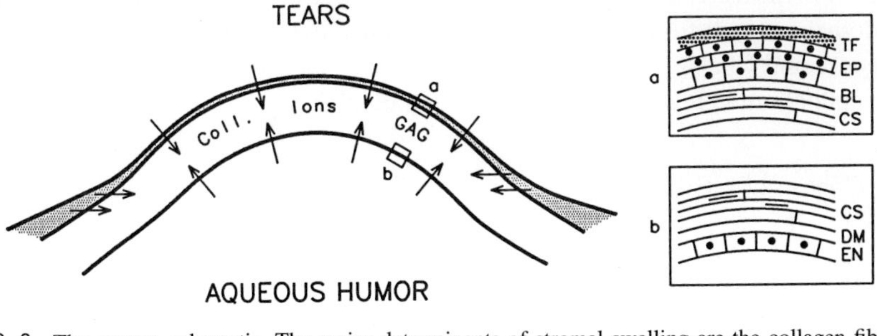

FIG. 2 The cornea, schematic. The major determinants of stromal swelling are the collagen fibers (Coll.), ions, and glycosaminoglycans (GAG); the arrows illustrate the tendency of passive inward movement of water from aqueous humor, tears, and limbal capillaries. The corneal ultrastructure is shown in the insets at right. TF, tear film; EP, epithelium; BL, Bowman's layer; CS, stroma; DM, Descemet's membrane; and EN, endothelium.

by Bowman's acellular layer (BL) and the epithelium (EP), and internally by Descemet's membrane (DM) and the endothelium (EN). The stroma is an extracellular matrix, which consists of proteoglycans and a lamellar arrangement of collagen fibrils running parallel to the corneal surface (Cogan, 1951). The collagen fibrils and extracellular matrix are maintained by flattened cells with limited cytoplasm called keratocytes lying between the collagen lamellae. From the keratocytes, fine tapering processes extend in several directions, some of which come in contact with processes of neighboring keratocytes to form tight junctions (Pepose and Ubels, 1992). The stroma is separated from the outside fluids, aqueous humor and tears, by membranes that permit movements of ions and water but not of the colloid. Accordingly, the stroma may be considered a gel with a tendency to take up water (Davson, 1990b). This uptake, if uncompensated, leads to swelling and loss of transparency. The swelling tendency is matched by active transport mechanisms in the corneal endothelium and epithelium, and to a smaller extent by osmotic water transport. This osmotic shrinkage tendency stems from differences in ionic activities, which define the diffusional and osmotic gradients for Na^+. While the concentrations of Na^+ and K^+ are higher in the stroma than in the aqueous humor (Otori, 1967), the ionic activities are smaller in the stroma than in the aqueous (Brubaker and Kupfer, 1962). This reflects cationic binding by anionic sites on stromal glycosaminoglycans (Hedbys, 1961). Accordingly, the effective osmolarity of the stroma suggests a passive exit of water from the stroma into the aqueous (or into the tear fluid). In corneal keratocytes, the water channel aquaporin-1 (AQP1) is significantly expressed in plasma membranes (Hamann *et al.,* 1998). Its role is as yet undefined.

Initially, only the corneal endothelium was considered responsible for active removal of fluid from the cornea; the corneal epithelium was regarded simply as a barrier to salt and water influx from the tears and limbal blood vessels (for reviews, see Davson, 1990b; Pepose and Ubels, 1992). This simplified concept gained support from the fact that the relative resistance of the corneal epithelium, stroma, and endothelium to diffusion of electrolytes was found to be in the ratio of 2000 : 1 : 10 (Pepose and Ubels, 1992). Recent work, however, points to an important role of the epithelium in regulation of stromal hydration.

A. Corneal Epithelium

1. Ion Transport

The corneal epithelium is multilayered and provides a high transepithelial resistance of 12 to 16 kohm cm^2 (Pepose and Ubels, 1992). Apparently, the zonulae occludentes of the superficial epithelial cells have low conductance to ion transport, and the epithelium can be characterized as tight. The epithelium forms a functional syncytium (Klyce, 1972). The rabbit corneal epithelium generates a

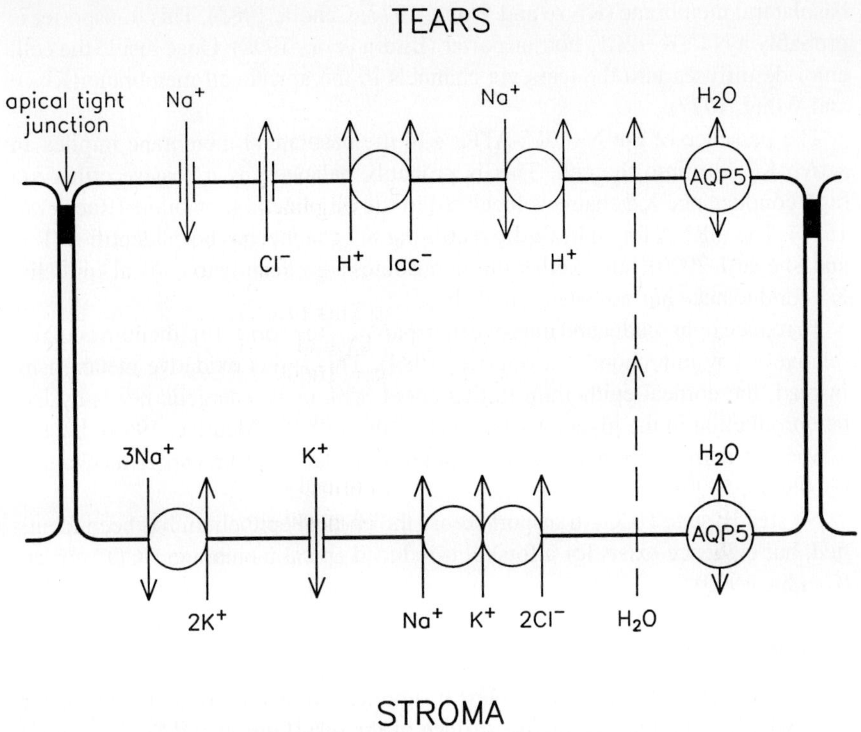

FIG. 3 Ion and water transport across the corneal epithelium. This multilayered epithelium forms a functional syncytium and is diagrammed as a monolayer for clarity. The epithelial cell has a Na^+/K^+-ATPase, a Na^+-K^+-$2Cl^-$, and a H^+-lactate cotransporter, a Na^+/H^+ exchanger, and Na^+, K^+, Cl^-, and H_2O channels (AQP5). The direction of active H_2O transport is denoted by the dotted line (see text).

transepithelial potential of 25 to 35 mV *in situ,* tear side negative to stromal side (Maurice, 1967; Klyce, 1972). This potential is maintained by transport of sodium and chloride ions (Fig. 3).

The net Na^+ flux is directed from tears to stroma, while the net Cl^- flux is directed from stroma into the tears. Na^+ enters through channels in the apical membrane of the superficial epithelial cells (Klyce, 1973) and also through an apical membrane Na^+/H^+ exchanger (Bonanno and Machen, 1989). The overall Na^+ permeability of the membrane is low (Klyce and Marshall, 1982). No amiloride-sensitive channels have been identified (Marshall and Klyce, 1983). Na^+ is transported into the stroma by the ouabain-sensitive Na^+/K^+-ATPase in the basolateral membrane (Donn *et al.,* 1959; Green, 1970; Klyce and Crosson, 1985). This creates an electrical gradient between the stroma and the epithelial cells, which serves as the driving force for transport of chloride into the tears (Zadunaisky, 1966). Sodium is also transported into the epithelial cells by a Na^+-coupled Cl^--cotransporter in the basolateral membrane (Klyce and Wong, 1977; Candia, 1985). This transporter is probably a Na^+-K^+-$2Cl^-$ cotransporter (Bildin *et al.*, 1998). Once inside the cell, chloride diffuses into the tears via channels in the apical cell membrane (Klyce and Wong, 1977).

The presence of the Na^+/K^+-ATPase in the basolateral membrane implies an active K^+ flux into the cell. This is probably balanced by a passive efflux via high conductance K^+ channels localized to the basolateral membrane (Rae *et al.*, 1990). The mRNA for an inwardly rectifying K^+ channel has been identified (Rae and Shepard, 2000), but whether this contributes significantly to corneal epithelial K^+ conductance has not been established.

To reduce light scatter and improve transparency, the corneal epithelium contains relatively few mitochondria (Maurice, 1984). This limits oxidative metabolism. Instead, the corneal epithelium fulfills about 85% of its energetic needs by lactate production in the glycolytic pathway (Riley, 1969b; Maurice, 1984). Lactate leaves the corneal epithelial cells through a H^+-lactate cotransport mechanism (Bonanno, 1990).

No steady-state HCO_3^- transport across the corneal epithelium has been identified, but evidence exists for a forskolin-induced apical membrane HCO_3^- current (Candia, 1990).

2. Water Transport

The corneal epithelium has two roles: it supplies a moist, transparent, apical surface, which is the main refractive surface of the eye (Gipson and Sugrue, 1994), and it participates in the regulation of stromal volume.

Outward fluid transport in the frog corneal epithelium *in vitro* can take place at a rate of about 7.5 $\mu l\ cm^{-2}\ hr^{-1}$ (calculated from the data of Candia and Zamudio, 1995). This rate is almost twice that of steady-state fluid transport across the corneal endothelium (see Table I). A recent study in cultured rabbit corneal epithelial cells gave fluid transport rates from the stromal to tear side of about 5.2 $\mu l\ cm^{-2}\ hr^{-1}$. This flux was inhibited by bumetanide as well as ouabain (Yang *et al.*, 2000b).

The water permeabilities of the epithelium are given in Table I: A diffusional water permeability of 6 $\mu m/sec$ (Parisi *et al.*, 1980) and a high osmotic permeability of 92 $\mu m/sec$ (Mishima and Hedbys, 1967) and 137 $\mu m/sec$ (Fischbarg and Montoreano, 1982). Interestingly, the water permeabilities have been suggested to be controlled by Cl^- on the tear side (Candia and Zamudio, 1995). The corneal epithelial Cl^- secretory pathway is coupled to the transport of water out of the cornea into the tears. When epithelial Cl^- secretion was stimulated pharmacologically or when the epithelial perfusion solutions were made Cl^- free, the previously swollen cornea was dehydrated (Klyce, 1977).

In normal eyes, the tear film is slightly hypertonic (approximately 304 mosm $liter^{-1}$), and increases in tear osmolarity are frequently seen in dry-eye patients (Farris, 1994). Recently it was found that long-term adaptation by corneal

TABLE I
Parameters for Ocular Water Transport

	J_v (μL cm^{-2} h^{-1})				P_f (μm sec^{-1})				L_p (μm sec^{-1})			P_d (μm sec^{-1})	Tonicity of transported fluid
	Frog	Rabbit	Human (*in vivo*)	Bovine	Frog	Rabbit	Human (cultures)	Bovine (cultures)	Monkey	Dog	Rabbit	frog	
Corneal epithelium	7.5[a]	5.2[b]	—	—	137[c]	92[d]	—	—	—	—	—	6.2[e]	Slightly hypertonic[f]
Corneal endothelium	—	3.7[g] 3.2[k] 6.7[j]	—	4.0[h]	711[c]	211[d] 162[l]	—	93[i] 460[m]	—	—	273[j] 114[n]	—	Isotonic[o]
Ciliary epithelium	—	2[p]	25[q]	—	—	—	98[r]	—	—	—	210[p]	—	Isotonic[s]
Retinal pigment epithelium	4.8[t] 7.6[cc]	4[u]	11[v]	1.4[x]	210[y]	—	—	—	869[z]	2235[aa]	—	—	Isotonic[bb]

J_v = rate of fluid transport.

P_f = osmotic water permeability. If P_f was given in cm sec^{-1} (osm liter^{-1})$^{-1}$ it has been multiplied by the partial molal volume of water of 55 mol liter^{-1} to give the P_f in units of 10^4 μm sec^{-1}.

L_p = hydraulic water permeability. The unit of L_p is usually μl min^{-1} cm^{-2} mmHg^{-1}. For comparison with the other permeabilities, this is converted to 10^4 μm sec^{-1} by multiplying by 17.74.

P_d = diffusional water permeability.

[a] (Candia and Zamudio, 1995)
[b] (Yang *et al.*, 2000b).
[c] (Fischbarg and Montoreano, 1982).
[d] (Mishima and Hedbys, 1967).
[e] (Parisi *et al.*, 1980).
[f] (Farris, 1994), about 1% hypertonic.
[g] (Fischbarg *et al.*, 1977a).
[h] (Narula *et al.*, 1992).
[i] (Echevarria *et al.*, 1993).
[j] (Baum *et al.*, 1984) .
[k] (Mayes and Hodson, 1978a).
[l] (Fischbarg *et al.*, 1980).
[m] (Sun *et al.*, 2001).
[n] (Fischbarg *et al.*, 1977b).
[o] (Davson, 1990b).
[p] (Brodwall and Fischbarg, 1982).
[q] (Brubaker, 1991), mean of value from several investigators.
[r] (Han *et al.*, 2000), permeability value of NPE (nonpigmented ciliary epithelium) only. Temperature = 10°C.
[s] (Gaasterland *et al.*, 1979).
[t] (Miller *et al.*, 1982).
[u] (Wolfensberger *et al.*, 2000).
[v] (Chihara and Nao-i, 1985). Rate of subretinal fluid resorption in the postoperative course following surgery of retinal detachment.
[x] (Edelman and Miller, 1991).
[y] (la Cour and Zeuthen, 1993), permeability value of retinal membrane only. .
[z] (Tsuboi and Pederson, 1988).
[aa] (Tsuboi, 1987). This value may possibly be an overestimate due to edge-damage of the tissue.
[bb] (Hughes *et al.*, 1998).
[cc] (Frambach *et al.*, 1985).

epithelial cells to a hypertonic medium depends on upregulation of the Na^+-K^+-$2Cl^-$ cotransporter gene, protein expression, and transport activity (Bildin *et al.*, 2000). This is consistent with a role of corneal epithelial Na^+-K^+-$2Cl^-$ cotransport in the maintenance of stromal hydration.

A functional water channel has been found in the frog corneal epithelium (Horwich *et al.*, 1995) and identified as AQP5 in rat corneal epithelium (Raina *et al.*, 1995). The subcellular localization to both apical and basolateral plasma membranes has been determined by high-resolution immunocytochemistry (Hamann *et al.*, 1998), Fig. 3.

Summing up, in both native and cultured corneal epithelial cells the functional elements for fluid transport are present. They include: (1) high osmotic permeability, (2) Cl^- secretion in the same direction as fluid transport, (3) transport proteins involved in Cl^- secretion, and (4) expression of AQP5 (see Fig. 3). What remains to be established is the molecular mechanism of coupling between chloride (and/or other ions) and water flux, the relative contribution to corneal epithelial fluid transport from AQP5 and the Cl^- coupled pathway, and the contribution of epithelial fluid transport to stromal dehydration *in vivo*.

B. Corneal Endothelium

1. Ion Transport

In contrast to the epithelium, the junctions that interconnect the corneal endothelial cells are maculae occludentes rather than zonulae occludentes, in that they do not completely encircle the cells (McLaughlin *et al.*, 1985). This gives a leaky cell layer with transepithelial resistances in the range from 20 to 70 ohm cm^2 (references in Lim and Ussing, 1982) and the possibility of high rates of ion transport via the paracellular pathway.

The Na^+/K^+-ATPase is located in the basolateral membrane of the endothelial cell (Riley *et al.*, 1994; Wigham *et al.*, 1994) (Fig. 4), and is vital to maintenance of corneal state of hydration. Inhibition with ouabain stops Na^+ transport, causes corneal swelling, and eliminates the $-500\ \mu V$ (aqueous humor negative to stroma) potential across the endothelium. The basolateral membrane also contains a Na^+/H^+-exchanger, which moves Na^+ into the cell and H^+ outward (Bonanno and Giasson, 1992a), and a Na^+-K^+-$2Cl^-$ cotransporter (Diecke *et al.*, 1998; Kuang *et al.*, 2001) possibly involved in volume-regulatory processes in these cells (see Section II.B.2).

A variety of K^+ channels is present in corneal endothelium. Two major K^+ currents have been measured: one due to an anion- and temperature-stimulated channel, and another with properties similar to those of channels from excitable cells (Rae and Watsky, 1996); recently the molecular cloning and expression of an inwardly rectifying K^+ channel in cultured bovine corneal endothelial cells

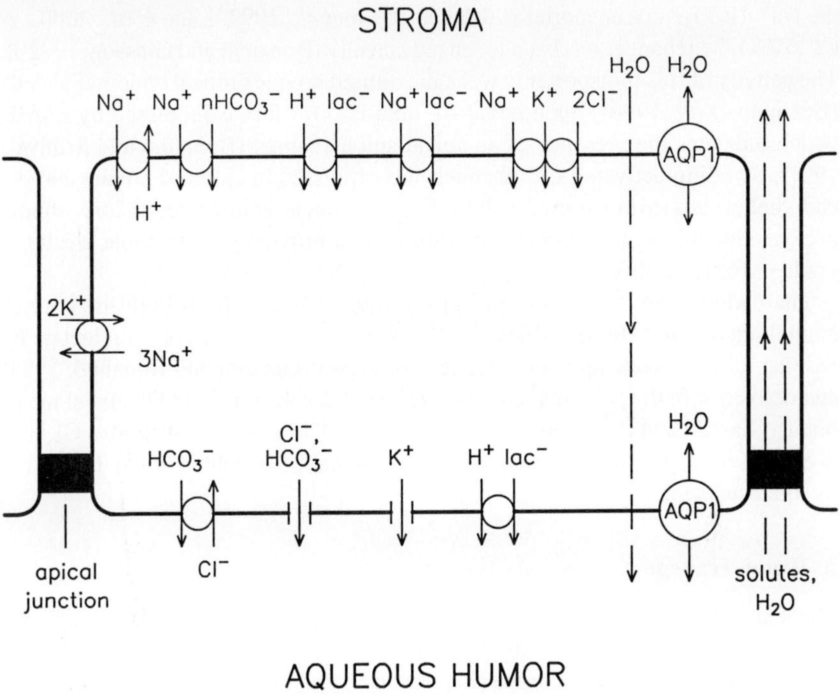

FIG. 4 Ion and water transport across the corneal endothelium. The endothelial cell has a Na^+/K^+–ATPase, an H^+-lactate, an Na^+-lactate, an Na^+-$nHCO_3^-$, and an Na^+-K^+-$2Cl^-$ cotransporter, a Na^+/H^+ exchanger, and channels for K^+, Cl^-, HCO_3^-, and H_2O (AQP1). The direction of active H_2O transport is denoted by the dotted line. The leaky intercellular apical junctions are the sites for an additional entry of solutes and H_2O from the aqueous humor (see text).

was described (Yang *et al.,* 2000a). It has been shown that fluoxetine (Prozac) at 20–1000 μM inhibits K^+ transport pathways (K^+ efflux, Na^+-K^+-$2Cl^-$ cotransport, and Na^+/K^+-ATPase) which are active in volume regulation in cultured bovine corneal endothelial cells (Hara *et al.,* 1999).

Bicarbonate is central to the maintenance of corneal hydration, as evidenced by the net flux of HCO_3^- from the stroma to the aqueous (Hodson and Miller, 1976; Hull *et al.,* 1977; Mayes and Hodson, 1978b; Hodson *et al.,* 1981). This transport is responsible, at least in part, for the transendothelial potential difference mentioned above (Kuang *et al.,* 1990). Electrophysiological evidence indicates the existence of an electrogenic, stilbene-inhibitable Na^+-HCO_3^- cotransporter, which probably moves these ions in a 1 : 2 ratio (Jentsch *et al.,* 1984). Recent studies have confirmed the presence of this cotransporter in the basolateral membrane of both cultured and intact fetal bovine corneal endothelial cells (Bonanno and Giasson, 1992b; Sun *et al.,* 2000) (see Fig. 4). Whether Cl^- has a role in the activation of the Na^+-HCO_3^- cotransporter is debated (Usui *et al.*, 1999; Lane *et al.*, 2000). A Cl^-/HCO_3^- exchanger has been localized apically (Bonanno and Giasson, 1992b). The activity of this transporter is weak in cultured bovine corneal endothelial cells (Bonanno *et al.*, 1998). Endothelial Cl^- and HCO_3^- flux is increased by cAMP, which indicates the presence of an apical anion channel (Bonanno and Srinivas, 1997). Swelling-activated Cl^- channels are expressed in cultured bovine corneal endothelial cells (Srinivas *et al.*, 1999). These channels facilitate regulatory volume decrease by enhancing anionic permeability and providing a favorable electrical gradient for K^+ efflux.

The endothelium is responsible for removal of lactate from both the corneal epithelium and the stroma (Riley, 1969a,b). The endothelium takes up lactate by basolateral H^+-lactate and Na^+-lactate cotransport (Giasson and Bonanno, 1994) and delivers it to the aqueous humor by H^+-lactate cotransport in the apical membrane (Giasson and Bonanno, 1994) (see Fig. 4). The glucose transporter GLUT1 is expressed in cultured bovine corneal endothelial cells (Ishida *et al.*, 1995), but its role is unclear.

2. Water Transport

The leaky junctions between corneal endothelial cells are most likely the sites for a large passive component of fluid transport into the stroma providing metabolic substrates to the avascular cornea (Pepose and Ubels, 1992).

The passive influx into the stroma is probably balanced by two components of water efflux, one active and one passive. The metabolically dependent active fluid transport across the corneal endothelium occurs at a rate of 3.7 $\mu l\ cm^{-2}\ hr^{-1}$ in rabbit preparations *in vitro* (Fischbarg *et al.*, 1977a) and about 4 $\mu l\ cm^{-2}\ hr^{-1}$ in cultured bovine cells (Narula *et al.*, 1992), Table I. The passive efflux component is determined by the osmolarities of the two compartments, stroma and aqueous humor. The Na^+ concentration of aqueous humor is 140 mosm $liter^{-1}$, while the Na^+ concentration of the stroma is 165 mosm $liter^{-1}$ (Stiemke *et al.*, 1992). This would imply an osmotic gradient favoring aqueous to stroma movement of water. However, all of aqueous Na^+ is osmotically active, while a large portion of the stromal Na^+ is chemically bound, leaving an osmotically active Na^+ level of only 134 mosm $liter^{-1}$. This results in an osmotic pressure difference, from Na^+ alone, of about 5–10 mosm $liter^{-1}$, driving water in the stroma to aqueous direction (Stiemke *et al.*, 1992). Given the passive water permeability of the endothelium (Table I) of about 200 $\mu m\ sec^{-1}$, it can be calculated that this osmotic mechanism supplies a water flux similar to the active one.

There is good evidence that the fluid entering the cornea is isotonic to plasma, and the same is true for that expelled by the endothelium (Davson, 1990b). The general mechanism of isotonic transport is debated. Yet, it is generally agreed that water transport is coupled to ion transport. The role of various ion transport mechanisms for water transport has been investigated. Removal of Na^+ from the medium bathing the corneal endothelial cells abolished corneal hydration regulation, resulting in pronounced swelling of the tissue (Dikstein and Maurice, 1972). Identical results were obtained with ouabain. Accordingly, this effect has been attributed to intact Na^+/K^+-ATPase activity (Trenberth and Mishima, 1968; Riley, 1982). Both HCO_3^- and Cl^- are critical to the fluid-transporting capacity of these cells, most likely via their effect on Na^+/H^+ and Cl^-/HCO_3^- exchangers, Na^+-HCO_3^- cotransporters, and apical anion channels (Dikstein and Maurice, 1972; Fischbarg and Lim, 1974; Kuang *et al.,* 1990; Winkler *et al.,* 1992; Riley *et al.,* 1995). Recently, the involvement of Na^+-K^+-$2Cl^-$ cotransport has also been proposed on the basis of experiments in cultured bovine corneal endothelial cells (Diecke *et al.,* 1998) and intact rabbit corneal endothelium (Kuang *et al.,* 2001); one study on intact rabbit endothelium showed no Na^+-K^+-$2Cl^-$ cotransport activity (Riley *et al.,* 1997).

Aquaporin-1 is expressed in both apical and basolateral plasma membranes of the corneal endothelium (Nielsen *et al.,* 1993; Stamer *et al.,* 1994; Hamann *et al.,* 1998) (Fig. 4). An osmotic water transport is likely to take place via this water channel. However, there is no report yet of stromal swelling, loss of corneal transparency, or impaired vision in AQP1-deficient knock-out mice (Verkman *et al.,* 2000). Thus, the precise role and mechanism for water transport through AQP1 in corneal endothelium remain unclear. It has recently been suggested that the passive water transport from aqueous to stroma could have an electroosmosis component (Fischbarg *et al.,* 2001).

III. Aqueous Humor

Aqueous humor is formed by the ciliary epithelium, which consists of two layers of cells, a pigmented layer (CPE) facing the stroma and a nonpigmented layer (NPE) facing the aqueous humor (Figs. 1 and 5). The apical membranes of the two cell layers are in tight apposition. Only between NPE cells are the zonulae occludentes observed (Raviola and Raviola, 1978). The cells within each layer and between the two layers are connected by gap junctions (Raviola and Raviola, 1978; Coca-Prados *et al.,* 1992), forming a functional syncytium (Green *et al.,* 1985; Carré *et al.,* 1992; Oh *et al.,* 1994; Edelman *et al.,* 1994c).

The aqueous humor bathes the structures within the eye and supplies oxygen and metabolites for the avascular tissues: the posterior cornea, trabecular meshwork, crystalline lens, and anterior vitreous. Aqueous humor is formed in three serial processes (Brubaker, 1991):

1. Blood flows to the ciliary processes in the anterior uvea.
2. Some of the plasma that reaches this vascular bed is filtered through the fenestrated capillaries into the interstitial spaces between the vessels and the ciliary epithelium.

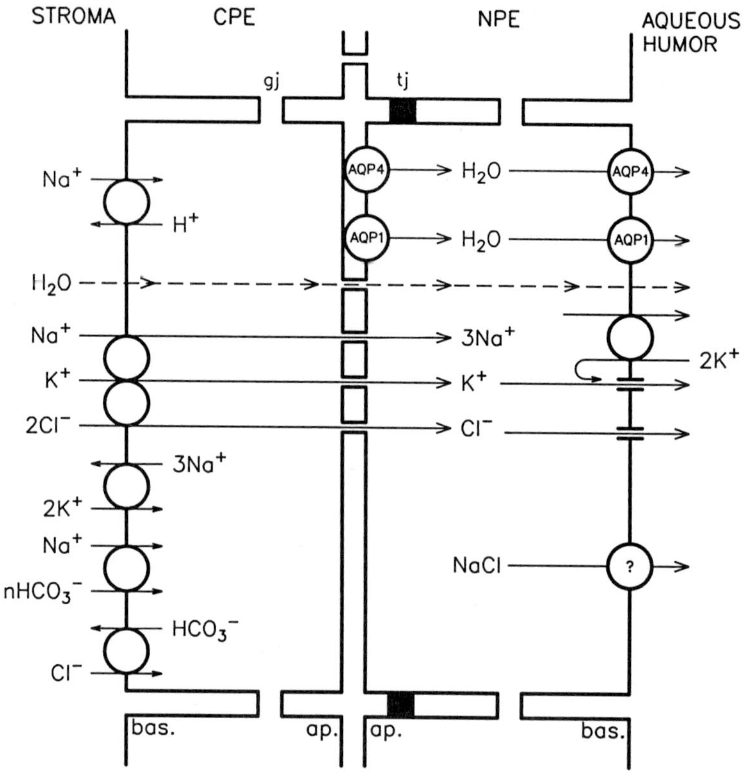

FIG. 5 Ion and water transport across the ciliary epithelium. Gap junctions (gj) connect pigmented (CPE) and nonpigmented (NPE) ciliary epithelial cells, forming a functional syncytium. Tight junctions (tj) are only developed by the NPE. The CPE cell contains a Na^+/K^+ATPase, a Na^+/H^+ and a Cl^-/HCO_3^- exchanger, and a Na^+-K^+-$2Cl^-$ and a Na^+-HCO_3^- cotransporter, all localized to the basolateral membrane. The NPE cell has a Na^+/K^+-ATPase, a Cl^- channel, and an electroneutral unidentified exit pathway for NaCl (symbolized by the question mark) localized to the basolateral membrane, as well as H_2O channels (AQP1 and –4) localized to both the apical and basolateral membrane. The net direction of H_2O transport across the entire epithelium is shown by the dashed line. For clarity, a series of ion channels, cotransporters, and exchangers with no clear role in transepithelial secretion are omitted from this simplified model.

3. The ciliary epithelium extracts from the filtered fluid the various ions and molecules that are ultimately secreted together with water into the posterior chamber.

The rate of aqueous humor production in the human varies markedly over the course of the day, with a peak value during the waking hours and a nadir during the early morning (Brubaker, 1998). Each minute, between 1 and 3 μL is secreted into the posterior chamber (Brubaker, 1998). Since the volume of the posterior chamber in the human eye is approximately 60 μl, the posterior chamber is replaced with fresh secretion approximately every 30 min. The aqueous humor flows through the pupillary aperture into the anterior chamber, and leaves the eye primarily through the trabecular meshwork leading to Schlemm's canal, where it then passes through a series of collector channels into the network of episcleral veins. Other routes of minor importance have been described (Bill and Phillips, 1971; Pederson, 1994); for further reading on drainage, see Hart (1992); Bill (1993); Kaufman (1996).

A. Ciliary Epithelia

1. Ion Transport

A transepithelial potential of 1 mV, aqueous humor negative, has been suggested to provide the electrochemical driving force for paracellular Na^+ secretion into the aqueous humor (Coca-Prados *et al.,* 1995; Jacob and Civan, 1996). Since the transepithelial potential depends on the presence of HCO_3^-, this ion may influence the rate of aqueous humor secretion (Krupin and Civan, 1995; Coca-Prados *et al.,* 1995) (Fig. 5).

a. Ion Transport in CPE. The Na^+/K^+-ATPase of the CPE couples the entry of Na^+, K^+, and Cl^- to the metabolism. The ions enter from the stroma through a furosemide-sensitive (Wiederholt and Zadunaisky, 1986) and bumetanide-sensitive (Edelman *et al.,* 1994c) Na^+-K^+-$2Cl^-$ cotransporter. The cotransporter has been localized to the basolateral membrane (Dunn *et al.,* 2001) (Fig. 5). It is activated by shrinkage of both CPE (Edelman *et al.,* 1994c) and NPE cells (Civan *et al.,* 1996). Bumetanide causes CPE cell shrinkage in isoosmotic bathing solutions (Edelman *et al.,* 1994c). Recent studies suggest that the Na^+-K^+-$2Cl^-$ cotransporter is the major Cl^- influx pathway, and may provide the driving force for aqueous humor formation (Do and To, 2000; Crook *et al.,* 2000). An additional pathway for NaCl uptake is mediated by a Cl^-/HCO_3^--exchanger (Wiederholt *et al.,* 1991; Counillon *et al.,* 2000) and a Na^+/H^+ exchanger (Counillon *et al.,* 2000). Recently a Na^+-HCO_3^- cotransporter of unknown stoichiometry has been localized to the basolateral membrane as well (Cilluffo *et al.,* 2001).

b. Ion Transport in NPE. From the CPE, the Na^+, K^+, and Cl^- ions diffuse through the gap junctions to the basal membranes of the NPE cells and are expelled into the aqueous humor through the Na^+/K^+-ATPase, K^+ channels, and Cl^- channels. The NPE contains the main fraction of Na^+/K^+-ATPase activity of the ciliary epithelium (Riley and Kishida, 1986; Ghosh *et al.,* 1990). The highest concentrations are along the lateral cellular interdigitations (Cole, 1964; Kaye and Pappas, 1965; Shiose and Sears, 1966). This would support the standing gradient hypothesis of aqueous humor secretion (see Section III.A.2).

The K^+ channels of the NPE cells probably constitute the main pathway for K^+ transport into the aqueous humor. These channels determine the voltage across the basal membranes of the NPE and CPE cells and thus across the entire epithelium (for a review see Jacob and Civan, 1996). The electrical potential provides the driving force for Cl^- secretion through Cl^- channels in the NPE basolateral membrane. This is the main pathway for Cl^- secretion from the NPE into the posterior chamber, and it is considered critical in the formation of aqueous humor (Coca-Prados *et al.,* 1995), and for volume regulation of the NPE cells (Jacob and Civan, 1996). Several candidates for the volume-activated Cl^- channel (or regulator) in NPE cells have been presented (Coca-Prados *et al.,* 1995, 1996; Wu *et al.,* 1996; Chen *et al.,* 1998; Wang *et al.,* 1998; Shi *et al.,* 1999).

In addition to Cl^- exit through Cl^- channels, an electroneutral exit pathway for NaCl (Fig. 5) has been suggested (Chu and Candia, 1987), possibly in the form of a Na^+-K^+-$2Cl^-$ cotransporter (McLaughlin *et al.,* 1998). However, the expression of Na^+-K^+-$2Cl^-$ cotransport protein in NPE is approximately 10 times lower than in CPE (Crook *et al.,* 2000). In a recent study, no expression of the cotransporter was found on the NPE basolateral membrane (Dunn *et al.,* 2001).

2. Water Transport

The rate of aqueous formation is a function of the rate of active solute transport in the ciliary epithelia (Cole, 1977). The predominant theory is that active solute pumping sets up a concentration gradient that forces an osmotic flow of water into the posterior chamber. In analogy to many other epithelia, the Na^+/K^+-ATPase is the primary point of investment of metabolic energy. Accordingly, ouabain, an inhibitor of Na^+/K^+-ATPase, decreases aqueous formation by some 70% (Cole, 1977). This has lead to the assumptions, that (1) the major pathway for water transport is cellular, and that (2) the site primarily responsible for aqueous humor secretion is the site with the major Na^+/K^+-ATPase activity, i.e., the NPE. This will be discussed further in Section V. In keeping with this, the water channels AQP1 and (to a lesser extent) AQP4 are expressed in apical and basolateral membranes of the NPE in contrast with the CPE, where no aquaporins have been found (Hamann *et al.,* 1998) (see Fig. 5). The presence of AQP1 (and AQP4) in the NPE is consistent with a specific osmotic role in aqueous humor secretion. Whether aquaporins are essential for the production of aqueous humor is as yet unclear, but transgenic mice lacking AQP1 or AQP4 did not reveal any ocular problems (Verkman *et al.,* 2000).

The ciliary epithelial water transport is isotonic. Rhesus monkey aqueous humor and serum are isomolar, with osmolalities about 303 mosm liter^{-1} (Gaasterland *et al.,* 1979). In some mammalian species, aqueous humor is slightly hyperosmotic to plasma (references in Caprioli, 1992). It is generally agreed that ion transport drives water transport, but it is still unclear how the ion transport mechanisms integrate to form the isotonic aqueous humor. The standing gradient model (Diamond and Bossert, 1967, 1968) adapted for the ciliary epithelia by Cole (1977) is still the only working model. In this theory, the NPE cells absorb Na^+ from the stroma via the CPE and transport it into the intercellular clefts, which are closed to the stromal side by the tight junctions but open at the aqueous humor side. The putative hyperosmolarity in the clefts causes osmotic flow of water into the clefts. Transport of Cl^-, HCO_3^-, and K^+ as well as sugars and amino acids may contribute to the process (Davson, 1990a). The existence of hyperosmotic intercellular spaces is yet to be demonstrated (see Zeuthen, this volume and Shachar-Hill and Hill, this volume). Recently, a direct coupling between water and ion transport has been demonstrated for several types of cotransporters (for a review see Zeuthen, 2000, and Cotransporters as Molecular Water Pumps by Zeuthen and MacAulay, this volume). For example, it has been suggested that the Na^+-K^+-$2Cl^-$ cotransporter may also mediate secondary active transport of H_2O (Zeuthen and Stein, 1994). This could contribute to the water transport in the ciliary epithelium. For a general discussion on this topic, see Section V.

IV. Retina

The retina is a complex multilayered structure, which functionally divides into the neural retina facing the vitreous body and the RPE facing the capillaries of the choroid (Fig. 6). The retina is an outgrowth of the developing brain. As such, its

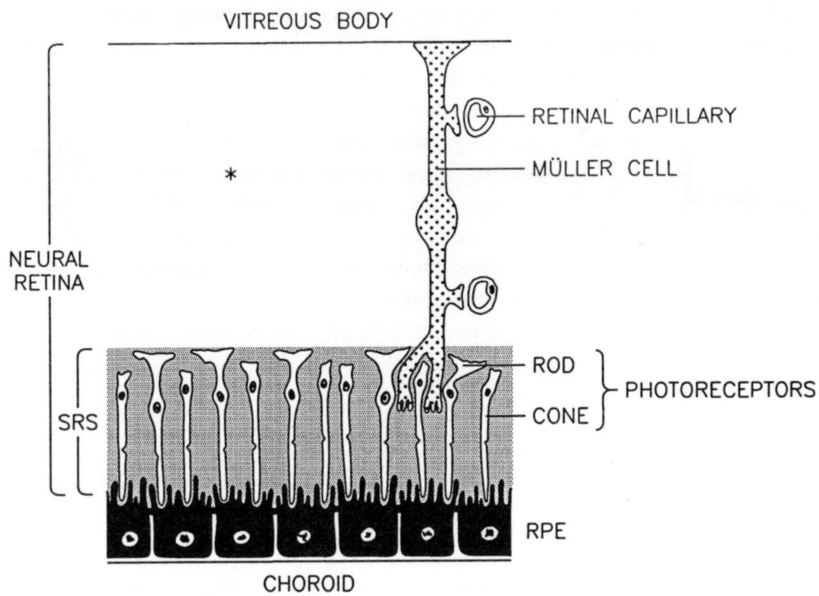

FIG. 6 A diagram of the structure of the human retina, with the elements relevant for water transport (see text for details). *, Space occupied by retinal neurons; SRS, subretinal space; RPE, retinal pigment epithelium.

neural constituents demand the same protection as brain tissue. The blood–brain barrier in the retina is formed by tight junctions between the endothelial cells of the capillaries and between the cells of the RPE. This barrier prevents leakage of fluid and protein from the vasculature into the retina (Marmor, 1999).

Water is transported from the vitreous body across the neural retina and RPE into the choroid and the systemic circulation (Fig. 1). Retina as a whole impedes the free flow of water (Fatt and Shantinath, 1971) to an extent that limits the rate at which water leaves the eye across the RPE. Accordingly, water transport by the RPE helps adherence of the neural retina by creating a negative pressure in the subretinal space (Marmor, 1998b). In isolation, however, the RPE is capable of pumping a lot more water per minute than leaves under *in vivo* conditions (Table I).

In situ, there are three mechanisms for water transport across the RPE, one active and two passive. Active solute-linked water transport is responsible for the major part, as evidenced by the high rates of volume transport by *in vitro* preparations in isoosmolar bathing solutions (Table I). With a typical rate of RPE fluid transport of 6 μl cm^{-2} hr^{-1}, the rate for the entire area of human RPE of approximately 10 cm^{2} would be 1.0 μl min^{-1}. Osmotic and hydrostatic forces determine the two passive mechanisms of RPE water transport. The osmotic pressure is due to a high protein concentration in the extravascular space of the choroid, where the albumin concentration is approximately 10% of that of the plasma (Toris *et al.,* 1990). This passive force is probably of minor importance, since the transport mechanisms of the RPE within physiological ranges are capable of overcoming whatever osmotic difference imposed to the epithelium *in vivo* (Marmor, 1998a). Intraocular hydrostatic pressure is continually pushing water into the retina, although the resulting flowthrough is very small, due to the flow resistance of the retinal tissue (Marmor, 1999). The pressure difference between the vitreous and suprachoroidal space is approximately 4 mmHg (Emi *et al.,* 1989). Using an average RPE hydraulic water permeability value of 0.009 μl min^{-1} cm^{-2} mmHg^{-1} (Table I), it is calculated that only about 0.4 μl min^{-1} could be transported across an RPE area of 10 cm^{2}. Thus, hydrostatic forces are of minor importance under normal conditions.

It should be recalled that the transport situation is different between the light- and dark-adapted eye. The retina is metabolically much more active in the dark-adapted situation. This means that the RPE in the dark faces a subretinal space high in lactate and K^+.

A. Retinal Pigment Epithelium

1. Ion Transport

The transepithelial potential is of the order of 5 to 15 mV, retinal side positive (references in Hughes *et al.,* 1998). This potential provides a driving force for Na^+ transport in the retina-to-choroid direction through the paracellular pathway.

In the RPE cell, the Na^+/K^+-ATPase is located to the retinal membrane (Miller and Steinberg, 1977a,b; Miller *et al.,* 1978; Joseph and Miller, 1991; Quinn and Miller, 1992) (Fig. 7). The secretion of Na^+ by the Na^+/K^+-ATPase contributes directly to Na^+ homeostasis in the subretinal space, which is important for a number of transport mechanisms in photoreceptor outer segments (Schnetkamp *et al.,* 1989), in retinal astrocytes and Müller cells (Newman, 1999), and in the RPE retinal membrane (Miller and Edelman, 1990; Edelman and Miller, 1991; Bialek

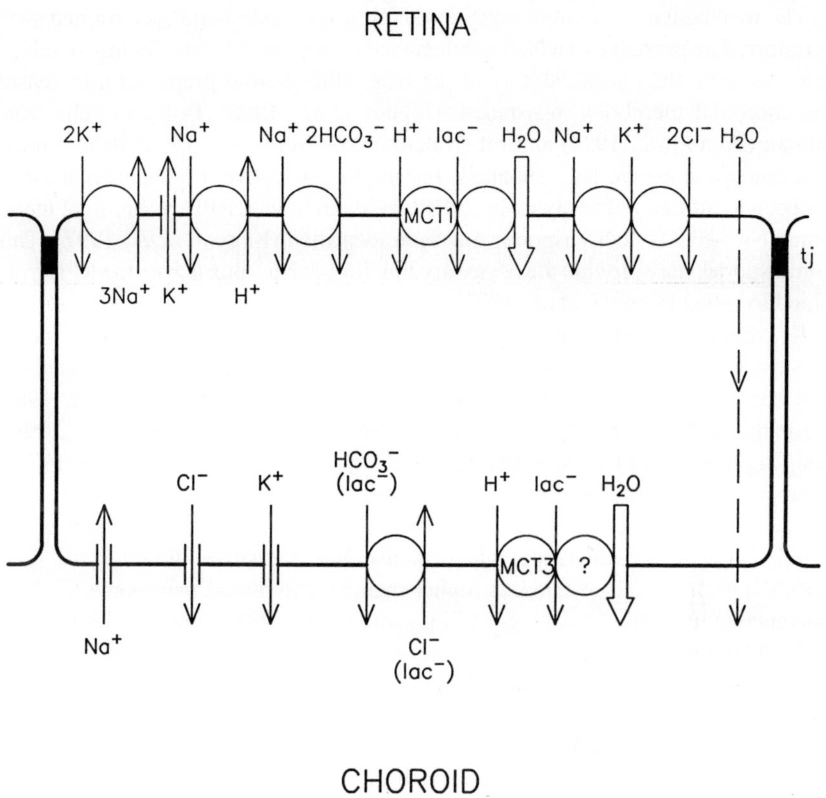

FIG. 7 Ion and water transport across the retinal pigment epithelium (RPE). The Na^+/K^+-ATPase is located in the retinal membrane. In addition, a variety of channels, cotransporters, and exchangers are shown (for details see text). Please note that there are two types of H^+-lactate cotransporters, MCT1 and MCT3 (MCT, monocarboxylate transporter). The RPE transports H_2O in the retina-to-choroid direction, which is depicted by the dotted line. Under normal conditions, the tight junctions (tj) connecting adjacent RPE cells hinder H_2O transport by the paracellular route. RPE fluid absorption is mediated in part by MCT1 in the retinal membrane, which transports H_2O together with the ions. Whether MCT3 in the choroidal membrane also transports H_2O is currently unknown, as symbolized by the question mark. For clarity, a series of ion channels, cotransporters, and exchangers with no clear role in RPE fluid absorption, are omitted from this simplified model.

and Miller, 1994). One of these, the Na^+/H^+ exchanger, has been located to the retinal membrane of human (Lin *et al.,* 1992), bullfrog (Lin and Miller, 1991), bovine RPE (Kenyon *et al.,* 1997), and cultured bovine (Keller *et al.,* 1988) and human (Hamann *et al.,* 2000) RPE cells. In excised human RPE preparations, this transport protein is the dominant acid extrusion mechanism in both the absence and the presence of HCO_3^- (Lin *et al.,* 1992). Under normal conditions, the passive absorptive flux of Na^+ through the paracellular pathway probably exceeds the active secretory Na^+ flux through the transcellular route, leading to net Na^+ absorption (Hughes *et al.,* 1998).

The mechanism of sodium entry at the choroidal side is not determined with certainty. The presence of a Na^+ conductance is suggested by the finding that Na^+ removal from the choroidal side of the frog RPE-choroid preparation increases the choroidal membrane resistance (Hughes *et al.,* 1988). Cultured cells from human (Wen *et al.,* 1994) and rat (Botchkin and Matthews, 1994) RPE express tetrodotoxin-sensitive Na^+-channels, but no Na^+-selective ion-channels have so far been identified in freshly isolated RPE cells. In bovine RPE, a choroidal membrane Na^+-HCO_3^- cotransporter has been identified (Kenyon *et al.,* 1997). This cotransporter may provide the necessary link for net Na^+ flux across the RPE from blood to retina (Kenyon *et al.,* 1997).

K^+ movement across the retinal membrane occurs by at least three mechanisms: (1) the electrogenic Na^+/K^+ pump; (2) an electroneutral furosemide- and bumetanide-sensitive Na^+-K^+-$2Cl^-$ cotransporter (Frambach *et al.,* 1989; Adorante and Miller, 1990; Kennedy, 1990; la Cour, 1992; Hu *et al.,* 1996); and (3) a Ba^{2+}- and Cs^+-sensitive K^+ conductance (Miller and Steinberg, 1977b; Griff *et al.,* 1985; la Cour *et al.,* 1986; Joseph and Miller, 1991; Hughes *et al.,* 1995a; Takahira and Hughes, 1997). The choroidal membrane contains a relatively large K^+ conductance, which is also Ba^{2+}-sensitive, although the Ba^{2+} concentration needed to block is higher than for the retinal membrane K^+ conductance (Miller and Steinberg, 1977a; Griff *et al.,* 1985; Immel and Steinberg, 1986; Gallemore *et al.,* 1993; Hernandez *et al.,* 1995). Two voltage-dependent K^+ channels have been studied in detail: a weakly voltage-dependent inward rectifier, which is blocked by Ba^{2+} and Cs^+ (Hughes and Steinberg, 1990; Wen *et al.,* 1993; Segawa and Hughes, 1994; Hughes and Takahira, 1996), and a sustained outward rectifier inhibitable by millimolar concentrations of Ba^{2+} (Hughes *et al.,* 1995b; Takahira and Hughes, 1997). Studies have shown that the retinal membrane K^+ conductance is mainly composed of inward rectifier K^+ channels (Hughes *et al.,* 1995a; Maruiwa *et al.,* 1996; Kusaka *et al.,* 1999). The channels responsible for the choroidal membrane K^+ conductance have not yet been determined.

HCO_3^- enters the RPE cell via an electrogenic Na^+-HCO_3^- contransporter in the retinal membrane (Hughes *et al.,* 1989; la Cour, 1989a,b, 1991; Lin and Miller, 1991) and exits at the choroidal membrane via a Cl^-/HCO_3^- exchanger

(Lin and Miller, 1991, 1994; Edelman *et al.,* 1994a,b). An additional exit pathway for HCO_3^-, an electrogenic Na^+-HCO_3^- cotransporter, has been found in the choroidal membrane of fresh bovine RPE (Kenyon *et al.,* 1997). The transport rate of HCO_3^- across the RPE is closely related to changes in retinal K^+ concentration that occur as a result of dark–light transition. Reducing retinal concentration of K^+ from 5 to 2 mmol liter^{-1} (mimicking the transition from dark to light) hyperpolarizes the retinal cell membrane. This shifts the retinal membrane potential toward the reversal potential of the Na^+-HCO_3^- cotransporter, slowing the transporter and resulting in an intracellular acidification. Reducing retinal K^+ concentration from 5 to 2 mmol liter^{-1} acidified the frog RPE cytoplasm by nearly 0.2 pH units (Lin and Miller, 1991). Similar results were obtained in human (Lin *et al.,* 1992) and in bovine RPE (Kenyon *et al.,* 1997).

Cl^- enters the RPE cell across the retinal membrane via the Na^+-K^+-$2Cl^-$ cotransporter. The inwardly directed Na^+ gradient maintained by the Na^+/K^+ pump is normally sufficient to drive the uphill influx of K^+ and Cl^-. The Cl^- exit at the choroidal side takes place via Cl^- channels. Ca^{2+}- and cAMP-activated Cl^- channels have been studied extensively (Joseph and Miller, 1991; la Cour, 1992; Fujii *et al.,* 1992; Gallemore *et al.,* 1993; Hughes and Segawa, 1993; Ueda and Steinberg, 1994; Strauss *et al.,* 1996). There may also be a swelling-activated Cl^- channel (Botchkin and Matthews, 1993; Ueda and Steinberg, 1994). Movement of Cl^- across the choroidal membrane is influenced by the activity of the Cl^-/HCO_3^- exchanger which, under resting conditions, transports Cl^- into the cell from the choroid in exchange for intracellular HCO_3^-. The rate of this exchanger is increased by cell alkalinization and reduced by cell acidification (Lin and Miller, 1994). Several conditions reverse the direction of Cl^- transport from transport into the choroid (absorption) to transport into the retina (secretion). Elevated retinal concentrations of K^+ or HCO_3^- or lowered bath pCO_2 all stimulated active Cl^- secretion by increasing the choroidal Cl^-/HCO_3^- exchange rate, probably implemented by an intracellular alkalinization (Edelman *et al.,* 1994b).

In retina, the oxygen tension is generally low due to the special location of the capillaries (Linsenmeier, 1986). Accordingly, anaerobic glycolysis, with lactate as an end product, plays a pivotal role in providing the energy (Winkler, 1981, 1989; Miceli *et al.,* 1990); furthermore, photoreceptors use lactate as a fuel for oxidative phosphorylation (Poitry-Yamate *et al.,* 1995; Tsacopoulos and Magistretti, 1996). Lactate concentration in retina is high; values between 3.8 and 13 mmol liter^{-1} in the subretinal space and 19 mmol liter^{-1} in the RPE cells have been determined in bovine tissue under conditions of light adaptation (Adler and Southwick, 1992). Thus, under these conditions, there is a gradient for lactate transport across the retinal membrane into the subretinal space. The rather high lactate concentration in the RPE reflects a high rate of aerobic glycolysis, which has also been described in RPE from other mammals (Berman, 1979; Miceli *et al.,* 1990).

In the more metabolically active, dark-adapted state, lactate concentration in the subretinal space increases (Wang *et al.,* 1997), lactate concentration in RPE decreases (Kenyon *et al.,* 1994), pH in the subretinal space decreases (Yamamoto *et al.,* 1992), and pH in RPE increases (Lin and Miller, 1991). In the dark-adapted eye, therefore, the gradients for both lactate and protons across the retinal membrane are from subretinal space into RPE. Lactate accumulation in the subretinal space would impose an osmotic load and a serious decrease in extracellular pH. Consequently, lactate is transported out of the subretinal space into the choroidal circulation by the RPE (Törnquist and Alm, 1979; Alm and Törnquist, 1985; Adler and Southwick, 1992).

Lactate entry into the RPE cells is mediated by a H^+-lactate contransport protein in the retinal membrane (la Cour *et al.,* 1994; Lin *et al.,* 1994; Kenyon *et al.,* 1994; Zeuthen *et al.,* 1996; Hamann *et al.,* 2000), probably by the monocarboxylate transporter 1 (MCT1) (Philp *et al.,* 1998; Bergersen *et al.,* 1999). In addition to this transporter, a Na^+-dependent transporter of organic anions has been found in bovine RPE (Kenyon *et al.,* 1994). Lactate exit across the choroidal membrane is mediated by an RPE-specific monocarboxylate transporter, MCT3 (Philp *et al.,* 1995, 1998; Yoon *et al.,* 1997). Supplementary or alternative exit pathways for lactate could be an electrogenic Na^+-lactate cotransporter (Kenyon *et al.,* 1994) and the choroidal membrane Cl^-/HCO_3^- exchanger, which may carry lactate instead of Cl^- (Lin and Miller, 1994).

2. Water Transport

Water transport rates in RPE are relatively large (Table I). The average rates estimated *in vitro* range from 1.4 to 7.6 μl cm^{-2} hr^{-1} (Miller *et al.,* 1982; Frambach *et al.,* 1985; Edelman and Miller, 1991; Wolfensberger *et al.,* 2000).

In vivo, the postoperative decrease of subretinal fluid after surgery of retinal detachments has been used as a measure of RPE fluid transport (Chihara and Nao-i, 1985). Given a total RPE area of the human eye of 10 cm^2, the rates measured in the human eye *in vivo* correspond approximately to a fluid transport rate of 1.8 μl min^{-1} across the entire RPE or 11 μl cm^{-2} hr^{-1}. Thus the capacity of RPE for fluid removal exceeds the amount of fluid leaving the intact eye by this route (roughly 0.4 μl/min). Apparently, under conditions of retinal detachment, the flow resistance of the retina is absent, and fluid is available for transport at a maximal rate (Marmor, 1994). For the same reasons, fluid absorption by the RPE has been suggested to play a direct role in normal retinal adhesion and recovery from retinal detachment. The RPE probably also participates in fluid outflow from the eye under normal conditions (Pederson, 1994; Marmor, 1994).

What is the molecular mechanism of RPE water transport? There is evidence that RPE transports ions and fluid isotonically, as shown in the frog (Hughes *et al.,* 1984; Miller and Farber, 1984; Edelman *et al.,* 1994a). When water transport rates and passive water permeabilities are considered, it seems that the isotonic fluid transport capacity of the RPE cannot be accounted for by osmosis. Given a typical permeability value of 0.1 cm sec^{-1} and a water transport rate of 6 $\mu l\ cm^{-2}\ hr^{-1}$ (Table I), it can be calculated that it would require a transepithelial osmotic difference of about 10 mosm $liter^{-1}$ to move water across the epithelium from the retinal compartment by osmosis. In many *in vivo* situations, however, the retinal side is actually hyperosmolar relative to the choroidal compartment by some 10 mosm $liter^{-1}$ mainly due to the build up of lactate (see above). Accordingly, the osmotic gradient is in the wrong direction for absorption. In addition, no known aquaporins have been localized to the RPE (Hamann *et al.,* 1998). To resolve this dilemma, it is essential first to determine the nature of the ion transport systems responsible for net transport and to study the link between these mechanisms and volume changes.

The ion transport proteins that have been associated specifically with water transport are the retinal membrane Na^+-K^+-$2Cl^-$ cotransporter, the choroidal membrane Cl^- channel, the retinal and choroidal membrane K^+ channel, the retinal membrane Na^+-$2HCO_3^-$ contransporter, the choroidal membrane Cl^-/HCO_3^- exchanger, and the retinal (and probably also choroidal) membrane H^+-lactate contransporter. The retinal membrane Na^+-K^+-$2Cl^-$ contransporter is involved in chloride and fluid absorption from retina to choroid (DiMattio *et al.,* 1983; Frambach and Misfeldt, 1983; Tsuboi *et al.,* 1986; Tsuboi, 1987; Miller and Edelman, 1990; Edelman and Miller, 1991; Li *et al.,* 1994), and has been suggested to play a role in the hydration of the subretinal space. The contransporter also plays a major role in regulatory volume increase (Adorante and Miller, 1990; Kennedy, 1990). Isoosmotic removal of retinal Na^+ resulted in significant cell shrinkage (Adorante and Miller, 1990) consistent with reversal of the driving forces, now directed out of the cell, for Na^+-K^+-$2Cl^-$ and Na^+-HCO_3^- cotransport.

A small isosmotic decrease in retinal K^+ concentration from 5 to 2 mmol $liter^{-1}$ mimics the changes observed when the RPE is adapted from dark to light. When this change was performed in bovine RPE, the Na^+-K^+-$2Cl^-$ cotransporter transiently reversed its direction from absorption to secretion (Bialek and Miller, 1994). Consistent with these findings, it was shown that light caused an increase in the size of the subretinal space correlating to RPE cell shrinkage (Huang and Karwoski, 1992), and Li *et al.* (1994) found that a similar light-induced RPE cell shrinkage was also produced by lowering the concentrations of Cl^- or K^+ in the retinal bath, and that the effect was bumetanide-, furosemide-, and Ba^{2+}-inhibitable.

When the opposite change in retinal bath K^+ concentration was performed, that is, a change from 2 to 5 mmol $liter^{-1}$ (mimicking the transition from light to dark), several effects on the distribution of water and ions were observed: The net fluid absorption was significantly reduced (Edelman *et al.,* 1994b), the RPE cells swelled (Adorante and Miller, 1990), and active Cl^- transport across the RPE was reversed from absorption to secretion (Edelman *et al.*, 1994b). Edelman *et al.*, proposed a model with a complicated interplay between transporters, involving a swelling-activated Cl^- channel located to the retinal membrane, to explain the findings. In accordance with this, a swelling-activated Cl^- current has been found in rat RPE cells (Botchkin and Matthews, 1993; Ueda and Steinberg, 1994), but it has not yet been determined whether the conductance is localized to the retinal or choroidal surface of the RPE cell. A retinal membrane Cl^- channel (or any other chloride exit mechanism) has not yet been found. The model proposed by Edelman *et al.* (1994b) dismissed the possibility that the Na^+-K^+-$2Cl^-$ contransporter may reverse its direction of transport as a result of the increase in intracellular Cl^- concentration and cell volume.

The results from the experiments with changes in extracellular K^+ concentration point to a direct role of the Na^+-K^+-$2Cl^-$ cotransporter in water transport consistent with the possibility that this cotransporter pulls water, as proposed by Zeuthen and Stein (1994).

Both Cl^- and HCO_3^- transport have been suggested to form the direct link to fluid transport. It has been shown that fluid exit across the choroidal membrane is coupled to Cl^- exit (DiMattio *et al.*, 1983; Miller and Edelman, 1990; Edelman and Miller, 1991; Quinn and Miller, 1992; Bialek and Miller, 1994; Edelman *et al.*, 1994a,b). The Cl^- exit pathway across the choroidal membrane is a Ca^{2+}-sensitive Cl^- channel (Joseph and Miller, 1992). Taken together with the results from studies on Na^+-K^+-$2Cl^-$ cotransport, fluid absorption across the RPE is closely coupled to the Cl^- transport pathway. The K^+ transport may interact with the Cl^- transport. The retinal membrane K^+ conductance has been implicated in the generation of volume changes in the subretinal space and RPE cells described above (Huang and Karwoski, 1992; Bialek and Miller, 1994; Li *et al.*, 1994). The regulatory volume decrease of RPE cells following hypotonicity-induced cell swelling was mediated by choroidal membrane K^+ channels in the frog RPE-choroid preparation (Adorante, 1995). Cl^- channels (Civan *et al.*, 1994) and KCl cotransport (Kennedy, 1994) have also been implicated.

In the frog, it has been shown that a large fraction of the transepithelial fluid transport is coupled to the HCO_3^- transport pathway. The transporters involved may be the retinal membrane Na^+-HCO_3^- cotransporter (Hughes *et al.*, 1984, 1987) and the choroidal membrane Cl^-/HCO_3^- exchanger (Lin and Miller, 1994; Edelman *et al.*, 1994a,b). The molecular mechanism behind the coupled transport of water and bicarbonate has not yet been determined. The transport is isotonic (Hughes *et al.*, 1984, 1987). In bovine RPE, however, net HCO_3 transport is not an important determinant of fluid absorption (Edelman and Miller, 1991), and there are pathways for both active HCO_3^- absorption and secretion across the epithelium (Kenyon *et al.*, 1997). The reason for this interspecific difference is unknown.

Electroneutral H^+-lactate cotransport in the retinal membrane of the RPE is well established (see above). We have studied the relationship between H^+-lactate cotransport and water transport using microelectrodes in bullfrog RPE

(Zeuthen *et al.*, 1996), and fluorescence techniques in cultured human (Hamann *et al.*, 2000) and porcine RPE cells (Hamann, unpublished). Experiments were performed with abrupt increases in lactate concentrations. This caused immediate influxes of water, which were independent of opposing osmotic gradients. We concluded that a downhill influx of H^+ and lactate ions in the cotransport protein could energize an uphill influx of water. In all types of experiments, the ratio of the H^+ and lactate fluxes to water fluxes was 109 mmol liter^{-1}, close to the coupling ratio for other cotransporters (see Cotransporters as Molecular Water Pumps by Zeuthen and MacAulay, this volume).

Water transport by the H^+-lactate cotransporter may contribute significantly to total RPE fluid absorption, particularly in the dark when the subretinal lactate concentrations are largest (Yamamoto *et al.*, 1992). If we assume that the lactate concentration in the subretinal space increases by 7 mmol liter^{-1} and the cells contain 1 mmol liter^{-1} or smaller, our data would predict that the activity of the cotransporter increases the water flux by about 23 μl cm^{-2} hr^{-1} (Zeuthen *et al.*, 1996). This is more than four times the spontaneous water flux of 4.8 μ cm^{-2} hr^{-1} measured under lactate-free conditions in frog RPE (Miller *et al.*, 1982).

Transepithelial movement of lactate and water would require additional transport proteins on the choroidal membrane of the RPE. The monocarboxylate transporter MCT3 is located in the choroidal membrane of rat RPE (Philp *et al.*, 1998). The concentration of lactate in blood plasma is usually about 1 mmol liter^{-1}, indicating a chemical gradient for a flux of H^+ and lactate from the RPE cytoplasm into the choroidal blood. Whether MCT3 also cotransports water has not yet been demonstrated.

V. General Discussion and Summary

Fluid transport in the eye encompasses several distinct functions, each maintained by a specialized epithelial system. The hydration of the cornea is controlled by the corneal epi- and endothelium to maintain transparency. The ciliary epithelia provide a constant secretion of nutritionally important aqueous humor and maintain an adequate intraocular pressure. The RPE provides a stable extracellular environment for proper neuronal function and probably helps keep the retina attached.

All ocular epithelia have high water permeabilities and are capable of high rates of water transport (Table I). Based on measurements in lipid bilayers and cell systems, P_f greater than 100 μm sec^{-1} is considered to be high and suggests the involvement of proteinaceous pathways, whereas P_f less than 50 μm sec^{-1} is consistent with diffusion through the lipid bilayer (Verkman, 2000).

The aquaporins are obvious candidates for the passive osmotic pathway across the cell membrane. To test for a general correlation between ocular fluid transport

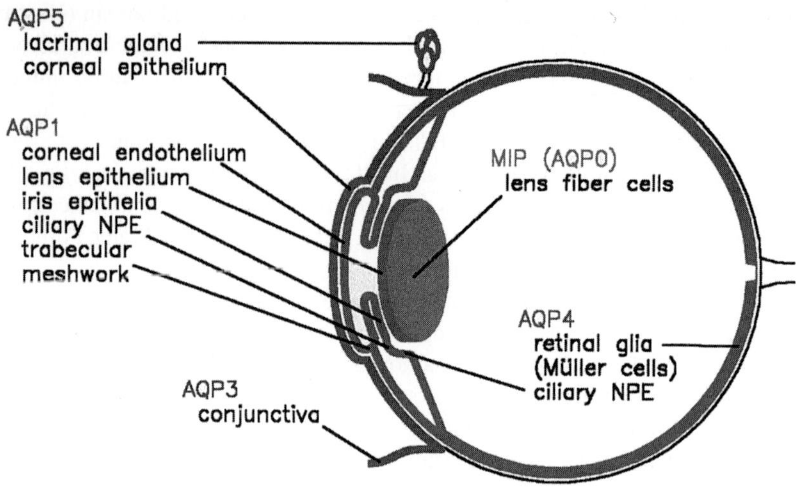

FIG. 8 Expression of aquaporins in eye. MIP, major intrinsic protein of lens or AQP0. (From Hamann *et al.,* 1998, with modifications.) (See also color insert.)

and the distribution of aquaporins, we have performed an immunohistochemical study to map the cellular and subcellular localization of the aquaporins 1 to 5 in the eye (Hamann *et al.,* 1998) (Fig. 8). There was no clear correlation between the localization of the aquaporins and the transport properties of the four epithelia discussed in this review. Both the corneal epithelium and endothelium have aquaporins at both faces of the cell layer (Figs. 3 and 4). Accordingly, water could in principle be transported across each cell layer entirely by aquaporins. In the ciliary epithelium, only one of the cell layers has aquaporins (Fig. 5), and the retinal pigment epithelium possesses no known aquaporins. Accordingly, inspection of Table I shows no marked differences between water permeability of membranes with or without aquaporins. At present, therefore, aquaporins cannot form the foundation of a general theory of fluid transport across ocular epithelia.

No ocular abnormalities have been reported from studies of transgenic mice lacking aquaporins 1, 3, 4, and 5 (Verkman *et al.,* 2000). Clinical eye problems associated with aquaporin dysfunction have been suggested. Naturally occurring mice mutations in AQP0, the major intrinsic protein of the lens (Mulders *et al.,* 1995), result in congenital cataracts (Shiels and Bassnett, 1996). AQP5, which normally is localized to the apical membrane of lacrimal glands (Hamann *et al.,* 1998), is localized to the cytoplasm in patients with Sjögren's syndrome (Tsubota *et al.,* 2001). This defect in lacrimal gland AQP5 trafficking might contribute to decreased lacrimation and dry eye in these patients. Deletion of AQP4 in mice reduces experimentally induced cerebral edema (Manley *et al.,* 2000). In the brain, AQP4 is expressed in glial cells bordering the subarachnoidal space, ventricles, and blood vessels (Nielsen *et al.,* 1997). A similar distribution pattern is seen in the retina, where AQP4 is expressed in the primary glial cell, the Müller cell, with extensive expression in processes facing the retinal capillaries and the vitreous body (Nagelhus *et al.,* 1998; Hamann *et al.,* 1998). In analogy, glial swelling could be a component of retinal edema (Marmor, 1999), determined by water transport through AQP4. In fact, recent studies have pointed to a role of AQP4 in the pathogenesis of surgery-related macular edema (Kitaoka *et al.,* 2001).

There is no fixed relationship between the ultrastructure of the epithelia and the direction of fluid transport. In corneal epithelium and endothelium, the direction is from the basal to the apical membrane, and in the retinal pigment epithelium, the direction is opposite, from apical to basal. In the ciliary epithelium, the direction is from basal to apical in the pigmented cell layer, and opposite in the nonpigmented layer (Fig. 9).

The lateral intercellular spaces play a pivotal role for the standing gradient hypothesis (Diamond and Bossert, 1967). The idea is that the Na^+/K^+-ATPase mediates a hypertonic solution in these spaces. Water would follow osmotically, leave the spaces by the open end, and enter the systemic circulation. A summary of the location of the Na^+/K^+-ATPase, the orientation of lateral spaces, and the direction of the fluid transport is given in Fig. 9. For the corneal epi- and endothelium, the direction of water transport is directly opposite that expected from the standing gradient hypothesis, given the orientation of the spaces and the direction of the active water transport. In the nonpigmented part of the ciliary epithelia, the orientation of the spaces and Na^+ pumping do comply with that of the standing gradient. In the retinal pigment epithelium, however, the active transport of Na^+ is again opposite that expected from the standing gradient theory. It seems that the standing gradient model cannot form a general concept for ocular water transport.

The direction of water transport correlates with the direction of cotransport by symporters in all four cell layers studied (Fig. 9). In the corneal epi- and endothelium, the Na^+-K^+-$2Cl^-$ and the H^+-lactate cotransporters transport in the same direction as the net water flux. In the ciliary epithelium, the dominant Na^+-K^+-$2Cl^-$cotransporter transports in the same direction as the net water flux. In the retinal pigment epithelium, the direction of Na^+-K^+-$2Cl^-$ and H^+-lactate transport coincides with the water transport. It seems that the ocular water transporting cell layers conform to the suggestion that cotransporters of the symport type are the site of coupling between salt and water transport across cell membranes (Zeuthen and Stein, 1994, and Zeuthen, this volume). The cotransport theory would suggest a working model based on a pump-leak system for aqueous humor secretion: Water is actively pumped into the CPE cells via Na^+-K^+-$2Cl^-$ cotransport in the basolateral membrane, transferred to the NPE cells through gap junctions, and

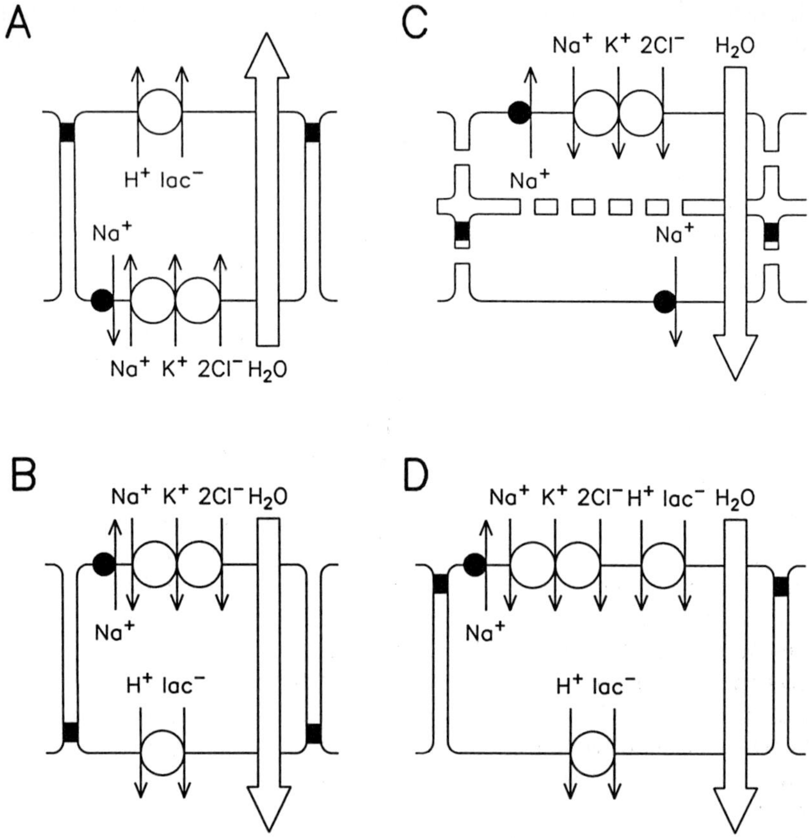

FIG. 9 Water-transporting epithelia are characterized by the location of tight junctions (black bars), direction of water transport (open arrows), and location of Na^+/K^+-ATPase (filled circles) and cotransporters (e.g., Na^+-K^+-$2Cl^-$ and H^+-lactate). A, Corneal epithelium; B, corneal endothelium; C, ciliary epithelia; and D, retinal pigment epithelium. The direction of cotransport and of water transport coincides. On the other hand, there is no correlation between the direction of the water transport and the location of the primary active Na^+/K^+-ATPase and the ultrastructure, i.e., the orientation of the lateral intercellular spaces.

subsequently leaks out of the NPE cells into the aqueous humor via aquaporins (Fig. 5).

Acknowledgments

It is a pleasure to thank Dr. Thomas Zeuthen for very helpful discussions and constructive suggestions as well as Dr. Jan Ulrik Prause for a critical reading of the manuscript. Svend Christoffersen is thanked for producing most of the figures. Financial support was received from Øjenforeningen Værn Om Synet.

References

Adler, A. J., and Southwick, R. E. (1992). Distribution of glucose and lactate in the interphotoreceptor matrix. *Ophthal. Res.* **24,** 243–252.

Adorante, J. S. (1995). Regulatory volume decrease in frog retinal pigment epithelium. *Am. J. Physiol.* **268,** C89–C100.

Adorante, J. S., and Miller, S. S. (1990). Potassium-dependent volume regulation in retinal pigment epithelium is mediated by Na,K,Cl cotransport. *J. Gen. Physiol.* **96,** 1153–1176.

Alm, A., and Törnquist, P. (1985). Lactate transport through the blood–retinal and the blood–brain barrier in rats. *Ophthal. Res.* **17,** 181–184.

Baum, J. P., Maurice, D. M., and McCarey, B. E. (1984). The active and passive transport of water across the corneal endothelium. *Exp. Eye Res.* **39,** 335–342.

Bergersen, L., Johannsson, E., Veruki, M. L., Nagelhus, E. A., Halestrap, A., Sejersted, O. M., and Ottersen, O. P. (1999). Cellular and subcellular expression of monocarboxylate transporters in the pigment epithelium and retina of the rat. *Neuroscience* **90,** 319–331.

Berman, E. R. (1979). Biochemistry of the retinal pigment epithelium. *In* "The Retinal Pigment Epithelium" (K. M. Zinn, and M. F. Marmor, Eds.), pp. 83–102. Harvard Univ. Press, Cambridge.

Bialek, S., and Miller, S. S. (1994). K^+ and Cl^- transport mechanisms in bovine pigment epithelium that could modulate subretinal space volume and composition. *J. Physiol.* (*Lond*). **475,** 401–417.

Bildin, V. N., Yang, H., Crook, R. B., Fischbarg, J., and Reinach, P. S. (2000). Adaptation by corneal epithelial cells to chronic hypertonic stress depends on upregulation of Na : K : 2Cl cotransporter gene and protein expression and ion transport activity [In Process Citation]. *J. Membr. Biol.* **177,** 41–50.

Bildin, V. N., Yang, H., Fischbarg, J., and Reinach, P. S. (1998). Effects of chronic hypertonic stress on regulatory volume increase and Na-K-2Cl cotransporter expression in cultured corneal epithelial cells. *Adv. Exp. Med. Biol.* **438,** 637–642.

Bill, A. (1993). Some aspects of aqueous humour drainage. *Eye* **7 (Pt. 1),** 14–19.

Bill, A., and Phillips, C. I. (1971). Uveoscleral drainage of aqueous humour in human eyes. *Exp. Eye Res.* **12,** 275–281.

Bonanno, J. A. (1990). Lactate–proton cotransport in rabbit corneal epithelium. *Curr. Eye Res.* **9,** 707–712.

Bonanno, J. A., and Giasson, C. (1992a). Intracellular pH regulation in fresh and cultured bovine corneal endothelium. I. Na^+/H^+ exchange in the absence and presence of HCO3−. *Invest. Ophthalmol. Vis. Sci.* **33,** 3058–3067.

Bonanno, J. A., and Giasson, C. (1992b). Intracellular pH regulation in fresh and cultured bovine corneal endothelium. II. $Na^+:HCO_3^-$ cotransport and Cl^-/HCO_3^- exchange. *Invest. Ophthalmol. Vis. Sci.* **33,** 3068–3079.

Bonanno, J. A., and Machen, T. E. (1989). Intracellular pH regulation in basal corneal epithelial cells measured in corneal explants: Characterization of Na/H exchange. *Exp. Eye Res.* **49,** 129–142.

Bonanno, J. A., and Srinivas, S. P. (1997). Cyclic AMP activates anion channels in cultured bovine corneal endothelial cells. *Exp. Eye Res.* **64,** 953–962.

Bonanno, J. A., Yi, G., Kang, X. J., and Srinivas, S. P. (1998). Reevaluation of Cl^-/HCO3-exchange in cultured bovine corneal endothelial cells. *Invest. Ophthalmol. Vis. Sci.* **39,** 2713–2722.

Botchkin, L. M., and Matthews, G. (1993). Chloride current activated by swelling in retinal pigment epithelium cells. *Am. J. Physiol.* **265,** C1037–C1045.

Botchkin, L. M., and Matthews, G. (1994). Voltage-dependent sodium channels develop in rat retinal pigment epithelium cells in culture. *Proc. Natl. Acad. Sci. USA* **91,** 4564–4568.

Brodwall, J., and Fischbarg, J. (1982). The hydraulic conductivity of rabbit ciliary epithelium in vitro. *Exp. Eye Res.* **34,** 121–129.

Brubaker, R. F. (1991). Flow of aqueous humor in humans [The Friedenwald Lecture]. *Invest. Ophthalmol. Vis. Sci.* **32,** 3145–3166.

Brubaker, R. F. (1998). Clinical measurement of aqueous dynamics: Implications for addressing glaucoma. *In* "Eye's Aqueous Humor: From Secretion to Glaucoma. Current Topics in Membranes," Vol. 45. (M. M. Civan, Ed.), pp. 233–284. Academic Press, San Diego.

Brubaker, R. F., and Kupfer, C. (1962). Microcryoscopic determination of the osmolality of interstitial fluid in the living rabbit cornea. *Invest. Ophthalmol.* **1,** 653.

Candia, O. A. (1985). The flux ratio of the Na–Cl cotransport mechanism in the frog corneal epithelium. *Curr. Eye Res.* **4,** 333–338.

Candia, O. A. (1990). Forskolin-induced HCO_3^- current across apical membrane of the frog corneal epithelium. *Am. J. Physiol.* **259,** C215–C223.

Candia, O. A., and Zamudio, A. C. (1995). Chloride-activated water permeability in the frog corneal epithelium. *J. Membr. Biol.* **143,** 259–266.

Caprioli, J. (1992). The ciliary epithelia and aqueous humor. *In* "Adler's Physiology of the Eye" (W. M., Jr., Hart, Ed.), pp. 228–247. Mosby-Year Book, St. Louis, Missouri.

Carré, D. A., Tang, C. S., Krupin, T., and Civan, M. M. (1992). Effect of bicarbonate on intracellular potential of rabbit ciliary epithelium. *Curr. Eye Res.* **11,** 609–624.

Chen, S., Wan, X. L., and Sears, M. (1998). pICln can regulate swelling-induced Cl^- currents in either layer of rabbit ciliary epithelium. *Biochem. Biophys. Res. Commun.* **246,** 59–63.

Chihara, E., and Nao-i, N. (1985). Resorption of subretinal fluid by transepithelial flow of the retinal pigment epithelium. *Graefes Arch. Clin. Exp. Ophthalmol.* **223,** 202–204.

Chu, T. C., and Candia, O. A. (1987). Electrically silent Na^+ and Cl^- fluxes across the rabbit ciliary epithelium. *Invest. Ophthalmol. Vis. Sci.* **28,** 445–450.

Cilluffo, M. C., Schmitt, B. M., Boron, W. F., and Farahbakhsh, N. A. (2001). An electrogenic Na/HCO_3 cotransporter is present in pigmented ciliary body epithelial cells of the mammalian eye. *Invest. Ophtalmol. Vis. Sci.* **42,** S317 (abstract).

Civan, M. M., Coca-Prados, M., and Peterson-Yantorno, K. (1996). Regulatory volume increase of human non-pigmented ciliary epithelial cells. *Exp. Eye Res.* **62,** 627–640.

Civan, M. M., Marano, C. W., Matschinsky, F. W., and Peterson-Yantorno, K. (1994). Prolonged incubation with elevated glucose inhibits the regulatory response to shrinkage of cultured human retinal pigment epithelial cells. *J. Membr. Biol.* **139,** 1–13.

Coca-Prados, M., Anguita, J., Chalfant, M. L., and Civan, M. M. (1995). PKC-sensitive Cl^- channels associated with ciliary epithelial homologue of pICln. *Am. J. Physiol.* **268,** C572–C579.

Coca-Prados, M., Ghosh, S., Gilula, N. B., and Kumar, N. M. (1992). Expression and cellular distribution of the alpha 1 gap junction gene product in the ocular pigmented ciliary epithelium. *Curr. Eye Res.* **11,** 113–122.

Coca-Prados, M., Sanchez-Torres, J., Peterson-Yantorno, K., and Civan, M. M. (1996). Association of ClC-3 channel with Cl^- transport by human nonpigmented ciliary epithelial cells. *J. Membr. Biol.* **150,** 197–208.

Cogan, D. G. (1951). Applied anatomy and physiology of the cornea. *Trans. Am. Acad. Ophthalmol.* **55,** 329.

Cole, D. F. (1964). Location of ouabain-sensitive adenosine triphosphatase in ciliary epithelium. *Exp. Eye Res.* **3,** 72.

Cole, D. F. (1977). Secretion of the aqueous humour. *Exp Eye Res.* **25 (Suppl.),** 161–176.

Counillon, L., Touret, N., Bidet, M., Peterson-Yantorno, K., Coca-Prados, M., Stuart-Tilley, A., Wilhelm, S., Alper, S. L., and Civan, M. M. (2000). Na^+/H^+ and Cl^-/HCO3-antiporters of bovine pigmented ciliary epithelial cells [In Process Citation]. *Pflugers Arch.* **440,** 667–678.

Crook, R. B., Takahashi, K., Mead, A., Dunn, J. J., and Sears, M. L. (2000). The role of NaKCl cotransport in blood-to-aqueous chloride fluxes across rabbit ciliary epithelium. *Invest. Ophthalmol. Vis. Sci.* **41,** 2574–2583.

Davson, H. (1990a). The aqueous humor and the intraocular pressure. *In* "Physiology of the Eye" pp. 3–95. MacMillan Press, London.

Davson, H. (1990b). The cornea. *In* "Physiology of the Eye" pp. 105–138. MacMillan Press, London.

Diamond, J. M., and Bossert, W. H. (1967). Standing-gradient osmotic flow. A mechanism for coupling of water and solute transport in epithelia. *J. Gen. Physiol.* **50,** 2061–2083.

Diamond, J. M., and Bossert, W. H. (1968). Functional consequences of ultrastructural geometry in "backwards" fluid transporting epithelia. *J. Cell Biol.* **37,** 694.

Diecke, F. P., Zhu, Z., Kang, F., Kuang, K., and Fischbarg, J. (1998). Sodium, potassium, two chloride cotransport in corneal endothelium: characterization and possible role in volume regulation and fluid transport. *Invest. Ophthalmol. Vis. Sci.* **39,** 104–110.

Dikstein, S., and Maurice, D. M. (1972). The metabolic basis to the fluid pump in the cornea. *J. Physiol.* (*Lond*). **221,** 29–41.

DiMattio, J., Degnan, K. G., and Zadunaisky, J. A. (1983). A model for transepithelial ion transport across the isolated retinal pigment epithelium of the frog. *Exp. Eye Res.* **37,** 409–420.

Do, C. W., and To, C. H. (2000). Chloride secretion by bovine ciliary epithelium: A model of aqueous humor formation. *Invest. Ophthalmol. Vis. Sci.* **41,** 1853–1860.

Donn, A., Maurice, D. M., and Mills, N. L. (1959). Studies in the living cornea *in vitro.* II, The active transport of sodium across the epithelium. *Arch. Ophthalmol.* **62,** 748.

Dunn, J. J., Lytle, C., and Crook, R. B. (2001). Immunolocalization of the Na-K-Cl cotransporter in bovine ciliary epithelium. *Invest. Ophthalmol. Vis. Sci.* **42,** 343–353.

Echevarria, M., Kuang, K., Iserovich, P., Li, J., Preston, G. M., Agre, P., and Fischbarg, J. (1993). Cultured bovine corneal endothelial cells express CHIP28 water channels. *Am. J. Physiol.* **265,** C1349–C1355.

Edelman, J. L., Lin, H., and Miller, S. S. (1994a). Acidification stimulates chloride and fluid absorption across frog retinal pigment epithelium. *Am. J. Physiol.* **266,** C946–C956.

Edelman, J. L., Lin, H., and Miller, S. S. (1994b). Potassium-induced chloride secretion across the frog retinal pigment epithelium. *Am. J. Physiol.* **266,** C957–C966.

Edelman, J. L., and Miller, S. S. (1991). Epinephrine stimulates fluid absorption across bovine retinal pigment epithelium. *Invest. Ophthalmol. Vis. Sci.* **32,** 3033–3040.

Edelman, J. L., Sachs, G., and Adorante, J. S. (1994c). Ion transport asymmetry and functional coupling in bovine pigmented and nonpigmented ciliary epithelial cells. *Am. J. Physiol.* **266,** C1210–C1221.

Emi, K., Pederson, J. E., and Toris, C. B. (1989). Hydrostatic pressure of the suprachoroidal space. *Invest. Ophthalmol. Vis. Sci.* **30,** 233–238.

Farris, R. L. (1994). Tear osmolarity—A new gold standard? *Adv. Exp. Med. Biol.* **350,** 495–503.

Fatt, I., and Shantinath, K. (1971). Flow conductivity of retina and its role in retinal adhesion. *Exp. Eye Res.* **12,** 218–226.

Fischbarg, J., Hofer, G., and Koatz, R. A. (1980). Priming of the fluid pump by osmotic gradients across rabbit corneal endothelium. *Biochim. Biophys. Acta* **603,** 198–206.

Fischbarg, J., and Lim, J. J. (1974). Role of cations, anions and carbonic anhydrase in fluid transport across rabbit corneal endothelium. *J. Physiol.* (*Lond*). **241,** 647–675.

Fischbarg, J., Lim, J. J., and Bourguet, J. (1977a). Adenosine stimulation of fluid transport across rabbit corneal endothelium. *J. Membr. Biol.* **35,** 95–112.

Fischbarg, J., and Montoreano, R. (1982). Osmotic permeabilities across corneal endothelium and antidiuretic hormone-stimulated toad urinary bladder structures. *Biochim. Biophys. Acta* **690,** 207–214.

Fischbarg, J., Sanchez, J. M., Li, Y., Wen, Q., Iserovich, P., and Kuang, K. (2001). Evidence that corneal endothelial fluid transport is due to electroosmosis across the intercellular junctions. *Invest Ophthalmol. Vis. Sci.* **42,** S318 (abstract).

Fischbarg, J., Warshavsky, C. R., and Lim, J. J. (1977b). Pathways for hydraulically and osmotically-induced water flows across epithelia. *Nature* **266,** 71–74.

Frambach, D. A., and Misfeldt, D. S. (1983). Furosemide-sensitive Cl^- transport in embryonic chicken retinal pigment epithelium. *Am. J. Physiol.* **244,** F679–F685.

Frambach, D. A., Valentine, J. L., and Weiter, J. J. (1989). Furosemide-sensitive Cl transport in bovine retinal pigment epithelium. *Invest. Ophthalmol. Vis. Sci.* **30,** 2271–2274.

Frambach, D. A., Weiter, J. J., and Adler, A. J. (1985). A photogrammetric method to measure fluid movement across isolated frog retinal pigment epithelium. *Biophys. J.* **47,** 547–552.

Fujii, S., Gallemore, R. P., Hughes, B. A., and Steinberg, R. H. (1992). Direct evidence for a basolateral membrane Cl^- conductance in toad retinal pigment epithelium. *Am. J. Physiol.* **262,** C374–C383.

Gaasterland, D. E., Pederson, J. E., MacLellan, H. M., and Reddy, V. N. (1979). Rhesus monkey aqueous humor composition and a primate ocular perfusate. *Invest. Ophthalmol. Vis. Sci.* **18,** 1139–1150.

Gallemore, R. P., Hernandez, E., Tayyanipour, R., Fujii, S., and Steinberg, R. H. (1993). Basolateral membrane Cl^- and K^+ conductances of the dark-adapted chick retinal pigment epithelium. *J. Neurophysiol.* **70,** 1656–1668.

Ghosh, S., Freitag, A. C., Martin-Vasallo, P., and Coca-Prados, M. (1990). Cellular distribution and differential gene expression of the three alpha subunit isoforms of the Na, K-ATPase in the ocular ciliary epithelium. *J. Biol. Chem.* **265,** 2935–2940.

Giasson, C., and Bonanno, J. A. (1994). Facilitated transport of lactate by rabbit corneal endothelium. *Exp. Eye Res.* **59,** 73–81.

Gipson, I. K., and Sugrue, S. P. (1994). Cell biology of the corneal epithelium. *In* "Principles and Practice of Ophthalmology. Basic Sciences" (D. M. Albert, and F. A. Jacobiec, Eds.), pp. 3–16. Saunders, Philadelphia, PA.

Green, K. (1970). Stromal cation binding after inhibition of epithelial transport in the cornea. *Am. J. Physiol.* **218,** 1642–1649.

Green, K., Bountra, C., Georgiou, P., and House, C. R. (1985). An electrophysiologic study of rabbit ciliary epithelium. *Invest. Ophthalmol. Vis. Sci.* **26,** 371–381.

Griff, E. R., Shirao, Y., and Steinberg, R. H. (1985). Ba^{2+} unmasks K^+ modulation of the Na^+-K^+ pump in the frog retinal pigment epithelium. *J. Gen. Physiol.* **86,** 853–876.

Hamann, S., la Cour, M., Lui, G. M., Bundgaard, M., and Zeuthen, T. (2000). Transport of protons and lactate in cultured human fetal retinal pigment epithelial cells. *Pflügers Arch.* **440,** 84–92.

Hamann, S., Zeuthen, T., la Cour, M., Nagelhus, E. A., Ottersen, O. P., Agre, P., and Nielsen, S. (1998). Aquaporins in complex tissues: Distribution of aquaporins 1–5 in human and rat eye. *Am. J. Physiol.* **274,** C1332–C1345.

Han, Z. B., Yang, J. B., Wax, M. B., and Patil, R. V. (2000). Molecular identification of functional water channel protein in cultured human nonpigmented ciliary epithelial cells. *Curr. Eye Res.* **20,** 242–247.

Hara, E., Reinach, P. S., Wen, Q., Iserovich, P., and Fischbarg, J. (1999). Fluoxetine inhibits K(+) transport pathways (K(+) efflux, Na(+)-K(+)-2Cl(−) cotransport, and Na(+) pump) underlying volume regulation in corneal endothelial cells. *J. Membr. Biol.* **171,** 75–85.

Hart, W. M., Jr. (1992). Intraocular pressure. *In* "Adler's Physiology of the Eye" (W. M. Jr., Hart, Ed.), pp. 248–267. Mosby-Year Book, St. Louis, Missouri.

Hedbys, B. O. (1961). The role of polysaccharides in corneal swelling. *Exp. Eye Res.* **1,** 81.

Hernandez, E. V., Hu, J. G., Frambach, D. A., and Gallemore, R. P. (1995). Potassium conductances in cultured bovine and human retinal pigment epithelium. *Invest. Ophthalmol. Vis. Sci.* **36,** 113–122.

Hodson, S., and Miller, F. (1976). The bicarbonate ion pump in the endothelium which regulates the hydration of rabbit cornea. *J. Physiol.(Lond).* **263,** 563–577.

Hodson, S., Wigham, C., Williams, L., Mayes, K. R., and Graham, M. V. (1981). Observation on the human cornea in vitro. *Exp. Eye Res.* **32,** 353–360.

Horwich, T., Ibarra, C., Ford, P., Zamudio, A., Parisi, M., and Candia, O. A. (1995). mRNA from frog corneal epithelium increases water permeability in Xenopus oocytes. *Invest. Ophthalmol. Vis. Sci.* **36,** 2772–2774.

Hu, J. G., Gallemore, R. P., Bok, D., and Frambach, D. A. (1996). Chloride transport in cultured fetal human retinal pigment epithelium. *Exp. Eye Res.* **62,** 443–448.

Huang, B., and Karwoski, C. J. (1992). Light-evoked expansion of subretinal space volume in the retina of the frog. *J. Neurosci.* **12,** 4243–4252.

Hughes, B. A., Adorante, J. S., Miller, S. S., and Lin, H. (1989). Apical electrogenic $NaHCO_3$ cotransport. A mechanism for HCO_3 absorption across the retinal pigment epithelium. *J. Gen. Physiol.* **94,** 125–150.

Hughes, B. A., Gallemore, R. P., and Miller, S. S. (1998). Transport mechanisms in the RPE. *In* "The Retinal Pigment Epithelium: Current Aspects of Function and Disease" (M. F. Marmor, and T. J. Wolfensberger, Eds.), pp. 103–134. Oxford Univ. Press, New York.

Hughes, B. A., Miller, S. S., and Farber, D. B. (1987). Adenylate cyclase stimulation alters transport in frog retinal pigment epithelium. *Am. J. Physiol.* **252,** C385–C395.

Hughes, B. A., Miller, S. S., Joseph, D. P., and Edelman, J. L. (1988). cAMP stimulates Na^+-K^+ pump in frog retinal pigment epithelium. *Am. J. Physiol.* **254,** C84–C98.

Hughes, B. A., Miller, S. S., and Machen, T. E. (1984). Effect of cyclic AMP on fluid absorption and ion transport across frog retinal pigment epithelium. Measurements in the open-circuit state. *J. Gen. Physiol.* **83,** 875–899.

Hughes, B. A., and Segawa, Y. (1993). cAMP-activated chloride currents in amphibian retinal pigment epithelial cells. *J. Physiol.(Lond).* **466,** 749–766.

Hughes, B. A., Shaikh, A., and Ahmad, A. (1995a). Effects of Ba^{2+} and Cs^+ on apical membrane K^+ conductance in toad retinal pigment epithelium. *Am. J. Physiol.* **268,** C1164–C1172.

Hughes, B. A., and Steinberg, R. H. (1990). Voltage-dependent currents in isolated cells of the frog retinal pigment epithelium. *J. Physiol.* **428,** 273–297.

Hughes, B. A., and Takahira, M. (1996). Inwardly rectifying K^+ currents in isolated human retinal pigment epithelial cells. *Invest. Ophthalmol. Vis. Sci.* **37,** 1125–1139.

Hughes, B. A., Takahira, M., and Segawa, Y. (1995b). An outwardly rectifying K^+ current active near resting potential in human retinal pigment epithelial cells. *Am. J. Physiol.* **269,** C179–C187.

Hull, D. S., Green, K., Boyd, M., and Wynn, H. R. (1977). Corneal endothelium bicarbonate transport and the effect of carbonic anhydrase inhibitors on endothelial permeability and fluxes and corneal thickness. *Invest. Ophthalmol. Vis. Sci.* **16,** 883–892.

Immel, J., and Steinberg, R. H. (1986). Spatial buffering of K^+ by the retinal pigment epithelium in frog. *J. Neurosci.* **6,** 3197–3204.

Ishida, K., Yamashita, H., Katagiri, H., and Oka, Y. (1995). Regulation of glucose transporter 1 (GLUT1) gene expression by epidermal growth factor in bovine corneal endothelial cells. *Jpn. J. Ophthalmol.* **39,** 225–232.

Jacob, T. J., and Civan, M. M. (1996). Role of ion channels in aqueous humor formation. *Am. J. Physiol.* **271,** C703–C720.

Jentsch, T. J., Keller, S. K., Koch, M., and Wiederholt, M. (1984). Evidence for coupled transport of bicarbonate and sodium in cultured bovine corneal endothelial cells. *J. Membr. Biol.* **81,** 189–204.

Joseph, D. P., and Miller, S. S. (1991). Apical and basal membrane ion transport mechanisms in bovine retinal pigment epithelium. *J. Physiol.* **435,** 439–463.

Joseph, D. P., and Miller, S. S. (1992). Alpha-1 adrenergic modulation of K and Cl transport in bovine retinal pigment epithelium. *J. Gen. Physiol.* **99,** 263–290.

Kaufman, P. L. (1996). Pressure-dependent outflow. *In* "The Glaucomas. Vol I: Basic Sciences" (R. Ritch, M. B. Shields, and T. Krupin, Eds.), pp. 307–335. Mosby-Year Book, St. Louis, Missouri.

Kaye, G. I. (1969). Stereologic measurement of cell volume fraction of rabbit corneal stroma. *Arch. Ophthalmol.* **82,** 792–794.

Kaye, G. L., and Pappas, G. D. (1965). Studies on the ciliary epithelium and zonule, III, The fine structure of the rabbit ciliary epithelium in relation to the localization of ATPase activity. *J Microsc.* **4,** 497.

Keller, S. K., Jentsch, T. J., and Wiederholt, M. (1988). Regulation of intracellular pH in cultured bovine retinal pigment epithelial cells. *Pflügers Arch.* **411,** 47–52.

Kennedy, B. G. (1990). Na^+-K^+-Cl^- cotransport in cultured cells derived from human retinal pigment epithelium. *Am. J. Physiol.* **259,** C29–C34.

Kennedy, B. G. (1994). Volume regulation in cultured cells derived from human retinal pigment epithelium. *Am. J. Physiol.* **266,** C676–C683.

Kenyon, E., Maminishkis, A., Joseph, D. P., and Miller, S. S. (1997). Apical and basolateral membrane mechanisms that regulate pH_i in bovine retinal pigment epithelium. *Am. J. Physiol.* **273,** C456–C472.

Kenyon, E., Yu, K., la Cour, M., and Miller, S. S. (1994). Lactate transport mechanisms at apical and basolateral membranes of bovine retinal pigment epithelium. *Am. J. Physiol.* **267,** C1561–C1573.

Kitaoka, T., Taniguchi, H., Oba, K, Mishima, K., and Amemiya, T. (2001). Aquaporin 4 in tissues obtained from internal limiting membrane peeling. *Invest. Ophthalmol. Vis. Sci.* **42,** S202 (abstract).

Klyce, S. D. (1972). Electrical profiles in the corneal epithelium. *J. Physiol.* (*Lond*). **226,** 407–429.

Klyce, S. D. (1973). Relationship of epithelial membrane potentials to corneal potential. *Exp. Eye Res.* **15,** 567–575.

Klyce, S. D. (1977). Enhancing fluid secretion by the corneal epithelium. *Invest. Ophthalmol. Vis. Sci.* **16,** 968–973.

Klyce, S. D., and Crosson, C. E. (1985). Transport processes across the rabbit corneal epithelium: A review. *Curr. Eye Res.* **4,** 323–331.

Klyce, S. D., and Marshall, W. S. (1982). Effects of Ag^+ on ion transport by the corneal epithelium of the rabbit. *J. Membr. Biol.* **66,** 133–144.

Klyce, S. D., and Wong, R. K. S. (1977). Site and mode of adrenaline action on chloride transport across the rabbit corneal epithelium. *J. Physiol.* (*Lond*). **266,** 777–799.

Krupin, T., and Civan, M. M. (1995). The physiologic basis of aqueous humor formation. *In* "The Glaucomas" (R. Ritch, M. B. Shields, and T. Krupin, Eds.), St. Louis, CV Mosby.

Kuang, K., Li, Y., Wen, Q., Wang, Z., Li, J., Yang, Y., Iserovich, P., Reinach, P. S., Sparrow, J., Diecke, F. P., and Fischbarg, J. (2001). Corneal endothelial NKCC: Molecular identification, location, and contribution to fluid transport. *Am. J. Physiol. Cell Physiol.* **280,** C491–C499.

Kuang, K. Y., Xu, M., Koniarek, J. P., and Fischbarg, J. (1990). Effects of ambient bicarbonate, phosphate and carbonic anhydrase inhibitors on fluid transport across rabbit corneal endothelium. *Exp. Eye Res.* **50,** 487–493.

Kusaka, S., Horio, Y., Fujita, A., Matsushita, K., Inanobe, A., Gotow, T., Uchiyama, Y., Tano, Y., and Kurachi, Y. (1999). Expression and polarized distribution of an inwardly rectifying K^+ channel, Kir4.1, in rat retinal pigment epithelium. *J. Physiol.*(*Lond*). **520 (Pt. 2),** 373–381.

la Cour, M. (1989a). Coupled transport of Na^+ and HCO_3 across the retinal membrane in frog RPE. *In* "Retinal Pigment Epithelium" (M. Zingirian and F. C. Piccolino, Eds.), pp. 29–31. Kugler & Ghedini, Amsterdam

la Cour, M. (1989b). Rheogenic sodium-bicarbonate co-transport across the retinal membrane of the frog retinal pigment epithelium. *J. Physiol.* **419,** 539–553.

la Cour, M. (1991). Kinetic properties and Na^+ dependence of rheogenic Na^+:$HCO_3{}^-$ co-transport in frog retinal pigment epithelium. *J. Physiol.* **439,** 59–72.

la Cour, M. (1992). Cl^- transport in frog retinal pigment epithelium. *Exp. Eye Res.* **54,** 921–931.

la Cour, M., Lin, H., Kenyon, E., and Miller, S. S. (1994). Lactate transport in freshly isolated human fetal retinal pigment epithelium. *Invest. Ophthalmol. Vis. Sci.* **35,** 434–442.

la Cour, M., Lund-Andersen, H., and Zeuthen, T. (1986). Potassium transport of the frog retinal pigment epithelium: Autoregulation of potassium activity in the subretinal space. *J. Physiol.* **375,** 461–479.

la Cour, M., and Zeuthen, T. (1993). Osmotic properties of the frog retinal pigment epithelium. *Exp. Eye Res.* **56,** 521–530.

Lane, J., Wigham, C. G., and Hodson, S. A. (2000). A chloride-activated Na(+)/HCO(3)(−)-coupled transport activity in corneal endothelial membranes. *Biophys. J.* **78,** 2493–2498.

Li, J. D., Gallemore, R. P., Dmitriev, A., and Steinberg, R. H. (1994). Light-dependent hydration of the space surrounding photoreceptors in chick retina. *Invest. Ophthalmol. Vis. Sci.* **35,** 2700–2711.

Lim, J. J., and Ussing, H. H. (1982). Analysis of presteady-state Na^+ fluxes across the rabbit corneal endothelium. *J. Membr. Biol.* **65,** 197–204.

Lin, H., Kenyon, E., and Miller, S. S. (1992). Na-dependent pH_i regulatory mechanisms in native human retinal pigment epithelium. *Invest. Ophthalmol. Vis. Sci.* **33,** 3528–3538.

Lin, H., la Cour, M., Andersen, M. V. N., and Miller, S. S. (1994). Proton-lactate cotransport in the apical membrane of frog retinal pigment epithelium. *Exp. Eye Res.* **59,** 679–688.

Lin, H., and Miller, S. S. (1991). pH_i regulation in frog retinal pigment epithelium: Two apical membrane mechanisms. *Am. J. Physiol.* **261,** C132–C142.

Lin, H., and Miller, S. S. (1994). pH_i-dependent Cl-HCO_3 exchange at the basolateral membrane of frog retinal pigment epithelium. *Am. J. Physiol.* **266,** C935–C945.

Linsenmeier, R. A. (1986). Effects of light and darkness on oxygen distribution and consumption in the cat retina. *J. Gen. Physiol.* **88,** 521–542.

Manley, G. T., Fujimura, M., Ma, T., Noshita, N., Filiz, F., Bollen, A. W., Chan, P., and Verkman, A. S. (2000). Aquaporin-4 deletion in mice reduces brain edema after acute water intoxication and ischemic stroke. *Nat. Med.* **6,** 159–163.

Marmor, M. F. (1994). Mechanisms of normal retinal adhesion. *In* "Retina" (S. J. Ryan, Ed.), Vol. III, pp. 1931–1953. Mosby, St. Louis.

Marmor, M. F. (1998a). Control of subretinal fluid and mechanisms of serous detachment. *In* "The Retinal Pigment Epithelium: Current Aspects of Function and Disease" (M. F. Marmor, and T. J. Wolfensberger, Eds.), pp. 420–438. Oxford Univ. Press, New York.

Marmor, M. F. (1998b). Mechanisms of retinal adhesiveness. *In* "The Retinal Pigment Epithelium: Current Aspects of Function and Disease" (M. F. Marmor, and T. J. Wolfensberger, Eds.), pp. 392–405. Oxford Univ. Press, New York.

Marmor, M. F. (1999). Mechanisms of fluid accumulation in retinal edema. *Doc. Ophthalmol.* **97,** 239–249.

Marshall, W. S., and Klyce, S. D. (1983). Cellular and paracellular pathway resistances in the "tight" Cl^- -secreting epithelium of rabbit cornea. *J. Membr. Biol.* **73,** 275–282.

Maruiwa, F., Gallemore, R. P., Takahira, M., Hughes, B. A., Hu, J., and Bok, D. (1996). Membrane polarization of potassium conductances in cultured retinal pigment epithelium. *Investigative Ophthalmol. Vis. Sci.* **37,** S229 (abstract).

Maurice, D. M. (1967). Epithelial potential of the cornea. *Exp. Eye Res.* **6,** 138–140.

Maurice, D. M. (1984). The cornea and the sclera. *In* "Vegetative Physiology of the Eye" (H. Davson, Ed.), pp. 1–158. Academic Press, Orlando.

Mayes, K. R., and Hodson, S. (1978a). Local osmotic coupling to the active trans-endothelial bicarbonate flux in the rabbit cornea. *Biochim. Biophys. Acta* **514,** 286–293.

Mayes, K. R., and Hodson, S. (1978b). Some effects of hydrostatic pressure on corneal hydration during specular microscopy. *Exp Eye Res.* **26,** 141–145.

McLaughlin, B. J., Caldwell, R. B., Sasaki, Y., and Wood, T. O. (1985). Freeze-fracture quantitative comparison of rabbit corneal epithelial and endothelial membranes. *Curr. Eye. Res.* **4,** 951–961.

McLaughlin, C. W., Peart, D., Purves, R. D., Carré, D. A., Macknight, A. D., and Civan, M. M. (1998). Effects of bicarbonate on cell composition of rabbit ciliary epithelium: A new model for aqueous humor formation. *Invest. Ophthalmol. Vis. Sci.* **39,** 1631–1641.

Miceli, M. V., Newsome, D. A., and Schriver, G. W. (1990). Glucose uptake, hexose monophosphate shunt activity, and oxygen consumption in cultured human retinal pigment epithelial cells. *Invest. Ophthalmol. Vis. Sci.* **31,** 277–283.

Miller, S. S., and Edelman, J. L. (1990). Active ion transport pathways in the bovine retinal pigment epithelium. *J. Physiol.* **424,** 283–300.

Miller, S. S., and Farber, D. (1984). Cyclic AMP modulation of ion transport across frog retinal pigment epithelium. Measurements in the short circuit state. *J. Gen. Physiol.* **83,** 853–874.

Miller, S. S., Hughes, B. A., and Machen, T. E. (1982). Fluid transport across retinal pigment epithelium is inhibited by cyclic AMP. *Proc. Natl. Acad Sci USA* **79,** 2111–2115.

Miller, S. S., and Steinberg, R. H. (1977a). Active transport of ions across frog retinal pigment epithelium. *Exp. Eye Res.* **25,** 235–248.

Miller, S. S., and Steinberg, R. H. (1977b). Passive ionic properties of frog retinal pigment epithelium. *J. Membr. Biol.* **36,** 337–372.

Miller, S. S., Steinberg, R. H., and Oakley, B. II. (1978). The electrogenic sodium pump of the frog retinal pigment epithelium. *J. Membr. Biol.* **44,** 259–279.

Mishima, S., and Hedbys, B. O. (1967). The permeability of the corneal epithelium and endothelium to water. *Exp. Eye Res.* **6,** 10–32.

Mulders, S. M., Preston, G. M., Deen, P. M., Guggino, W. B., van Os, C. H., and Agre, P. (1995). Water channel properties of major intrinsic protein of lens. *J. Biol. Chem.* **270,** 9010–9016.

Nagelhus, E. A., Veruki, M. L., Torp, R., Haug, F. M., Laake, J. H., Nielsen, S., Agre, and Ottersen, O. P. (1998). Aquaporin-4 water channel protein in the rat retina and optic nerve: Polarized expression in Muller cells and fibrous astrocytes. *J. Neurosci.* **18,** 2506–2519.

Narula, P., Xu, M., Kuang, K. Y., Akiyama, R., and Fischbarg, J. (1992). Fluid transport across cultured bovine corneal endothelial cell monolayers. *Am. J. Physiol.* **262,** C98–C103.

Newman, E. A. (1999). Sodium-bicarbonate cotransport in retinal astrocytes and Muller cells of the rat. *GLIA* **26,** 302–308.

Nielsen, S., Nagelhus, E. A., Amiry-Moghaddam, M., Bourque, C., Agre, P., and Ottersen, O. P. (1997). Specialized membrane domains for water transport in glial cells: High-resolution immunogold cytochemistry of aquaporin-4 in rat brain. *J. Neurosci.* **17,** 171–180.

Nielsen, S., Smith, B. L., Christensen, E. I., and Agre, P. (1993). Distribution of the aquaporin CHIP in secretory and resorptive epithelia and capillary endothelia. *Proc. Natl. Acad. Sci. USA* **90,** 7275–7279.

Oh, J., Krupin, T., Tang, L. Q., Sveen, J., and Lahlum, R. A. (1994). Dye coupling of rabbit ciliary epithelial cells in vitro. *Invest. Ophthalmol. Vis. Sci.* **35,** 2509–2514.

Otori, T. (1967). Electrolyte content of the rabbit corneal stroma. *Exp. Eye Res.* **6,** 356–367.

Parisi, M., Candia, O., and Alvarez, L. (1980). Water permeability of the toad corneal epithelium: The effects of pH and amphotericin B. *Pflügers Arch.* **383,** 131–136.

Pederson, J. E. (1994). Fluid physiology of the subretinal space. *In* "Retina" (S. J. Ryan, Ed.), Vol. III, pp. 1955–1967. Mosby, St. Louis.

Pepose, J. S., and Ubels, J. L. (1992). The cornea. *In* "Adler's Physiology of the Eye" (W. M. Jr., Hart, Ed.), pp. 29–70. Mosby-Year Book, St. Louis, Missouri.

Philp, N., Chu, P., Pan, T. C., Zhang, R. Z., Chu, M. L., Stark, K., Boettiger, D., Yoon, H., and Kieber-Emmons, T. (1995). Developmental expression and molecular cloning of REMP, a novel retinal epithelial membrane protein. *Exp. Cell. Res.* **219,** 64–73.

Philp, N. J., Yoon, H., and Grollman, E. F. (1998). Monocarboxylate transporter MCT1 is located in the apical membrane and MCT3 in the basal membrane of rat RPE. *Am. J. Physiol.* **274,** R1824–R1828.

Poitry-Yamate, C. L., Poitry, S., and Tsacopoulos, M. (1995). Lactate released by Muller glial cells is metabolized by photoreceptors from mammalian retina. *J. Neurosci.* **15,** 5179–5191.

Quinn, R. H., and Miller, S. S. (1992). Ion transport mechanisms in native human retinal pigment epithelium. *Invest. Ophthalmol. Vis. Sci.* **33,** 3513–3527.

Rae, J. L., Dewey, J., Rae, J. S., Nesler, M., and Cooper, K. (1990). Single potassium channels in corneal epithelium. *Invest. Ophthalmol. Vis. Sci.* **31,** 1799–1809.

Rae, J. L., and Shepard, A. R. (2000). Kir2.1 Potassium channels and corneal epithelia. *Curr. Eye Res.* **20,** 144–152.

Rae, J. L., and Watsky, M. A. (1996). Ionic channels in corneal endothelium. *Am. J. Physiol.* **270,** C975–C989.

Raina, S., Preston, G. M., Guggino, W. B., and Agre, P. (1995). Molecular cloning and characterization of an aquaporin cDNA from salivary, lacrimal, and respiratory tissues. *J. Biol. Chem.* **270,** 1908–1912.

Raviola, G., and Raviola, E. (1978). Intercellular junctions in the ciliary epithelium. *Invest. Ophthalmol. Vis. Sci.* **17,** 958–981.

Riley, M. V. (1969a). Aerobic glycolysis in the ox cornea. *Exp. Eye Res.* **8,** 201–204.

Riley, M. V. (1969b). Glucose and oxygen utilization by the rabbit cornea. *Exp. Eye Res.* **8,** 193–200.

Riley, M. V. (1982). Corneal endothelial transport. *In* "Cell Biology of the Eye" (D. S. McDevitt, Ed.), pp. 53–95. Academic Press, New York.

Riley, M. V., and Kishida, K. (1986). ATPases of ciliary epithelium: Cellular and subcellular distribution and probable role in secretion of aqueous humor. *Exp. Eye Res.* **42,** 559–568.

Riley, M. V., Winkler, B. S., Czajkowski, C. A., and Peters, M. I. (1995). The roles of bicarbonate and CO_2 in transendothelial fluid movement and control of corneal thickness. *Invest. Ophthalmol. Vis. Sci.* **36,** 103–112.

Riley, M. V., Winkler, B. S., Peters, M. I., and Czajkowski, C. A. (1994). Relationship between fluid transport and in situ inhibition of Na(+)-K^+ adenosine triphosphatase in corneal endothelium. *Invest. Ophthalmol. Vis. Sci.* **35,** 560–567.

Riley, M. V., Winkler, B. S., Starnes, C. A., and Peters, M. I. (1997). Fluid and ion transport in corneal endothelium: Insensitivity to modulators of Na(+)-K(+)-2Cl-cotransport. *Am. J. Physiol.* **273,** C1480–C1486.

Schnetkamp, P. P., Basu, D. K., and Szerencsei, R. T. (1989). Na^+-$Ca2^+$ exchange in bovine rod outer segments requires and transports K^+. *Am. J. Physiol.* **257,** C153–C157.

Segawa, Y., and Hughes, B. A. (1994). Properties of the inwardly rectifying K^+ conductance in the toad retinal pigment epithelium. *J. Physiol. (Lond).* **476,** 41–53.

Shi, C., Ryan, J. S., French, A. S., Coca-Prados, M., and Kelly, M. E. (1999). Hyposmotically activated chloride channels in cultured rabbit non-pigmented ciliary epithelial cells. *J. Physiol. (Lond).* **521 (Pt. 1),** 57–67.

Shiels, A., and Bassnett, S. (1996). Mutations in the founder of the MIP gene family underlie cataract development in the mouse. *Nat. Genet.* **12,** 212–215.

Shiose, Y., and Sears, M. L. (1966). Fine structural localization of nucleotide activity in the ciliary epithelium of albino rabbits. *Invest Ophthalmol.* **5,** 152.

Srinivas, S. P., Guan, Y., and Bonanno, J. A. (1999). Swelling activated chloride channels in cultured bovine corneal endothelial cells. *Exp. Eye Res.* **68,** 165–177.

Stamer, W. D., Snyder, R. W., Smith, B. L., Agre, P., and Regan, J. W. (1994). Localization of aquaporin CHIP in the human eye: Implications in the pathogenesis of glaucoma and other disorders of ocular fluid balance. *Invest. Ophthalmol. Vis. Sci.* **35,** 3867–3872.

Stiemke, M. M., Roman, R. J., Palmer, M. L., and Edelhauser, H. F. (1992). Sodium activity in the aqueous humor and corneal stroma of the rabbit. *Exp. Eye Res.* **55,** 425–433.

Strauss, O., Wiederholt, M., and Wienrich, M. (1996). Activation of Cl^- currents in cultured rat retinal pigment epithelial cells by intracellular applications of inositol-1,4,5-triphosphate: Differences between rats with retinal dystrophy (RCS) and normal rats. *J. Membr. Biol.* **151,** 189–200.

Sun, X. C., Allen, K. T., Xie, Q., Stamer, W. D., and Bonanno, J. A. (2001). Effect of AQP1 expression level on CO(2) permeability in bovine corneal endothelium. *Invest. Ophthalmol. Vis. Sci.* **42,** 417–423.

Sun, X. C., Bonanno, J. A., Jelamskii, S., and Xie, Q. (2000). Expression and localization of Na(+)-HCO(3)(−) cotransporter in bovine corneal endothelium [In Process Citation]. *Am. J. Physiol. Cell Physiol.* **279,** C1648–C1655.

Takahira, M., and Hughes, B. A. (1997). Isolated bovine retinal pigment epithelial cells express delayed rectifier type and M-type K^+ currents. *Am. J. Physiol.* **273,** C790–C803.

Toris, C. B., Pederson, J. E., Tsuboi, S., Gregerson, D. S., and Rice, T. J. (1990). Extravascular albumin concentration of the uvea. *Invest. Ophthalmol. Vis. Sci.* **31,** 43–53.

Törnquist, P., and Alm, A. (1979). Retinal and choroidal contribution to retinal metabolism in vivo. A study in pigs. *Acta Physiol. Scand.* **106,** 351–357.

Trenberth, S. M., and Mishima, S. (1968). The effect of ouabain on the rabbit corneal endothelium. *Invest. Ophthalmol.* **7,** 44–52.

Tsacopoulos, M., and Magistretti, P. J. (1996). Metabolic coupling between glia and neurons. *J. Neurosci.* **16,** 877–885.

Tsuboi, S. (1987). Measurement of the volume flow and hydraulic conductivity across the isolated dog retinal pigment epithelium. *Invest. Ophthalmol. Vis. Sci.* **28,** 1776–1792.

Tsuboi, S., Manabe, R., and Iizuka, S. (1986). Aspects of electrolyte transport across isolated dog retinal pigment epithelium. *Am. J. Physiol.* **250,** F781–F784.

Tsuboi, S., and Pederson, J. E. (1988). Volume flow across the isolated retinal pigment epithelium of cynomolgus monkey eyes Invest. *Ophthalmol. Vis. Sci.* **29,** 1652–1655.

Tsubota, K., Hirai, S., King, L. S., Agre, P., and Ishida, N. (2001). Defective cellular trafficking of lacrimal gland aquaporin-5 in Sjögren's syndrome. *Lancet* **357,** 688–689.

Ueda, Y., and Steinberg, R. H. (1994). Chloride currents in freshly isolated rat retinal pigment epithelial cells. *Exp. Eye Res.* **58,** 331–342.

Usui, T., Seki, G., Amano, S., Oshika, T., Miyata, K., Kunimi, M., Taniguchi, S., Uwatoko, S., Fujita, T., and Araie, M. (1999). Functional and molecular evidence for Na(+)-HCO_3^- contransporter in human corneal endothelial cells. *Pflügers Arch.* **438,** 458–462.

Verkman, A. S. (2000). Water permeability measurement in living cells and complex tissues. *J. Membr. Biol.* **173,** 73–87.

Verkman, A. S., Yang, B., Song, Y., Manley, G. T., and Ma, T. (2000). Role of water channels in fluid transport studied by phynotype analysis of aquaporin knockout mice. *Exp. Physiol.* **85(Spec No.),** 233S–241S.

Wang, L., Chen, L., Walker, V., and Jacob, T. J. (1998). Antisense to MDR1 mRNA reduces P-glycoprotein expression, swelling-activated $C1^-$current and volume regulation in bovine ciliary epithelial cells. *J. Physiol. (Lond).* **511 (Pt. 1),** 33–44.

Wang, L., Tornquist, P., and Bill, A. (1997). Glucose metabolism in pig outer retina in light and darkness. *Acta Physiol. Scand.* **160,** 75–81.

Wen, R., Lui, G. M., and Steinberg, R. H. (1993). Whole-cell K^+ currents in fresh and cultured cells of the human and monkey retinal pigment epithelium. *J Physiol. (Lond).* **465,** 121–147.

Wen, R., Lui, G. M., and Steinberg, R. H. (1994). Expression of a tetrodotoxin-sensitive Na^+ current in cultured human retinal pigment epithelial cells. *J. Physiol. (Lond).* **476,** 187–196.

Wiederholt, M., Helbig, H., and Korbmacher, C. (1991). Ion transport across the ciliary epithelium: Lessons from cultured cells and proposed role of the carbonic anhydrase. *In* "Carbonic Anhydrase" (F. Botre, G. Gross, and B. Storey, Eds.), pp. 232–244. VCH Verlagsgesellschaft, Weinheim, Germany.

Wiederholt, M., and Zadunaisky, J. A. (1986). Membrane potentials and intracellular chloride activity in the ciliary body of the shark. *Pflügers Arch.* **407(Suppl 2),** S112–S115.

Wigham, C. G., Guggenheim, J. A., and Hodson, S. A. (1994). Sodium movement into and out of corneal endothelium. *Pflugers Arch.* **428,** 577–582.

Winkler, B. S. (1981). Glycolytic and oxidative metabolism in relation to retinal function. *J. Gen. Physiol.* **77,** 667–692.

Winkler, B. S. (1989). Retinal aerobic glycolysis revisited. *Ivest. Ophthalmol. Vis. Sci.* **30,** 1023.

Winkler, B. S., Riley, M. V., Peters, M. I., and Williams, F. J. (1992). Chloride is required for fluid transport by the rabbit corneal endothelium. *Am. J. Physiol.* **262,** C1167–C1174.

Wolfensberger, T. J., Chiang, R. K., Takeuchi, A., and Marmor, M. F. (2000). Inhibition of membrane-bound carbonic anhydrase enhances subretinal fluid absorption and retinal adhesisveness. *Graefes Arch. Clin. Exp. Ophthalmol.* **238,** 76–80.

Wu, J., Zhang, J. J., Koppel, H., and Jacob, T. J. (1996). P-glycoprotein regulates a volume-activated chloride current in bovine non-pigmented ciliary epithelial cells. *J. Physiol. (Lond).* **491 (Pt. 3),** 743–755.

Yamamoto, F., Borgula, G. A., and Steinberg, R. H. (1992). Effects of light and darkness on pH outside rod photoreceptors in the cat retina. *Exp. Eye Res.* **54,** 685–697.

Yang, D., Sun, F., Thomas, L. L., Offord, J., MacCallum, D. K., Dawson, D. C., Hughes, B. A., and Ernst, S. A. (2000a). Molecular cloning and expression of an inwardly rectifying K(+) channel from bovine corneal endothelial cells. *Invest. Ophthalmol. Vis. Sci.* **41,** 2936–2944.

Yang, H., Reinach, P. S., Koniarek, J. P., Wang, Z., Iserovich, P., and Fischbarg, J. (2000b). Fluid transport by cultured corneal epithelial cell layers. *Br. J. Ophthalmol.* **84,** 199–204.

Yoon, H., Fanelli, A., Grollman, E. F., and Philp, N. J. (1997). Identification of a unique monocarboxylate transporter (MCT3) in retinal pigment epithelium. *Biochem. Biophys. Res. Commun.* **234,** 90–94.

Zadunaisky, J. A. (1966). Active transport of chloride across the cornea. *Nauture* **209,** 1136–1137.

Zeuthen, T. (2000). Molecular water pumps. *Rev. Physiol. Biochem. Pharmacol.* **141,** 97–151.

Zeuthen, T., Hamann, S., and la, C. M. (1996). Cotransport of H^+, lactate and H_2O by membrane proteins in retinal pigment epithelium of bullfrog. *J. Physiol.* **497,** 3–17.

Zeuthen, T., and Stein, W. D. (1994). Contransport of salt and water in membrane proteins: Membrane proteins as osmotic engines. *J. Membr. Biol.* **137,** 179–195.

INDEX

A

B

C

D

E

F

G

H

I

K

L

N

O

P

R

S

T

U

V

W

X

Y